Student's Solutions Manual

Elementary Algebra

Concepts and Applications

FOURTH EDITION

Student's Solutions Manual

Elementary Algebra

Concepts and Applications

FOURTH EDITION

BITTINGER/ KEEDY/ ELLENBOGEN

Judith A. Penna

ADDISON-WESLEY PUBLISHING COMPANY

Reading, Massachusetts • Menlo Park, California • New York
Don Mills, Ontario • Wokingham, England • Amsterdam • Bonn
Sydney • Singapore • Tokyo • Madrid • San Juan • Milan • Paris

Reproduced by Addison-Wesley from camera-ready copy supplied by the authors.

ISBN: 0-201-53784-2

1 2 3 4 5 6 7 8 9 10-VG-9796959493

TABLE OF CONTENTS

Special thanks are extended to Patsy Hammond for her excellent typing and to Pam Smith for her careful proofreading. Their patience, efficiency, and good humor made the author's work much easier.

Student's Solutions Manual

Elementary Algebra

Concepts and Applications

FOURTH EDITION

Exercise Set 1.1

1. Substitute 7 for x and multiply.
$6x = 6 \cdot 7 = 42$

2. 49

3. Substitute 7 for a and add.
$9 + a = 9 + 7 = 16$

4. 4

5. $\frac{3p}{q} = \frac{3 \cdot 2}{6} = \frac{6}{6} = 1$

6. 3

7. $\frac{x + y}{5} = \frac{10 + 20}{5} = \frac{30}{5} = 6$

8. 9

9. $\frac{x - y}{8} = \frac{20 - 4}{8} = \frac{16}{8} = 2$

10. 2

11. $\frac{x}{y} = \frac{3}{6} = \frac{1}{2}$

12. $\frac{1}{4}$

13. $\frac{5z}{y} = \frac{5 \cdot 8}{2} = \frac{40}{2} = 20$

14. 2

15. $rt = 55(4 \text{ mi}) = 220 \text{ mi}$

16. $\frac{13}{25}$, or 0.52

17. $bh = (6.5 \text{ cm})(15.4 \text{ cm})$
$= (6.5)(15.4)(\text{cm})(\text{cm})$
$= 100.1 \text{ cm}^2$, or 100.1 square centimeters

18. 2.7 hr

19. $5t = 5(30 \text{ sec}) = 150 \text{ sec}$;
$5t = 5(90 \text{ sec}) = 450 \text{ sec}$;
$5t = 5(2 \text{ min}) = 10 \text{ min}$

20. 23; 28; 41

21. $b + 6$, or $6 + b$

22. $t + 8$, or $8 + t$

23. $c - 9$

24. $d - 4$

25. $q + 6$, or $6 + q$

26. $z + 11$, or $11 + z$

27. $a + b$, or $b + a$

28. $d + c$, or $c + d$

29. $y - x$

30. $h - c$

31. $x \div w$, or $\frac{x}{w}$

32. $s \div t$, or $\frac{s}{t}$

33. $n - m$

34. $q - p$

35. $r + s$, or $s + r$

36. $d + f$, or $f + d$

37. $2x$

38. $3p$

39. $\frac{1}{3}t$, or $\frac{t}{3}$

40. $\frac{1}{4}d$, or $\frac{d}{4}$

41. Let n represent "some number." Then we have 97%n, or 0.97n.

42. 43%n, or 0.43n

43. \$d - \$29.95

44. 65t mi

45.

$x + 17 = 32$		Writing the equation
$15 + 17$?	32	Substituting 15 for x
32	32	32 = 32 is TRUE.

Since the left-hand and right-hand sides are the same, 15 is a solution.

46. No

47.

$x - 7 = 12$		Writing the equation
$21 - 7$?	12	Substituting 21 for x
14	12	14 = 12 is FALSE.

Since the left-hand and right-hand sides are not the same, 21 is not a solution.

48. Yes

49. $\dfrac{6x = 54}{6 \cdot 7 \;?\; 54}$

$42 \mid 54 \qquad 42 = 54$ is FALSE.

7 is not a solution.

50. Yes

51. $\dfrac{\frac{x}{6} = 5}{\frac{30}{6} \;?\; 5}$

$5 \mid 5 \qquad 5 = 5$ is TRUE.

5 is a solution.

52. No

53. $\dfrac{5x + 7 = 107}{5 \cdot 19 + 7 \;?\; 107}$

$95 + 7$

$102 \mid 107 \qquad 102 = 107$ is FALSE.

19 is not a solution.

54. Yes

55. Let x represent the number.

	What number	added to	60	is	112?
	↓	↓	↓	↓	↓
Translating:	x	+	60	=	112

$x + 60 = 112$

56. $7w = 2233$

57. Let y represent the number.

Rewording:	42	times	what number	is	2352?
	↓	↓	↓	↓	↓
Translating:	42	$\cdot$	y	=	2352

$42y = 2352$

58. $x + 345 = 987$

59. Let s represent the number of squares your opponent gets.

Rewording:	What number	added to	35	is	64?
	↓	↓	↓	↓	↓
Translating:	s	+	35	=	64

$s + 35 = 64$

60. $\$80y = \$53{,}400$

61. Let c = the cost of one box.

Rewording:	4	times	what number	is	\$7.96?
	↓	↓	↓	↓	↓
Translating:	4	$\cdot$	c	=	\$7.96

$4c = \$7.96$

62. $d + \$0.2 = \$6.5$, where d is in billions of dollars.

63.

64.

65. $y + 2x$

66. $a + 2 + b$

67. $2x - 3$

68. $a + 5$

69. $b - 2$

70. $x + x$, or $2x$

71. $s + s + s + s$, or $4s$

72. $\ell + w + \ell + w$, or $2\ell + 2w$

73. $y = 8$, $x = 2y = 2 \cdot 8 = 16$;

$\dfrac{x + y}{4} = \dfrac{8 + 16}{4} = \dfrac{24}{4} = 6$

74. 5

75. $x = 9$, $y = 3x = 3 \cdot 9 = 27$;

$\dfrac{y - x}{3} = \dfrac{27 - 9}{3} = \dfrac{18}{3} = 6$

76. 9

77. The next whole number is one more than $w + 3$:

$w + 3 + 1 = w + 4$

78. d

79. If t is the larger number, the other number is 3 less than t, or $t - 3$.
If t is the smaller number, the other number is 3 more than t, or $t + 3$.

80. $n + 10\%n$, or $n + 0.1n$, or $1.01n$

Exercise Set 1.2

1. $5 + y$ Changing the order

2. $6 + x$

3. $ab + 5$

4. $3y + x$

5. $3y + 9x$

6. $7b + 3a$

7. $2(3 + a)$

8. $9(5 + x)$

9. tr Changing the order

10. nm

11. $a5$

12. $b7$

13. $5 + ba$

14. $x + y3$

15. $(a + 3)2$

16. $(x + 5)9$

17. $(x + y) + 2$

18. $a + (3 + b)$

19. $9 + (m + 2)$

20. $(x + 2) + y$

21. $ab + (c + d)$

22. $m + (np + r)$

23. $(5a)b$

24. $7(xy)$

25. $6(mn)$

26. $(9r)p$

27. $(3\cdot2)(a + b)$

28. $(5x)(2 + y)$

29. a) $(a + b) + 2 = 2 + (a + b)$ Using the commutative law
$= 2 + (b + a)$ Using the commutative law again

b) $(a + b) + 2 = 2 + (a + b)$ Using the commutative law
$= (2 + a) + b$ Using the associative law

Answers may vary.

30. $v + (w + 5)$; $(v + 5) + w$; answers may vary.

31. a) $7(ab) = (ab)7$ Using the commutative law
$= (ba)7$ Using the commutative law again

b) $7(ab) = (ab)7$ Using the commutative law
$= a(b7)$ Using the associative law
$= a(7b)$ Using the commutative law again

Answers may vary.

32. $(3x)y$; $x(3y)$; answers may vary.

33. $(3a)4 = 4(3a)$ Commutative law
$= (4\cdot3)a$ Associative law
$= 12a$ Simplifying

34. $(2 + m) + 3 = (m + 2) + 3$ Commutative law
$= m + (2 + 3)$ Associative law
$= m + 5$ Simplifying

35. $5 + (2 + x) = (5 + 2) + x$ Associative law
$= x + (5 + 2)$ Commutative law
$= x + 7$ Simplifying

36. $(a3)5 = a(3\cdot5)$ Associative law
$= (3\cdot5)a$ Commutative law
$= 15a$ Simplifying

37. $2(b + 5) = 2\cdot b + 2\cdot5 = 2b + 10$

38. $4x + 12$

39. $7(1 + t) = 7\cdot1 + 7\cdot t = 7 + 7t$

40. $6v + 24$

41. $3(x + 1) = 3\cdot x + 3\cdot1 = 3x + 3$

42. $7x + 56$

43. $4(1 + y) = 4\cdot1 + 4\cdot y = 4 + 4y$

44. $9s + 9$

45. $6(5x + 2) = 6\cdot5x + 6\cdot2 = 30x + 12$

46. $54m + 63$

47. $7(x + 4 + 6y) = 7\cdot x + 7\cdot4 + 7\cdot6y = 7x + 28 + 42y$

48. $20x + 32 + 12p$

49. $(a + b)2 = a(2) + b(2) = 2a + 2b$

50. $7x + 14$

51. $(x + y + 2)5 = x(5) + y(5) + 2(5) = 5x + 5y + 10$

52. $12 + 6a + 6b$

53. $2x + 2y = 2(x + y)$ The common factor is 2.

54. $5(y + z)$

55. $5 + 5y = 5\cdot 1 + 5\cdot y$ The common factor is 5.
$= 5(1 + y)$ Using the distributive law

56. $13(1 + x)$

57. $3x + 12y = 3\cdot x + 3\cdot 4y = 3(x + 4y)$

58. $5(x + 4y)$

59. $5x + 10 + 15y = 5\cdot x + 5\cdot 2 + 5\cdot 3y = 5(x + 2 + 3y)$

60. $3(1 + 9b + 2c)$

61. $9x + 9 = 9\cdot x + 9\cdot 1 = 9(x + 1)$
Check: $9(x + 1) = 9\cdot x + 9\cdot 1 = 9x + 9$

62. $6(x + 1)$

63. $9x + 3y = 3\cdot 3x + 3\cdot y = 3(3x + y)$
Check: $3(3x + y) = 3\cdot 3x + 3\cdot y = 9x + 3y$

64. $5(3x + y)$

65. $2a + 16b + 64 = 2\cdot a + 2\cdot 8b + 2\cdot 32 = 2(a + 8b + 32)$
Check: $2(a + 8b + 32) = 2\cdot a + 2\cdot 8b + 2\cdot 32 =$
$2a + 16b + 64$

66. $5(1 + 4x + 7y)$

67. $11x + 44y + 121 = 11\cdot x + 11\cdot 4y + 11\cdot 11 =$
$11(x + 4y + 11)$
Check: $11(x + 4y + 11) = 11\cdot x + 11\cdot 4y + 11\cdot 11 =$
$11x + 44y + 121$

68. $7(1 + 2b + 8w)$

69. $t - 9$

70. $\frac{1}{2}m$, or $\frac{m}{2}$

71. ◆

72. ◆

73. Yes; commutative law of addition

74. Yes; associative and distributive laws

75. $axy + ax = ax(y + 1)$ Distributive law
$= xa(y + 1)$ Commutative law of multiplication
$= xa(1 + y)$ Commutative law of addition
The expressions are equivalent.

76. $3a(b + c) = 3[a(b + c)]$ Associative law of multiplication
$= 3(ab + ac)$ Distributive law
$= (ba + ca)3$ Using the commutative law of multiplication 3 times
$= (ca + ba)3$ Commutative law of addition

77. ◆

78. $P(1 + rt)$

79. ◆

Exercise Set 1.3

1. We write several factorizations of 56. There are other correct answers.
$4\cdot 14,\ 7\cdot 8,\ 2\cdot 4\cdot 7$

2. $2\cdot 51,\ 6\cdot 17$; there are other correct answers.

3. $1\cdot 93,\ 3\cdot 31$

4. $4\cdot 36,\ 12\cdot 12$; there are other correct answers.

5. $14 = 2\cdot 7$

6. $3\cdot 5$

7. $33 = 3\cdot 11$

8. $5\cdot 11$

9. $9 = 3\cdot 3$

10. $5\cdot 5$

11. $49 = 7\cdot 7$

12. $11\cdot 11$

13. We begin by factoring 18 in any way that we can and continue factoring until each factor is prime.
$18 = 2\cdot 9 = 2\cdot 3\cdot 3$

14. $2\cdot 2\cdot 2\cdot 3$

15. We begin by factoring 40 in any way that we can and continue factoring until each factor is prime.

$$40 = 4\cdot10 = 2\cdot2\cdot2\cdot5$$

16. $2\cdot2\cdot2\cdot7$

17. We begin by factoring 90 in any way that we can and continue factoring until each factor is prime.

$$90 = 2\cdot45 = 2\cdot9\cdot5 = 2\cdot3\cdot3\cdot5$$

18. $2\cdot2\cdot2\cdot3\cdot5$

19. $210 = 2\cdot105 = 2\cdot3\cdot35 = 2\cdot3\cdot5\cdot7$

20. $2\cdot3\cdot5\cdot11$

21. 79 is prime.

22. $11\cdot13$

23. $119 = 7\cdot17$

24. $13\cdot17$

25. $\frac{18}{45} = \frac{2\cdot9}{5\cdot9}$ Factoring numerator and denominator

$= \frac{2}{5}\cdot\frac{9}{9}$ Rewriting as a product of two fractions

$= \frac{2}{5}\cdot1$ $9/9 = 1$

$= \frac{2}{5}$ Using the identity property of 1

26. $\frac{2}{7}$

27. $\frac{49}{14} = \frac{7\cdot7}{2\cdot7} = \frac{7}{2}\cdot\frac{7}{7} = \frac{7}{2}\cdot1 = \frac{7}{2}$

28. $\frac{8}{3}$

29. $\frac{6}{42} = \frac{1\cdot6}{7\cdot6}$ Factoring and using the identity property of 1 to write 6 as $1\cdot6$

$= \frac{1}{7}\cdot\frac{6}{6}$

$= \frac{1}{7}\cdot1 = \frac{1}{7}$

30. $\frac{1}{8}$

31. $\frac{56}{7} = \frac{8\cdot7}{1\cdot7} = \frac{8}{1}\cdot\frac{7}{7} = \frac{8}{1}\cdot1 = 8$

32. 12

33. $\frac{19}{76} = \frac{1\cdot19}{4\cdot19}$ Factoring and using the identity property of 1 to write 19 as $1\cdot19$

$= \frac{1\cdot\cancel{19}}{4\cdot\cancel{19}}$ Removing a factor of 1: $\frac{19}{19} = 1$

$= \frac{1}{4}$

34. $\frac{1}{3}$

35. $\frac{100}{20} = \frac{5\cdot20}{1\cdot20}$ Factoring and using the identity property of 1 to write 20 as $1\cdot20$

$= \frac{5\cdot\cancel{20}}{1\cdot\cancel{20}}$ Removing a factor of 1: $\frac{20}{20} = 1$

$= \frac{5}{1}$

$= 5$ Simplifying

36. 6

37. $\frac{425}{525} = \frac{17\cdot25}{21\cdot25}$ Factoring the numerator and the denominator

$= \frac{17\cdot\cancel{25}}{21\cdot\cancel{25}}$ Removing a factor of 1: $\frac{25}{25} = 1$

$= \frac{17}{21}$

38. $\frac{25}{13}$

39. $\frac{2600}{1400} = \frac{2\cdot13\cdot100}{2\cdot7\cdot100}$ Factoring

$= \frac{13\cdot\cancel{2\cdot100}}{7\cdot\cancel{2\cdot100}}$ Removing a factor of 1: $\frac{2\cdot100}{2\cdot100} = 1$

$= \frac{13}{7}$

40. 3

41. $\frac{8\cdot x}{6\cdot x} = \frac{2\cdot4\cdot x}{2\cdot3\cdot x}$ Factoring

$= \frac{4\cdot\cancel{2\cdot x}}{3\cdot\cancel{2\cdot x}}$ Removing a factor of 1: $\frac{2\cdot x}{2\cdot x} = 1$

$= \frac{4}{3}$

42. $\frac{1}{3}$

43. $\frac{1}{4}\cdot\frac{1}{2} = \frac{1\cdot1}{4\cdot2}$ Multiplying numerators and denominators

$= \frac{1}{8}$

44. $\frac{44}{25}$

45. $\frac{17}{2}\cdot\frac{3}{4} = \frac{17\cdot3}{2\cdot4} = \frac{51}{8}$

46. 1

47. $\frac{1}{2} + \frac{1}{2} = \frac{1+1}{2}$ Adding numerators; keeping the common denominator

$= \frac{2}{2} = 1$

48. $\frac{3}{4}$

49. $\frac{4}{9} + \frac{13}{18} = \frac{4}{9} \cdot \frac{2}{2} + \frac{13}{18}$ Using 18 as the common denominator

$= \frac{8}{18} + \frac{13}{18}$

$= \frac{21}{18}$

$= \frac{7 \cdot \cancel{3}}{6 \cdot \cancel{3}} = \frac{7}{6}$ Simplifying

50. $\frac{4}{3}$

51. $\frac{3}{a} \cdot \frac{b}{7} = \frac{3b}{7a}$ Multiplying numerators and denominators

52. $\frac{xy}{5z}$

53. $\frac{3}{x} + \frac{2}{x} = \frac{5}{x}$ Adding numerators; keeping the common denominator

54. $\frac{2}{a}$

55. $\frac{3}{10} + \frac{8}{15} = \frac{3}{10} \cdot \frac{3}{3} + \frac{8}{15} \cdot \frac{2}{2}$ Using 30 as the common denominator

$= \frac{9}{30} + \frac{16}{30}$

$= \frac{25}{30}$

$= \frac{5 \cdot \cancel{5}}{6 \cdot \cancel{5}} = \frac{5}{6}$ Simplifying

56. $\frac{41}{24}$

57. $\frac{5}{4} - \frac{3}{4} = \frac{2}{4}$

$= \frac{1 \cdot 2}{2 \cdot 2} = \frac{1}{2}$

58. 2

59. $\frac{13}{18} - \frac{4}{9} = \frac{13}{18} - \frac{4}{9} \cdot \frac{2}{2}$ Using 18 as a common denominator

$= \frac{13}{18} - \frac{8}{18}$

$= \frac{5}{18}$

60. $\frac{31}{45}$

61. $\frac{11}{12} - \frac{2}{5} = \frac{11}{12} \cdot \frac{5}{5} - \frac{2}{5} \cdot \frac{12}{12}$ Using 60 as a common denominator

$= \frac{55}{60} - \frac{24}{60}$

$= \frac{31}{60}$

62. $\frac{13}{48}$

63. $\frac{7}{6} \div \frac{3}{5} = \frac{7}{6} \cdot \frac{5}{3}$ Multiplying by the reciprocal of the divisor

$= \frac{35}{18}$

64. $\frac{28}{15}$

65. $\frac{8}{9} \div \frac{4}{15} = \frac{8}{9} \cdot \frac{15}{4} = \frac{2 \cdot \cancel{4} \cdot \cancel{3} \cdot 5}{\cancel{3} \cdot 3 \cdot \cancel{4}} = \frac{10}{3}$

66. $\frac{7}{4}$

67. $\frac{1}{4} \div \frac{1}{2} = \frac{1}{4} \cdot \frac{2}{1} = \frac{1 \cdot \cancel{2}}{\cancel{2} \cdot 2 \cdot 1} = \frac{1}{2}$

68. $\frac{1}{2}$

69. $\frac{\frac{13}{12}}{\frac{39}{5}} = \frac{13}{12} \div \frac{39}{5} = \frac{13}{12} \cdot \frac{5}{39} = \frac{\cancel{13} \cdot 5}{12 \cdot 3 \cdot \cancel{13}} = \frac{5}{36}$

70. $\frac{68}{9}$

71. $100 \div \frac{1}{5} = \frac{100}{1} \cdot \frac{5}{1} = \frac{500}{1} = 500$

72. 468

73. $\frac{3}{4} \div 10 = \frac{3}{4} \cdot \frac{1}{10} = \frac{3}{40}$

74. $\frac{1}{18}$

75. $\frac{5}{3} \div \frac{a}{b} = \frac{5}{3} \cdot \frac{b}{a} = \frac{5b}{3a}$

76. $\frac{xy}{28}$

77. $\frac{x}{6} - \frac{1}{3} = \frac{x}{6} - \frac{1}{3} \cdot \frac{2}{2}$ Using 6 as a common denominator

$= \frac{x}{6} - \frac{2}{6}$

$= \frac{x-2}{6}$

78. $\frac{9 + 5x}{10}$

79. $5(x + 3) = 5(3 + x)$ Commutative law of addition
Answers may vary.

80. $(a + b) + 7$; answers may vary.

81.

82.

83. $\frac{128}{192} = \frac{2\cdot 64}{3\cdot 64} = \frac{2}{3}$

84. $\frac{p}{t}$

85. $\frac{33sba}{2(11a)} = \frac{3\cdot 11\cdot s\cdot b\cdot a}{2\cdot 11\cdot a} = \frac{3sb}{2}$

86. $\frac{12}{5}$

87. $\frac{36\cdot(2rh)}{8\cdot(9hg)} = \frac{4\cdot 9\cdot 2\cdot r\cdot h}{2\cdot 4\cdot 9\cdot h\cdot g} = \frac{r}{g}$

88. $\frac{5}{2}$

89. We need to find the smallest number that has both 6 and 8 as factors. Starting with 6 we list some numbers with a factor of 6, and starting with 8 we also list some numbers with a factor of 8. Then we find the first number that is on both lists.

6, 12, 18, 24, 30, 36, . . .

8, 16, 24, 32, 40, 48, . . .

Since 24 is the smallest number that is on both lists, the carton should be 24 in. long.

90.

Product	56	63	36	72	140	96	48	168	110	90	432	63
Factor	7	7	2	36	14	8	6	21	11	9	24	3
Factor	8	9	18	2	10	12	8	8	10	10	18	21
Sum	15	16	20	38	24	20	14	29	21	19	42	24

91. $A = \ell w = \left(\frac{4}{5}\text{ m}\right)\left(\frac{7}{9}\text{ m}\right)$

$= \left(\frac{4}{5}\right)\left(\frac{7}{9}\right)(\text{m})(\text{m})$

$= \frac{28}{45}\text{ m}^2$, or $\frac{28}{45}$ square meters

92. $\frac{25}{28}\text{ m}^2$, or $\frac{25}{28}$ square meters

93. $P = 4s = 4\left(\frac{5}{9}\text{ m}\right) = \frac{20}{9}\text{ m}$

94. $\frac{142}{45}$ m

95.

Exercise Set 1.4

1. The integer 5 corresponds to winning 5 points, and the integer -12 corresponds to losing 12 points.

2. 18, -2

3. The integer -170 corresponds to owing \$170, and the integer 950 corresponds to having \$950 in a bank account.

4. 1200, -560

5. The integer -1286 corresponds to 1286 ft below sea level. The integer 29,028 corresponds to 29,028 ft above sea level.

6. Jets: -34, Strikers: 34

7. The integer 750 corresponds to a \$750 deposit, and the integer -125 corresponds to a \$125 withdrawal.

8. -3,000,000

9. The integers 20, -150, and 300 correspond to the interception of the missile, the loss of the starship, and the capture of the base, respectively.

10. -10, 235

11. Since $\frac{10}{3} = 3\frac{1}{3}$, its graph is $\frac{1}{3}$ of a unit to the right of 3.

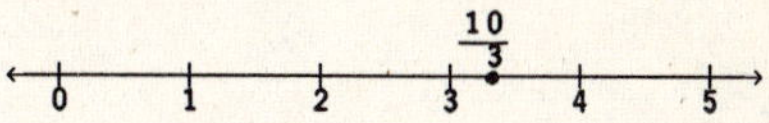

12.
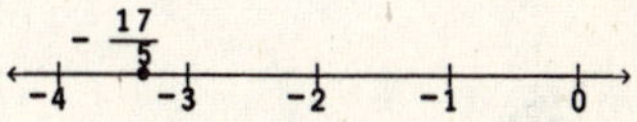

13. The graph of -4.3 is $\frac{3}{10}$ of a unit to the left of -4.

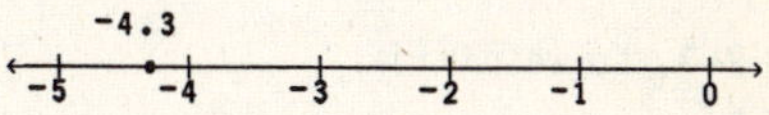

14.
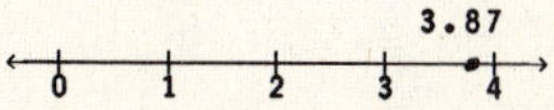

15.
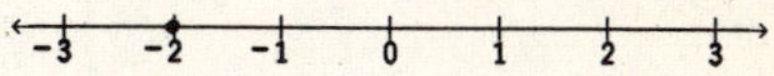

16.

17. We first find decimal notation for $\frac{3}{8}$. Since $\frac{3}{8}$ means $3 \div 8$, we divide.

```
   0.3 7 5
 8)3.0 0 0
   2 4
     6 0
     5 6
       4 0
       4 0
         0
```

Thus $\frac{3}{8} = 0.375$, so $-\frac{3}{8} = -0.375$.

18. -0.125

19. $\frac{5}{3}$ means $5 \div 3$, so we divide.

```
   1.6 6. . .
 3)5.0 0
   3
   2 0
   1 8
     2 0
     1 8
       2
```

We have $\frac{5}{3} = 1.\overline{6}$.

20. $0.8\overline{3}$

21. $\frac{7}{6}$ means $7 \div 6$, so we divide.

```
   1.1 6 6. . .
 6)7.0 0 0
   6
   1 0
     6
     4 0
     3 6
       4 0
       3 6
         4
```

We have $\frac{7}{6} = 1.1\overline{6}$.

22. $0.41\overline{6}$

23. $\frac{2}{3}$ means $2 \div 3$, so we divide.

```
   0.6 6 6. . .
 3)2.0 0 0
   1 8
     2 0
     1 8
       2 0
       1 8
         2
```

We have $\frac{2}{3} = 0.\overline{6}$

24. 0.25

25. We first find decimal notation for $\frac{1}{2}$. Since $\frac{1}{2}$ means $1 \div 2$, we divide.

```
   0.5
 2)1.0
   1 0
     0
```

Thus, $\frac{1}{2} = 0.5$, so $-\frac{1}{2} = -0.5$.

26. 0.625

27. $\frac{1}{10}$ means $1 \div 10$, so we divide.

```
    0.1
 10)1.0
    1 0
      0
```

We have $\frac{1}{10} = 0.1$

28. -0.35

29. Since 5 is to the right of 0, we have $5 > 0$.

30. $9 > 0$

31. Since -9 is to the left of 5, we have $-9 < 5$.

32. $8 > -8$

33. Since -6 is to the left of 6, we have $-6 < 6$.

34. $0 > -7$

35. Since -8 is to the left of -5, we have $-8 < -5$.

36. $-4 < -3$

37. Since -5 is to the right of -11, we have $-5 > -11$.

38. $-3 > -4$

39. Since -12.5 is to the left of -9.4, we have $-12.5 < -9.4$.

40. $-10.3 > -14.5$

41. Since 2.14 is to the right of 1.24, we have $2.14 > 1.24$.

42. $-3.3 < -2.2$

43. We convert to decimal notation.

$\frac{5}{12} = 0.41\overline{6}$ and $\frac{11}{25} = 0.44$. Thus, $\frac{5}{12} < \frac{11}{25}$.

44. $-\frac{14}{17} < -\frac{27}{35}$

45. $x < -6$ has the same meaning as $-6 > x$.

46. $8 > x$

47. $y \geqslant -10$ has the same meaning as $-10 \leqslant y$.

48. $t \leqslant 12$

49. $-3 \geqslant -11$ is true, since $-3 > -11$ is true.

50. False

51. $0 \geqslant 8$ is false, since neither $0 > 8$ nor $0 = 8$ is true.

52. True

53. $-8 \leqslant -8$ is true because $-8 = -8$ is true.

54. True

55. English: -5 is greater than some number.
Translation: $-5 > x$

56. $x < -1$

57. English: A score of 120 is better than a score of -20.
Translation: $120 > -20$

58. $20 > -25$

59. English: A deficit of \$500,000 is worse than an excess of \$1,000,000.
Translation: $-500,000 < 1,000,000$

60. $60 > 20$

61. English: Alicia's test score was at most 95.
Translation: $s \leqslant 95$

62. $n \leqslant -9$

63. English: Profits considered poor don't exceed \$15,000.
Translation: $p \leqslant 15,000$

64. $n \leqslant 0$

65. $|-3| = 3$ since -3 is 3 units from 0.

66. 7

67. $|10| = 10$ since 10 is 10 units from 0.

68. 11

69. $|0| = 0$ since 0 is 0 units from itself.

70. 4

71. $|-24| = 24$ since -24 is 24 units from 0.

72. 325

73. $\left|-\frac{2}{3}\right| = \frac{2}{3}$ since $-\frac{2}{3}$ is $\frac{2}{3}$ of a unit from 0.

74. $\frac{10}{7}$

75. $|43.9| = 43.9$ since 43.9 is 43.9 units from 0.

76. 14.8

77. When $x = 5$, $|x| = |5| = 5$.

78. $\frac{7}{8}$

79. Answers may vary. $-\frac{9}{7}$, 0, $4\frac{1}{2}$, -1.97, -491, 128, $\frac{3}{11}$, $-\frac{1}{7}$, 0.000011, $-26\frac{1}{3}$

80. Answers may vary. 1.26, $9\frac{1}{5}$, $\frac{3}{2}$, 0.17, $\frac{6}{11}$, $-\frac{1}{10,000}$, -0.1, -5.6283, -8.3, $-47\frac{1}{2}$

81. Answers may vary. $-\pi$, $\sqrt{42}$, 8.4262262226...

82. Answers may vary. -67,892, -356, -2

83. $\frac{21}{5} \cdot \frac{1}{7} = \frac{21 \cdot 1}{5 \cdot 7}$ Multiplying numerators and denominators

$= \frac{3 \cdot 7 \cdot 1}{5 \cdot 7}$ Factoring the numerator

$= \frac{3}{5}$ Removing a factor of 1

84. 42

85. $5 + ab$ is equivalent to $ab + 5$ by the commutative law of addition.

$ba + 5$ is equivalent to $ab + 5$ by the commutative law of multiplication.

$5 + ba$ is equivalent to $ab + 5$ by both commutative laws.

86. $3(x + 3 + 4y)$

87.

88.

89. List the numbers as they occur on the number line, from left to right: -17, -12, 5, 13

90. -23, -17, 0, 4

91. Converting to decimal notation, we can write

$\frac{4}{5}, \frac{4}{3}, \frac{4}{8}, \frac{4}{6}, \frac{4}{9}, \frac{4}{2}, -\frac{4}{3}$ as

$0.8, 1.\overline{33}, 0.5, 0.\overline{66}, 0.\overline{44}, 2, -1.\overline{33}$, respectively. List the numbers (in fractional form) as they occur on the number line, from left to right:

$-\frac{4}{3}, \frac{4}{9}, \frac{4}{8}, \frac{4}{6}, \frac{4}{5}, \frac{4}{3}, \frac{4}{2}$

92. $-\frac{5}{6}, -\frac{3}{4}, -\frac{2}{3}, \frac{1}{6}, \frac{3}{8}, \frac{1}{2}$

93. $|-5| = 5$ and $|-2| = 2$, so $|-5| > |-2|$.

94. $|4| < |-7|$

95. $|-8| = 8$ and $|8| = 8$, so $|-8| = |8|$.

96. $|23| = |-23|$

97. $|-3| = 3$ and $|5| = 5$, so $|-3| < |5|$.

98. $|-19| < |-27|$

99. $|x| = 7$

x represents a number whose distance from 0 is 7. Thus, x = 7 or x = -7.

100. -1, 0, 1

101. a) $0.5555... = \frac{5}{6}(0.6666)....) = \frac{5}{6} \cdot \frac{2}{3} = \frac{5\cdot 2}{6\cdot 3} = \frac{5\cdot 2}{2\cdot 3\cdot 3} = \frac{5}{9}$

b) $0.1111... = \frac{0.3333...}{3} = \frac{\frac{1}{3}}{3} = \frac{1}{9}$

c) $0.2222... = \frac{0.6666...}{3} = \frac{\frac{2}{3}}{3} = \frac{2}{9}$

d) $0.9999... = 3(0.3333....) = 3 \cdot \frac{1}{3} = \frac{3}{3} = 1$

Exercise Set 1.5

1. Start at -9. Move 2 units to the right.

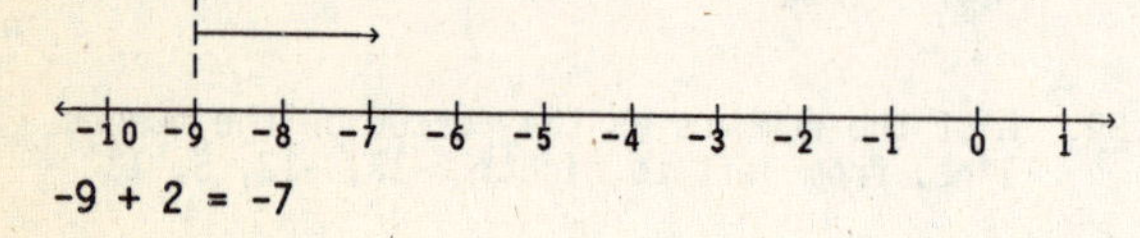

-9 + 2 = -7

2. -3

3. Start at -10. Move 6 units to the right.

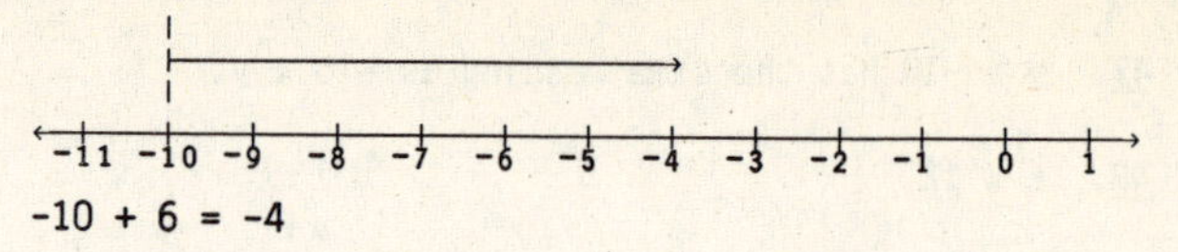

-10 + 6 = -4

4. 5

5. Start at -8. Move 8 units to the right.

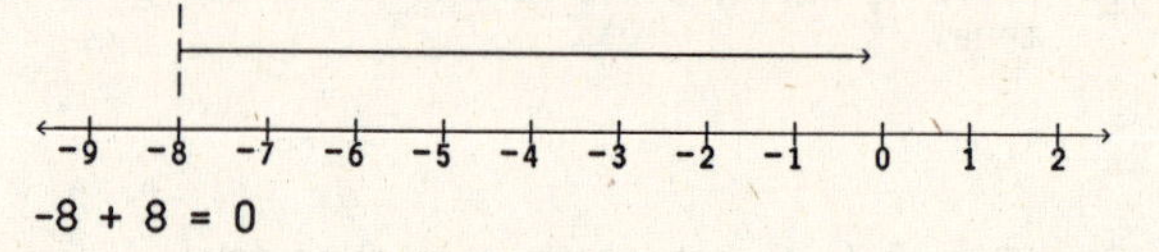

-8 + 8 = 0

6. 0

7. Start at -3. Move 5 units to the left.

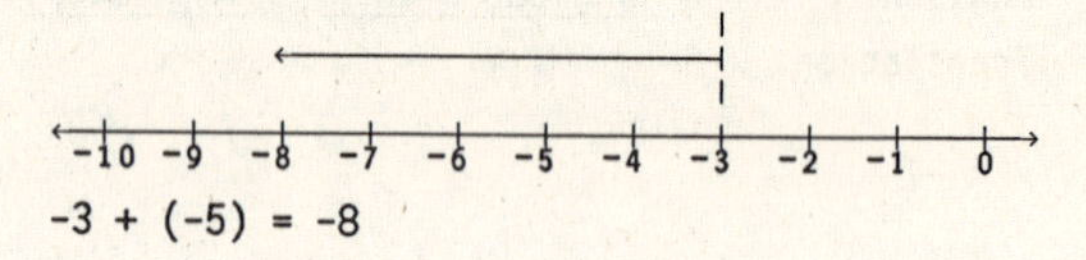

-3 + (-5) = -8

8. -10

9. -7 + 0 One number is 0. The answer is the other number. -7 + 0 = -7

10. -13

11. 0 + (-27) One number is 0. The answer is the other number. 0 + (-27) = -27

12. -35

13. 17 + (-17) The numbers have the same absolute value. The sum is 0. 17 + (-17) = 0

14. 0

15. -17 + (-25) Two negatives. Add the absolute values, getting 42. Make the answer negative. -17 + (-25) = -42

16. -41

17. -18 + 18 The numbers have the same absolute value. The sum is 0. -18 + 18 = 0

18. 0

19. 8 + (-5) The absolute values are 8 and 5. The difference is 8 - 5, or 3. The positive number has the larger absolute value, so the answer is positive. 8 + (-5) = 3

20. 1

21. -4 + (-5) Two negatives. Add the absolute values, getting 9. Make the answer negative. -4 + (-5) = -9

22. -2

23. 13 + (-6) The absolute values are 13 and 6. The difference is 13 - 6, or 7. The positive number has the larger absolute value, so the answer is positive. 13 + (-6) = 7

24. 11

25. 11 + (-9) The absolute values are 11 and 9. The difference is 11 - 9, or 2. The positive number has the larger absolute value, so the answer is positive. 11 + (-9) = 2

26. -33

27. -20 + (-6) Two negatives. Add the absolute values, getting 26. Make the answer negative. -20 + (-6) = -26

28. 0

29. -15 + (-7) Two negatives. Add the absolute values, getting 22. Make the answer negative. -15 + (-7) = -22

30. 18

31. 40 + (-8) The absolute values are 40 and 8. The difference is 40 - 8, or 32. The positive number has the larger absolute value, so the answer is positive. 40 + (-8) = 32

32. -32

33. -25 + 25 The numbers have the same absolute value. The sum is 0. -25 + 25 = 0

34. 0

35. 63 + (-18) The absolute values are 63 and 18. The difference is 63 - 18, or 45. The positive number has the larger absolute value, so the answer is positive. 63 + (-18) = 45

36. 20

37. -6.5 + 4.7 The absolute values are 6.5 and 4.7. The difference is 6.5 - 4.7, or 1.8. The negative number has the larger absolute value, so the answer is negative. -6.5 + 4.7 = -1.8

38. -1.7

39. -2.8 + (-5.3) Two negatives. Add the absolute values, getting 8.1. Make the answer negative. -2.8 + (-5.3) = -8.1

40. -14.4

41. $-\frac{3}{5} + \frac{2}{5}$ The absolute values are $\frac{3}{5}$ and $\frac{2}{5}$. The difference is $\frac{3}{5} - \frac{2}{5}$, or $\frac{1}{5}$. The negative number has the larger absolute value, so the answer is negative. $-\frac{3}{5} + \frac{2}{5} = -\frac{1}{5}$

42. $-\frac{2}{3}$

43. $-\frac{3}{7} + \left(-\frac{5}{7}\right)$ Two negatives. Add the absolute values, getting $\frac{8}{7}$. Make the answer negative. $-\frac{3}{7} + \left(-\frac{5}{7}\right) = -\frac{8}{7}$

44. $-\frac{10}{9}$

45. $-\frac{5}{8} + \frac{1}{4}$ The absolute values are $\frac{5}{8}$ and $\frac{1}{4}$. The difference is $\frac{5}{8} - \frac{2}{8}$, or $\frac{3}{8}$. The negative number has the larger absolute value, so the answer is negative. $-\frac{5}{8} + \frac{1}{4} = -\frac{3}{8}$

46. $-\frac{1}{6}$

47. $-\frac{3}{7} + \left(-\frac{2}{5}\right)$ Two negatives. Add the absolute values, getting $\frac{15}{35} + \frac{14}{35}$, or $\frac{29}{35}$. Make the answer negative. $-\frac{3}{7} + \left(-\frac{2}{5}\right) = -\frac{29}{35}$

48. $-\frac{23}{24}$

49. 75 + (-14) + (-17) + (-5)
= 75 + [(-14) + (-17) + (-5)] Using the associative law of addition
= 75 + (-36) Adding the negatives
= 39 Adding a positive and a negative

50. -62

51. $-44 + \left(-\frac{3}{8}\right) + 95 + \left(-\frac{5}{8}\right)$
$= \left[-44 + \left(-\frac{3}{8}\right) + \left(-\frac{5}{8}\right)\right] + 95$ Using the associative law of addition
= -45 + 95 Adding the negatives
= 50 Adding a negative and a positive

52. 37.9

53. 98 + (-54) + 113 + (-998) + 44 + (-612) + (-18) + 334
= (98 + 113 + 44 + 334) + [-54 + (-998) + (-612) + (-18)]
= 589 + (-1682) Adding the positives; adding the negatives
= -1093

54. -1021

55. Rewording: First try plus second try plus third try plus

Translating: 13 + 0 + (-12) +

fourth try plus fifth try is the total gain (or loss).

21 + (-14) = Total gain (or loss).

Since $13+0+(-12)+21+(-14) = 13+(-12)+21+(-14)$
$= 1+21+(-14)$
$= 22+(-14)$
$= 8,$

the total gain was 8 yd.

56. $77,320 profit

57. Rewording: First change plus second change plus third change plus

Translating: -6 + 3 + (-14) +

fourth change is total change.

4 = Total change.

Since $-6 + 3 + (-14) + 4 = [-6 + (-14)] + (3 + 4)$
$= -20 + 7$
$= -13,$

the total change in pressure is a 13 mb drop.

58. $0

59. Rewording: Amount owed plus amount paid plus

Translating: -470 + 45 +

additional charge plus additional payment is amount owed.

(-160) + 500 = amount owed.

Since $-470 + 45 + (-160) + 500$
$= [-470 + (-160)] + (45 + 500)$
$= -630 + 545$
$= -85,$

Kyle owes the company $85.

60. $85 overdrawn

61. $3a + 8a = (3 + 8)a$ Using the distributive law
$= 11a$

62. $12x$

63. $-2x + 15x = (-2 + 15)x$ Using the distributive law
$= 13x$

64. $-5m$

65. $4x + 7x = (4 + 7)x = 11x$

66. $14a$

67. $7m + (-9m) = [7 + (-9)]m = -2m$

68. $5x$

69. $-6a + 10a = (-6 + 10)a = 4a$

70. $-7n$

71. $-3 + 8x + 4 + (-10x)$
$= -3 + 4 + 8x + (-10x)$ Using the commutative law of addition
$= (-3 + 4) + [8 + (-10)]x$ Using the distributive law
$= 1 - 2x$ Adding

72. $7a + 2$

73. Perimeter $= 7 + 5x + 9 + 6x$
$= 7 + 9 + 5x + 6x$
$= (7 + 9) + (5 + 6)x$
$= 16 + 11x$

74. $10a + 13$

75. Perimeter $= 9 + 3m + 7 + 4m + 2m$
$= 9 + 7 + 3m + 4m + 2m$
$= (9 + 7) + (3 + 4 + 2)m$
$= 16 + 9m$

76. $19n + 11$

77. $7(3z + y + 2) = 7\cdot 3z + 7\cdot y + 7\cdot 2 = 21z + 7y + 14$

78. $\frac{28}{3}$

79.

80.

81. Starting with the final value, we "undo" the rise and drop in value by adding their opposites. The result is the original value.

Rewording: Final value plus opposite of rise plus

Translating: $64\frac{3}{8}$ + $\left(-2\frac{3}{8}\right)$ +

opposite of drop is original value.

$3\frac{1}{4}$ is original value.

Since $64\frac{3}{8} + \left[-2\frac{3}{8}\right] + 3\frac{1}{4} = 62 + 3\frac{1}{4}$

$= 65\frac{1}{4}$,

the stock's original value was $\$65\frac{1}{4}$.

82. \$40.80

83. $7x + \_\_\_\_ + (-9x) + (-2y)$
$= 7x + (-9x) + \_\_\_\_ + (-2y)$
$= [7 + (-9)]x + \_\_\_\_ + (-2y)$
$= -2x + \_\_\_\_ + (-2y)$
This expression is equivalent to -2x - 7y, so the missing term is the term which yields -7y when added -2y. Since -5y + (-2y) = -7y, the missing term is -5y.

84. -11b

85. $3m + 2n + \_\_\_\_ + (-2m) = 2n + \_\_\_\_ + (-2m) + 3m$
$= 2n + \_\_\_\_ + (-2 + 3)m$
$= 2n + \_\_\_\_ + m$
This expression is equivalent to 2n + (-6m), so the missing term is the term which yields -6m when added to m. Since -7m + m = -6m, the missing term is -7m.

86. -3y

87. $P = 2\ell + 2w = 6x + 10$
We know $2\ell = 2\cdot 5 = 10$, so 2w is 6x. Then the width is a number which yields 6x when added to itself. Since 3x + 3x = 6x, the width is 3x.

88. 1 under par

Exercise Set 1.6

1. The opposite of 24 is -24 because 24 + (-24) = 0.

2. 64

3. The opposite of -9 is 9 because -9 + 9 = 0.

4. $-\frac{7}{2}$

5. The opposite of -26.9 is 26.9 because -26.9 + 26.9 = 0.

6. -48.2

7. If x = 9, then -x = -(9) = -9. (The opposite of 9 is -9.)

8. 26

9. If $x = -\frac{14}{3}$, then $-x = -\left[-\frac{14}{3}\right] = \frac{14}{3}$.
$\left[\text{The opposite of } -\frac{14}{3} \text{ is } \frac{14}{3}.\right]$

10. $-\frac{1}{328}$

11. If x = 0.101, then -x = -(0.101) = -0.101. (The opposite of 0.101 is -0.101.)

12. 0

13. If x = -65, then -(-x) = -[-(-65)] = -65 (The opposite of the opposite of -65 is -65.)

14. 29

15. If $x = \frac{5}{3}$, then $-(-x) = -\left[-\frac{5}{3}\right] = \frac{5}{3}$.
$\left[\text{The opposite of the opposite of } \frac{5}{3} \text{ is } \frac{5}{3}.\right]$

16. -9.1

17. When we change the sign of -1 we obtain 1.

18. 7

19. When we change the sign of 7 we obtain -7.

20. -10

21. 3 - 7 = 3 + (-7) = -4

22. -5

23. 0 - 7 = 0 + (-7) = -7

24. -10

25. -8 - (-2) = -8 + 2 = -6

26. 2

27. -10 - (-10) = -10 + 10 = 0

28. 0

29. 12 - 16 = 12 + (-16) = -4

30. -5

31. 20 - 27 = 20 + (-27) = -7

32. 26

33. -9 - (-3) = -9 + 3 = -6

34. 2

35. -40 - (-40) = -40 + 40 = 0

36. 0

37. $7 - 7 = 7 + (-7) = 0$

38. 0

39. $7 - (-7) = 7 + 7 = 14$

40. 8

41. $8 - (-3) = 8 + 3 = 11$

42. -11

43. $-6 - 8 = -6 + (-8) = -14$

44. 16

45. $-4 - (-9) = -4 + 9 = 5$

46. -16

47. $-6 - (-5) = -6 + 5 = -1$

48. -1

49. $8 - (-10) = 8 + 10 = 18$

50. 11

51. $0 - 5 = 0 + (-5) = -5$

52. -6

53. $-5 - (-2) = -5 + 2 = -3$

54. -2

55. $-7 - 14 = -7 + (-14) = -21$

56. -25

57. $0 - (-5) = 0 + 5 = 5$

58. 1

59. $-8 - 0 = -8 + 0 = -8$

60. -9

61. $7 - (-5) = 7 + 5 = 12$

62. 35

63. $2 - 25 = 2 + (-25) = -23$

64. -45

65. $-42 - 26 = -42 + (-26) = -68$

66. -81

67. $-71 - 2 = -71 + (-2) = -73$

68. -52

69. $24 - (-92) = 24 + 92 = 116$

70. 121

71. $-50 - (-50) = -50 + 50 = 0$

72. 0

73. $\frac{3}{8} - \frac{5}{8} = \frac{3}{8} + \left(-\frac{5}{8}\right) = -\frac{2}{8} = -\frac{1}{4}$

74. $-\frac{2}{3}$

75. $\frac{3}{4} - \frac{2}{3} = \frac{9}{12} - \frac{8}{12} = \frac{9}{12} + \left(-\frac{8}{12}\right) = \frac{1}{12}$

76. $-\frac{1}{8}$

77. $-\frac{3}{4} - \frac{2}{3} = -\frac{9}{12} - \frac{8}{12} = -\frac{9}{12} + \left(-\frac{8}{12}\right) = -\frac{17}{12}$

78. $-\frac{11}{8}$

79. $-2.8 - 0 = -2.8 + 0 = -2.8$

80. 4.94

81. $0.99 - 1 = 0.99 + (-1) = -0.01$

82. -0.13

83. $\frac{1}{6} - \frac{2}{3} = \frac{1}{6} - \frac{4}{6} = \frac{1}{6} + \left(-\frac{4}{6}\right) = -\frac{3}{6} = -\frac{1}{2}$

84. $\frac{1}{8}$

85. $-\frac{4}{7} - \left(-\frac{10}{7}\right) = -\frac{4}{7} + \frac{10}{7} = \frac{6}{7}$

86. 0

87. We subtract the smaller number from the larger.
Translate: $1.5 - (-3.5)$
Simplify: $1.5 - (-3.5) = 1.5 + 3.5 = 5$

88. $-2.1 - (-5.9)$; 3.8

89. We subtract the smaller number from the larger.
Translate: $114 - (-79)$
Simplify: $114 - (-79) = 114 + 79 = 193$

90. $23 - (-17)$; 40

91. $-13 - 41 = -13 + (-41) = -54$

92. -26

93. $9 - (-25) = 9 + 25 = 34$

94. 26

95. -3.2 - 5.8 is read "negative three point two minus five point eight."

$-3.2 - 5.8 = -3.2 + (-5.8) = -9$

96. Negative two point seven minus five point nine; -8.6

97. -230 - (-500) is read "negative two hundred thirty minus negative five hundred."

$-230 - (-500) = -230 + 500 = 270$

98. Negative three hundred fifty minus negative one thousand; 650

99. $18 - (-15) - 3 - (-5) + 2 =$

$18 + 15 + (-3) + 5 + 2 = 37$

100. -22

101. $-31 + (-28) - (-14) - 17 =$

$(-31) + (-28) + 14 + (-17) = -62$

102. 22

103. $-34 - 28 + (-33) - 44 =$

$(-34) + (-28) + (-33) + (-44) = -139$

104. 5

105. $-93 - (-84) - 41 - (-56) =$

$(-93) + 84 + (-41) + 56 = 6$

106. 4

107. $3x - 2y = 3x + (-2y)$, so the terms are $3x$ and $-2y$.

108. $7a$, $-9b$

109. $-5 + 3m - 6mn = -5 + 3m + (-6mn)$, so the terms are -5, $3m$, and $-6mn$.

110. -9, $-4t$, $10rt$

111. $5 - a - 6b + 2 = 5 + (-a) + (-6b) + 2$, so the terms are 5, $-a$, $-6b$, and 2.

112. -2, $3x$, $-y$, -8

113. $7a - 12a = 7a + (-12a)$ Adding the opposite

$= (7 + (-12))a$ Using the distributive law

$= -5a$

114. $-12x$

115. $-3m - 5 + m = -3m + (-5) + m$ Rewriting as addition

$= -3m + m + (-5)$ Using the commutative law of addition

$= -2m + (-5)$ Adding like terms mentally

$= -2m - 5$ Rewriting as subtraction

116. $9n - 15$

117. $3x + 5 - 9x = 3x + 5 + (-9x)$

$= 3x + (-9x) + 5$

$= -6x + 5$

118. $3a - 5$

119. $2 - 6t - 9 - 2t = 2 + (-6t) + (-9) + (-2t)$

$= 2 + (-9) + (-6t) + (-2t)$

$= -7 -8t$

120. $-2b - 12$

121. $-5 - (-3x) + 3x + 4x - (-12) =$

$-5 + 3x + 3x + 4x + 12 =$

$3x + 3x + 4x + (-5) + 12 = 10x + 7$

122. $7x + 46$

123. $13x - (-2x) + 45 - (-21) = 13x + 2x + 45 + 21 =$

$15x + 66$

124. $15x + 39$

125. We subtract the amount borrowed from the total assets:

\$619.46 - \$950 = \$619.46 + (-\$950) = -\$330.54

Your total assets are now -\$330.54.

126. \$264

127. We subtract the lower average temperature from the higher average temperature:

$19 - (-31) = 19 + 31 = 50$

The difference in the average daily temperature is 50° C.

128. 17° C

129. We draw a picture of the situation.

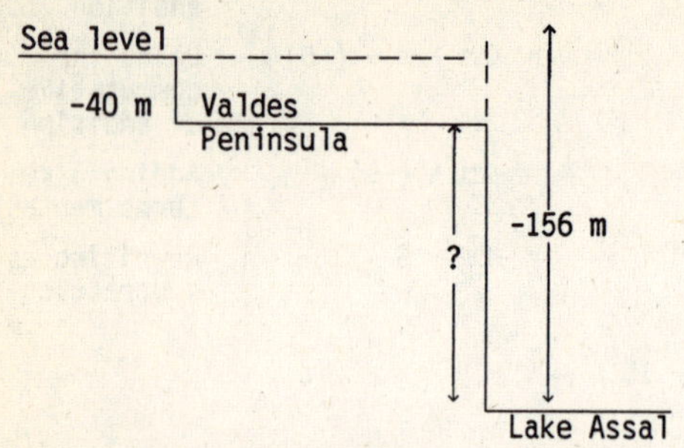

We subtract the lower altitude from the higher altitude:

$-40 - (-156) = -40 + 156 = 116$

Lake Assal is 116 m lower than the Valdes Peninsula.

130. 1767 m

131. Area = ℓw = (36 ft)(12 ft) = 432 ft^2

132. $2\cdot2\cdot2\cdot2\cdot2\cdot3\cdot3\cdot3$

133.

134.

135. See the answer section in the text.

136. False. For example, let $m = -3$ and $n = -5$. Then $-3 > -5$, but $-3 + (-5) = -8 \not> 0$.

137. See the answer section in the text.

138. True. For example, for $m = 4$ and $n = -4$, $4 = -(-4)$ and $4 + (-4) = 0$; for $m = -3$ and $n = 3$, $-3 = -3$ and $-3 + 3 = 0$.

139.

140.

Exercise Set 1.7

1. -16

2. -10

3. -42

4. -18

5. -24

6. -45

7. -72

8. -30

9. 16 Multiplying absolute values

10. 10

11. 42 Multiplying absolute values

12. 18

13. -120

14. 120

15. -238

16. 195

17. 1200

18. -1677

19. 98

20. -203.7

21. -72

22. -63

23. 21.7

24. 12.8

25. $\frac{2}{3}\cdot\left(-\frac{3}{5}\right) = -\left(\frac{2\cdot3}{3\cdot5}\right) = -\left(\frac{2}{5}\cdot\frac{3}{3}\right) = -\frac{2}{5}$

26. $-\frac{10}{21}$

27. $-\frac{3}{8}\cdot\left(-\frac{2}{9}\right) = \frac{\cancel{3}\cdot\cancel{2}\cdot1}{4\cdot\cancel{2}\cdot\cancel{3}\cdot3} = \frac{1}{12}$

28. $\frac{1}{4}$

29. -17.01

30. -38.95

31. $-\frac{5}{9}\cdot\frac{3}{4} = -\frac{5\cdot\cancel{3}}{\cancel{3}\cdot3\cdot4} = -\frac{5}{12}$

32. -6

33. $7\cdot(-4)\cdot(-3)\cdot5 = 7\cdot12\cdot5 = 7\cdot60 = 420$

34. 756

35. $-\frac{2}{3}\cdot\frac{1}{2}\cdot\left(-\frac{6}{7}\right) = -\frac{2}{6}\cdot\left(-\frac{6}{7}\right) = \frac{2\cdot\cancel{6}}{7\cdot\cancel{6}} = \frac{2}{7}$

36. $-\frac{3}{160}$

37. $-3\cdot(-4)\cdot(-5) = 12\cdot(-5) = -60$

38. -70

39. $-2\cdot(-5)\cdot(-3)\cdot(-5) = 10\cdot 15 = 150$

40. 30

41. 0, The product of 0 and any real number is 0.

42. 0

43. $(-8)(-9)(-10) = 72(-10) = -720$

44. 5040

45. $(-6)(-7)(-8)(-9)(-10) = 42\cdot 72\cdot(-10) =$
$3024\cdot(-10) = -30,240$

46. 151,200

47. $36 \div (-6) = -6$ Check: $-6\cdot(-6) = 36$

48. -4

49. $\frac{26}{-2} = -13$ Check: $-13\cdot(-2) = 26$

50. -2

51. $\frac{-16}{8} = -2$ Check: $-2\cdot 8 = -16$

52. 11

53. $\frac{-48}{-12} = 4$ Check: $4(-12) = -48$

54. 7

55. $\frac{-72}{9} = -8$ Check: $-8\cdot 9 = -72$

56. -2

57. $-100 \div (-50) = 2$ Check: $2(-50) = -100$

58. -25

59. $-108 \div 9 = -12$ Check: $-12\cdot 9 = -108$

60. $\frac{64}{7}$

61. $\frac{200}{-25} = -8$ Check: $-8(-25) = 200$

62. $\frac{300}{13}$

63. Undefined

64. 0

65. $\frac{88}{-9} = -\frac{88}{9}$ Check: $-\frac{88}{9}\cdot(-9) = 88$

66. Indeterminate

67. $\frac{0}{-9} = 0$

68. Undefined

69. $0 \div 0$ is indeterminate.

70. 0

71. $\frac{9}{-5} = \frac{-9}{5}$ and $\frac{9}{-5} = -\frac{9}{5}$

72. $\frac{12}{-7}, -\frac{12}{7}$

73. $\frac{-36}{11} = \frac{36}{-11}$ and $\frac{-36}{11} = -\frac{36}{11}$

74. $\frac{-9}{14}, -\frac{9}{14}$

75. $-\frac{7}{3} = \frac{-7}{3}$ and $-\frac{7}{3} = \frac{7}{-3}$

76. $\frac{-4}{15}, \frac{4}{-15}$

77. $\frac{-x}{2} = \frac{x}{-2}$ and $\frac{-x}{2} = -\frac{x}{2}$

78. $\frac{-9}{a}, -\frac{9}{a}$

79. The reciprocal of $\frac{-3}{7}$ is $\frac{7}{-3}$ $\left(\text{or equivalently, } -\frac{7}{3}\right)$ because $\frac{-3}{7}\cdot\frac{7}{-3} = 1$.

80. $\frac{-9}{2}$, or $-\frac{9}{2}$

81. The reciprocal of $-\frac{47}{13}$ is $-\frac{13}{47}$ because $-\frac{47}{13}\left(-\frac{13}{47}\right) = 1$.

82. $-\frac{12}{31}$

83. The reciprocal of -10 is $\frac{1}{-10}$ $\left(\text{or equivalently, } -\frac{1}{10}\right)$ because $-10\left(\frac{1}{-10}\right) = 1$.

84. $\frac{1}{13}$

85. The reciprocal of 4.3 is $\frac{1}{4.3}$ because $4.3\left(\frac{1}{4.3}\right) = 1$.

86. $\frac{1}{-8.5}$, or $-\frac{1}{8.5}$

87. The reciprocal of $\frac{5}{-3}$ is $\frac{-3}{5}$ $\left(\text{or equivalently, } -\frac{3}{5}\right)$ because $\frac{5}{-3}\left(\frac{-3}{5}\right) = 1$.

88. $\frac{11}{-6}$, or $-\frac{11}{6}$

89. The reciprocal of -1 is $\frac{1}{-1}$, or -1 because $(-1)(-1) = 1$.

90. $\frac{1}{2}$

91. $\left(-\frac{3}{7}\right)\left(\frac{2}{-5}\right) = \left(\frac{3}{-7}\right)\left(\frac{2}{-5}\right)$ Rewriting $-\frac{3}{7}$ as $\frac{3}{-7}$

$= \frac{6}{35}$

92. $\frac{8}{27}$

93. $\left(\frac{7}{-2}\right)\left(\frac{-5}{6}\right) = \left(\frac{-7}{2}\right)\left(\frac{-5}{6}\right) = \frac{35}{12}$

94. $\frac{12}{55}$

95. $\frac{-4}{5} + \frac{7}{-5} = \frac{-4}{5} + \frac{-7}{5}$ Rewriting $\frac{7}{-5}$ with a denominator of 5

$= \frac{-11}{5}$, or $-\frac{11}{5}$

96. -1

97. $\left(-\frac{2}{7}\right)\left(\frac{5}{-8}\right) = \left(\frac{2}{-7}\right)\left(\frac{5}{-8}\right) = \frac{10}{56} = \frac{2\cdot 5}{2\cdot 28} = \frac{5}{28}$

98. $\frac{18}{7}$

99. $\frac{-9}{7} + \left(-\frac{4}{7}\right) = \frac{-9}{7} + \frac{-4}{7} = \frac{-13}{7}$, or $-\frac{13}{7}$

100. $-\frac{8}{11}$

101. $\frac{3}{4} \div \left(-\frac{2}{3}\right) = \frac{3}{4} \cdot \left(-\frac{3}{2}\right) = -\frac{9}{8}$

102. $-\frac{7}{4}$

103. $\frac{-5}{12} \cdot \frac{7}{15} = -\frac{5}{12} \cdot \frac{7}{15} = -\frac{5\cdot 7}{12\cdot 15} = -\frac{\cancel{5}\cdot 7}{12\cdot\cancel{5}\cdot 3} = -\frac{7}{36}$

104. -12

105. $\left(-\frac{12}{5}\right) + \left(-\frac{3}{5}\right) = -\frac{15}{5} = -3$

106. -3

107. $-\frac{5}{4} \div \left(-\frac{3}{4}\right) = -\frac{5}{4} \cdot \left(-\frac{4}{3}\right) = \frac{5\cdot\cancel{4}}{\cancel{4}\cdot 3} = \frac{5}{3}$

108. $\frac{2}{3}$

109. $-6.6 \div 3.3 = -2$ Do the long division. The answer is negative.

110. 7

111. $\frac{-3}{7} - \frac{2}{7} = -\frac{3}{7} - \frac{2}{7} = -\frac{3}{7} + \left(-\frac{2}{7}\right) = -\frac{5}{7}$

112. $-\frac{7}{9}$

113. $\frac{-5}{9} + \frac{2}{-3} = \frac{-5}{9} + \frac{-2}{3}$

$= \frac{-5}{9} + \frac{-2}{3} \cdot \frac{3}{3}$ Using a common denominator of 6

$= \frac{-5}{9} + \frac{-6}{9}$

$= \frac{-11}{9}$, or $-\frac{11}{9}$

114. $-\frac{7}{10}$

115. $\left(\frac{-3}{5}\right) \div \frac{6}{15} = \left(-\frac{3}{5}\right) \cdot \frac{15}{6}$ Rewriting $\frac{-3}{5}$ as $-\frac{3}{5}$; multiplying by the reciprocal of the divisor

$= -\frac{3\cdot 15}{5\cdot 6}$

$= -\frac{\cancel{3}\cdot 3\cdot\cancel{5}}{\cancel{5}\cdot\cancel{3}\cdot 2}$

$= -\frac{3}{2}$

116. $-\frac{7}{6}$

117. $\frac{4}{9} - \frac{1}{-9} = \frac{4}{9} - \left(-\frac{1}{9}\right) = \frac{4}{9} + \frac{1}{9} = \frac{5}{9}$

118. $\frac{6}{7}$

119. $\frac{3}{-10} + \frac{-1}{5} = \frac{-3}{10} + \frac{-1}{5} = \frac{-3}{10} + \frac{-1}{5} \cdot \frac{2}{2} = \frac{-3}{10} + \frac{-2}{10} =$

$\frac{-5}{10} = \frac{-1\cdot\cancel{5}}{2\cdot\cancel{5}} = \frac{-1}{2}$, or $-\frac{1}{2}$

120. $-\frac{14}{15}$

121. $\frac{-2}{3} - \frac{1}{-6} = \frac{-2}{3} - \left(-\frac{1}{6}\right) = \frac{-2}{3} + \frac{1}{6} = \frac{-2}{3} \cdot \frac{2}{2} + \frac{1}{6} =$

$\frac{-4}{6} + \frac{1}{6} = \frac{-3}{6} = \frac{-1\cdot\cancel{3}}{2\cdot\cancel{3}} = \frac{-1}{2}$, or $-\frac{1}{2}$

122. $-\frac{1}{2}$

123. $\frac{264}{468} = \frac{\cancel{2}\cdot\cancel{2}\cdot 2\cdot\cancel{3}\cdot 11}{\cancel{2}\cdot\cancel{2}\cdot\cancel{3}\cdot 3\cdot 13} = \frac{22}{39}$

124. $12x - 2y - 9$

125.

126.

127. There are none. A reciprocal has the same sign as the number. Zero has no reciprocal.

128. -1 and 1

129. When n is negative, -n is positive so $\frac{-n}{m}$ is the quotient of a positive and a negative number and, thus, is negative.

130. Positive

131. $\frac{-n}{m}$ is negative (see Exercise 129), so $-\left(\frac{-n}{m}\right)$ is the opposite, or additive inverse, of a negative number and, thus, is positive.

132. Positive

133. When n and m are negative, -n and -m are positive, so $\frac{-n}{-m}$ is the quotient of two positive numbers and, thus, is positive. Then $-\left(\frac{-n}{-m}\right)$ is the opposite, or additive inverse, of a positive number and, thus, is negative.

134. a) m and n have different signs;

b) either m or n is zero;

c) m and n have the same sign

135. $a(-b) + ab = a[-b + b]$ Distributive law

$= a(0)$ Law of opposites

$= 0$ Multiplicative property of zero

Therefore, $a(-b) = -(ab)$. Law of opposites

136. 

Exercise Set 1.8

1. $\underbrace{10 \times 10 \times 10}_{3 \text{ factors}} = 10^3$

2. 6^4

3. $\underbrace{x \cdot x \cdot x \cdot x \cdot x \cdot x \cdot x}_{7 \text{ factors}} = x^7$

4. y^6

5. $3y \cdot 3y \cdot 3y \cdot 3y = (3y)^4$

6. $(5m)^5$

7. $2^4 = 2 \cdot 2 \cdot 2 \cdot 2 = 4 \cdot 4 = 16$

8. 125

9. $(-3)^2 = (-3)(-3) = 9$

10. 49

11. $1^5 = 1 \cdot 1 \cdot 1 \cdot 1 \cdot 1 = 1 \cdot 1 \cdot 1 = 1 \cdot 1 = 1$

12. -1

13. $4^3 = 4 \cdot 4 \cdot 4 = 16 \cdot 4 = 64$

14. 9

15. $(-4)^3 = (-4)(-4)(-4) = 16(-4) = -64$

16. 625

17. $7^1 = 7$ (1 factor)

18. 1

19. $(4a)^2 = (4a)(4a) = 4 \cdot 4 \cdot a \cdot a = 16a^2$

20. $9x^2$

21. $(-7x)^3 = (-7x)(-7x)(-7x) = (-7)(-7)(-7)(x)(x)(x) = -343x^3$

22. $625x^4$

23. $7 + 2 \times 6 = 7 + 12$ Multiplying

$= 19$ Adding

24. 27

25. $8 \times 7 + 6 \times 5 = 56 + 30$ Multiplying

$= 86$ Adding

26. 51

27. $19 - 5 \times 3 + 3 = 19 - 15 + 3$ Multiplying

$= 4 + 3$ Subtracting and adding from left to right

$= 7$

28. 9

29. $9 \div 3 + 16 \div 8 = 3 + 2$ Dividing

$= 5$ Adding

30. 28

31. $7 + 10 - 10 \div 2 = 7 + 10 - 5$ Dividing

$= 17 - 5$ Adding and subtracting from left to right

$= 12$

32. 9

33. $(3 - 5)^3 = (-2)^3$ Working within parentheses first

$= -8$ Simplifying the exponential expression

34. 24

35. $8 - 2\cdot 3 - 9 = 8 - 6 - 9$ Multiplying
$= 2 - 9$ Adding and subtracting from left to right
$= -7$

36. 11

37. $(8 - 2\cdot 3) - 9 = (8 - 6) - 9$ Multiplying inside the parentheses
$= 2 - 9$ Subtracting inside the parentheses
$= -7$

38. -36

39. $(-24) \div (-3)\cdot\left(-\frac{1}{2}\right) = 8\cdot\left(-\frac{1}{2}\right) = -\frac{8}{2} = -4$

40. 32

41. $16\cdot(-24) + 50 = -384 + 50 = -334$

42. -160

43. $2^4 + 2^3 - 10 = 16 + 8 - 10 = 24 - 10 = 14$

44. 23

45. $5^3 + 26\cdot 71 - (16 + 25\cdot 3) =$
$5^3 + 26\cdot 71 - (16 + 75) = 5^3 + 26\cdot 71 - 91 =$
$125 + 26\cdot 71 - 91 = 125 + 1846 - 91 =$
$1971 - 91 = 1880$

46. 305

47. $[2\cdot(5 - 3)]^2 = [2\cdot 2]^2 = 4^2 = 16$

48. 76

49. $\frac{7 + 2}{5^2 - 4^2} = \frac{9}{25 - 16} = \frac{9}{9} = 1$

50. 2

51. $8(-7) + |6(-5)| = -56 + |-30| = -56 + 30 = -26$

52. 49

53. $19 - 5(-3) + 3 = 19 + 15 + 3 = 34 + 3 = 37$

54. 33

55. $9 \div (-3)\cdot 16 \div 8 = -3\cdot 16 \div 8 = -48 \div 8 = -6$

56. -28

57. $20 + 4^3 \div (-8)\cdot 2 = 20 + 64 \div (-8)\cdot 2 =$
$20 + (-8)\cdot 2 = 20 + (-16) = 4$

58. -3000

59. $8|(6 - 13) - 11| = 8|-7 - 11| = 8|-18| =$
$8\cdot 18 = 144$

60. 60

61. $256 \div (-32) \div (-4) = -8 \div (-4)$ Doing the divisions in order from left to right
$= 2$

62. 1

63. $\frac{5^2 - 4^3 - 3}{9^2 - 2^2 - 1^5} = \frac{25 - 64 - 3}{81 - 4 - 1} = \frac{-39 - 3}{77 - 1} = \frac{-42}{76} =$
$-\frac{\cancel{2}\cdot 21}{\cancel{2}\cdot 38} = -\frac{21}{38}$

64. $-\frac{23}{18}$

65. $\frac{20(8 - 3) - 4(10 - 3)}{10(2 - 6) - 2(5 + 2)} = \frac{20\cdot 5 - 4\cdot 7}{10(-4) - 2\cdot 7} =$
$\frac{100 - 28}{-40 - 14} = \frac{72}{-54} = -\frac{\cancel{18}\cdot 4}{\cancel{18}\cdot 3} = -\frac{4}{3}$

66. -118

67. $7 - 3x = 7 - 3\cdot 5 = 7 - 15 = -8$

68. -7

69. $a \div 6\cdot 2 = 12 \div 6\cdot 2 = 2\cdot 2 = 4$

70. 25

71. $-20 \div t^2 - 3(t - 1) = -20 \div (-4)^2 - 3((-4) - 1) =$
$-20 \div (-4)^2 - 3\cdot(-5) = -20 \div 16 - 3\cdot(-5) =$
$\frac{-20}{16} + 15 = \frac{-5}{4} + 15 = \frac{-5}{4} + \frac{60}{4} = \frac{55}{4}$

72. 20

73. $-x^2 - 5x = -(-3)^2 - 5\cdot(-3) = -9 - 5\cdot(-3) =$
$-9 + 15 = 6$

74. 24

75. $-(2x + 7) = -2x - 7$ Removing parentheses and changing the sign of each term

76. $-3x - 5$

77. $-(5x - 8) = -5x + 8$ Removing parentheses and changing the sign of each term

78. $-6x + 7$

79. $-(4a - 3b + 7c) = -4a + 3b - 7c$

80. $-5x + 2y + 3z$

81. $-(3x^2 + 5x - 1) = -3x^2 - 5x + 1$

82. $-8x^3 + 6x - 5$

83. $9x - (4x + 3) = 9x - 4x - 3$ Removing parentheses and changing the sign of every term
$= 5x - 3$ Collecting like terms

84. $5y - 9$

85. $2a - (5a - 9) = 2a - 5a + 9 = -3a + 9$

86. $8n + 7$

87. $2x + 7x - (4x + 6) = 2x + 7x - 4x - 6 = 5x - 6$

88. $a - 7$

89. $2x - 4y - 3(7x - 2y) = 2x - 4y - 21x + 6y = -19x + 2y$

90. $-a - 4b$

91. $15x - y - 5(3x - 2y + 5z)$
$= 15x - y - 15x + 10y - 25z$ Multiplying each term in parentheses by -5
$= 9y - 25z$

92. $-16a + 27b - 32c$

93. $3x^2 + 7 - (2x^2 + 5) = 3x^2 + 7 - 2x^2 - 5$
$= x^2 + 2$

94. $2x^4 + 6x$

95. $9x^3 + x - 2(x^3 + 3x) = 9x^3 + x - 2x^3 - 6x$
$= 7x^3 - 5x$

96. $-10x^2 + 17x$

97. $12a^2 - 3ab + 5b^2 - 5(-5a^2 + 4ab - 6b^2)$
$= 12a^2 - 3ab + 5b^2 + 25a^2 - 20ab + 30b^2$
$= 37a^2 - 23ab + 35b^2$

98. $-20a^2 + 29ab + 48b^2$

99. $-7t^3 - t^2 - 3(5t^3 - 3t) = -7t^3 - t^2 - 15t^3 + 9t$
$= -22t^3 - t^2 + 9t$

100. $9t^4 - 45t^3 + 17t$

101. $[10(x + 3) - 4] + [2(x - 1) + 6]$
$= [10x + 30 - 4] + [2x - 2 + 6]$
$= [10x + 26] + [2x + 4]$
$= 10x + 26 + 2x + 4$
$= 12x + 30$

102. $13x - 1$

103. $[7(x^2 + 5) - 19] - [4(x^2 - 6) + 10]$
$= [7x^2 + 35 - 19] - [4x^2 - 24 + 10]$
$= [7x^2 + 16] - [4x^2 - 14]$
$= 7x^2 + 16 - 4x^2 + 14$
$= 3x^2 + 30$

104. $x^3 + 41$

105. $3\{[7(x - 2) + 4] - [2(2x - 5) + 6]\}$
$= 3\{[7x - 14 + 4] - [4x - 10 + 6]\}$
$= 3\{[7x - 10] - [4x - 4]\}$
$= 3\{7x - 10 - 4x + 4\}$
$= 3\{3x - 6\}$
$= 9x - 18$

106. $-16x + 44$

107. $4\{[5(x^3 - 3) + 2] - 3[2(x^3 + 5) - 9]\}$
$= 4\{[5x^3 - 15 + 2] - 3[2x^3 + 10 - 9]\}$
$= 4\{[5x^3 - 13] - 3[2x^3 + 1]\}$
$= 4\{5x^3 - 13 - 6x^3 - 3\}$
$= 4\{-x^3 - 16\}$
$= -4x^3 - 64$

108. $-12x^2 - 237$

109. $2x + 9$

110. $\frac{1}{2}(x + y)$

111.

112.

113. $z - \{2z - [3z - (4z - 5z) - 6z] - 7z\} - 8z$
$= z - \{2z - [3z - (-z) - 6z] - 7z\} - 8z$
$= z - \{2z - [3z + z - 6z] - 7z\} - 8z$
$= z - \{2z - [-2z] - 7z\} - 8z$
$= z - \{2z + 2z - 7z\} - 8z$
$= z - \{-3z\} - 8z$
$= z + 3z - 8z$
$= -4z$

114. $-2x - f$

115. $x-\{x-1-[x-2-(x-3-\{x-4-[x-5-(x-6)]\})]\}$
$= x-\{x-1-[x-2-(x-3-\{x-4-[x-5-x+6]\})]\}$
$= x - \{x - 1 - [x - 2 - (x - 3 - \{x - 4 - 1\})]\}$
$= x - \{x - 1 - [x - 2 - (x - 3 - \{x - 5\})]\}$
$= x - \{x - 1 - [x - 2 - (x - 3 - x + 5)]\}$
$= x - \{x - 1 - [x - 2 - 2]\}$
$= x - \{x - 1 - [x - 4]\}$
$= x - \{x - 1 - x + 4\}$
$= x - 3$

116.

117.

118. False

119. False; $-n + m = -(n - m) \neq -(n + m)$ for $m > 0$

120. True

121. False; $-n - m = -(n + m) \neq -(n - m)$ for $m > 0$

122. False

123. False; $-m(n - m) = -mn + m^2 = -(mn - m^2) \neq -(mn + m^2)$ for $m > 0$

124. True

125. True; $-n(-n - m) = n^2 + nm = n(n + m)$

Exercise Set 2.1

1. $x + 2 = 6$

$x + 2 + (-2) = 6 + (-2)$ Adding -2 on both sides

$x = 4$ Simplifying

Check: $x + 2 = 6$

$4 + 2 \ ? \ 6$

$6 \mid 6$ TRUE

2. 3

3. $x + 15 = -5$

$x + 15 + (-15) = -5 + (-15)$ Adding -15 on both sides

$x = -20$

Check: $x + 15 = -5$

$-20 + 15 \ ? \ -5$

$-5 \mid -5$ TRUE

4. 34

5. $x + 6 = -8$

$x + 6 + (-6) = -8 + (-6)$

$x = -14$

Check: $x + 6 = -8$

$-14 + 6 \ ? \ -8$

$-8 \mid -8$ TRUE

6. -21

7. $-2 = x + 16$

$-2 + (-16) = x + 16 + (-16)$

$-18 = x$

Check: $-2 = x + 16$

$-2 \ ? \ -18 + 16$

$-2 \mid -2$ TRUE

8. -31

9. $x - 9 = 6$

$x - 9 + 9 = 6 + 9$

$x = 15$

Check: $x - 9 = 6$

$15 - 9 \ ? \ 6$

$6 \mid 6$ TRUE

10. 13

11. $x - 7 = -21$

$x - 7 + 7 = -21 + 7$

$x = -14$

Check: $x - 7 = -21$

$-14 - 7 \ ? \ -21$

$-21 \mid -21$ TRUE

12. -11

13. $5 + t = 7$

$-5 + 5 + t = -5 + 7$

$t = 2$

Check: $5 + t = 7$

$5 + 2 \ ? \ 7$

$7 \mid 7$ TRUE

14. 4

15. $13 = -7 + y$

$7 + 13 = 7 + (-7) + y$

$20 = y$

Check: $13 = -7 + y$

$13 \ ? \ -7 + 20$

$13 \mid 13$ TRUE

16. 24

17. $-3 + t = -9$

$3 + (-3) + t = 3 + (-9)$

$t = -6$

Check: $-3 + t = -9$

$-3 + (-6) \ ? \ -9$

$-9 \mid -9$ TRUE

18. -15

19. $r + \frac{1}{3} = \frac{8}{3}$

$r + \frac{1}{3} + \left(-\frac{1}{3}\right) = \frac{8}{3} + \left(-\frac{1}{3}\right)$

$r = \frac{7}{3}$

Check: $r + \frac{1}{3} = \frac{8}{3}$

$\frac{7}{3} + \frac{1}{3} \ ? \ \frac{8}{3}$

$\frac{8}{3} \mid \frac{8}{3}$ TRUE

20. $\frac{1}{4}$

21. $m + \frac{5}{6} = -\frac{11}{12}$

$m + \frac{5}{6} + \left(-\frac{5}{6}\right) = -\frac{11}{12} + \left(-\frac{5}{6}\right)$

$m = -\frac{11}{12} + \left(-\frac{5}{6}\right)\left(\frac{2}{2}\right)$

$m = -\frac{11}{12} + \left(-\frac{10}{12}\right)$

$m = -\frac{21}{12} = -\frac{3 \cdot 7}{3 \cdot 4}$

$m = -\frac{7}{4}$

Check: $m + \frac{5}{6} = -\frac{11}{12}$

$-\frac{7}{4} + \frac{5}{6} \ ? \ -\frac{11}{12}$

$-\frac{21}{12} + \frac{10}{12}$

$-\frac{11}{12} \mid -\frac{11}{12}$ TRUE

22. $-\frac{3}{2}$

23. $x - \frac{5}{6} = \frac{7}{8}$

$x - \frac{5}{6} + \frac{5}{6} = \frac{7}{8} + \frac{5}{6}$

$x = \frac{7}{8} \cdot \frac{3}{3} + \frac{5}{6} \cdot \frac{4}{4}$

$x = \frac{21}{24} + \frac{20}{24}$

$x = \frac{41}{24}$

Check: $x - \frac{5}{6} = \frac{7}{8}$

$\frac{41}{24} - \frac{5}{6} \ ? \ \frac{7}{8}$

$\frac{41}{24} - \frac{20}{24} \mid \frac{21}{24}$

$\frac{21}{24} \mid \frac{21}{24}$ TRUE

24. $\frac{19}{12}$

25. $-\frac{1}{5} + z = -\frac{1}{4}$

$\frac{1}{5} - \frac{1}{5} + z = \frac{1}{5} - \frac{1}{4}$

$z = \frac{1}{5} \cdot \frac{4}{4} - \frac{1}{4} \cdot \frac{5}{5}$

$z = \frac{4}{20} - \frac{5}{20}$

$z = -\frac{1}{20}$

Check: $-\frac{1}{5} + z = -\frac{1}{4}$

$-\frac{1}{5} + \left(-\frac{1}{20}\right) \ ? \ -\frac{1}{4}$

$-\frac{4}{20} + \left(-\frac{1}{20}\right) \mid -\frac{5}{20}$

$-\frac{5}{20} \mid -\frac{5}{20}$ TRUE

26. $-\frac{5}{8}$

27. $x + 2.3 = 7.4$

$x + 2.3 + (-2.3) = 7.4 + (-2.3)$

$x = 5.1$

Check: $x + 2.3 = 7.4$

$5.1 + 2.3 \ ? \ 7.4$

$7.4 \mid 7.4$ TRUE

28. 4.7

29. $-9.7 = -4.7 + y$

$4.7 + (-9.7) = 4.7 + (-4.7) + y$

$-5 = y$

Check: $-9.7 = -4.7 + y$

$-9.7 \ ? \ -4.7 + (-5)$

$-9.7 \mid -9.7$ TRUE

30. -10.6

31. $6x = 36$

$\frac{6x}{6} = \frac{36}{6}$

$1 \cdot x = 6$

$x = 6$

Check: $6x = 36$

$6 \cdot 6 \ ? \ 36$

$36 \mid 36$ TRUE

32. 13

33. $5x = 45$

$\frac{5x}{5} = \frac{45}{5}$

$1 \cdot x = 9$

$x = 9$

Check: $5x = 45$

$5 \cdot 9 \ ? \ 45$

$45 \mid 45$ TRUE

34. 8

35. $84 = 7x$

$\frac{84}{7} = \frac{7x}{7}$

$12 = x$

Check: $84 = 7x$

$84 \ ? \ 7 \cdot 12$

$84 \mid 84$ TRUE

36. 7

37. $-x = 40$

$-1 \cdot x = 40$

$-1 \cdot (-1 \cdot x) = -1 \cdot 40$

$1 \cdot x = -40$

$x = -40$

Check: $-x = 40$

$-(-40) \ ? \ 40$

$40 \mid 40$ TRUE

38. -100

39. $-x = -1$

$-1 \cdot x = -1$

$-1 \cdot (-1 \cdot x) = -1 \cdot (-1)$

$1 \cdot x = 1$

$x = 1$

Check: $-x = -1$

$-(1) \ ? \ -1$

$-1 \mid -1$ TRUE

40. 68

41. $7x = -49$

$\frac{7x}{7} = \frac{-49}{7}$

$x = -7$

Check: $7x = -49$

$7(-7) \ ? \ -49$

$-49 \mid -49$ TRUE

42. -4

43. $-12x = 72$

$\frac{-12x}{-12} = \frac{72}{-12}$

$x = -6$

Check: $-12x = 72$

$-12(-6) \ ? \ 72$

$72 \mid 72$ TRUE

44. -7

45. $-21x = -126$

$\frac{-21x}{-21} = \frac{-126}{-21}$

$x = 6$

Check: $-21x = -126$

$-21\cdot 6 \ ?\ -126$

$-126 \mid -126$ TRUE

46. 8

47. $\frac{t}{7} = -9$

$7 \cdot \left[\frac{1}{7}t\right] = 7\cdot(-9)$

$t = -63$

Check: $\frac{t}{7} = -9$

$\frac{-63}{7} \ ?\ -9$

$-9 \mid -9$ TRUE

48. -88

49. $\frac{3}{4}x = 27$

$\frac{4}{3} \cdot \frac{3}{4}x = \frac{4}{3} \cdot 27$

$x = \frac{4\cdot\cancel{3}\cdot 3\cdot 3}{\cancel{3}\cdot 1}$

$x = 36$

Check: $\frac{3}{4}x = 27$

$\frac{3}{4} \cdot 36 \ ?\ 27$

$27 \mid 27$ TRUE

50. 20

51. $\frac{-t}{3} = 7$

$3 \cdot \frac{1}{3} \cdot (-t) = 3\cdot 7$

$-t = 21$

$-1\cdot(-1\cdot t) = -1\cdot 21$

$1\cdot t = -21$

$t = -21$

Check: $\frac{-t}{3} = 7$

$\frac{-(-21)}{3} \ ?\ 7$

$\frac{21}{3}$

$7 \mid 7$ TRUE

52. -54

53. $\frac{1}{5} = -\frac{m}{3}$

$\frac{1}{5} = -\frac{1}{3} \cdot m$

$-3 \cdot \frac{1}{5} = -3 \cdot \left[-\frac{1}{3} \cdot m\right]$

$-\frac{3}{5} = m$

Check: $\frac{1}{5} = -\frac{m}{3}$

$\frac{1}{5} \ ?\ -\frac{-\frac{3}{5}}{3}$

$-\left[-\frac{3}{5} \div 3\right]$

$-\left[-\frac{3}{5} \cdot \frac{1}{3}\right]$

$-\left[-\frac{1}{5}\right]$

$\frac{1}{5} \mid \frac{1}{5}$ TRUE

54. $-\frac{7}{9}$

55. $-\frac{3}{5}r = -\frac{9}{10}$

$-\frac{5}{3} \cdot \left[-\frac{3}{5}r\right] = -\frac{5}{3} \cdot \left[-\frac{9}{10}\right]$

$r = \frac{\cancel{5}\cdot\cancel{3}\cdot 3}{\cancel{3}\cdot\cancel{5}\cdot 2}$

$r = \frac{3}{2}$

Check: $-\frac{3}{5}r = -\frac{9}{10}$

$-\frac{3}{5} \cdot \frac{3}{2} \ ?\ -\frac{9}{10}$

$-\frac{9}{10} \mid -\frac{9}{10}$ TRUE

56. $\frac{2}{3}$

57. $\frac{-3r}{2} = -\frac{27}{4}$

$-\frac{3}{2}r = -\frac{27}{4}$

$-\frac{2}{3} \cdot \left[-\frac{3}{2}r\right] = -\frac{2}{3} \cdot \left[-\frac{27}{4}\right]$

$r = \frac{\cancel{2}\cdot\cancel{3}\cdot 3\cdot 3}{\cancel{3}\cdot\cancel{2}\cdot 2}$

$r = \frac{9}{2}$

Check: $\frac{-3r}{2} = -\frac{27}{4}$

$-\frac{3}{2} \cdot \frac{9}{2} \ ?\ -\frac{27}{4}$

$-\frac{27}{4} \mid -\frac{27}{4}$ TRUE

58. -1

59. $6.3x = 44.1$

$\frac{6.3x}{6.3} = \frac{44.1}{6.3}$

$x = 7$

Check: $6.3x = 44.1$

$6.3\cdot 7 \ ?\ 44.1$

$44.1 \mid 44.1$ TRUE

60. 20

61. $3.7 + t = 8.2$

$3.7 + t - 3.7 = 8.2 - 3.7$

$t = 4.5$

62. 24

63. $18 = -\frac{2}{3}x$

$-\frac{3}{2} \cdot 18 = -\frac{3}{2}\left[-\frac{2}{3}x\right]$

$-\frac{3\cdot 2\cdot 9}{2\cdot 1} = x$

$-27 = x$

64. -5.5

65. $17 = y + 29$

$17 - 29 = y + 29 - 29$

$-12 = y$

66. -128

67. $y - \frac{2}{3} = -\frac{1}{6}$

$y - \frac{2}{3} + \frac{2}{3} = -\frac{1}{6} + \frac{2}{3}$

$y = -\frac{1}{6} + \frac{2}{3} \cdot \frac{2}{2}$

$y = -\frac{1}{6} + \frac{4}{6}$

$y = \frac{3}{6}$

$y = \frac{1}{2}$

68. $-\frac{14}{9}$

69. $-24 = \frac{8x}{5}$

$-24 = \frac{8}{5}x$

$\frac{5}{8}(-24) = \frac{5}{8} \cdot \frac{8}{5}x$

$-\frac{5 \cdot 8 \cdot 3}{8 \cdot 1} = x$

$-15 = x$

70. $-\frac{1}{2}$

71. $-4.1t = 10.25$

$\frac{-4.1t}{-4.1} = \frac{10.25}{-4.1}$

$t = -2.5$

72. $-\frac{19}{23}$

73. $3x + 4x = (3 + 4)x = 7x$

74. $-x + 5$

75. $3x - (4 + 2x) = 3x - 4 - 2x = x - 4$

76. $-5x - 23$

77.

78.

79. $-356.788 = -699.034 + t$

$699.034 + (-356.788) = 699.034 + (-699.034) + t$

$342.246 = t$

80. -8655

81. For all x, $0 \cdot x = 0$. There is no solution to $0 \cdot x = 9$.

82. All real numbers

83. $4|x| = 48$

$|x| = 12$

x represents a number whose distance from 0 is 12. Thus, $x = -12$ or $x = 12$.

84. No solution

85. For all x, $0 \cdot x = 0$. Thus, the solution is all real numbers.

86. 0

87. $x + 4 = 5 + x$

$x + 4 - x = 5 + x - x$

$4 = 5$

Since $4 = 5$ is false, the equation has no solution.

88. -2, 2

89. $ax = 5a$

$\frac{ax}{a} = \frac{5a}{a}$

$x = 5$

90. $a + 4$

91. $3x = \frac{b}{a}$

$\frac{1}{3} \cdot 3x = \frac{1}{3} \cdot \frac{b}{a}$

$x = \frac{b}{3a}$

92. $\frac{a^2 + 1}{c}$

93. $1 - c = a + x$

$1 - c - a = a + x - a$

$1 - c - a = x$

94. -13, 13

95. $x - 4720 = 1634$

$x - 4720 + 4720 = 1634 + 4720$

$x = 6354$

$x + 4720 = 6354 + 4720$

$x + 4720 = 11{,}074$

96. 250

97. 

Exercise Set 2.2

1. $5x + 6 = 31$
$5x + 6 - 6 = 31 - 6$
$5x = 25$
$\frac{5x}{5} = \frac{25}{5}$
$x = 5$

Check: $5x + 6 = 31$
$5 \cdot 5 + 6 \ ? \ 31$
$25 + 6 \ | \ $
$31 \ | \ 31$ TRUE

2. 8

3. $8x + 4 = 68$
$8x + 4 - 4 = 68 - 4$
$8x = 64$
$\frac{8x}{8} = \frac{64}{8}$
$x = 8$

Check: $8x + 4 = 68$
$8 \cdot 8 + 4 \ ? \ 68$
$64 + 4 \ | \ $
$68 \ | \ 68$ TRUE

4. 9

5. $4x - 6 = 34$
$4x - 6 + 6 = 34 + 6$
$4x = 40$
$\frac{4x}{4} = \frac{40}{4}$
$x = 10$

Check: $4x - 6 = 34$
$4 \cdot 10 - 6 \ ? \ 34$
$40 - 6 \ | \ $
$34 \ | \ 34$ TRUE

6. 3

7. $3x - 9 = 33$
$3x - 9 + 9 = 33 + 9$
$3x = 42$
$\frac{3x}{3} = \frac{42}{3}$
$x = 14$

Check: $3x - 9 = 33$
$3 \cdot 14 - 9 \ ? \ 33$
$42 - 9 \ | \ $
$33 \ | \ 33$ TRUE

8. 11

9. $7x + 2 = -54$
$7x + 2 - 2 = -54 - 2$
$7x = -56$
$\frac{7x}{7} = \frac{-56}{7}$
$x = -8$

Check: $7x + 2 = -54$
$7(-8) + 2 \ ? \ -54$
$-56 + 2 \ | \ $
$-54 \ | \ -54$ TRUE

10. -9

11. $-45 = 6y + 3$
$-45 - 3 = 6y + 3 - 3$
$-48 = 6y$
$-\frac{48}{6} = \frac{6y}{6}$
$-8 = y$

Check: $-45 = 6y + 3$
$-45 \ ? \ 6(-8) + 3$
$\ | \ -48 + 3$
$-45 \ | \ -45$ TRUE

12. -11

13. $-4x + 7 = 35$
$-4x + 7 - 7 = 35 - 7$
$-4x = 28$
$\frac{4x}{-4} = \frac{28}{-4}$
$x = -7$

Check: $-4x + 7 = 35$
$-4(-7) + 7 \ ? \ 35$
$28 + 7 \ | \ $
$35 \ | \ 35$ TRUE

14. -23

15. $-7x - 24 = -129$
$-7x - 24 + 24 = -129 + 24$
$-7x = -105$
$\frac{-7x}{-7} = \frac{-105}{-7}$
$x = 15$

Check: $-7x - 24 = -129$
$-7 \cdot 15 - 24 \ ? \ -129$
$-105 - 24 \ | \ $
$-129 \ | \ -129$ TRUE

16. 19

17. $5x + 7x = 72$
$12x = 72$
$\frac{12x}{12} = \frac{72}{12}$
$x = 6$

Check: $5x + 7x = 72$
$5 \cdot 6 + 7 \cdot 6 \ ? \ 72$
$30 + 42 \ | \ $
$72 \ | \ 72$ TRUE

18. 5

19. $8x + 7x = 60$
$15x = 60$
$\frac{15x}{15} = \frac{60}{15}$
$x = 4$

Check: $8x + 7x = 60$
$8 \cdot 4 + 7 \cdot 4 \ ? \ 60$
$32 + 28 \ | \ $
$60 \ | \ 60$ TRUE

20. 8

21. $4x + 3x = 42$
$7x = 42$
$\frac{7x}{7} = \frac{42}{7}$
$x = 6$

Check: $4x + 3x = 42$
$4 \cdot 6 + 3 \cdot 6 \ ? \ 42$
$24 + 18 \ | \ $
$42 \ | \ 42$ TRUE

22. 4

23. $-6y - 3y = 27$
$-9y = 27$
$\frac{-9y}{-9} = \frac{27}{-9}$
$y = -3$

Check: $-6y - 3y = 27$
$-6(-3) - 3(-3) \ ? \ 27$
$18 + 9 \ | \ $
$27 \ | \ 27$ TRUE

24. -4

25. $-7y - 8y = -15$
$-15y = -15$
$\frac{-15y}{-15} = \frac{-15}{-15}$
$y = 1$

Check: $-7y - 8y = -15$
$-7 \cdot 1 - 8 \cdot 1 \ ? \ -15$
$-7 - 8 \ | \ $
$-15 \ | \ -15$ TRUE

26. 3

27. $10.2y - 7.3y = -58$
$2.9y = -58$
$\frac{2.9y}{2.9} = \frac{-58}{2.9}$
$y = -\frac{58}{2.9}$
$y = -20$

Check: $10.2y - 7.3y = -58$
$10.2(-20) - 7.3(-20) \ ? \ -58$
$-204 + 146 \ | \ $
$-58 \ | \ -58$ TRUE

28. -20

29. $x + \frac{1}{3}x = 8$
$\left(1 + \frac{1}{3}\right)x = 8$
$\frac{4}{3}x = 8$
$\frac{3}{4} \cdot \frac{4}{3}x = \frac{3}{4} \cdot 8$
$x = 6$

Check: $x + \frac{1}{3}x = 8$
$6 + \frac{1}{3} \cdot 6 \ ? \ 8$
$6 + 2 \ | \ $
$8 \ | \ 8$ TRUE

30. 8

31. $8y - 35 = 3y$
$8y = 3y + 35$
$8y - 3y = 35$
$5y = 35$
$y = \frac{35}{5}$
$y = 7$

Check: $8y - 35 = 3y$
$8 \cdot 7 - 35 \ ? \ 3 \cdot 7$
$56 - 35 \ | \ 21$
$21 \ | \ 21$ TRUE

32. -3

33. $8x - 1 = 23 - 4x$
$8x + 4x = 23 + 1$
$12x = 24$
$x = \frac{24}{12}$
$x = 2$

Check: $8x - 1 = 23 - 4x$
$8 \cdot 2 - 1 \ ? \ 23 - 4 \cdot 2$
$16 - 1 \ | \ 23 - 8$
$15 \ | \ 15$ TRUE

34. 5

35. $2x - 1 = 4 + x$
$2x - x = 4 + 1$
$x = 5$

Check: $2x - 1 = 4 + x$
$2 \cdot 5 - 1 \ ? \ 4 + 5$
$10 - 1 \ | \ 9$
$9 \ | \ 9$ TRUE

36. 2

37. $6x + 3 = 2x + 11$
$6x - 2x = 11 - 3$
$4x = 8$
$x = \frac{8}{4}$
$x = 2$

Check: $6x + 3 = 2x + 11$
$6 \cdot 2 + 3 \ ? \ 2 \cdot 2 + 11$
$12 + 3 \ | \ 4 + 11$
$15 \ | \ 15$ TRUE

38. 4

39. $5 - 2x = 3x - 7x + 25$
$5 - 2x = -4x + 25$
$4x - 2x = 25 - 5$
$2x = 20$
$x = \frac{20}{2}$
$x = 10$

Check: $5 - 2x = 3x - 7x + 25$
$5 - 2 \cdot 10 \ ? \ 3 \cdot 10 - 7 \cdot 10 + 25$
$5 - 20 \ | \ 30 - 70 + 25$
$-15 \ | \ -40 + 25$
$-15 \ | \ -15$ TRUE

40. 10

41. $4 + 3x - 6 = 3x + 2 - x$
$3x - 2 = 2x + 2$ Collecting like terms on each side
$3x - 2x = 2 + 2$
$x = 4$

Check:
$$\begin{array}{r|l}
\multicolumn{2}{c}{4 + 3x - 6 = 3x + 2 - x} \\
\hline
4 + 3\cdot 4 - 6 \ ? & 3\cdot 4 + 2 - 4 \\
4 + 12 - 6 & 12 + 2 - 4 \\
16 - 6 & 14 - 4 \\
10 & 10 \quad \text{TRUE}
\end{array}$$

42. 0

43. $4y - 4 + y + 24 = 6y + 20 - 4y$
$$5y + 20 = 2y + 20$$
$$5y - 2y = 20 - 20$$
$$3y = 0$$
$$y = 0$$

Check:
$$\begin{array}{r|l}
\multicolumn{2}{c}{4y - 4 + y + 24 = 6y + 20 - 4y} \\
\hline
4\cdot 0 - 4 + 0 + 24 \ ? & 6\cdot 0 + 20 - 4\cdot 0 \\
0 - 4 + 0 + 24 & 0 + 20 - 0 \\
20 & 20 \quad \text{TRUE}
\end{array}$$

44. 7

45. $\frac{7}{2}x + \frac{1}{2}x = 3x + \frac{3}{2} + \frac{5}{2}x$

The number 2 is the least common denominator, so we multiply by 2 on both sides.

$$2\left(\frac{7}{2}x + \frac{1}{2}x\right) = 2\left(3x + \frac{3}{2} + \frac{5}{2}x\right)$$
$$2 \cdot \frac{7}{2}x + 2 \cdot \frac{1}{2}x = 2\cdot 3x + 2 \cdot \frac{3}{2} + 2 \cdot \frac{5}{2}x$$
$$7x + x = 6x + 3 + 5x$$
$$8x = 11x + 3$$
$$8x - 11x = 3$$
$$-3x = 3$$
$$x = \frac{3}{-3}$$
$$x = -1$$

Check:
$$\begin{array}{r|l}
\multicolumn{2}{c}{\frac{7}{2}x + \frac{1}{2}x = 3x + \frac{3}{2} + \frac{5}{2}x} \\
\hline
\frac{7}{2}(-1) + \frac{1}{2}(-1) \ ? & 3(-1) + \frac{3}{2} + \frac{5}{2}(-1) \\
-\frac{7}{2} - \frac{1}{2} & -3 + \frac{3}{2} - \frac{5}{2} \\
-\frac{8}{2} & -\frac{3}{2} - \frac{5}{2} \\
-4 & -\frac{8}{2} \\
-4 & -4 \quad \text{TRUE}
\end{array}$$

46. $\frac{1}{2}$

47. $\frac{2}{3} + \frac{1}{4}t = 6$

The number 12 is the least common denominator, so we multiply by 12 on both sides.

$$12\left(\frac{2}{3} + \frac{1}{4}t\right) = 12 \cdot 6$$
$$12 \cdot \frac{2}{3} + 12 \cdot \frac{1}{4}t = 72$$
$$8 + 3t = 72$$
$$3t = 72 - 8$$
$$3t = 64$$
$$t = \frac{64}{3}$$

Check:
$$\begin{array}{r|l}
\multicolumn{2}{c}{\frac{2}{3} + \frac{1}{4}t = 6} \\
\hline
\frac{2}{3} + \frac{1}{4}\left(\frac{64}{3}\right) \ ? & 6 \\
\frac{2}{3} + \frac{16}{3} & \\
\frac{18}{3} & \\
6 & 6 \quad \text{TRUE}
\end{array}$$

48. $-\frac{2}{3}$

49. $\frac{2}{3} + 3y = 5y - \frac{2}{15}$

The number 15 is the least common denominator, so we multiply by 15 on both sides.

$$15\left(\frac{2}{3} + 3y\right) = 15\left(5y - \frac{2}{15}\right)$$
$$15 \cdot \frac{2}{3} + 15\cdot 3y = 15\cdot 5y - 15 \cdot \frac{2}{15}$$
$$10 + 45y = 75y - 2$$
$$10 + 2 = 75y - 45y$$
$$12 = 30y$$
$$\frac{12}{30} = y$$
$$\frac{2}{5} = y$$

Check:
$$\begin{array}{r|l}
\multicolumn{2}{c}{\frac{2}{3} + 3y = 5y - \frac{2}{15}} \\
\hline
\frac{2}{3} + 3 \cdot \frac{2}{5} \ ? & 5 \cdot \frac{2}{5} - \frac{2}{15} \\
\frac{2}{3} + \frac{6}{5} & 2 - \frac{2}{15} \\
\frac{10}{15} + \frac{18}{15} & \frac{30}{15} - \frac{2}{15} \\
\frac{28}{15} & \frac{28}{15} \quad \text{TRUE}
\end{array}$$

50. -3

51. $\frac{5}{3} + \frac{2}{3}x = \frac{25}{12} + \frac{5}{4}x + \frac{3}{4}$

The number 12 is the least common denominator, so we multiply by 12 on both sides.

$$12\left(\frac{5}{3} + \frac{2}{3}x\right) = 12\left(\frac{25}{12} + \frac{5}{4}x + \frac{3}{4}\right)$$
$$12 \cdot \frac{5}{3} + 12 \cdot \frac{2}{3}x = 12 \cdot \frac{25}{12} + 12 \cdot \frac{5}{4}x + 12 \cdot \frac{3}{4}$$
$$20 + 8x = 25 + 15x + 9$$
$$20 + 8x = 15x + 34$$
$$20 - 34 = 15x - 8x$$
$$-14 = 7x$$
$$\frac{-14}{7} = x$$
$$-2 = x$$

Check: $\frac{5}{3} + \frac{2}{3}x = \frac{25}{12} + \frac{5}{4}x + \frac{3}{4}$

$\frac{5}{3} + \frac{2}{3}(-2)$?	$\frac{25}{12} + \frac{5}{4}(-2) + \frac{3}{4}$	
$\frac{5}{3} - \frac{4}{3}$	$\frac{25}{12} - \frac{5}{2} + \frac{3}{4}$	
$\frac{1}{3}$	$\frac{25}{12} - \frac{30}{12} + \frac{9}{12}$	
	$\frac{4}{12}$	
$\frac{1}{3}$	$\frac{1}{3}$	TRUE

52. -3

53. $2.1x + 45.2 = 3.2 - 8.4x$ Greatest number of decimal places is 1

$10(2.1x + 45.2) = 10(3.2 - 8.4x)$ Multiplying by 10 to clear decimals

$$10(2.1x) + 10(45.2) = 10(3.2) - 10(8.4x)$$
$$21x + 452 = 32 - 84x$$
$$21x + 84x = 32 - 452$$
$$105x = -420$$
$$x = \frac{-420}{105}$$
$$x = -4$$

Check: $2.1x + 45.2 = 3.2 - 8.4x$

$2.1(-4) + 45.2$?	$3.2 - 8.4(-4)$	
$-8.4 + 45.2$	$3.2 + 33.6$	
36.8	36.8	TRUE

54. $\frac{5}{3}$

55. $1.03 - 0.6x = 0.71 - 0.2x$
Greatest number of decimal places is 2

$100(1.03 - 0.6x) = 100(0.71 - 0.2x)$
Multiplying by 100 to clear decimals

$$100(1.03) - 100(0.6x) = 100(0.71) - 100(0.2x)$$
$$103 - 60x = 71 - 20x$$
$$32 = 40x$$
$$\frac{32}{40} = x$$
$$\frac{4}{5} = x, \text{ or}$$
$$0.8 = x$$

Check: $1.03 - 0.6x = 0.71 - 0.2x$

$1.03 - 0.6(0.8)$?	$0.71 - 0.2(0.8)$	
$1.03 - 0.48$	$0.71 - 0.16$	
0.55	0.55	TRUE

56. 1

57. $\frac{2}{7}x - \frac{1}{2}x = \frac{3}{4}x + 1$

The least common denominator is 28.

$$28\left(\frac{2}{7}x - \frac{1}{2}x\right) = 28\left(\frac{3}{4}x + 1\right)$$
$$28 \cdot \frac{2}{7}x - 28 \cdot \frac{1}{2}x = 28 \cdot \frac{3}{4}x + 28 \cdot 1$$
$$8x - 14x = 21x + 28$$
$$-6x = 21x + 28$$
$$-6x - 21x = 28$$
$$-27x = 28$$
$$x = -\frac{28}{27}$$

Check: $\frac{2}{7}x - \frac{1}{2}x = \frac{3}{4}x + 1$

$\frac{2}{7}\left(-\frac{28}{27}\right) - \frac{1}{2}\left(-\frac{28}{27}\right)$?	$\frac{3}{4}\left(-\frac{28}{27}\right) + 1$	
$-\frac{8}{27} + \frac{14}{27}$	$-\frac{21}{27} + 1$	
$\frac{6}{27}$	$\frac{6}{27}$	TRUE

58. $\frac{32}{7}$

59. $3(2y - 3) = 27$
$$6y - 9 = 27$$
$$6y = 27 + 9$$
$$6y = 36$$
$$y = 6$$

Check: $3(2y - 3) = 27$

$3(2\cdot6 - 3)$?	27	
$3(12 - 3)$		
$3\cdot9$		
27	27	TRUE

60. 5

61. $40 = 5(3x + 2)$
$40 = 15x + 10$
$40 - 10 = 15x$
$30 = 15x$
$2 = x$

Check: $40 = 5(3x + 2)$
$40 \; ? \; 5(3 \cdot 2 + 2)$
$40 \mid 5(6 + 2)$
$5 \cdot 8$
$40 \mid 40$ TRUE

62. 1

63. $2(3 + 4m) - 9 = 45$
$6 + 8m - 9 = 45$
$8m - 3 = 45$
$8m = 45 + 3$
$8m = 48$
$m = 6$

Check: $2(3 + 4m) - 9 = 45$
$2(3 + 4 \cdot 6) - 9 \; ? \; 45$
$2(3 + 24) - 9$
$2 \cdot 27 - 9$
$54 - 9$
$45 \mid 45$ TRUE

64. 9

65. $5r - (2r + 8) = 16$
$5r - 2r - 8 = 16$
$3r - 8 = 16$
$3r = 16 + 8$
$3r = 24$
$r = 8$

Check: $5r - (2r + 8) = 16$
$5 \cdot 8 - (2 \cdot 8 + 8) \; ? \; 16$
$40 - (16 + 8)$
$40 - 24$
$16 \mid 16$ TRUE

66. 8

67. $6 - 2(3x - 1) = 2$
$6 - 6x + 2 = 2$
$8 - 6x = 2$
$8 - 2 = 6x$
$6 = 6x$
$1 = x$

Check: $6 - 2(3x - 1) = 2$
$6 - 2(3 \cdot 1 - 1) \; ? \; 2$
$6 - 2(3 - 1)$
$6 - 2 \cdot 2$
$6 - 4$
$2 \mid 2$ TRUE

68. 2

69. $5(d + 4) = 7(d - 2)$
$5d + 20 = 7d - 14$
$20 + 14 = 7d - 5d$
$34 = 2d$
$17 = d$

Check: $5(d + 4) = 7(d - 2)$
$5(17 + 4) \; ? \; 7(17 - 2)$
$5 \cdot 21 \mid 7 \cdot 15$
$105 \mid 105$ TRUE

70. -4

71. $8(2t + 1) = 4(7t + 7)$
$16t + 8 = 28t + 28$
$16t - 28t = 28 - 8$
$-12t = 20$
$t = -\frac{20}{12}$
$t = -\frac{5}{3}$

Check: $8(2t + 1) = 4(7t + 7)$
$8\left[2\left(-\frac{5}{3}\right) + 1\right] \; ? \; 4\left[7\left(-\frac{5}{3}\right) + 7\right]$
$8\left(-\frac{10}{3} + 1\right) \mid 4\left(-\frac{35}{3} + 7\right)$
$8\left(-\frac{7}{3}\right) \mid 4\left(-\frac{14}{3}\right)$
$-\frac{56}{3} \mid -\frac{56}{3}$ TRUE

72. -8

73. $3(r - 6) + 2 = 4(r + 2) - 21$
$3r - 18 + 2 = 4r + 8 - 21$
$3r - 16 = 4r - 13$
$13 - 16 = 4r - 3r$
$-3 = r$

Check: $3(r - 6) + 2 = 4(r + 2) - 21$
$3(-3 - 6) + 2 \; ? \; 4(-3 + 2) - 21$
$3(-9) + 2 \mid 4(-1) - 21$
$-27 + 2 \mid -4 - 21$
$-25 \mid -25$ TRUE

74. -12

75. $19 - (2x + 3) = 2(x + 3) + x$
$19 - 2x - 3 = 2x + 6 + x$
$16 - 2x = 3x + 6$
$16 - 6 = 3x + 2x$
$10 = 5x$
$2 = x$

Check: $19 - (2x + 3) = 2(x + 3) + x$

$$\begin{array}{r|l} 19 - (2\cdot 2 + 3) \; ? & 2(2 + 3) + 2 \\ 19 - (4 + 3) & 2\cdot 5 + 2 \\ 19 - 7 & 10 + 2 \\ 12 & 12 \end{array}$$ TRUE

76. 1

77. $\frac{1}{3}(6x + 24) - 20 = -\frac{1}{4}(12x - 72)$

$$2x + 8 - 20 = -3x + 18$$
$$2x - 12 = -3x + 18$$
$$5x = 30$$
$$x = 6$$

The check is left to the student.

78. 5

79. $2[4 - 2(3 - x)] - 1 = 4[2(4x - 3) + 7] - 25$

$$2[4 - 6 + 2x] - 1 = 4[8x - 6 + 7] - 25$$
$$2[-2 + 2x] - 1 = 4[8x + 1] - 25$$
$$-4 + 4x - 1 = 32x + 4 - 25$$
$$4x - 5 = 32x - 21$$
$$-5 + 21 = 32x - 4x$$
$$16 = 28x$$
$$\frac{16}{28} = x$$
$$\frac{4}{7} = x$$

The check is left to the student.

80. $-\frac{27}{19}$

81. $\frac{2}{3}(2x - 1) = 10$

$3 \cdot \frac{2}{3}(2x - 1) = 3\cdot 10$ Multiplying by 3 to clear the fraction

$$2(2x - 1) = 30$$
$$4x - 2 = 30$$
$$4x = 30 + 2$$
$$4x = 32$$
$$x = 8$$

Check: $\frac{2}{3}(2x - 1) = 10$

$$\begin{array}{r|l} \frac{2}{3}(2\cdot 8 - 1) \; ? & 10 \\ \frac{2}{3}(16 - 1) & \\ \frac{2}{3} \cdot 15 & \\ 10 & 10 \end{array}$$ TRUE

82. 7

83. $\frac{3}{4}\left(3x - \frac{1}{2}\right) - \frac{2}{3} = \frac{1}{3}$

$$\frac{9}{4}x - \frac{3}{8} - \frac{2}{3} = \frac{1}{3}$$

The number 24 is the least common denominator, so we multiply by 24 on both sides.

$$24\left(\frac{9}{4}x - \frac{3}{8} - \frac{2}{3}\right) = 24 \cdot \frac{1}{3}$$
$$24 \cdot \frac{9}{4}x - 24 \cdot \frac{3}{8} - 24 \cdot \frac{2}{3} = 8$$
$$54x - 9 - 16 = 8$$
$$54x - 25 = 8$$
$$54x = 8 + 25$$
$$54x = 33$$
$$x = \frac{33}{54}$$
$$x = \frac{11}{18}$$

The check is left to the student.

84. $-\frac{5}{32}$

85. $0.7(3x + 6) = 1.1 - (x + 2)$

$$2.1x + 4.2 = 1.1 - x - 2$$

$10(2.1x + 4.2) = 10(1.1 - x - 2)$ Clearing decimals

$$21x + 42 = 11 - 10x - 20$$
$$21x + 42 = -10x - 9$$
$$21x + 10x = -9 - 42$$
$$31x = -51$$
$$x = -\frac{51}{31}$$

The check is left to the student.

86. $\frac{39}{14}$

87. $a + (a - 3) = (a + 2) - (a + 1)$

$$a + a - 3 = a + 2 - a - 1$$
$$2a - 3 = 1$$
$$2a = 1 + 3$$
$$2a = 4$$
$$a = 2$$

Check: $a + (a - 3) = (a + 2) - (a + 1)$

$$\begin{array}{r|l} 2 + (2 - 3) \; ? & (2 + 2) - (2 + 1) \\ 2 - 1 & 4 - 3 \\ 1 & 1 \end{array}$$ TRUE

88. -7.4

89. Do the long division. The answer is negative.

```
          6. 5
3.4^ )2 2.1^0
       2 0 4
         1 7 0
         1 7 0
             0
```

$-22.1 \div 3.4 = -6.5$

90. $7(x - 3 - 2y)$

91. Since -15 is to the left of -13 on the number line, -15 is less than -13, so $-15 < -13$.

92. -14

93.

94.

95. Since we are using a calculator we will not clear the decimals.

$$\begin{aligned} 0.008 + 9.62x - 42.8 &= 0.944x + 0.0083 - x \\ 9.62x - 42.792 &= -0.056x + 0.0083 \\ 9.62x + 0.056x &= 0.0083 + 42.792 \\ 9.676x &= 42.8003 \\ x &= \frac{42.8003}{9.676} \\ x &\approx 4.4233464 \end{aligned}$$

96. -4

97.

$$\begin{aligned} 0 &= y - (-14) - (-3y) \\ 0 &= y + 14 + 3y \\ 0 &= 4y + 14 \\ -14 &= 4y \\ \frac{-14}{4} &= y \\ -\frac{7}{2} &= y \end{aligned}$$

98. All real numbers

99.

$$\begin{aligned} 475(54x + 7856) + 9762 &= 402(83x + 975) \\ 25{,}650x + 3{,}731{,}600 + 9762 &= 33{,}366x + 391{,}950 \\ 25{,}650x + 3{,}741{,}362 &= 33{,}366x + 391{,}950 \\ 3{,}741{,}362 - 391{,}950 &= 33{,}366x - 25{,}650x \\ 3{,}349{,}412 &= 7716x \\ \frac{3{,}349{,}412}{7716} &= x \\ \frac{837{,}353}{1929} &= x \end{aligned}$$

100. -0.000036365

101.

$$\begin{aligned} x(x - 4) &= 3x(x + 1) - 2(x^2 + x - 5) \\ x^2 - 4x &= 3x^2 + 3x - 2x^2 - 2x + 10 \\ x^2 - 4x &= x^2 + x + 10 \\ -x^2 + x^2 - 4x &= -x^2 + x^2 + x + 10 \\ -4x &= x + 10 \\ -4x - x &= 10 \\ -5x &= 10 \\ x &= -2 \end{aligned}$$

102. -2

103.

$$\begin{aligned} -2y + 5y &= 6y \\ 3y &= 6y \\ 0 &= 3y \\ 0 &= y \end{aligned}$$

104. 0

105. $\dfrac{5 + 2y}{3} = \dfrac{25}{12} + \dfrac{5y + 3}{4}$

The least common denominator is 12.

$$\begin{aligned} 4(5 + 2y) &= 25 + 3(5y + 3) \\ 20 + 8y &= 25 + 15y + 9 \\ -7y &= 14 \\ y &= -2 \end{aligned}$$

106. $\frac{52}{45}$

Exercise Set 2.3

1. $A = bh$

$\frac{A}{h} = b$ Dividing by h

2. $h = \frac{A}{b}$

3. $d = rt$

$\frac{d}{t} = r$ Dividing by t

4. $t = \frac{d}{r}$

5. $I = Prt$

$\frac{I}{rt} = P$ Dividing by rt

6. $t = \frac{I}{Pr}$

7. $F = ma$

$\frac{F}{m} = a$

8. $m = \frac{F}{a}$

9. $P = 2\ell + 2w$

$P - 2\ell = 2w$ Subtracting 2ℓ

$\frac{P - 2\ell}{2} = w$ Dividing by 2

10. $\ell = \frac{P - 2w}{2}$

11. $A = \pi r^2$

$\frac{A}{\pi} = r^2$

12. $\pi = \frac{A}{r^2}$

13. $A = \frac{1}{2}bh$

$2A = bh$ Multiplying by 2

$\frac{2A}{h} = b$ Dividing by h

14. $h = \frac{2A}{b}$

15. $E = mc^2$

$\frac{E}{c^2} = m$ Dividing by c^2

16. $c^2 = \frac{E}{m}$

17. $Q = \frac{c + d}{2}$

$2Q = c + d$ Multiplying by 2

$2Q - c = d$ Subtracting c

18. $p = 2Q + q$

19. $A = \frac{a + b + c}{3}$

$3A = a + b + c$ Multiplying by 3

$3A - a - c = b$ Subtracting a and c

20. $c = 3A - a - b$

21. $v = \frac{3k}{t}$

$tv = 3k$ Multiplying by t

$t = \frac{3k}{v}$ Dividing by v

22. $c = \frac{ab}{P}$

23. $Ax + By = C$

$By = C - Ax$ Subtracting Ax

$y = \frac{C - Ax}{B}$ Dividing by B

24. $x = \frac{C - By}{A}$

25. $A = \frac{1}{2}ah + \frac{1}{2}bh$

$2A = 2\left(\frac{1}{2}ah + \frac{1}{2}bh\right)$ Clearing the fractions

$2A = ah + bh$

$2A - ah = bh$ Subtracting ah

$\frac{2A - ah}{h} = b$ Dividing by h

26. $a = \frac{2A + bh}{h}$; $h = \frac{2A}{a - b}$

27. $Q = 3a + 5ca$

$Q = a(3 + 5c)$ Factoring

$\frac{Q}{3 + 5c} = a$ Dividing by 3 + 5c

28. $m = \frac{P}{4 + 7n}$

29. $A = P + Prt$

$A = P(1 + rt)$ Factoring

$\frac{A}{1 + rt} = P$ Dividing 1 + rt

30. $P = \frac{S}{1 - 0.01r}$

31. $A = \frac{\pi r^2 S}{360}$

$\frac{360}{\pi r^2} \cdot A = \frac{360}{\pi r^2} \cdot \frac{\pi r^2 S}{360}$

$\frac{360A}{\pi r^2} = S$

32. $r^2 = \frac{360A}{\pi S}$

33. $R = -0.0075t + 3.85$

$R - 3.85 = -0.0075t$

$\frac{R - 3.85}{-0.0075} = t$

34. $C = \frac{5F - 160}{9}$, or $\frac{5}{9}(F - 32)$

35. We multiply from left to right.

$7(-3)2 = (-21)2 = -42$

36. $-\frac{3}{5}$

37. $10 \div (-2)\cdot 5 - 4 = -5\cdot 5 - 4$ Dividing

$= -25 - 4$ Multiplying

$= -29$ Subtracting

38. 30

39. 

40. 

41. $\dfrac{\left(\frac{y}{z}\right)}{\left(\frac{z}{t}\right)} = 1$

$\frac{y}{z} = \frac{z}{t}$ Multiplying by $\frac{z}{t}$

$y = \frac{z^2}{t}$ Multiplying by z

42. $F = \frac{1 - GE}{G}$, or $\frac{1}{G} - E$

43. $q = r(s + t)$

$q = rs + rt$

$q - rs = rt$

$\frac{q - rs}{r} = t$

We could also solve for t as follows.

$q = r(s + t)$

$\frac{q}{r} = s + t$

$\frac{q}{r} - s = t$

44. $c = \frac{d}{a - b}$

45. $a = c(x + y) + bx$

$a = cx + cy + bx$

$a - cy = cx + bx$

$a - cy = x(c + b)$

$\frac{a - cy}{c + b} = x$

46. $a = \frac{c}{3 + b + d}$

47. 

48. 

Exercise Set 2.4

1. 76% = 76 × 0.01 Replacing % by × 0.01
   = 0.76

2. 0.54

3. 54.7% = 54.7 × 0.01 Replacing % by × 0.01
   = 0.547

4. 0.962

5. 100% = 100 × 0.01 = 1

6. 0.01

7. 0.61% = 0.61 × 0.01 = 0.0061

8. 1.25

9. 240% = 240 × 0.01 = 2.4

10. 0.0073

11. 3.25% = 3.25 × 0.01 = 0.0325

12. 0.023

13. 4.54

First move the decimal point two places to the right; 4.54.

then write a % symbol: 454%.

14. 100%

15. 0.998

First move the decimal point two places to the right; 0.99.8

then write a % symbol: 99.8%.

16. 73%

17. 2 (Note: 2 = 2.00)

First move the decimal point two places to the right; 2.00.

then write a % symbol: 200%.

18. 0.57%

19. 0.072

First move the decimal point two places to the right; 0.07.2

then write a % symbol: 7.2%.

20. 134%

21. 9.2 (Note: 9.2 = 9.20)

First move the decimal point two places to the right; 9.20.

then write a % symbol: 920%.

22. 1.3%

23. 0.0068

First move the decimal point two places to the right; 0.00.68

then write a % symbol: 0.68%.

24. 67.5%

25. $\frac{1}{8}$ [Note: $\frac{1}{8} = 0.125$]

First move the decimal point two places to the right; 0.12.5

then write a % symbol: 12.5%.

26. $33.\overline{3}\%$, or $33\frac{1}{3}\%$

27. $\frac{17}{25}$ [Note: $\frac{17}{25} = 0.68$]

First move the decimal point two places to the right; 0.68.

then write a % symbol: 68%.

28. 55%

29. $\frac{3}{4}$ [Note: $\frac{3}{4} = 0.75$]

First move the decimal point two places to the right; 0.75.

then write a % symbol: 75%.

30. 40%

31. $\frac{7}{10}$ [Note: $\frac{7}{10} = 0.7$, or 0.70]

First move the decimal point two places to the right; 0.70.

then write a % symbol: 70%.

32. 80%

33. $\frac{3}{5}$ [Note: $\frac{3}{5} = 0.6$, or 0.60]

First move the decimal point two places to the right; 0.60.

then write a % symbol: 60%.

34. 34%

35. $\frac{2}{3}$ [Note: $\frac{2}{3} = 0.66\overline{6}$]

First move the decimal point two places to the right; $0.66.\overline{6}$

then write a % symbol: $66.\overline{6}\%$.

Since $0.\overline{6} = \frac{2}{3}$, this can also be expressed as $66\frac{2}{3}\%$.

36. 37.5%

37. Translate.

What percent of 68 is 17?

$y \cdot 68 = 17$

We solve the equation and then convert to percent notation.

$$y \cdot 68 = 17$$
$$y = \frac{17}{68}$$
$$y = 0.25 = 25\%$$

The answer is 25%.

38. 48%

39. Translate.

What percent of 125 is 30?

$y \cdot 125 = 30$

We solve the equation and then convert to percent notation.

$$y \cdot 125 = 30$$
$$y = \frac{30}{125}$$
$$y = 0.24 = 24\%$$

The answer is 24%.

40. 19%

41. Translate.

45 is 30% of what number?

$45 = 30\% \cdot y$

We solve the equation.

$$45 = 0.3y \qquad (30\% = 0.3)$$
$$\frac{45}{0.3} = y$$
$$150 = y$$

The answer is 150.

42. 85

43. Translate.

0.3 is 12% of what number?

$0.3 = 12\% \cdot y$

We solve the equation.

$$0.3 = 0.12y \qquad (12\% = 0.12)$$
$$\frac{0.3}{0.12} = y$$
$$2.5 = y$$

The answer is 2.5.

44. 4

45. Translate.

What number is 65% of 840?

$y = 65\% \cdot 840$

$y = 0.65 \times 840$ (65% = 0.65)

$y = 546$ Multiplying

The answer is 546.

46. 10,000

47. Translate.

What percent of 80 is 100?

$y \cdot 80 = 100$

We solve the equation and then convert to percent notation.

$y \cdot 80 = 100$

$y = \frac{100}{80}$

$y = 1.25 = 125\%$

The answer is 125%.

48. 2050%

49. Translate.

What is 2% of 40?

$x = 2\% \cdot 40$

$x = 0.02 \times 40$ (2% = 0.02)

$x = 0.8$ Multiplying

The answer is 0.8.

50. 0.8

51. Translate.

2 is what percent of 40?

$2 = y \cdot 40$

We solve the equation.

$2 = y \cdot 40$

$\frac{2}{40} = y$

$0.05 = y$, or $5\% = y$

The answer is 5%.

52. 2000

53. We reword the problem. If y = the percent accepted, we have:

What percent of 16,000 is 600?

$y \cdot 16{,}000 = 600$

$y = \frac{600}{16{,}000}$

$y = 0.0375 = 3.75\%$

The FBI accepts 3.75% of the applicants.

54. 7410

55. We reword the problem. If y = the number of bowlers who are left-handed, we have:

What is 17% of 160?

$y = 17\% \cdot 160$

$y = 0.17 \times 160$

$y = 27.2$

You would expect 27 of the bowlers to be left-handed. (We round to the nearest one.)

56. 7%

57. We reword the problem. If y = the percent that were correct, we have:

What percent of 88 is 76?

$y \cdot 88 = 76$

$y = \frac{76}{88}$

$y = 0.864 = 86.4\%$

86.4% were correct.

58. 52%

59. When sales tax is 5%, the total paid is 105% of the price of the merchandise. If c = the cost of the merchandise, we have:

\$37.80 is 105% of c.

$37.80 = 1.05 \cdot c$

$\frac{37.80}{1.05} = c$

$36 = c$

The merchandise cost \$36 before tax.

60. \$940

61. We divide:

$$\begin{array}{r} .92 \\ 25\overline{)23.00} \\ \underline{225} \\ 50 \\ \underline{50} \\ 0 \end{array}$$

Decimal notation for $\frac{23}{25}$ is 0.92.

62. -90

63. $-45.8 - (-32.6) = -45.8 + 32.6 = -13.2$

64. $-21a + 12b$

65. ◆

66. ◆

67. At Rollie's Music, the total cost is 107% of the price of the disk. If R = the cost at Rollie's, we have:

$R = 107\%(\$11.99)$

$R = 1.07(\$11.99)$

$R = \$12.83$ (Rounding)

At Sound Warp, the total cost is \$13.99 - \$2.00 plus 7% of the original price of \$13.99. If S = the cost at Sound Warp, we have:

$S = (\$13.99 - \$2.00) + 7\%(\$13.99)$

$S = \$11.99 + 0.07(\$13.99)$

$S = \$11.99 + \$0.98$ (Rounding)

$S = \$12.97$

68. 40%; 70%; 95%

69. $x = 1.6y$, so $y = \frac{x}{1.6}$, or $\frac{1}{1.6}x$, or $0.625x$.

Since 0.625 = 62.5%, y is 62.5% of x.

70. 20%

71. Since the tax rate r is in decimal notation, the total cost is 1 + r times the cost of the merchandise. Then we have:

$T = (1 + r)c$

$\frac{T}{1 + r} = c$

Exercise Set 2.5

1. Let x = the number. Then "twice the number" translates to 2x, and "three less than 2x" translates to 2x - 3.

2. $\frac{x}{8} - 5$

3. Let x = the number. Then "the product of a number and 7" translates to $x \cdot 7$, or 7x, and "one half of 7x" translates to $\frac{1}{2} \cdot 7x$.

4. 10n - 2

5. Let a = the number. Then "the sum of 3 and some number" translates to a + 3, and "5 times a + 3" translates to 5(a + 3).

6. 6(x + y)

7. The longer piece is 2 ft longer than L, or L + 2.

8. $x \div 3$, or $\frac{x}{3}$

9. The amount of the reduction is 30%b. Then the sale price is b - 30%b, or b - 0.3b, or 0.7b.

10. p - 20%p, or p - 0.2p, or 0.8p

11. Each even integer is 2 greater than the one preceding it. If we let x = the first of the even integers, then x + 2 = the second and x + 4 = the third. We can express their sum as x + (x + 2) + (x + 4).

12. x + (x + 1) + (x + 2)

13. The cost is

(Initial charge) plus (mileage charge), or

\$34.95	plus	Cost per mile	times	Number of miles driven
\$34.95	+	\$0.27	$\cdot$	m

which is \$34.95 + \$0.27m.

14. $\frac{t}{2} + 2$

15. a) Twice the width is 2w.

b) The width is one-half the length, or $\frac{1}{2}\ell$.

16. 3(h + 5)

17. Second angle: Three times the first is 3x

Third angle: 30° more than the first is x + 30

18. 4x; 5x - 45

19. Familiarize. Let x = the number. Using the result of Exercise 1, we know that "three less than twice a number" translates to 2x - 3.

Translate.

Three less than twice a number is 25.

$2x - 3 = 25$

Carry out. We solve the equation.

$2x - 3 = 25$

$2x = 28$ Adding 3

$x = 14$ Dividing by 2

Check. Twice, or two times, 14 is 28. Three less than 28 is 25. The answer checks.

State. The number is 14.

20. 12

21. Familiarize. Let y = the number. Then "the sum of 3 and some number" translates to y + 3, and "five times y + 3" translates to 5(y + 3).

Translate.

Five times the sum of 3 and some number is 70.

$5(y + 3) = 70$

Carry out. We solve the equation.

$5(y + 3) = 70$

$5y + 15 = 70$ Using the distributive law

$5y = 55$ Subtracting 15

$y = 11$ Dividing by 5

Check. The sum of 3 and 11 is 14, and $5 \cdot 14 = 70$. The answer checks.

State. The number is 11.

22. 13

23. Familiarize. Let n = the number.

Translate. We reword the problem.

Six times a number less 18 is 96.

$6 \cdot n - 18 = 96$

Carry out. We solve the equation.

$6n - 18 = 96$

$6n = 114$ Adding 18

$n = 19$ Multiplying by $\frac{1}{6}$

Check. Six times 19 is 114. Subtracting 18 from 114, we get 96. This checks.

State. The number is 19.

24. 52

25. Familiarize. Let y = the number.

Translate. We reword the problem.

Two times a number plus 16 is $\frac{2}{5}$ of the number.

$2 \cdot y + 16 = \frac{2}{5} \cdot y$

Carry out. We solve the equation.

$2y + 16 = \frac{2}{5}y$

$5(2y + 16) = 5 \cdot \frac{2}{5}y$ Clearing the fraction

$10y + 80 = 2y$

$80 = -8y$ Adding $-10y$

$-10 = y$ Dividing by -8

Check. We double -10 and get -20. Adding 16, we get -4. Also, $\frac{2}{5}(-10) = -4$. The answer checks.

State. The number is -10.

26. -68

27. Familiarize. Let x = the number.

Translate. We reword the problem.

A number plus two-fifths of the number is 56.

$x + \frac{2}{5} \cdot x = 56$

Carry out. We solve the equation.

$x + \frac{2}{5}x = 56$

$\frac{7}{5}x = 56$ Collecting like terms

$x = \frac{5}{7} \cdot 56$ Multiplying by $\frac{5}{7}$

$x = 40$

Check. $\frac{2}{5} \cdot 40 = 16$, and $40 + 16 = 56$. The answer checks.

State. The number is 40.

28. 36

29. Familiarize. First draw a picture.

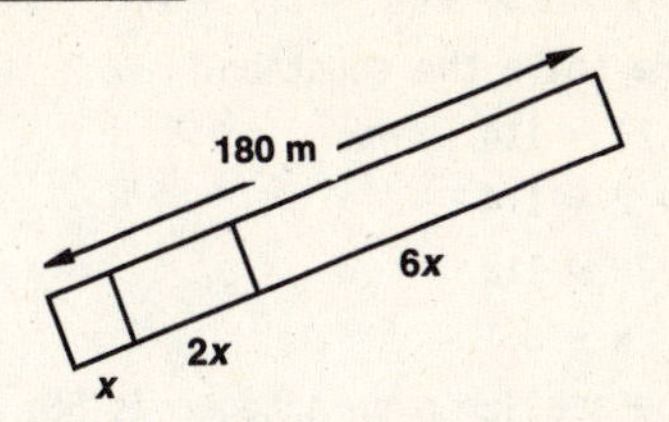

We use x for the first length, $2x$ for the second length, and $3 \cdot 2x$, or $6x$, for the third length.

Translate. The lengths of the three pieces add up to 180 m. This gives us the equation.

Length of 1st piece plus Length of 2nd piece plus Length of 3rd piece is 180

$x + 2x + 6x = 180$

Carry out. We solve the equation.

$x + 2x + 6x = 180$

$9x = 180$

$x = 20$

Check. If the first piece is 20 m long, then the second is $2 \cdot 20$ m, or 40 m and the third is $6 \cdot 20$ m, or 120 m. The lengths of these pieces add up to 180 m ($20 + 40 + 120 = 180$). This checks.

State. The first piece measures 20 m. The second measures 40 m, and the third measures 120 m.

30. 30 m, 90 m, 360 m

31. Familiarize. The page numbers are consecutive integers. (See Example 3.) If we let p = the smaller number, then $p + 1$ = the larger number.

Translate. We reword the problem.

First integer + Second integer = 273

$x + (x + 1) = 273$

Carry out. We solve the equation.

$x + (x + 1) = 273$

$2x + 1 = 273$ Collecting like terms

$2x = 272$ Adding -1

$x = 136$ Dividing by 2

Check. If $x = 136$, then $x + 1 = 137$. These are consecutive integers, and $136 + 137 = 273$. The answer checks.

State. The page numbers are 136 and 137.

32. 140 and 141.

33. Familiarize. We let x = the smaller even integer. Then $x + 2$ = the next even integer.

Translate.

Smaller even integer plus next even integer is 114.

$$x + (x + 2) = 114$$

Carry out. We solve the equation.

$$x + (x + 2) = 114$$
$$2x + 2 = 114$$
$$2x = 112$$
$$x = 56$$

Check. If the smaller even integer is 56, then the larger is 56 + 2, or 58. They are consecutive even integers. Their sum, 56 + 58, is 114. This checks.

State. The integers are 56 and 58.

34. 52 and 54

35. Familiarize. Let x = the first integer. Then $x + 1$ = the second integer, and $x + 2$ = the third.

Translate.

First integer plus Second integer plus Third integer is 108.

$$x + (x + 1) + (x + 2) = 108$$

Carry out. We solve the equation.

$$x + (x + 1) + (x + 2) = 108$$
$$3x + 3 = 108$$
$$3x = 105$$
$$x = 35$$

Check. If the first integer is 35, then the second is 35 + 1, or 36, and the third is 35 + 2, or 37. They are consecutive integers. Their sum, 35 + 36 + 37, is 108. This checks.

State. The integers are 35, 36, and 37.

36. 41, 42, and 43

37. Familiarize. Let x = the first odd integer. Then $x + 2$ = the second odd integer, and $x + 4$ = the third.

Translate.

First integer plus Second integer plus Third integer is 189.

$$x + (x + 2) + (x + 4) = 189$$

Carry out. We solve the equation.

$$x + (x + 2) + (x + 4) = 189$$
$$3x + 6 = 189$$
$$3x = 183$$
$$x = 61$$

Check. If the first odd integer is 61, then the second odd integer is 61 + 2, or 63, and the third is 61 + 4, or 65. They are consecutive odd integers. Their sum, 61 + 63 + 65, is 189. This checks.

State. The integers are 61, 63, and 65.

38. 130, 132, and 134

39. Familiarize. We draw a picture. Let w = the width of the rectangle. Then $w + 60$ = the length.

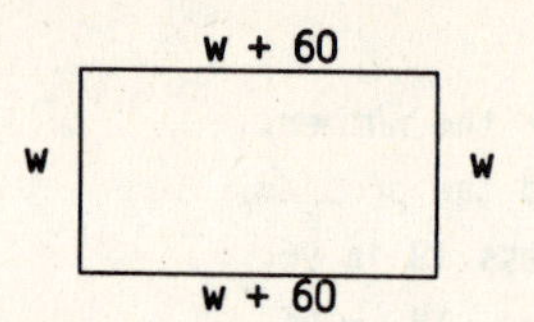

The perimeter of a rectangle is the sum of the lengths of the sides. The area is the product of the length and the width.

Translate. We use the definition of perimeter to write an equation that will allow us to find the width and length.

Width + Width + Length + Length = Perimeter

$$w + w + (w + 60) + (w + 60) = 520$$

To find the area we will compute the product of the length and width, or $(w + 60)w$.

Carry out. We solve the equation.

$$w + w + (w + 60) + (w + 60) = 520$$
$$4w + 120 = 520$$
$$4w = 400$$
$$w = 100$$

If $w = 100$, then $w + 60 = 100 + 60 = 160$, and the area is $160(100) = 16{,}000$.

Check. The length is 60 ft more than the width. The perimeter is 100 + 100 + 160 + 160 = 520 ft. This checks. To check the area we recheck the computation. This also checks.

State. The width is 100 ft, the length is 160 ft, and the area is 16,000 ft^2.

40. Width: 165 ft, length: 265 ft; area: 43,725 ft^2.

41. Familiarize. We draw a picture. Let ℓ = the length of the paper. Then $\ell - 6.3$ = the width.

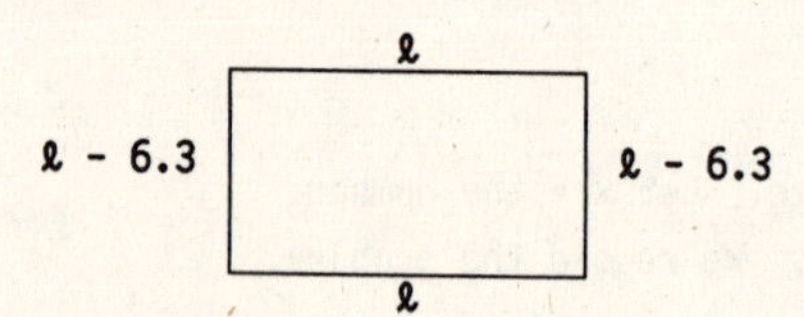

The perimeter is the sum of the lengths of the sides.

Translate. We use the definition of perimeter to write an equation.

Width + Width + Length + Length is 99.

$$(\ell - 6.3) + (\ell - 6.3) + \ell + \ell = 99$$

Carry out. We solve the equation.

$$(\ell - 6.3) + (\ell - 6.3) + \ell + \ell = 99$$
$$4\ell - 12.6 = 99$$
$$4\ell = 111.6$$
$$\ell = 27.9$$

Then $\ell - 6.3 = 21.6$.

Check. The width, 21.6 cm, is 6.3 cm less than the length, 27.9 cm. The perimeter is 21.6 cm + 21.6 cm + 27.9 cm + 27.9 cm, or 99 cm. This checks.

State. The length is 27.9 cm, and the width is 21.6 cm.

42. Length: 365 mi, width: 275 mi

43. Familiarize. We draw a picture. We let x = the measure of the first angle. Then 4x = the measure of the second angle, and (x + 4x) - 45, or 5x - 45 = the measure of the third angle.

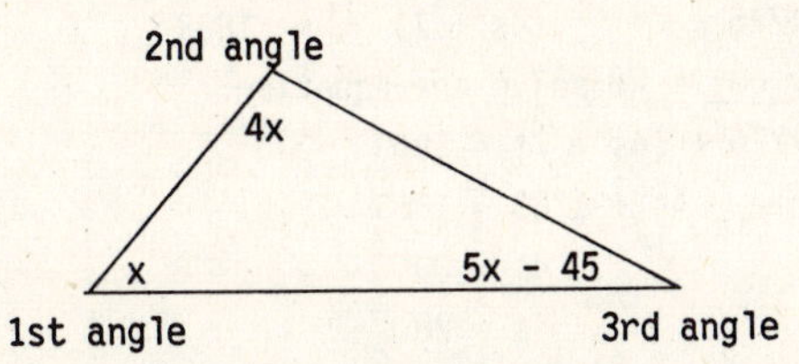

Recall that the measures of the angles of any triangle add up to 180°.

Translate.

$$\text{Measure of 1st angle} + \text{Measure of 2nd angle} + \text{Measure of 3rd angle} = 180.$$
$$x + 4x + (5x - 45) = 180$$

Carry out. We solve the equation.

$$x + 4x + (5x - 45) = 180$$
$$10x - 45 = 180$$
$$10x = 225$$
$$x = 22.5$$

Possible answers for the angle measures are as follows:

1st angle: $x = 22.5°$

2nd angle: $4x = 4(22.5) = 90°$

3rd angle: $5x - 45 = 5(22.5) - 45 = 112.5 - 45 = 67.5°$

Check. Consider 22.5°, 90°, and 67.5°. The second is four times the first, and the third is 45° less than five times the first. The sum is 180°. These numbers check.

State. The measure of the first angle is 22.5°.

44. 25.625°

45. Familiarize. Let x = the original price. Then, 40%x = the reduction. The sale price is found by subtracting the amount of reduction from the original price.

Translate.

Original price minus reduction is \$9.60.

$$x - 40\%x = 9.60$$

Carry out. We solve the equation.

$$x - 40\%x = 9.60$$
$$1 \cdot x - 0.40x = 9.60$$
$$(1 - 0.40)x = 9.60$$
$$0.6x = 9.60$$
$$x = \frac{9.60}{0.6}$$
$$x = 16$$

Check. 40% of \$16 is \$6.40. Subtracting this from \$16 we get \$9.60. This checks.

State. The original price was \$16.

46. \$14

47. Familiarize. Let x = the original investment. Interest earned in 1 year is found by taking 6% of the original investment. Then 6%x = the interest. The amount in the account at the end of the year is the sum of the original investment and the interest earned.

Translate.

Original investment plus interest earned is \$4664.

$$x + 6\%x = 4664$$

Carry out. We solve the equation.

$$x + 6\%x = 4664$$
$$1 \cdot x + 0.06x = 4664$$
$$1.06x = 4664$$
$$x = \frac{4664}{1.06}$$
$$x = 4400$$

Check. 6% of \$4400 is \$264. Adding this to \$4400 we get \$4664. This checks.

State. The original investment was \$4400.

48. \$6540

49. Familiarize. The total cost is the daily charge plus the mileage charge. The mileage charge is the cost per mile times the number of miles driven. Let m = the number of miles that can be driven for \$80.

Translate.

Daily rate plus Cost per mile times Number of miles driven is Amount

$$34.95 + 0.10 \cdot m = 80$$

Carry out. We solve the equation.

$$34.95 + 0.10m = 80$$
$$100(34.95 + 0.10m) = 100(80) \quad \text{Clearing the decimals}$$
$$3495 + 10m = 8000$$
$$10m = 4505$$
$$m = 450.5$$

Check. The mileage cost is found by multiplying 450.5 by \$0.10 obtaining \$45.05. Then we add \$45.05 to \$34.95, the daily rate, and get \$80.

State. The businessperson can drive 450.5 mi on the car-rental allotment.

50. 460.5 mi

51. Familiarize. From the drawing in the text, we see that x = the measure of the first angle, 3x = the measure of the second angle, and x + 40 = the measure of the third angle. Recall that the sum of the measures of the angles of a triangle is 180°.

Translate.

Measure of 1st angle + Measure of 2nd angle + Measure of 3rd angle = 180.

$$x + 3x + (x + 40) = 180$$

Carry out.

$$x + 3x + (x + 40) = 180$$
$$5x + 40 = 180$$
$$5x = 140$$
$$x = 28$$

Possible answers for the angle measures are as follows:

1st angle: $x = 28°$

2nd angle: $3x = 3(28) = 84°$

3rd angle: $x + 40 = 28 + 40 = 68°$

Check. Consider 28°, 84°, and 68°. The second angle is three times the first, and the third is 40° more than the first. The sum, 28° + 84° + 68°, is 180°. These numbers check.

State. The measures of the angles are 28°, 84°, and 68°.

52. 5°, 160°, 15°

53. Familiarize. We will use the equation R = -0.028t + 20.8 where R is in seconds and t is the number of years since 1920. We want to find t when R = 18.0 sec.

Translate.

Record is 18.0 sec.

$$-0.028t + 20.8 = 18.0$$

Carry out.

$$-0.028t + 20.8 = 18.0$$
$$1000(-0.028t + 20.8) = 1000(18.0) \quad \text{Clearing the decimals}$$
$$-28t + 20{,}800 = 18{,}000$$
$$-28t = -2800$$
$$t = 100$$

Check. Substitute 100 for t in the given equation:

$$R = -0.028(100) + 20.8 = -2.8 + 20.8 = 18.0$$

This checks.

State. The record will be 18.0 sec 100 years after 1920, or in 2020.

54. 160

55. $3x - 12y + 60 = 3\cdot x - 3\cdot 4y + 3\cdot 20$
$= 3(x - 4y + 20)$

56. $11x - 13$

57.

58.

59. Familiarize. Let s = one score. Then four score = 4s and four score and seven = 4s + 7.

Translate. We reword.

1776 plus four score and seven is 1863.

$$1776 + (4s + 7) = 1863$$

Carry out. We solve the equation.

$$1776 + (4s + 7) = 1863$$
$$4s + 1783 = 1863$$
$$4s = 80$$
$$s = 20$$

Check. If a score is 20, then four score and seven years represents 87 years. Adding 87 to 1776 we get 1863. This checks.

State. A score is 20.

60. 16 and 4

61. Familiarize. The cost of the rental is the daily charge plus the mileage charge. Let c = the cost per mile that will make the total cost equal to the budget amount. Then c is the highest price per mile the person can afford.

Translate.

Daily rate plus Cost per mile times Number of miles driven is Budget amount.

$$18.90 + c \cdot 190 = 55$$

Carry out.

$$18.90 + 190c = 55$$
$$10(18.90 + 190c) = 10(55) \quad \text{Clearing the decimal}$$
$$189 + 1900c = 550$$
$$1900c = 361$$
$$c = 0.19$$

Check. The mileage cost is found by multiplying 190 by $0.19 obtaining $36.10. Adding $36.10 to $18.90, the daily rate, we get $55.

State. The cost per mile cannot exceed $0.19 to stay within a $55 budget.

62. 19

63. Familiarize. We let x = the length of the original rectangle. Then $\frac{3}{4}x$ = the width. We draw a picture of the enlarged rectangle. Each dimension is increased by 2 cm, so x + 2 = the length of the enlarged rectangle and $\frac{3}{4}x + 2$ = the width.

x + 2

$\frac{3}{4}x + 2$ $\frac{3}{4}x + 2$

x + 2

Translate. We use the perimeter of the enlarged rectangle to write an equation.

Width + Width + Length + Length is Perimeter.

$$\left(\frac{3}{4}x + 2\right) + \left(\frac{3}{4}x + 2\right) + (x + 2) + (x + 2) = 50$$

Carry out.

$$\left(\frac{3}{4}x + 2\right) + \left(\frac{3}{4}x + 2\right) + (x + 2) + (x + 2) = 50$$
$$\frac{7}{2}x + 8 = 50$$
$$2\left(\frac{7}{2}x + 8\right) = 2\cdot 50$$
$$7x + 16 = 100$$
$$7x = 84$$
$$x = 12$$

Then $\frac{3}{4}x = \frac{3}{4}(12) = 9$.

Check. If the dimensions of the original rectangle are 12 cm and 9 cm, then the dimensions of the enlarged rectangle are 14 cm and 11 cm. The perimeter of the enlarged rectangle is $11 + 11 + 14 + 14 = 50$ cm. Also, 9 is $\frac{3}{4}$ of 12. These values check.

State. The length is 12 cm, and the width is 9 cm.

64. 120

65. Familiarize. Let x = the number of additional games the Falcons will have to play. Then $\frac{x}{2}$ = the number of those games they will win, $15 + \frac{x}{2}$ = the total number of games won, and $20 + x$ = the total number of games played.

Translate.

The number of games won is 60% of the number of games played.

$$\left(15 + \frac{x}{2}\right) = 60\% \cdot (20 + x)$$

Carry out.

$$15 + \frac{x}{2} = 60\%(20 + x)$$
$$15 + \frac{x}{2} = 0.6(20 + x)$$
$$15 + 0.5x = 12 + 0.6x \quad \text{Expressing } \frac{1}{2} \text{ as } 0.5$$
$$10(15 + 0.5x) = 10(12 + 0.6x) \quad \text{Clearing decimals}$$
$$150 + 5x = 120 + 6x$$
$$30 = x$$

Check. If 30 more games are played of which $15\left(\frac{30}{2} = 15\right)$ are won, then the total games played will be 20 + 30, or 50, and the total games won will be 15 + 15, or 30. 30 is 60% of 50. The numbers check.

State. The Falcons must play 30 more games.

66. $600

67. Familiarize. Let h = the height of the triangle. We know that the base is 8 in. Recall that the area of a triangle is given by the formula $A = \frac{1}{2}bh$.

Translate.

Area is 2.9047 in².

$$\frac{1}{2} \cdot 8 \cdot h = 2.9047$$

Carry out. We solve the equation.

$$\frac{1}{2} \cdot 8 \cdot h = 2.9047$$
$$4h = 2.9047$$
$$h = 0.726175$$

Check. The area of a triangle whose base is 8 in. and whose height is 0.726175 in. is $\frac{1}{2}(8)(0.726175)$, or 2.9047. The answer checks.

State. The height of the triangle is 0.726175 in.

68. 76

69.

Exercise Set 2.6

1. $x > -4$
   a) Since $4 > -4$ is true, 4 is a solution.
   b) Since $0 > -4$ is true, 0 is a solution.
   c) Since $-4.1 > -4$ is false, -4.1 is not a solution.
   d) Since $-3.9 > -4$ is true, -3.9 is a solution.
   e) Since $5.6 > -4$ is true, 5.6 is a solution.

2. a) Yes, b) No, c) Yes, d) Yes, e) No

3. $x \geq 6$
   a) Since $-6 \geq 6$ is false, -6 is not a solution.
   b) Since $0 \geq 6$ is false, 0 is not a solution.
   c) Since $6 \geq 6$ is true, 6 is a solution.
   d) Since $6.01 \geq 6$ is true, 6.01 is a solution.
   e) Since $-3\frac{1}{2} \geq 6$ is false, $-3\frac{1}{2}$ is not a solution.

4. a) Yes, b) Yes, c) Yes, d) No, e) Yes

5. The solutions of $x > 4$ are those numbers greater than 4. They are shown on the graph by shading all points to the right of 4. The open circle at 4 indicates that 4 is not part of the graph.

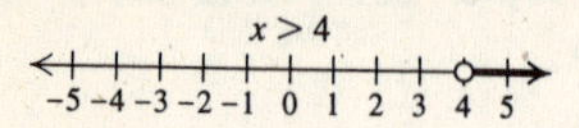

6. 0

7. The solutions of $t < -3$ are those numbers less than -3. They are shown on the graph by shading all points to the left of -3. The open circle at -3 indicates that -3 is not part of the graph.

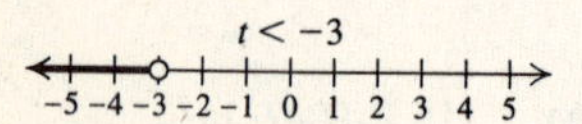

8. 0 5

9. The solutions of $m \geq -1$ are shown by shading the point for -1 and all points to the right of -1. The closed circle at -1 indicates that -1 is part of the graph.

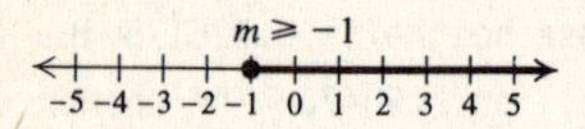

10. 0 3

11. In order to be a solution of the inequality $-3 < x \leq 4$, a number must be a solution of both $-3 < x$ and $x \leq 4$. The solution set is graphed as follows:

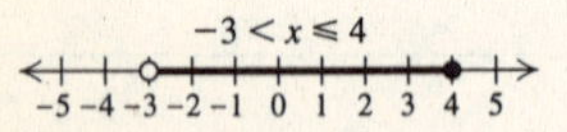

The open circle at -3 means that -3 is not part of the graph. The closed circle at 4 means that 4 is part of the graph.

12. -5 2

13. In order to be a solution of the inequality $0 < x < 3$, a number must be a solution of both $0 < x$ and $x < 3$. The solution set is graphed as follows:

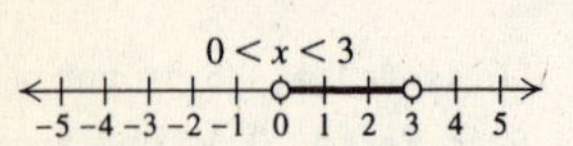

The open circles at 0 and 3 mean that 0 and 3 are not part of the graph.

14.

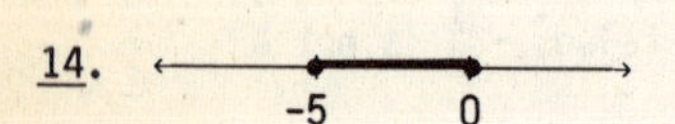

15. All points to the right of -1 are shaded. The open circle at -1 indicates that -1 is not part of the graph. Using set-builder notation we have $\{x \mid x > -1\}$.

16. $\{x \mid x < 3\}$

17. The point 2 and all points to the left of 2 are shaded. Using set-builder notation we have $\{x \mid x \leq 2\}$.

18. $\{x \mid x \geq -2\}$

19. All points to the left of -2 are shaded. The open circle at -2 indicates that -2 is not part of the graph. Using set-builder notation we have $\{x \mid x < -2\}$.

20. $\{x \mid x > 1\}$

21. The point 0 and all points to the right of 0 are shaded. Using set-builder notation we have $\{x \mid x \geq 0\}$.

22. $\{x \mid x \leq 0\}$

23. $y + 5 > 8$

$y + 5 - 5 > 8 - 5$ Adding -5

$y > 3$

The solution set is $\{y \mid y > 3\}$.

The graph is as follows:

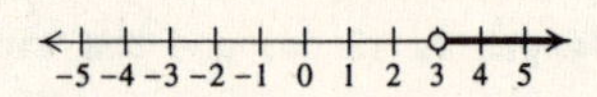

24. $\{y \mid y > 2\}$

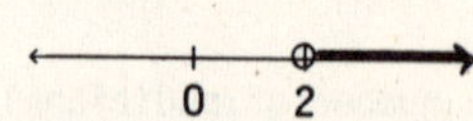

25. $x + 8 \leq -10$

$x + 8 - 8 \leq -10 - 8$

$x \leq -18$

The solution set is $\{x \mid x \leq -18\}$.

The graph is as follows:

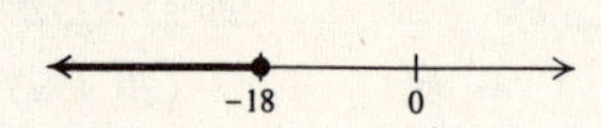

26. $\{x \mid x \leq -21\}$

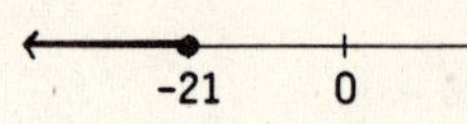

27. $x - 7 < 9$

$x - 7 + 7 < 9 + 7$

$x < 16$

The solution set is $\{x|x < 16\}$.

The graph is as follows:

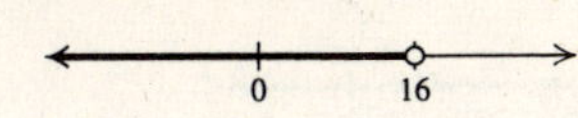

28. $\{x|x < 17\}$

29. $x - 6 \geq 2$

$x - 6 + 6 \geq 2 + 6$

$x \geq 8$

The solution set is $\{x|x \geq 8\}$.

The graph is as follows:

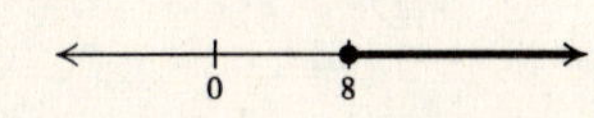

30. $\{x|x \geq 13\}$

31. $y - 7 > -12$

$y - 7 + 7 > -12 + 7$

$y > -5$

The solution set is $\{y|y > -5\}$.

The graph is as follows:

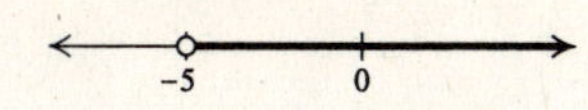

32. $\{y|y > -6\}$

33. $2x + 3 \leq x + 5$

$2x + 3 - 3 \leq x + 5 - 3$ Adding -3

$2x \leq x + 2$ Simplifying

$2x - x \leq x + 2 - x$ Adding -x

$x \leq 2$ Simplifying

The solution set is $\{x|x \leq 2\}$.

The graph is as follows:

34. $\{x|x \leq 3\}$

35. $3x - 6 \geq 2x + 7$

$3x - 6 + 6 \geq 2x + 7 + 6$ Adding 6

$3x \geq 2x + 13$

$3x - 2x \geq 2x + 13 - 2x$ Adding -2x

$x \geq 13$

The solution set is $\{x|x \geq 13\}$.

36. $\{x|x \geq 20\}$

37. $5x - 6 < 4x - 2$

$5x - 6 + 6 < 4x - 2 + 6$

$5x < 4x + 4$

$5x - 4x < 4x + 4 - 4x$

$x < 4$

The solution set is $\{x|x < 4\}$.

38. $\{x|x < -1\}$

39. $7 + c > 7$

$-7 + 7 + c > -7 + 7$

$c > 0$

The solution set is $\{c|c > 0\}$.

40. $\{c|c > 18\}$

41. $y + \frac{1}{4} \leq \frac{1}{2}$

$y + \frac{1}{4} - \frac{1}{4} \leq \frac{1}{2} - \frac{1}{4}$

$y \leq \frac{2}{4} - \frac{1}{4}$ Obtaining a common denominator

$y \leq \frac{1}{4}$

The solution set is $\left\{y \middle| y \leq \frac{1}{4}\right\}$.

42. $\left\{y \middle| y \leq \frac{1}{2}\right\}$

43. $x - \frac{1}{3} > \frac{1}{4}$

$x - \frac{1}{3} + \frac{1}{3} > \frac{1}{4} + \frac{1}{3}$

$x > \frac{3}{12} + \frac{4}{12}$ Obtaining a common denominator

$x > \frac{7}{12}$

The solution set is $\left\{x \middle| x > \frac{7}{12}\right\}$.

44. $\left\{x \middle| x > \frac{5}{8}\right\}$

45. $-14x + 21 > 21 - 15x$

$-14x + 21 + 15x > 21 - 15x + 15x$

$x + 21 > 21$

$x + 21 - 21 > 21 - 21$

$x > 0$

The solution set is $\{x|x > 0\}$.

46. $\{x|x > 3\}$

47. $5x < 35$

$\frac{1}{5} \cdot 5x < \frac{1}{5} \cdot 35$ Multiplying by $\frac{1}{5}$

$x < 7$

The solution set is $\{x|x < 7\}$. The graph is as follows:

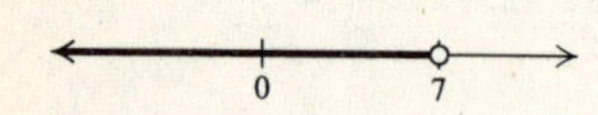

48. $\{x|x \geqslant 4\}$

49. $9y \leqslant 81$

$\frac{1}{9} \cdot 9y \leqslant \frac{1}{9} \cdot 81$ Multiplying by $\frac{1}{9}$

$y \leqslant 9$

The solution set is $\{y|y \leqslant 9\}$. The graph is as follows:

50. $\{x|x > 24\}$

51. $7x < 13$

$\frac{1}{7} \cdot 7x < \frac{1}{7} \cdot 13$

$x < \frac{13}{7}$

The solution set is $\left\{x\middle|x < \frac{13}{7}\right\}$. The graph is as follows:

52. $\left\{y\middle|y < \frac{17}{8}\right\}$

53. $12x > -36$

$\frac{1}{12} \cdot 12x > \frac{1}{12} \cdot (-36)$

$x > -3$

The solution set is $\{x|x > -3\}$. The graph is as follows:

54. $\{x|x < -4\}$

55. $5y \geqslant -2$

$\frac{1}{5} \cdot 5y \geqslant \frac{1}{5} \cdot (-2)$

$y \geqslant -\frac{2}{5}$

The solution set is $\left\{y\middle|y \geqslant -\frac{2}{5}\right\}$.

56. $\left\{x\middle|x > -\frac{4}{7}\right\}$

57. $-2x \leqslant 12$

$-\frac{1}{2} \cdot (-2x) \geqslant -\frac{1}{2} \cdot 12$ Multiplying by $-\frac{1}{2}$

The symbol has to be reversed.

$x \geqslant -6$ Simplifying

The solution set is $\{x|x \geqslant -6\}$.

58. $\{y|y \geqslant -5\}$

59. $-4y \geqslant -16$

$-\frac{1}{4} \cdot (-4y) \leqslant -\frac{1}{4} \cdot (-16)$ Multiplying by $-\frac{1}{4}$

The symbol has to be reversed.

$y \leqslant 4$

The solution set is $\{y|y \leqslant 4\}$.

60. $\{x|x > 3\}$

61. $-3x < -17$

$-\frac{1}{3} \cdot (-3x) > -\frac{1}{3} \cdot (-17)$ Multiplying by $-\frac{1}{3}$

The symbol has to be reversed.

$x > \frac{17}{3}$

The solution set is $\left\{x\middle|x > \frac{17}{3}\right\}$.

62. $\left\{y\middle|y < \frac{23}{5}\right\}$

63. $-2y > \frac{1}{7}$

$-\frac{1}{2} \cdot (-2y) < -\frac{1}{2} \cdot \frac{1}{7}$

The symbol has to be reversed.

$y < -\frac{1}{14}$

The solution set is $\left\{y \middle| y < -\frac{1}{14}\right\}$.

64. $\left\{x \middle| x \geqslant -\frac{1}{36}\right\}$

65. $-\frac{6}{5} \leqslant -4x$

$-\frac{1}{4} \cdot \left(-\frac{6}{5}\right) \geqslant -\frac{1}{4} \cdot (-4x)$

$\frac{6}{20} \geqslant x$

$\frac{3}{10} \geqslant x$, or $x \leqslant \frac{3}{10}$

The solution set is $\left\{x \middle| \frac{3}{10} \geqslant x\right\}$, or $\left\{x \middle| x \leqslant \frac{3}{10}\right\}$.

66. $\left\{t \middle| t > \frac{1}{64}\right\}$

67. $4 + 3x < 28$

$-4 + 4 + 3x < -4 + 28$ Adding -4

$3x < 24$ Simplifying

$\frac{1}{3} \cdot 3x < \frac{1}{3} \cdot 24$ Multiplying by $\frac{1}{3}$

$x < 8$ Simplifying

The solution set is $\{x|x < 8\}$.

68. $\{y|y < 8\}$

69. $6 + 5y \geqslant 36$

$-6 + 6 + 5y \geqslant -6 + 36$ Adding -6

$5y \geqslant 30$

$\frac{1}{5} \cdot 5y \geqslant \frac{1}{5} \cdot 30$ Multiplying by $\frac{1}{5}$

$y \geqslant 6$

The solution set is $\{y|y \geqslant 6\}$.

70. $\{x|x \geqslant 8\}$

71. $3x - 5 \leqslant 13$

$3x - 5 + 5 \leqslant 13 + 5$ Adding 5

$3x \leqslant 18$

$\frac{1}{3} \cdot 3x \leqslant \frac{1}{3} \cdot 18$ Multiplying by $\frac{1}{3}$

$x \leqslant 6$

The solution set is $\{x|x \leqslant 6\}$.

72. $\{y|y \leqslant 6\}$

73. $13x - 7 < -46$

$13x - 7 + 7 < -46 + 7$

$13x < -39$

$\frac{1}{13} \cdot 13x < \frac{1}{13} \cdot (-39)$

$x < -3$

The solution set is $\{x|x < -3\}$.

74. $\{y|y < -6\}$

75. $5x + 3 \geqslant -7$

$5x + 3 - 3 \geqslant -7 - 3$

$5x \geqslant -10$

$\frac{1}{5} \cdot 5x \geqslant \frac{1}{5} \cdot (-10)$

$x \geqslant -2$

The solution set is $\{x|x \geqslant -2\}$.

76. $\{y|y \geqslant -2\}$

77. $13 < 4 - 3y$

$13 - 4 < 4 - 3y - 4$ Adding -4

$9 < -3y$

$-\frac{1}{3} \cdot 9 > -\frac{1}{3} \cdot (-3y)$ Multiplying by $-\frac{1}{3}$

The symbol has to be reversed.

$-3 > y$

The solution set is $\{y|-3 > y\}$, or $\{y|y < -3\}$.

78. $\{x|x < -2\}$

79. $30 > 3 - 9x$

$30 - 3 > 3 - 9x - 3$ Adding -3

$27 > -9x$

$-\frac{1}{9} \cdot 27 < -\frac{1}{9} \cdot (-9x)$ Multiplying by $-\frac{1}{9}$

The symbol has to be reversed.

$-3 < x$

The solution set is $\{x|-3 < x\}$, or $\{x|x > -3\}$.

80. $\{y|y > -5\}$

81. $3 - 6y > 23$

$-3 + 3 - 6y > -3 + 23$

$-6y > 20$

$-\frac{1}{6} \cdot (-6y) < -\frac{1}{6} \cdot 20$

The symbol has to be reversed.

$y < -\frac{20}{6}$

$y < -\frac{10}{3}$

The solution set is $\left\{y \middle| y < -\frac{10}{3}\right\}$.

82. $\{y|y < -3\}$

83.
$$-3 < 8x + 7 - 7x$$
$$-3 < x + 7 \quad \text{Collecting like terms}$$
$$-3 - 7 < x + 7 - 7$$
$$-10 < x$$

The solution set is $\{x|-10 < x\}$, or $\{x|x > -10\}$.

84. $\{x|x > -13\}$

85.
$$6 - 4y > 4 - 3y$$
$$6 - 4y + 4y > 4 - 3y + 4y \quad \text{Adding } 4y$$
$$6 > 4 + y$$
$$-4 + 6 > -4 + 4 + y \quad \text{Adding } -4$$
$$2 > y, \text{ or } y < 2$$

The solution set is $\{y|2 > y\}$, or $\{y|y < 2\}$.

86. $\{y|y < 2\}$

87.
$$5 - 9y \leqslant 2 - 8y$$
$$5 - 9y + 9y \leqslant 2 - 8y + 9y$$
$$5 \leqslant 2 + y$$
$$-2 + 5 \leqslant -2 + 2 + y$$
$$3 \leqslant y, \text{ or } y \geqslant 3$$

The solution set is $\{y|3 \leqslant y\}$, or $\{y|y \geqslant 3\}$.

88. $\{y|y \geqslant 2\}$

89.
$$21 - 8y < 6y + 49$$
$$21 - 8y + 8y < 6y + 49 + 8y$$
$$21 < 14y + 49$$
$$21 - 49 < 14y + 49 - 49$$
$$-28 < 14y$$
$$\frac{1}{14} \cdot -28 < \frac{1}{14} \cdot 14y$$
$$-2 < y, \text{ or } y > -2$$

The solution set is $\{y|-2 < y\}$, or $\{y|y > -2\}$.

90. $\{x|x > -4\}$

91.
$$27 - 11x > 14x - 18$$
$$27 - 11x + 11x > 14x - 18 + 11x$$
$$27 > 25x - 18$$
$$27 + 18 > 25x - 18 + 18$$
$$45 > 25x$$
$$\frac{1}{25} \cdot 45 > \frac{1}{25} \cdot 25x$$
$$\frac{45}{25} > x$$
$$\frac{9}{5} > x, \text{ or } x < \frac{9}{5}$$

The solution set is $\left\{x\middle|\frac{9}{5} > x\right\}$, or $\left\{x\middle|x < \frac{9}{5}\right\}$.

92. $\left\{y\middle|y < \frac{61}{28}\right\}$

93.
$$2.1x + 45.2 > 3.2 - 8.4x$$
$$10(2.1x + 45.2) > 10(3.2 - 8.4x) \quad \text{Multiplying by 10 to clear decimals}$$
$$21x + 452 > 32 - 84x$$
$$21x + 84x > 32 - 452 \quad \text{Adding } 84x \text{ and } -452$$
$$105x > -420$$
$$x > -4 \quad \text{Multiplying by } \frac{1}{105}$$

The solution set is $\{x|x > -4\}$.

94. $\left\{y\middle|y \leqslant \frac{5}{3}\right\}$

95.
$$0.7n - 15 + n \geqslant 2n - 8 - 0.4n$$
$$1.7n - 15 \geqslant 1.6n - 8 \quad \text{Collecting like terms}$$
$$10(1.7n - 15) \geqslant 10(1.6n - 8) \quad \text{Multiplying by 10}$$
$$17n - 150 \geqslant 16n - 80$$
$$17n - 16n \geqslant -80 + 150 \quad \text{Adding } -16n \text{ and } 150$$
$$n \geqslant 70$$

The solution set is $\{n|n \geqslant 70\}$.

96. $\{t|t > 1\}$

97.
$$\frac{x}{3} - 2 \leqslant 1$$
$$3\left(\frac{x}{3} - 2\right) \leqslant 3\cdot 1 \quad \text{Multiplying by 3 to clear the fraction}$$
$$x - 6 \leqslant 3 \quad \text{Simplifying}$$
$$x \leqslant 9 \quad \text{Adding 6}$$

The solution set is $\{x|x \leqslant 9\}$.

98. $\{x|x > 2\}$

99.
$$\frac{y}{5} + 1 \leqslant \frac{2}{5}$$
$$5\left(\frac{y}{5} + 1\right) \leqslant 5 \cdot \frac{2}{5} \quad \text{Clearing fractions}$$
$$y + 5 \leqslant 2$$
$$y \leqslant -3 \quad \text{Adding } -5$$

The solution set is $\{y|y \leqslant -3\}$.

100. $\{x|x \geqslant -25\}$

101.
$$3(2y - 3) < 27$$
$$6y - 9 < 27 \quad \text{Removing parentheses}$$
$$6y < 36 \quad \text{Adding 9}$$
$$y < 6 \quad \text{Multiplying by } \frac{1}{6}$$

The solution set is $\left\{y\middle|y < 6\right\}$.

102. $\{y|y > 5\}$

103. $5(d + 4) \leq 7(d - 2)$

$5d + 20 \leq 7d - 14$ Removing parentheses

$5d - 7d \leq -14 - 20$ Adding $-7d$ and -20

$-2d \leq -34$

$d \geq 17$ Multiplying by $-\frac{1}{2}$

The symbol has to be reversed.

The solution set is $\{d | d \geq 17\}$.

104. $\{t | t \leq -4\}$

105. $8(2t + 1) > 4(7t + 7)$

$16t + 8 > 28t + 28$

$16t - 28t > 28 - 8$

$-12t > 20$

$t < -\frac{20}{12}$ Multiplying by $-\frac{1}{12}$ and reversing the symbol

$t < -\frac{5}{3}$

The solution set is $\left\{t \middle| t < -\frac{5}{3}\right\}$.

106. $\{x | x > -8\}$

107. $3(r - 6) + 2 < 4(r + 2) - 21$

$3r - 18 + 2 < 4r + 8 - 21$

$3r - 16 < 4r - 13$

$-16 + 13 < 4r - 3r$

$-3 < r$, or $r > -3$

The solution set is $\{r | r > -3\}$.

108. $\{t | t > -12\}$

109. $\frac{2}{3}(2x - 1) \geq 10$

$\frac{3}{2} \cdot \frac{2}{3}(2x - 1) \geq \frac{3}{2} \cdot 10$ Multiplying by $\frac{3}{2}$

$2x - 1 \geq 15$

$2x \geq 16$

$x \geq 8$

The solution set is $\{x | x \geq 8\}$.

110. $\{x | x \leq 7\}$

111. $\frac{3}{4}\left(3x - \frac{1}{2}\right) - \frac{2}{3} < \frac{1}{3}$

$\frac{3}{4}\left(3x - \frac{1}{2}\right) < 1$ Adding $\frac{2}{3}$

$\frac{9}{4}x - \frac{3}{8} < 1$ Removing parentheses

$8 \cdot \left(\frac{9}{4}x - \frac{3}{8}\right) < 8 \cdot 1$ Clearing fractions

$18x - 3 < 8$

$18x < 11$

$x < \frac{11}{18}$

The solution set is $\left\{x \middle| x < \frac{11}{18}\right\}$.

112. $\left\{x \middle| x > -\frac{5}{32}\right\}$

113. $10 \div 2 \cdot 5 - 3^2 + (-4)^2$

$= 10 \div 2 \cdot 5 - 9 + 16$ Evaluating the exponential notation

$= 5 \cdot 5 - 9 + 16$ Dividing

$= 25 - 9 + 16$ Multiplying

$= 32$ Subtracting and adding

114. 98

115.

116.

117. $2[4 - 2(3 - x)] - 1 \geq 4[2(4x - 3) + 7] - 25$

$2[4 - 6 + 2x] - 1 \geq 4[8x - 6 + 7] - 25$

$2[-2 + 2x] - 1 \geq 4[8x + 1] - 25$

$-4 + 4x - 1 \geq 32x + 4 - 25$

$-5 + 4x \geq 32x - 21$

$4x - 32x \geq -21 + 5$

$-28x \geq -16$

$x \leq \frac{-16}{-28}$

$x \leq \frac{4}{7}$

The solution set is $\left\{x \middle| x \leq \frac{4}{7}\right\}$.

118. $\left\{t \middle| t > -\frac{27}{19}\right\}$

119. $-(x + 5) \geq 4a - 5$

$-x - 5 \geq 4a - 5$

$-x \geq 4a - 5 + 5$

$-x \geq 4a$

$-1(-x) \leq -1 \cdot 4a$

$x \leq -4a$

The solution set is $\{x | x \leq -4a\}$.

120. $\{x|x > 7\}$

121. $y < ax + b$ (Assume $a > 0$.)

$y - b < ax$

$\frac{y-b}{a} < x$ (Since $a > 0$, the inequality symbol stays the same.)

The solution set is $\left\{x|x > \frac{y-b}{a}\right\}$.

122. $\left\{x\middle|x < \frac{y-b}{a}\right\}$

123. $|x| < 3$

a) Since $|0| = 0$, and $0 < 3$ is true, 0 is a solution.

b) Since $|-2| = 2$ and $2 < 3$ is true, -2 is a solution.

c) Since $|-3| = 3$ and $3 < 3$ is false, -3 is not a solution.

d) Since $|4| = 4$ and $4 < 3$ is false, 4 is not a solution.

e) Since $|3| = 3$ and $3 < 3$ is false, 3 is not a solution.

f) Since $|1.7| = 1.7$ and $1.7 < 3$ is true, 1.7 is a solution.

g) Since $|-2.8| = 2.8$ and $2.8 < 3$ is true, -2.8 is a solution.

124. 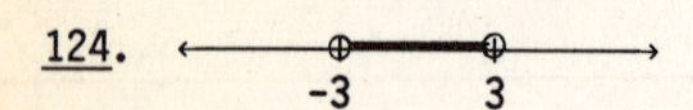

Exercise Set 2.7

1. $x > 4$

2. $x < 7$

3. $x \leq -6$

4. $y \geq 13$

5. $t \leq 80$

6. $w \geq 2$

7. $75 < a < 100$

8. $90 < s < 110$

9. $p \geq 1200$

10. $c \leq \$3457.95$

11. $y \leq 500$

12. $c \geq \$0.99$

13. $3x + 2 < 13$

14. $\frac{1}{2}n - 5 > 17$

15. Familiarize. The average of the five scores is their sum divided by the number of quizzes, 5. We let s represent the student's score on the last quiz.

Translate. The average of the five scores is given by

$$\frac{73 + 75 + 89 + 91 + s}{5}.$$

Since this average must be at least 85, this means that it must be greater than or equal to 85. Thus, we can translate the problem to the inequality

$$\frac{73 + 75 + 89 + 91 + s}{5} \geq 85.$$

Carry out. We first multiply by 5 to clear fractions.

$$5\left(\frac{73 + 75 + 89 + 91 + s}{5}\right) \geq 5\cdot 85$$

$$73 + 75 + 89 + 91 + s \geq 425$$

$$328 + s \geq 425$$

$$s \geq 425 - 328$$

$$s \geq 97$$

Check. Suppose s is a score greater than or equal to 97. Then by successively adding 73, 75, 89, and 91 on both sides of the inequality we get

$$73 + 75 + 89 + 91 + s \geq 425$$

so

$$\frac{73 + 75 + 89 + 91 + s}{5} \geq \frac{425}{5}, \text{ or } 85.$$

State. Any score which is at least 97 will give an average quiz grade of 85. The solution set is $\{s|s \geq 97\}$.

16. $\{s|s \geq 84\}$

17. Familiarize. Let m represent the number of miles per day. Then the cost per day for those miles is \$0.46m. The total cost is the daily rate plus the daily mileage cost. The total cost cannot exceed \$200. In other words the total cost must be less than or equal to \$200, the daily budget.

Translate.

Daily rate	+	Mileage cost	$\leq$	Budget
42.95	+	0.46 m	$\leq$	200

Carry out.

$42.95 + 0.46m \leq 200$

$4295 + 46m \leq 20{,}000$ Clearing decimals

$46m \leq 15{,}705$

$m \leq \frac{15{,}705}{46}$

$m \leq 341.4$ Rounding to the nearest tenth

Check. We can check to see if the solution set seems reasonable.

When $m = 342$, the total cost is

$42.95 + 0.46(342)$, or \$200.27.

When $m = 341.4$, the total cost is

$42.95 + 0.46(341.4)$, or \$199.99.

When $m = 341$, the total cost is

$42.95 + 0.46(341)$, or \$199.81.

From these calculations it would appear that $m \leqslant 341.4$ is the correct solution.

State. To stay within the budget, the number of miles the family drives must not exceed 341.4. The solution set is $\{m | m \leqslant 341.4 \text{ mi}\}$.

18. $\{m | m \leqslant 525.8 \text{ mi}\}$

19. Familiarize. We first make a drawing. We let ℓ represent the length.

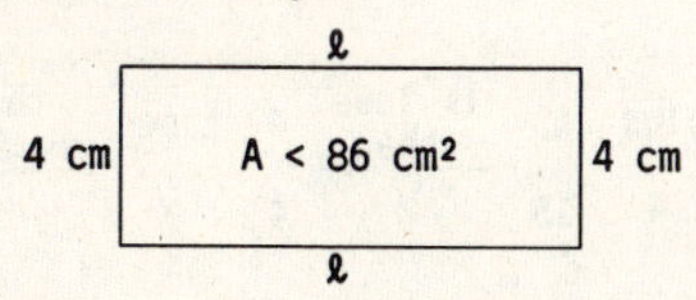

The area is the length times the width, or 4ℓ.

Translate.

Area	is less than	86 cm².
4ℓ	<	86

Carry out.

$4\ell < 86$

$\ell < 21.5$

Check. We check to see if the solution seems reasonable.

When $\ell = 22$, the area is $22 \cdot 4$, or 88 cm².

When $\ell = 21.5$, the area is $21.5(4)$, or 86 cm².

When $\ell = 21$, the area is $21 \cdot 4$, or 84 cm².

From these calculations, it would appear that the solution is correct.

State. The area will be less than 86 cm² for lengths less than 21.5 cm. The solution set is $\{\ell | \ell < 21.5 \text{ cm}\}$.

20. $\{\ell | \ell \geqslant 16.5 \text{ yd}\}$

21. Familiarize. We let n = the number of half hours that Laura's car is parked. Then the total parking time t, in hours, will be $t = n/2$. We will express all costs in dollars.

Translate.

\$0.45 charge	plus	charge for parking time	is at least	\$2.20
0.45	+	0.25n	$\geqslant$	2.20

Carry out. We solve the inequality.

$0.45 + 0.25n \geqslant 2.20$

$45 + 25n \geqslant 220$ Clearing decimals

$25n \geqslant 175$

$n \geqslant 7$

Note that when $n \geqslant 7$, $n/2 \geqslant 7/2$, or 3.5.

Check. We check to see if the solution seems reasonable.

When $n = 6$, the charge is $\$0.45 + \$0.25(6)$, or \$1.95.

When $n = 7$, the charge is $\$0.45 + \$0.25(7)$, or \$2.20.

When $n = 8$, the charge is $\$0.45 + \$0.25(8)$, or \$2.45.

From these calculations, it would appear that the solution is correct.

State. The charge is at least \$2.20 when the car is parked for at least 3.5 hr. The solution set is $\{t | t \geqslant 3.5 \text{ hr}\}$.

22. $\{m | m \geqslant 5 \text{ min}\}$

23. Familiarize.

$R = -0.075t + 3.85$

In the formula R represents the world record and t represents the years since 1930. When $t = 0$ (1930), the record was $-0.075 \cdot 0 + 3.85$, or 3.85 minutes. When $t = 2$ (1932), the record was $-0.075(2) + 3.85$, or 3.7. For what values of t will $-0.075t + 3.85$ be less than 3.5?

Translate. The record is to be less than 3.5. We have the inequality

$R < 3.5$.

To find the t values which satisfy this condition we substitute $-0.075t + 3.85$ for R.

$-0.075 + 3.85 < 3.5$

Carry out.

$-0.075t + 3.85 < 3.5$

$-0.075t < 3.5 - 3.85$

$-0.075t < -0.35$

$t > \frac{-0.35}{-0.075}$

$t > 4\frac{2}{3}$

Check. We check to see if the solution set we obtained seems reasonable.

When $t = 4\frac{1}{2}$, $R = -0.075(4.5) + 3.85$, or 3.5125.

When $t = 4\frac{2}{3}$, $R = -0.075\left(\frac{14}{3}\right) + 3.85$, or 3.5.

When $t = 4\frac{3}{4}$, $R = -0.075(4.75) + 3.85$, or 3.49375.

Since $R = 3.5$ when $t = 4\frac{2}{3}$ and R decreases as t increases, R will be less than 3.5 when t is greater than $4\frac{2}{3}$.

State. Thus, the world record will be less than 3.5 minutes when t is greater than $4\frac{2}{3}$ years $\left[\text{more than } 4\frac{2}{3} \text{ years since 1930}\right]$. The solution set is $\{t|t > 1934\}$.

24. $\{t|t > 1984\}$

25. Familiarize. Let w = the number of weeks it takes for the puppy's weight to exceed $22\frac{1}{2}$ lb.

Translate.

$$\underbrace{\text{Initial weight}}\ \text{plus}\ \underbrace{\text{amount gained in w weeks}}\ \text{exceeds}\ 22\tfrac{1}{2}\text{ lb.}$$

$$9 \quad + \quad \frac{3}{4}w \quad > \quad 22\frac{1}{2}$$

Carry out. We solve the inequality.

$$9 + \frac{3}{4}w > 22\frac{1}{2}$$

$$\frac{3}{4}w > 13\frac{1}{2}$$

$$\frac{3}{4}w > \frac{27}{2} \qquad \left[13\frac{1}{2} = \frac{27}{2}\right]$$

$$w > 18 \qquad \text{Multiplying by } \frac{4}{3}$$

Check. We check to see if the solution seems reasonable.

When w = 17, $9 + \frac{3}{4}\cdot 17 = 21\frac{3}{4}$.

When w = 18, $9 + \frac{3}{4}\cdot 18 = 22\frac{1}{2}$.

When w = 19, $9 + \frac{3}{4}\cdot 19 = 23\frac{1}{4}$.

It would appear that the solution is correct.

State. The puppy's weight will exceed $22\frac{1}{2}$ lb after 18 weeks. The solution set is $\{w|w > 18 \text{ wk}\}$.

26. $\{w|w \geq 6$ wk after July 1$\}$

27. Familiarize. We will use the formula $F = \frac{9}{5}C + 32$.

Translate.

$$\underbrace{\text{Fahrenheit temperature}}\ \underbrace{\text{is below}}\ 88°.$$

$$F \quad < \quad 88$$

Substituting $\frac{9}{5}C + 32$ for F, we have

$$\frac{9}{5}C + 32 < 88.$$

Carry out. We solve the inequality.

$$\frac{9}{5}C + 32 < 88$$

$$\frac{9}{5}C < 56$$

$$C < \frac{280}{9}$$

$$C < 31.1 \qquad \text{Rounding}$$

Check. We check to see if the solution seems reasonable.

When C = 31, $\frac{9}{5}\cdot 31 + 32 = 87.8$.

When C = 31.1, $\frac{9}{5}(31.1) + 32 = 87.98$.

When C = 31.2, $\frac{9}{5}(31.2) + 32 = 88.16$.

It would appear that the solution is correct, considering that rounding occurred.

State. Butter stays solid at Celsuis temperatures below about 31.1°. The solution set is $\{C|C < 31.1°\}$.

28. $\{C|C > 37°\}$

29. Familiarize. Let n represent the number.

Translate.

$$\underbrace{\text{The number}}\ \text{plus}\ 15\ \underbrace{\text{is less than}}\ 4\ \text{times}\ \underbrace{\text{the number.}}$$

$$n \quad + \quad 15 \quad < \quad 4 \quad \cdot \quad n$$

Carry out.

$$n + 15 < 4n$$

$$15 < 3n$$

$$5 < n, \text{ or } n > 5$$

Check. We check to see if the solution seems reasonable.

When n = 4, we have $4 + 15 < 4\cdot 4$, or $19 < 16$. This is false.

When n = 5, we have $5 + 15 < 4\cdot 5$, or $20 < 20$. This is false.

When n = 6, we have $6 + 15 < 4\cdot 6$, or $21 < 24$. This is true.

Since the inequality is false for the numbers less than or equal to 5 that we tried and true for the number greater than 5, it would appear that n > 5 is correct.

State. All numbers greater than 5 are solutions. The solution set is $\{n|n > 5\}$.

30. $\{n|n \leq 0\}$

31. Familiarize. We first make a drawing. We let w represent the width.

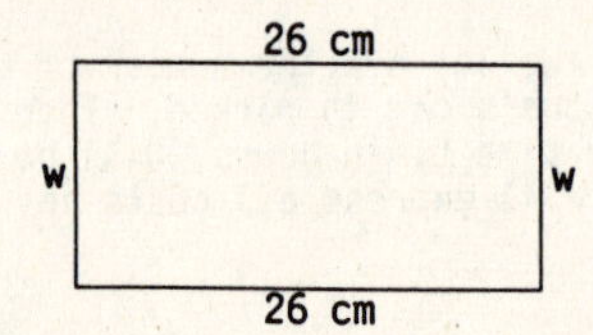

The perimeter is $P = 2\ell + 2w$, or $2\cdot 26 + 2w$, or $52 + 2w$.

Translate.

$$\underbrace{\text{The perimeter}}\ \underbrace{\text{is greater than}}\ 80\text{ cm.}$$

$$52 + 2w \quad > \quad 80$$

Carry out.

$52 + 2w > 80$

$2w > 28$

$w > 14$

Check. We check to see if the solution seems reasonable.

When $w = 13$, $P = 2\cdot26 + 2\cdot13$, or 78 cm.

When $w = 14$, $P = 2\cdot26 + 2\cdot14$, or 80 cm.

When $w = 15$, $P = 2\cdot26 + 2\cdot15$, or 82 cm.

From these calculations, it appears that the solution is correct.

State. Widths greater than 14 cm will make the perimeter greater than 80 cm. The solution set is $\{w|w > 14 \text{ cm}\}$.

32. $\{\ell|\ell \geqslant 92 \text{ ft}\}$; $\{\ell|\ell \leqslant 92 \text{ ft}\}$

33. Familiarize. We first make a drawing. Let b represent the length of the base. Then the lengths of the other sides are b - 2 and b + 3.

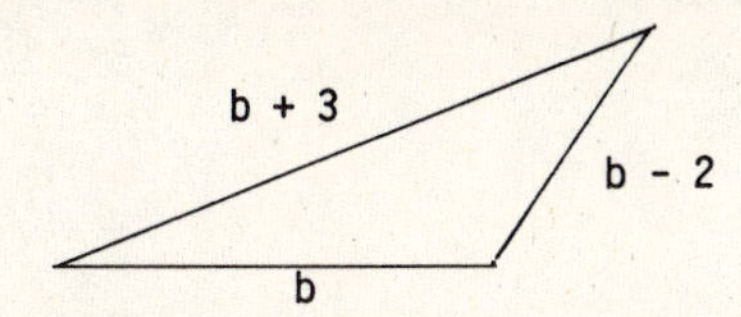

The perimeter is the sum of the lengths of the sides or $b + b - 2 + b + 3$, or $3b + 1$.

Translate.

The perimter	is greater than	19 cm.
$3b + 1$	$>$	19

Carry out.

$3b + 1 > 19$

$3b > 18$

$b > 6$

Check. We check to see if the solution seems reasonable.

When $b = 5$, the perimeter is $3\cdot5 + 1$, or 16 cm.

When $b = 6$, the perimeter is $3\cdot6 + 1$, or 19 cm.

When $b = 7$, the perimeter is $3\cdot7 + 1$, or 22 cm.

From these calculations, it would appear that the solution is correct.

State. For lengths of the base greater than 6 cm the perimeter will be greater than 19 cm. The solution set is $\{b|b > 6 \text{ cm}\}$.

34. $\left\{w \middle| w \leqslant \frac{35}{3} \text{ ft}\right\}$

35. Familiarize. The average number of calls per week is the sum of the calls for the three weeks divided by the number of weeks, 3. We let c represent the number of calls made during the third week.

Translate. The average of the three weeks is given by

$$\frac{17 + 22 + c}{3}.$$

Since the average must be at least 20, this means that it must be greater than or equal to 20. Thus, we can translate the problem to the inequality

$$\frac{17 + 22 + c}{3} \geqslant 20.$$

Carry out. We first multiply by 3 to clear of fractions.

$$3\left[\frac{17 + 22 + c}{3}\right] \geqslant 3\cdot20$$

$17 + 22 + c \geqslant 60$

$39 + c \geqslant 60$

$c \geqslant 21$

Check. Suppose c is a number greater than or equal to 21. Then by adding 17 and 22 on both sides of the inequality we get

$17 + 22 + c \geqslant 17 + 22 + 21$

$17 + 22 + c \geqslant 60$

so

$$\frac{17 + 22 + c}{3} \geqslant \frac{60}{3}, \text{ or } 20.$$

State. Any number of calls which is at least 21 will maintain an average of at least 20 for the three-week period. The solution set is $\{c|c \geqslant 21\}$.

36. George: more than 12 hours, Joan: more than 15 hours

37. Familiarize. Let s represent the amount Angelo can spend on each sweater. Then the total amount he can spend is represented by \$21.95 + 2s.

Translate.

Total spent	is less than or equal to	\$120.00
$21.95 + 2s$	$\leqslant$	120

Carry out.

$21.95 + 2s \leqslant 120$

$2195 + 200s \leqslant 12{,}000$ Clearing decimals

$200s \leqslant 9805$

$s \leqslant 49.02$ Rounding

Check. We check to see if the solution seems reasonable.

When s = \$49.01, the student spends

$21.95 + 2(49.01)$, or \$119.97.

When s = \$49.02, the student spends

$21.95 + 2(49.02)$, or \$119.99.

When s = \$49.03, the student spends

$21.95 + 2(49.03)$, or \$120.01.

From these calculations, it would appear that the solution is correct.

State. Angelo can spend at most \$49.02 for each sweater. The solution set is $\{s|s \leqslant \$49.02\}$.

38. $\{s|0 \text{ lb} \leqslant s \leqslant 9 \text{ lb}\}$

39. **Familiarize**. We first make a drawing. Let ℓ represent the length.

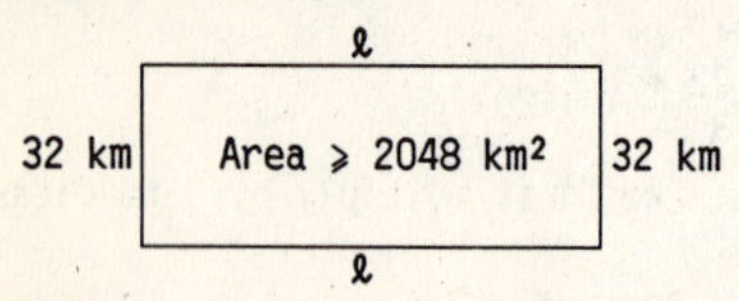

The area is length times width, or 32ℓ.

Translate.

The area is at least 2048 km².

$$32\ell \qquad \geqslant \qquad 2048$$

Carry out.

$$32\ell \geqslant 2048$$
$$\ell \geqslant 64$$

Check. We check to see if the solution seems reasonable.

When $\ell = 63$, the area is $32 \cdot 63$, or 2016 km².

When $\ell = 64$, the area is $32 \cdot 64$, or 2048 km².

When $\ell = 65$, the area is $32 \cdot 65$, or 2080 km².

From these calculations, it would appear that the solution is correct.

State. Lengths of 64 km or more will make the area at least 2048 km². The solution set is $\{\ell|\ell \geqslant 64 \text{ km}\}$.

40. $\{b|b \leqslant 4 \text{ cm}\}$

41. **Familiarize**. We will use the formula

$$P = 0.1522Y - 298.592.$$

Translate. We have the inequality $P \geqslant 6$. To find the years that satisfy this condition we substitute $0.1522Y - 298.592$ for P:

$$0.1522Y - 298.592 \geqslant 6$$

Carry out.

$$0.1522Y - 298.592 \geqslant 6$$
$$0.1522Y \geqslant 304.592$$
$$Y \geqslant 2001 \quad \text{Rounding}$$

Check. We check to see if the solution seems reasonable.

When $Y = 2000$, $P = 0.1522(2000) - 298.592$, or about \$5.81.

When $Y = 2001$, $P = 0.1522(2001) - 298.592$, or about \$5.96.

When $Y = 2002$, $P = 0.1522(2002) - 298.592$, or about \$6.11.

From these calculations, it would appear that the solution is correct considering that rounding occurred.

State. From about 2001 on, the average price of a movie ticket will be at least \$6. The solution set is $\{Y|Y \geqslant 2001\}$.

42. $\left\{x \middle| x \leqslant 215\frac{5}{27} \text{ mi}\right\}$

43.
$$-3 + 2(-5)^2(-3) - 7$$
$= -3 + 2(25)(-3) - 7$ Evaluating the exponential expression
$= -3 - 150 - 7$ Multiplying
$= -160$ Subtracting

44. $4a^2 - 2$

45.
$$9x - 5 + 4x^2 - 2 - 13x$$
$$= 4x^2 + 9x - 13x - 5 - 2$$
$$= 4x^2 + (9 - 13)x - 5 - 2$$
$$= 4x^2 - 4x - 7$$

46. $-17x + 18$

47.

48.

49. **Familiarize**. We make a drawing. Let s represent the length of a side of the square.

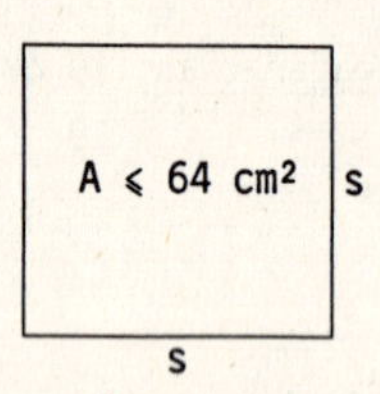

The area s is the square of the length of a side, or s^2.

Translate.

The area is no more than 64 cm².

$$s^2 \qquad \leqslant \qquad 64$$

Carry out.

$$s^2 \leqslant 64$$
$$s^2 - 64 \leqslant 0$$
$$(s + 8)(s - 8) \leqslant 0$$

We know that $(s + 8)(s - 8) = 0$ for $s = -8$ or $s = 8$. Now $(s + 8)(s - 8) < 0$ when the two factors have opposite signs. That is:

$s+8 > 0$ and $s-8 < 0$ or $s+8 < 0$ and $s-8 > 0$

$s > -8$ and $s < 8$ or $s < -8$ and $s > 8$

This can be expressed as $-8 < s < 8$. This is not possible.

Then $(s + 8)(s - 8) \leqslant 0$ for $-8 \leqslant s \leqslant 8$.

Check. Since the length of a side cannot be negative we only consider positive values of s, or $0 < s \leqslant 8$. We check to see if this solution seems reasonable.

When $s = 7$, the area is 7^2, or 49 cm^2.

When $s = 8$, the area is 8^2, or 64 cm^2.

When $s = 9$, the area is 9^2, or 81 cm^2.

From these calculations, it appears that the solution is correct.

State. Sides of length 8 or less will allow an area of no more than 64 cm^2. The solution set is $\{s \mid s \leq 8 \text{ cm and } s \text{ is positive}\}$, or $\{s \mid 0 \text{ cm} < s \leq 8 \text{ cm}\}$.

50. 47 and 49

51. Familiarize. Let h = the number of hours the car has been parked. Then $h - 1$ = the number of hours after the first hour.

Translate.

Charge for first hour	plus	charge for additional hours	exceeds	\$16.50.
4.00	+	2.50(h - 1)	>	16.50

Carry out. We solve the inequality.

$$4.00 + 2.50(h - 1) > 16.50$$
$$40 + 25(h - 1) > 165 \quad \text{Multiplying by 10 to clear decimals}$$
$$40 + 25h - 25 > 165$$
$$25h + 15 > 165$$
$$25h > 150$$
$$h > 6$$

Check. We check to see if this solution seems reasonable.

When $h = 5$, $4.00 + 2.50(5 - 1) = 14.00$.

When $h = 6$, $4.00 + 2.50(6 - 1) = 16.50$.

When $h = 7$, $4.00 + 2.50(7 - 1) = 19.00$.

It appears that the solution is correct.

State. The charge exceeds \$16.50 when the car has been parked for more than 6 hr.

52. $\{s \mid s < \$20{,}000\}$

53. Familiarize. We define h as in Exercise 51. Note that the parking charge must be more than \$14 and also less than \$24. Then the charge can be at most \$24 - \$2.50, or \$21.50. We will solve two inequalities and find the solutions that are in both solution sets.

Translate.

Charge for first hour	plus	charge for additional hours	is more than	\$14.
4.00	+	2.50(h - 1)	>	14

and

Charge for first hour	plus	charge for additional hours	is at most	\$21.50.
4.00	+	2.50(h - 1)	≤	21.50

Carry out. We solve both inqualities. We get $h > 5$ and $h \leq 8$. The solutions that are in both solution sets are $\{h \mid 5 < h \leq 8\}$.

Check. The check is left to the student.

State. The car has been parked more than 5 hr but at most 8 hr.

54. Between $-15°$ and $-9\frac{4}{9}°$

55. ◈

56. ◈

Chapter 2 [illegible]

14. [illegible]

21. [illegible]

61. [illegible] [illegible] [illegible] [illegible] [illegible] [illegible] [illegible] [illegible] the [illegible]

Tests trends

[illegible]

[illegible] [illegible]

[illegible]

[illegible]

[illegible]

[illegible]

[illegible]

Exercise Set 3.1

1. We go to the top of the bar that is above the body weight 100 lb. Then we move horizontally from the top of the bar to the vertical scale listing numbers of drinks. It appears approximately 3 drinks will give a 100 lb person a blood-alcohol level of 0.10%.

2. Approximately 5 drinks

3. From $3\frac{1}{2}$ on the vertical scale we move horizontally until we reach a bar whose top is above the horizontal line on which we are moving. The first such bar corresponds to a body weight of 120 lb. Thus, an individual weighs at least 120 lb if $3\frac{1}{2}$ drinks are consumed without reaching a blood-alcohol level of 0.10%.

4. 160 lb

5. The longest bar represents boredom. Thus this was the reason given most often.

6. Work/Military service

7. We go to the right end of the bar representing grades and then go down to the percent scale. We find that approximately 5% dropped out because of grades.

8. Approximately 40%

9. We locate 85 on the vertical scale and then move right until the line is reached. At that point we move down to the horizontal scale and read the information we are seeking. We see that the pulse rate was 85 beats per minute after 1 month of regular exercise.

10. 6 months

11. By observation or by computing the decrease/increase in beats per minute between successive pairs of points, we find that the greatest drop is about 15 beats per minute. This occurs between the second and third points from the left. We move down the horiziontal scale and see that these points correspond to 1 month and 2 months. Thus, the greatest drop in pulse rate occurred during the second month.

12. The fifth month

13. We find the portion of the graph labeled medical care and read that 12% of the income was spent on medical expenses.

14. \$220

15. Familiarize. The graph tells us that 28% of the income is spent on food. We let y = the amount spent on food.

Translate. We reword and translate.

What is 28% of income?

$y = 28\% \cdot \$2400$

Carry out. We do the computation.

$y = 0.28 \cdot \$2400 = \$672$

Check. We go over the computation.

State. The family would spend \$672 on food.

16. 55%

17. We locate 1965 on the horizontal scale and then move up to the line representing public education expenditures. At that point we move left to the vertical scale and read the information we are seeking. Approximately 4% of the GNP was spent on public education in 1965.

18. Approximately 8%

19. We locate 10% on the vertical scale and then move right until the line representing health care expenditures is reached. At that point we can move down to the horizontal scale and read the information we are seeking. Health care costs represented about 10% of the GNP in 1982.

20. 1990

21. We locate 8% on the vertical scale and then move right, noting the three points at which we hit the line representing defense expenditures. From each of the points we move down to the horizontal scale and read the information we are seeking. Defense expenditures were approximately 8% of the GNP in 1953, 1963, and 1970.

22. 1989

23. The highest point on the line representing defense expenditures occurs above 1955 on the horizontal scale. Defense expenditures peaked in 1955.

24. 1966, 1980

25. Familiarize. From the graph we read that health care expenditures were about 7% of GNP in 1970 and about 12% of GNP in 1990. We let y = the growth over the years 1970-1990.

Translate. We reword and translate.

Growth is 1990 percentage less 1970 percentage.

$y = 12\% - 7\%$

Carry out. We do the computation.

$y = 12\% - 7\% = 5\%$

Check. We go over the computation.

State. Health care expenditures (as a percentage of GNP) grew approximately 5% over the years 1970-1990.

26. Approximately 1%

27. We find the portion of the graph labeled Jazz and read that 3.7% of all recordings sold are jazz.

28. 9.0%

29. Familiarize. Let p = the percent of all recordings sold that are either soul or pop/rock. We will use the graph to find the percent for each type of music and then find their sum.

Translate. We reword the problem.

Percent together	is	soul percent	plus	pop/rock percent.
p	=	12.0%	+	58.1%

Carry out. We do the computation.

$p = 12.0\% + 58.1\% = 70.1\%$

Check. We go over the computation. The answer checks.

State. Together, 70.1% of all recordings sold are either soul or pop/rock.

30. 10.5%

31. Familiarize. The graph tells us that 58.1% of all recordings sold are pop/rock, 12.0% are soul, and 9.0% are country. Let p, s, and c represent the number of pop/rock, soul, and country recordings sold, respectively.

Translate. We reword the problem and write an equation for each type of music.

Number of pop/rock recordings sold	is	58.1%	of	total number sold.
p	=	58.1%	$\cdot$	3000

Number of soul recordings sold	is	12.0%	of	total number sold.
s	=	12.0%	$\cdot$	3000

Number of country recordings sold	is	9.0%	of	total number sold.
c	=	9.0%	$\cdot$	3000

Carry out. We do the three computations.

$p = 0.581 \cdot 3000 = 1743$

$s = 0.12 \cdot 3000 = 360$

$c = 0.09 \cdot 3000 = 270$

Check. We go over the computations. The answers check.

State. The store sells 1743 pop/rock, 360 soul, and 270 country recordings.

32. Pop/rock: 1452; classical: 170; gospel: 30

33. (2,5) is 2 units right and 5 units up.
(-1,3) is 1 unit left and 3 units up.
(3,-2) is 3 units right and 2 units down.
(-2,-4) is 2 units left and 4 units down.
(0,4) is 0 units left or right and 4 units up.
(0,-5) is 0 units left or right and 5 units down.
(5,0) is 5 units right and 0 units up or down.
(-5,0) is 5 units left and 0 units up or down.

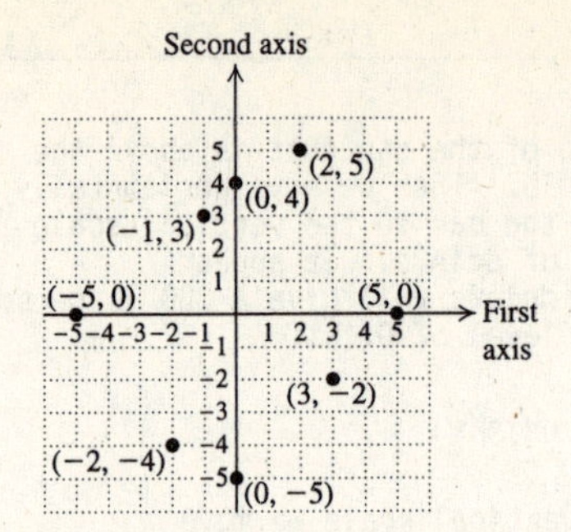

34.

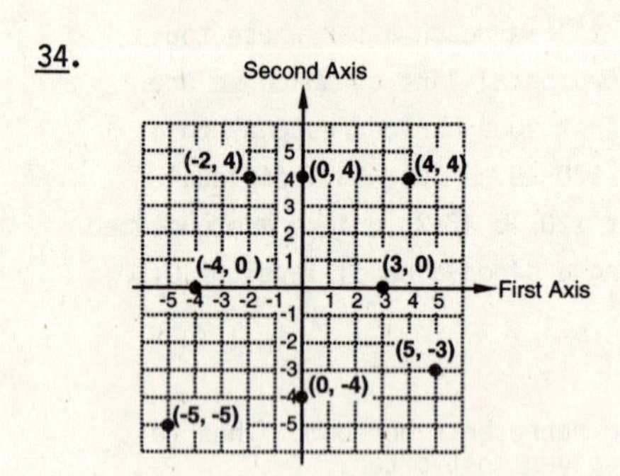

35. Since the first coordinate is negative and the second coordinate positive, the point (-5,3) is located in the second quadrant.

36. II

37. Since the first coordinate is positive and the second coordinate negative, the point (100,-1) is in the fourth quadrant.

38. IV

39. Since both coordinates are negative, the point (-6,-29) is in the third quadrant.

40. III

41. Since both coordinates are positive, the point (3.8,9.2) is the first quadrant.

42. I

43. In quadrant III, first coordinates are always negative and second coordinates are always negative.

44. Second, first

45.

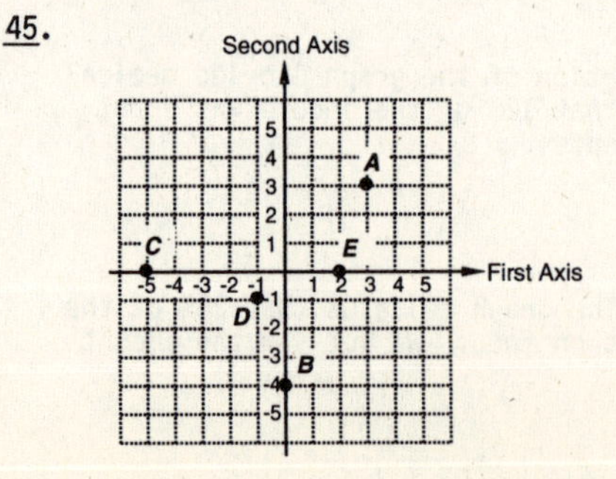

Point A is 3 units right and 3 units up. The coordinates of A are (3,3).

Point B is 0 units left or right and 4 units down. The coordinates of B are (0,-4).

Point C is 5 units left and 0 units up or down. The coordinates of C are (-5,0).

Point D is 1 unit left and 1 unit down. The coordinates of D are (-1,-1).

Point E is 2 units right and 0 units up or down. The coordinates of E are (2,0).

46. A: (4,1), B: (0,-5), C: (-4,0), D: (-3,-2), E: (3,0)

47. $\frac{3}{5} \cdot \frac{10}{9} = \frac{3 \cdot 10}{5 \cdot 9}$

$= \frac{3 \cdot 2 \cdot 5}{5 \cdot 3 \cdot 3}$

$= \frac{\cancel{3} \cdot 2 \cdot \cancel{5}}{\cancel{5} \cdot \cancel{3} \cdot 3}$

$= \frac{2}{3}$

48. $\frac{13}{15}$

49. $\frac{3}{7} - \frac{4}{5}$

$= \frac{3}{7} \cdot \frac{5}{5} - \frac{4}{5} \cdot \frac{7}{7}$ Using 35 as a common denominator

$= \frac{15}{35} - \frac{28}{35}$

$= -\frac{13}{35}$

50. $-\frac{2}{15}$

51.

52.

53.

Second Axis; First Axis; II (–, +); I (+, +); III (–, –); IV (+, –)

If the first coordinate is positive, then the point must be in either I or IV.

54. III or IV

55. If the first and second coordinates are equal, they must either be both positive or both negative. The point must be in either I (both positive) or III (both negative).

56. II or IV

57.

Second Axis; First Axis; (-1, 1); (4, 1); (-1, -5); (4, -5)

The coordinates of the fourth vertex are (-1,-5).

58. (5,2), (-7,2), or (3,-8)

59. Answers may vary.

We select eight points such that the sum of the coordinates for each point is 6.

(-1,7)	-1 + 7 = 6
(0,6)	0 + 6 = 6
(1,5)	1 + 5 = 6
(2,4)	2 + 4 = 6
(3,3)	3 + 3 = 6
(4,2)	4 + 2 = 6
(5,1)	5 + 1 = 6
(6,0)	6 + 0 = 6

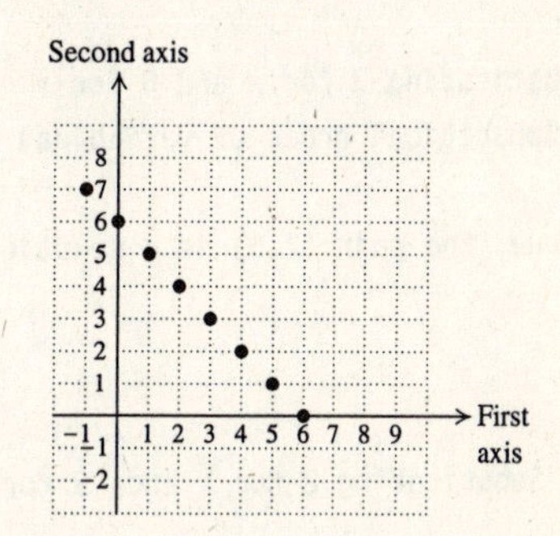

60. Answers may vary

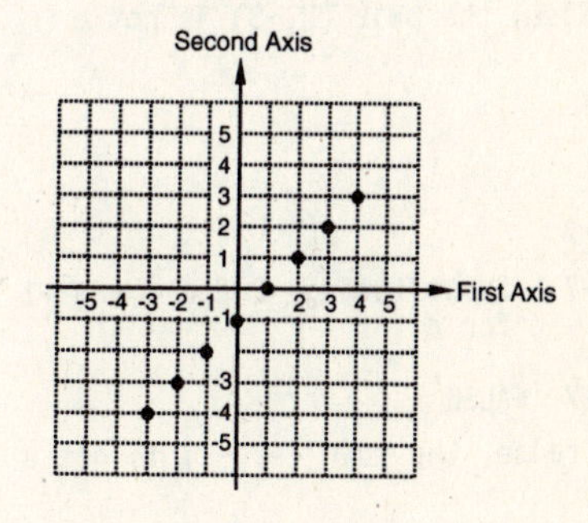

61.

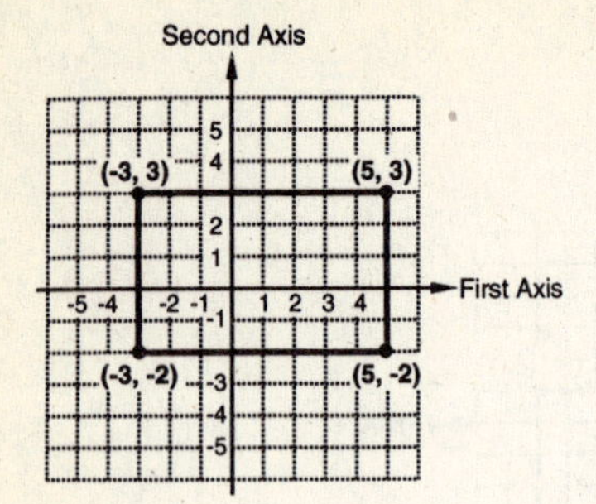

The length is 8, and the width is 5.

$P = 2\ell + 2w$

$P = 2\cdot 8 + 2\cdot 5 = 16 + 10 = 26$

62. $32\frac{1}{2}$

63. Latitude 32.5° North,
Longitude 64.5° West

64. Latitude 27° North,
Longitude 81° West

Exercise Set 3.2

1. $y = 3x - 1$

$5 \; ? \; 3\cdot 2 - 1$ Substituting 2 for x and 5 for y

$\quad 6 - 1$ (alphabetical order of variables)

$5 \mid 5$ TRUE

Since 5 = 5 is true, the pair (2,5) is a solution.

2. Yes

3. $3x - y = 4$

$3\cdot 2 - (-3) \; ? \; 4$ Substituting 2 for x and -3 for y

$6 + 3$

$9 \mid 4$ FALSE

Since 9 = 4 is false, the pair (2,-3) is not a solution.

4. No

5. $2c + 2d = -7$

$2(-2) + 2(-1) \; ? \; -7$ Substituting -2 for c and -1 for d

$-4 - 2$

$-6 \mid -7$ FALSE

Since -6 = -7 is false, the pair (-2,-1) is not a solution.

6. No

7. $y = x$

We first make a table of values. We choose any number for x and then determine y by substitution.

When $x = 0$, $y = 0$.

When $x = -2$, $y = -2$.

When $x = 3$, $y = 3$.

x	y
0	0
-2	-2
3	3

Since two points determine a line, that is all we really need to graph a line, but you may plot a third point as a check.

Plot these points, draw the line they determine, and label the graph y = x.

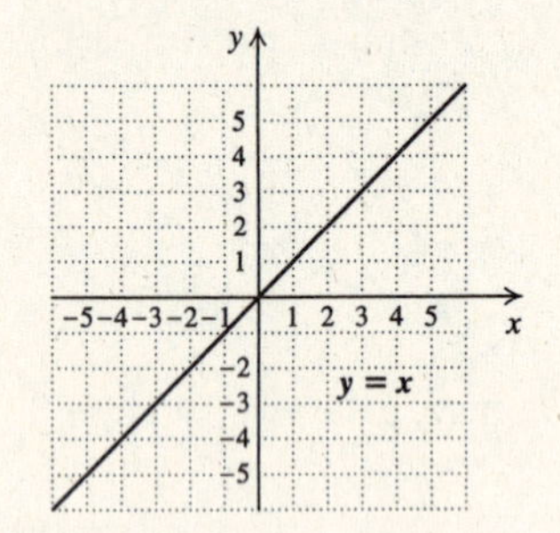

8.

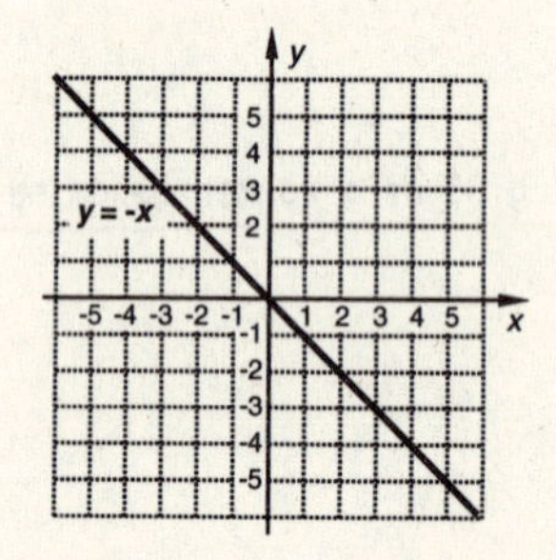

9. $y = -2x$

We first make a table of values.

When $x = 0$, $y = -2\cdot 0 = 0$.

When $x = 2$, $y = -2\cdot 2 = -4$.

When $x = -1$, $y = -2(-1) = 2$.

x	y
0	0
2	-4
-1	2

Plot these points, draw the line they determine, and label the graph y = -2x.

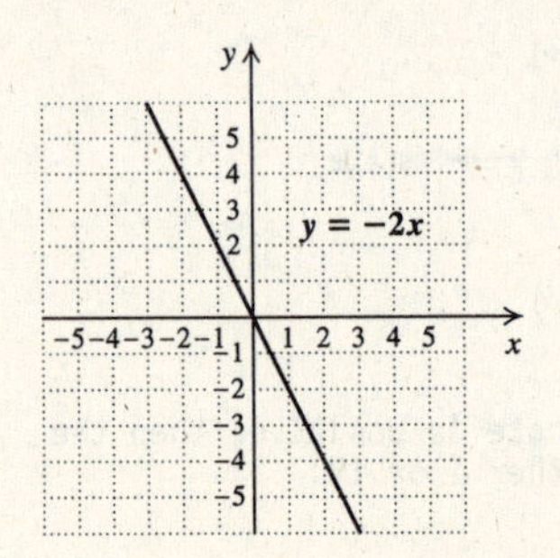

10.

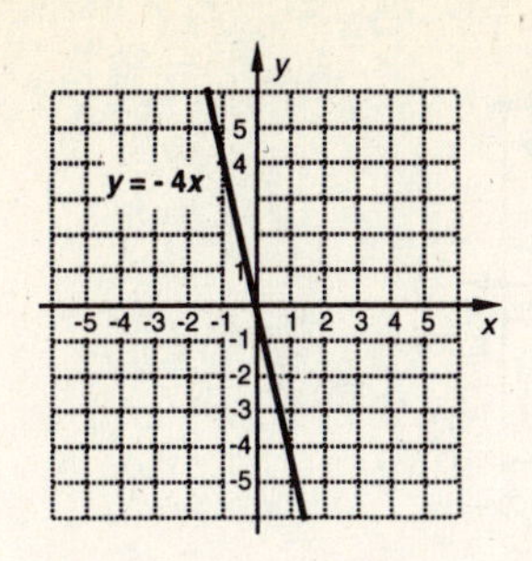

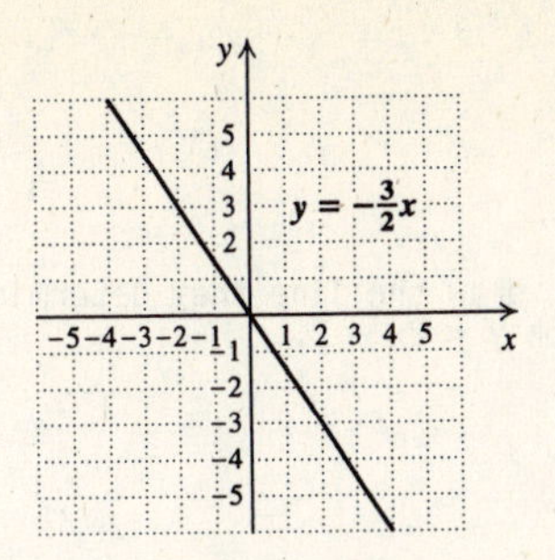

11. $y = \frac{1}{3}x$

We first make a table of values. Using multiples of 3 for x avoids fractions.

When x = 0, $y = \frac{1}{3} \cdot 0 = 0$.

When x = 6, $y = \frac{1}{3} \cdot 6 = 2$.

When x = -3, $y = \frac{1}{3}(-3) = -1$.

x	y
0	0
6	2
-3	-1

Plot these points, draw the line they determine, and label the graph $y = \frac{1}{3}x$.

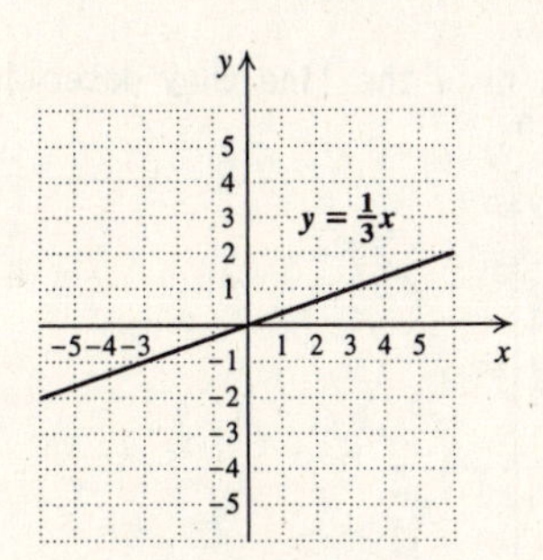

12.

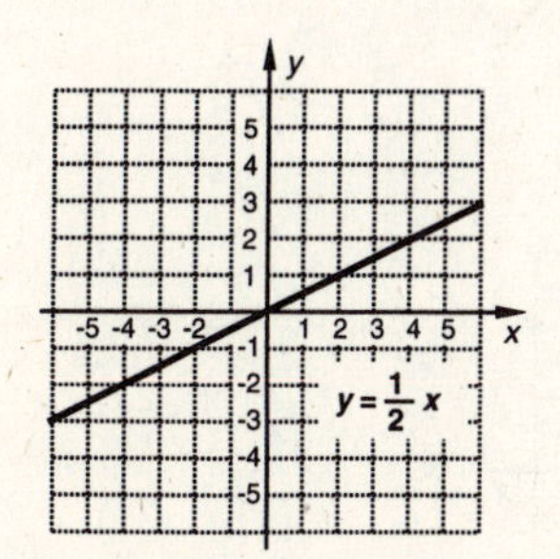

13. $y = -\frac{3}{2}x$

We first make a table of values. Using multiples of 2 for x avoids fractions.

When x = 0, $y = -\frac{3}{2} \cdot 0 = 0$.

When x = 2, $y = -\frac{3}{2} \cdot 2 = -3$.

When x = -2, $y = -\frac{3}{2}(-2) = 3$.

x	y
0	0
2	-3
-2	3

Plot these points, draw the line they determine, and label the graph $y = -\frac{3}{2}x$.

14.

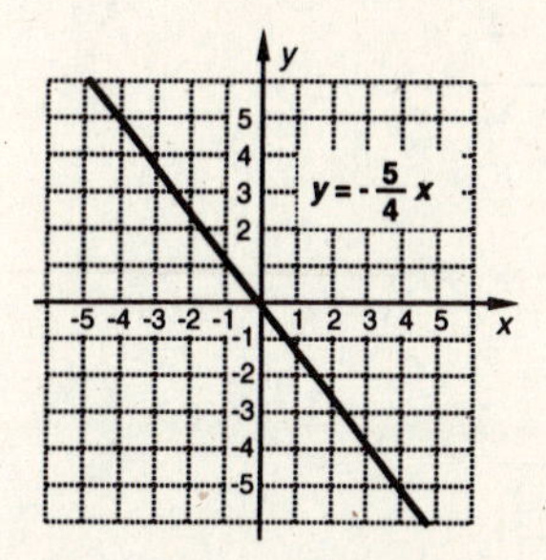

15. y = x + 1

We first make a table of values. We choose <u>any</u> number for x and then determine y by substitution.

When x = 0, y = 0 + 1 = 1.

When x = 3, y = 3 + 1 = 4.

When x = -5, y = -5 + 1 = -4.

x	y
0	1
3	4
-5	-4

Plot these points, draw the line they determine, and label the graph y = x + 1.

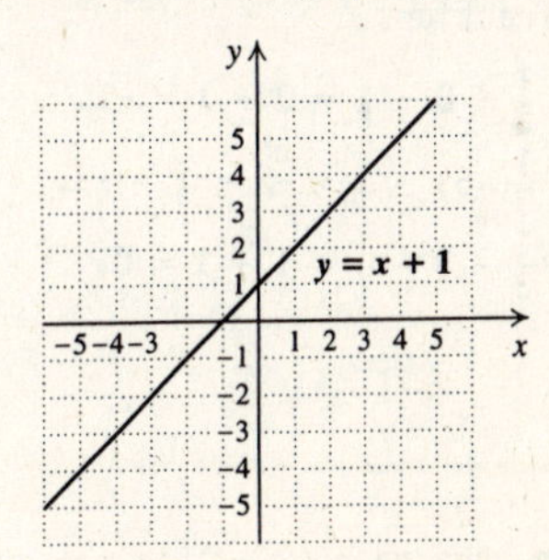

16.

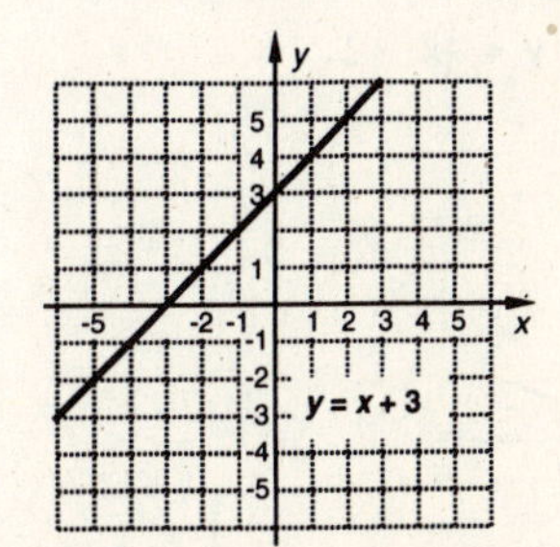

17. y = 2x + 2

We first make a table of values.

When x = 0, y = 2·0 + 2 = 0 + 2 = 2.

When x = -3, y = 2(-3) + 2 = -6 + 2 = -4.

When x = 1, y = 2·1 + 2 = 2 + 2 = 4.

x	y
0	2
-3	-4
1	4

Plot these points, draw the line they determine, and label the graph y = 2x + 2.

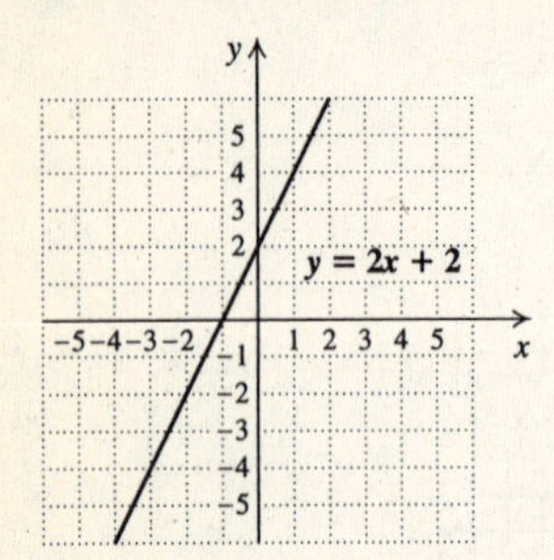

18.

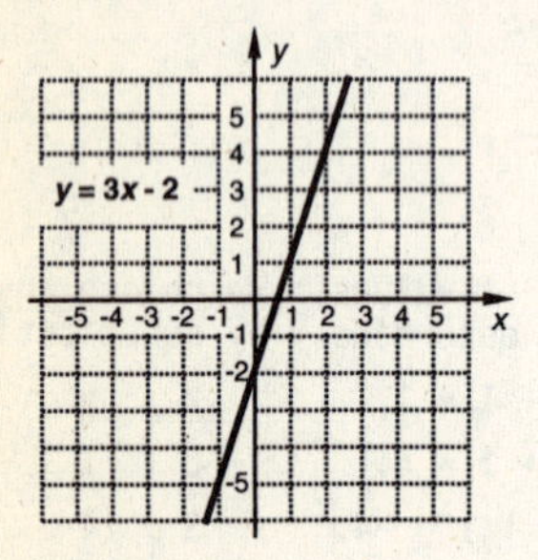

19. $y = \frac{1}{3}x - 1$

We first make a table of values. Using multiples of 3 for x avoids fractions.

When x = 0, $y = \frac{1}{3} \cdot 0 - 1 = 0 - 1 = -1.$

When x = -6, $y = \frac{1}{3}(-6) - 1 = -2 - 1 = -3.$

When x = 3, $y = \frac{1}{3} \cdot 3 - 1 = 1 - 1 = 0.$

x	y
0	-1
-6	-3
3	0

Plot these points, draw the line they determine, and label the graph $y = \frac{1}{3}x - 1$.

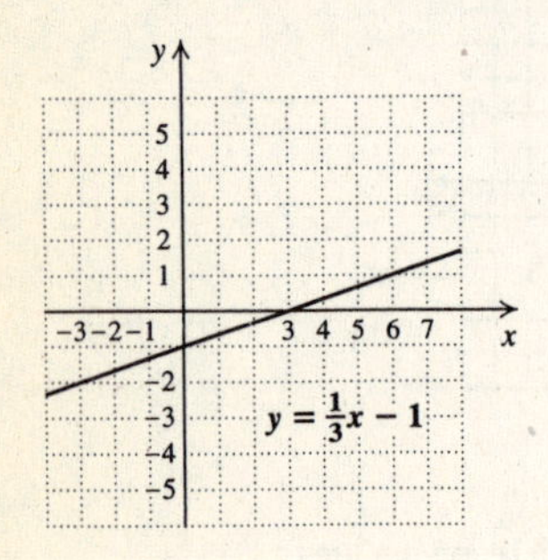

20.

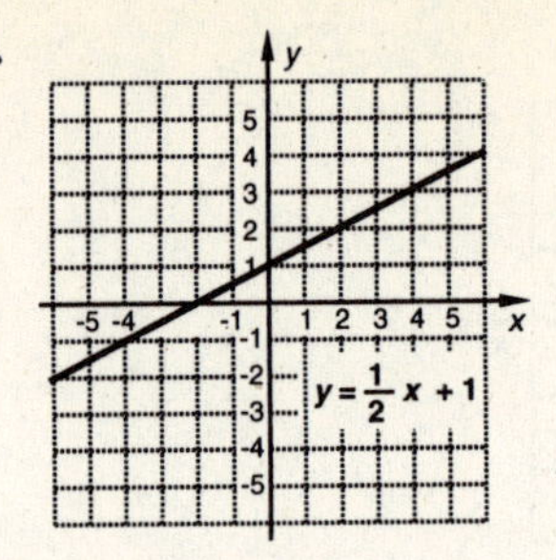

21. y + x = -3

y = -x - 3 Solving for y

We first make a table of values.

When x = 0, y = -0 - 3 = -3.

When x = 1, y = -1 - 3 = -4.

When x = -5, y = -(-5) - 3 = 5 - 3 = 2.

x	y
0	-3
1	-4
-5	2

Plot these points, draw the line they determine, and label the graph.

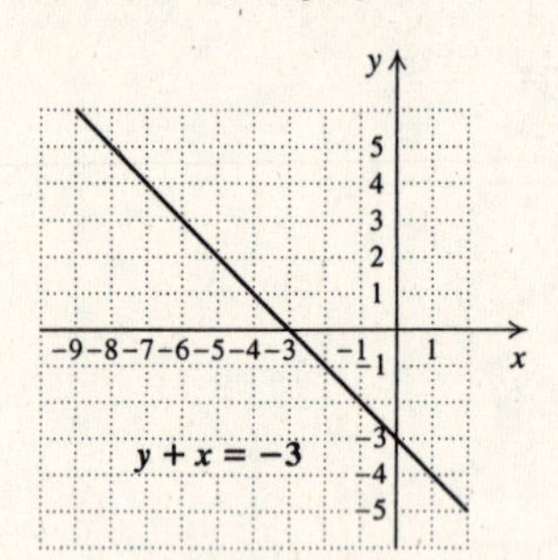

22.

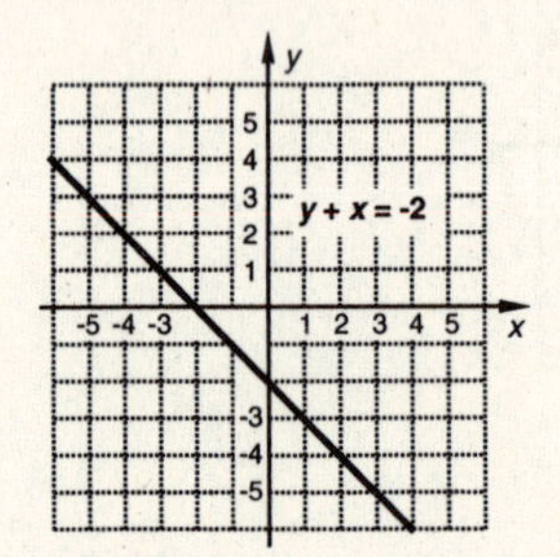

23. $y = \frac{5}{2}x + 3$

We first make a table of values. Using multiples of 2 for x avoids fractions.

When x = 0, $y = \frac{5}{2} \cdot 0 + 3 = 0 + 3 = 3.$

When x = -2, $y = \frac{5}{2}(-2) + 3 = -5 + 3 = -2.$

When x = -4, $y = \frac{5}{2}(-4) + 3 = -10 + 3 = -7.$

x	y
0	3
-2	-2
-4	-7

Plot these points, draw the line they determine, and label the graph.

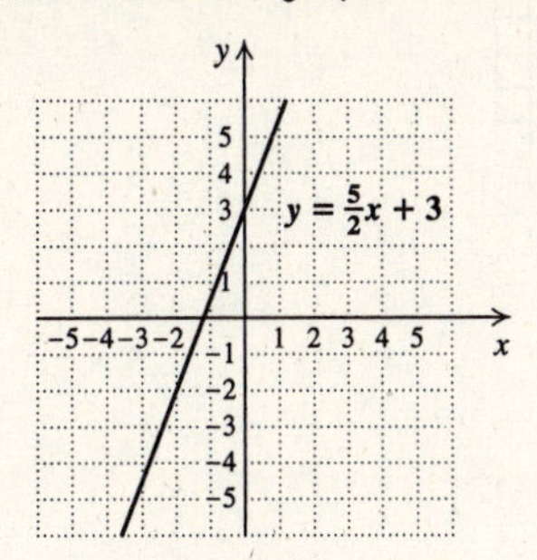

24.

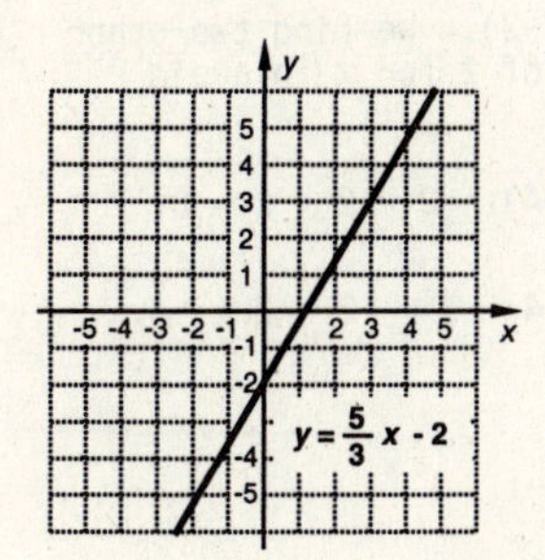

25. $y = -\frac{5}{2}x - 2$

We first make a table of values. Using multiples of 2 for x avoids fractions.

When x = 0, $y = -\frac{5}{2}\cdot 0 - 2 = 0 - 2 = -2.$

When x = -2, $y = -\frac{5}{2}(-2) - 2 = 5 - 2 = 3.$

When x = 2, $y = -\frac{5}{2}(2) - 2 = -5 - 2 = -7.$

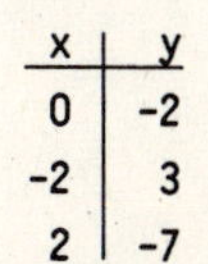

x	y
0	-2
-2	3
2	-7

Plot these points, draw the line they determine, and label the graph.

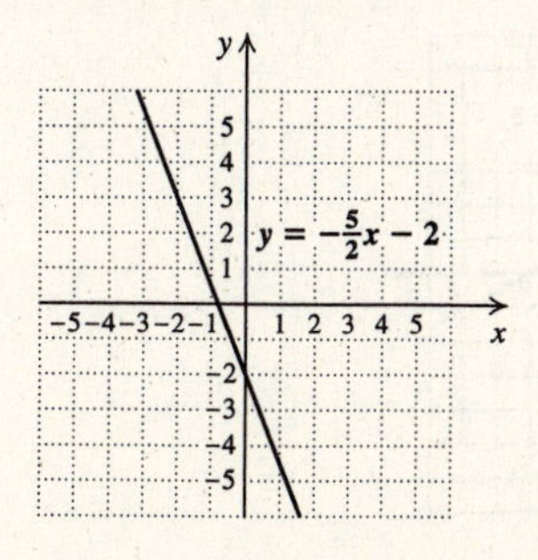

26.

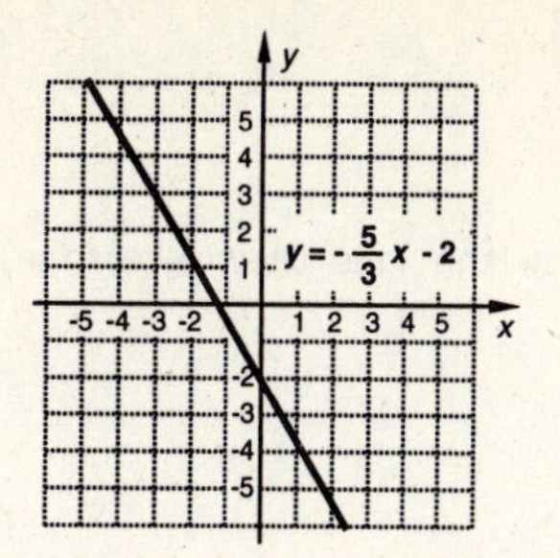

27. $y = \frac{1}{2}x - 5$

The y-intercept is (0,-5). We find two other pairs using multiples of 2 for x to avoid fractions.

When x = 2, $y = \frac{1}{2}\cdot 2 - 5 = 1 - 5 = -4.$

When x = 4, $y = \frac{1}{2}\cdot 4 - 5 = 2 - 5 = -3.$

x	y
0	-5
2	-4
4	-3

Plot these points, draw the line they determine, and label the graph.

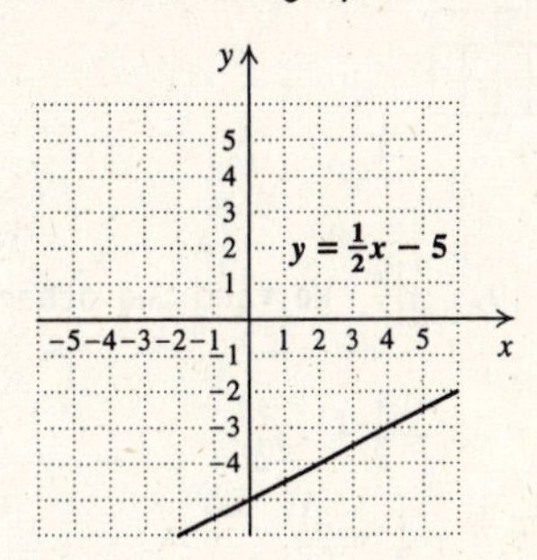

28.

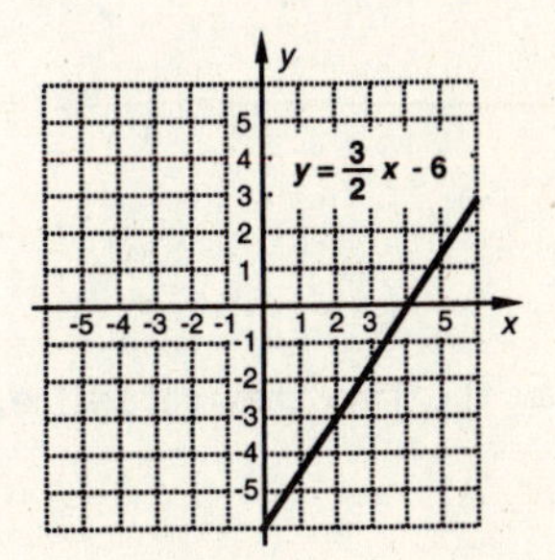

29. $2x + y = 3$

$y = -2x + 3$ Solving for y

The y-intercept is (0,3). We find two other pairs.

When x = -1, $y = -2(-1) + 3 = 2 + 3 = 5.$

When x = 3, $y = -2\cdot 3 + 3 = -6 + 3 = -3.$

x	y
0	3
-1	5
3	-3

Plot these points, draw the line they determine, and label the graph.

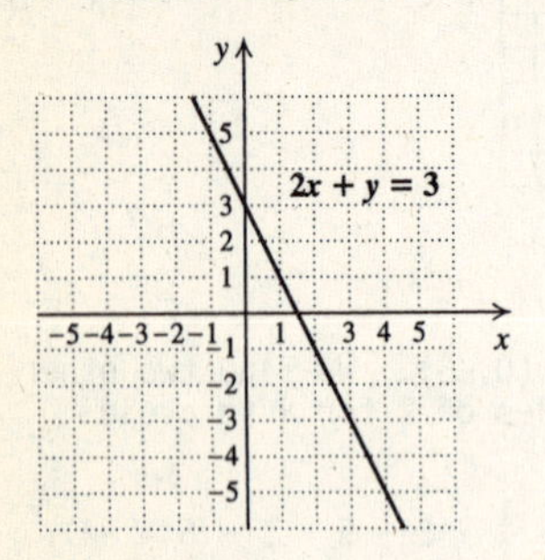

30.

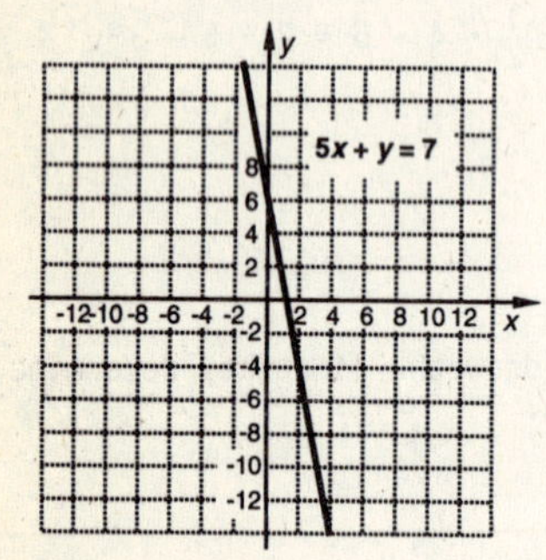

31. $y = x - \frac{1}{2}$

The y-intercept is $\left(0,-\frac{1}{2}\right)$. We find two other pairs.

When $x = -3$, $y = -3 - \frac{1}{2} = -3\frac{1}{2}$.

When $x = 4$, $y = 4 - \frac{1}{2} = 3\frac{1}{2}$.

x	y
0	$-\frac{1}{2}$
-3	$-3\frac{1}{2}$
4	$3\frac{1}{2}$

Plot these points, draw the line they determine, and label the graph.

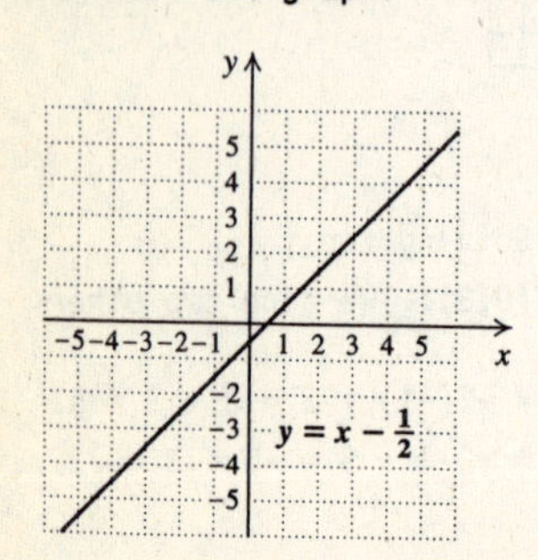

32.

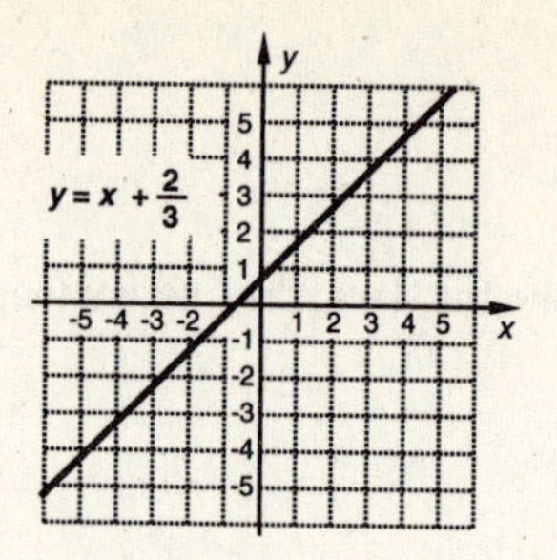

33. We solve for y.

$x + 2y = -4$

$2y = -x - 4$

$y = \frac{1}{2}(-x - 4)$

$y = -\frac{1}{2}x - 2$

The y-intercept is (0,-2). We find two other pairs using multiples of 2 for x to avoid fractions.

When $x = -4$, $y = -\frac{1}{2}(-4) - 2 = 2 - 2 = 0$.

When $x = 4$, $y = -\frac{1}{2} \cdot 4 - 2 = -2 - 2 = -4$.

x	y
0	-2
-4	0
4	-4

Plot these points, draw the line they determine, and label the graph.

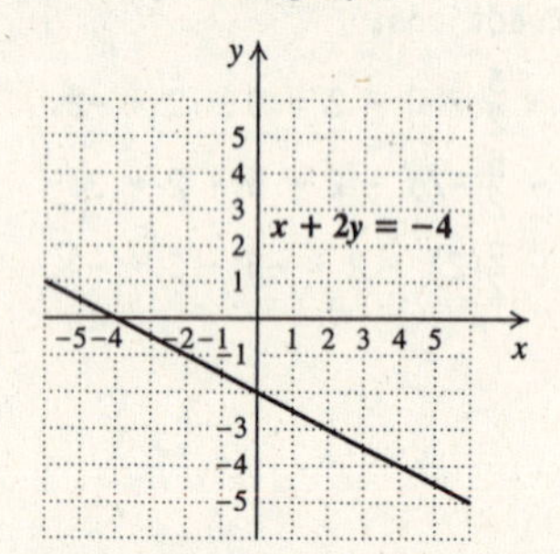

34.

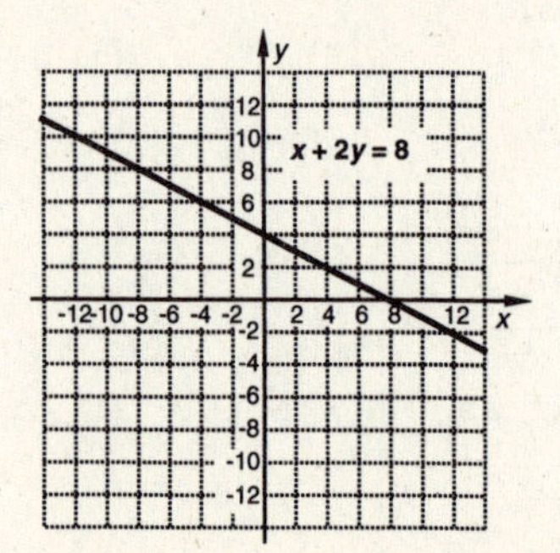

35. We solve for y.

$6x - 3y = 9$

$-3y = -6x + 9$

$y = -\frac{1}{3}(-6x + 9)$

$y = 2x - 3$

The y-intercept is (0,-3). We find two other pairs.

When x = -1, $y = 2(-1) - 3 = -2 - 3 = -5$.

When x = 3, $y = 2\cdot 3 - 3 = 6 - 3 = 3$.

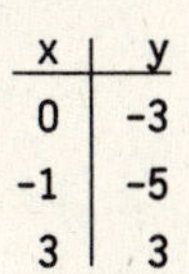

x	y
0	-3
-1	-5
3	3

Plot these points, draw the line they determine, and label the graph.

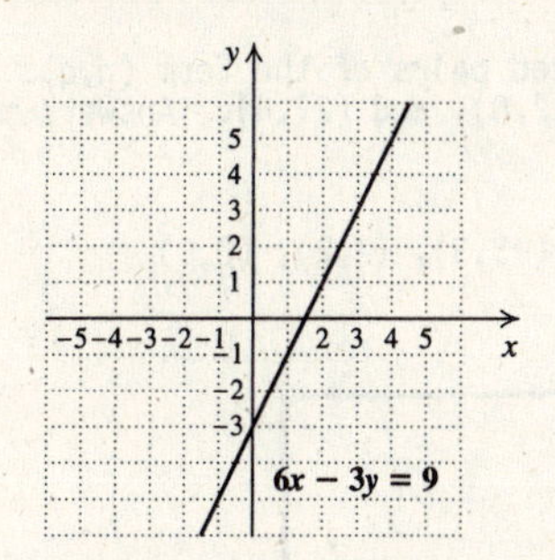

36.

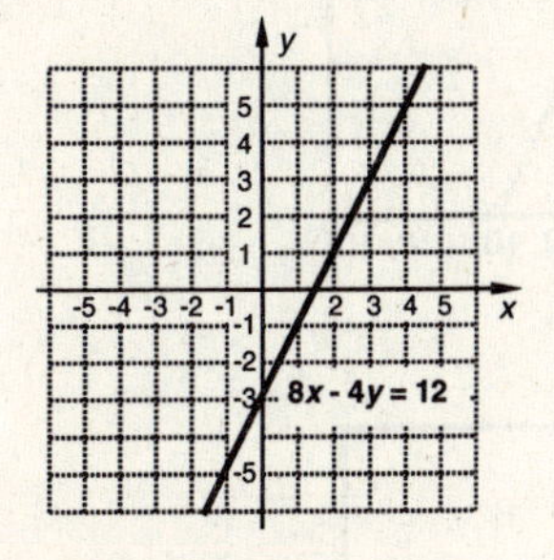

37. We solve for y.

$6y + 2x = 8$

$6y = -2x + 8$

$y = \frac{1}{6}(-2x + 8)$

$y = -\frac{1}{3}x + \frac{4}{3}$

The y-intercept is $\left(0,\frac{4}{3}\right)$. We find two other pairs.

When x = -2, $y = -\frac{1}{3}(-2) + \frac{4}{3} = \frac{2}{3} + \frac{4}{3} = \frac{6}{3} = 2$.

When x = 4, $y = -\frac{1}{3}\cdot 4 + \frac{4}{3} = -\frac{4}{3} + \frac{4}{3} = 0$.

x	y
0	$\frac{4}{3}$
-2	2
4	0

Plot these points, draw the line they determine, and label the graph.

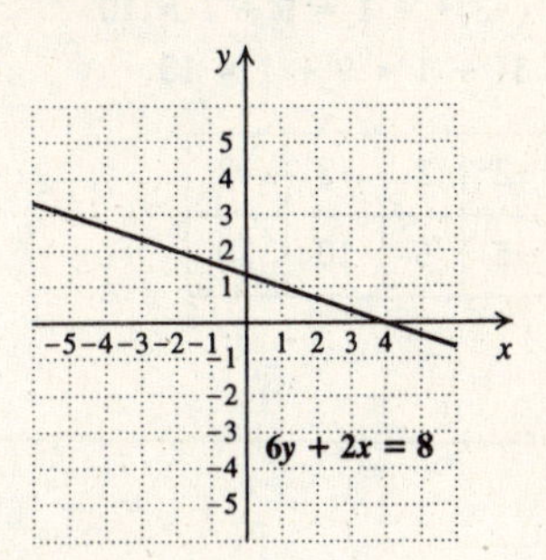

38.

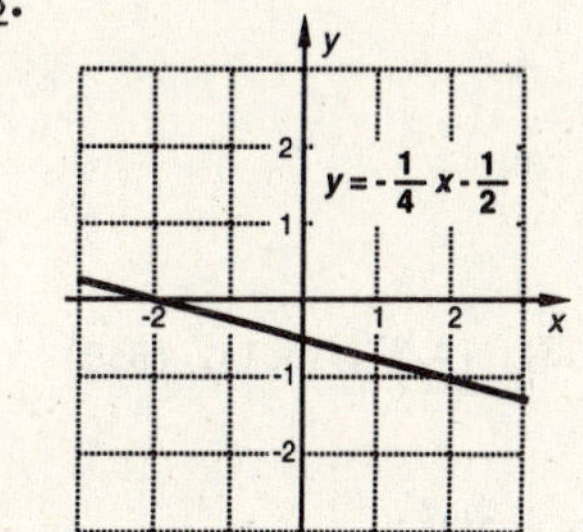

39. $3x - 7 = -34$

$3x = -27$ Adding 7

$x = -9$ Dividing by 3

Check:

$3x - 7 = -34$	
$3(-9) - 7$?	-34
$-27 - 7$	
-34	-34 TRUE

The solution is -9.

40. $\frac{16}{5}$

41. $Ax + By = C$

$By = C - Ax$ Subtracting Ax

$y = \frac{C - Ax}{B}$ Dividing by B

42. $Q = 2A - T$

43. ◈

44. ◈

45. $y = x^2 + 1$

When $x = 0$, $y = 0^2 + 1 = 0 + 1 = 1$.

When $x = -1$, $y = (-1)^2 + 1 = 1 + 1 = 2$.

When $x = 1$, $y = 1^2 + 1 = 1 + 1 = 2$.

When $x = -2$, $y = (-2)^2 + 1 = 4 + 1 = 5$.

When $x = 2$, $y = 2^2 + 1 = 4 + 1 = 5$.

When $x = -3$, $y = (-3)^2 + 1 = 9 + 1 = 10$.

When $x = 3$, $y = 3^2 + 1 = 9 + 1 = 10$.

x	0	-1	1	-2	2	-3	3
y	1	2	2	5	5	10	10

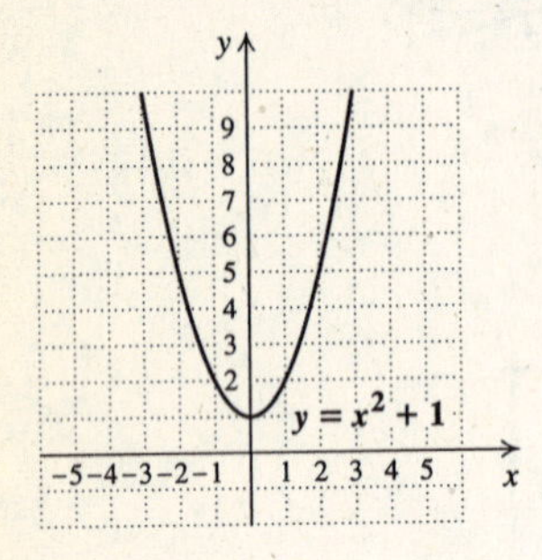

46. (0,6), (1,5), (2,4), (3,3), (4,2), (5,1), (6,0)

47. $x + 3y = 15$

Since 3y and 15 are both multiples of 3, x must also be a multiple of 3.

When $x = 0$, $0 + 3y = 15$
$3y = 15$
$y = 5$

When $x = 3$, $3 + 3y = 15$
$3y = 12$
$y = 4$

When $x = 6$, $6 + 3y = 15$
$3y = 9$
$y = 3$

When $x = 9$, $9 + 3y = 15$
$3y = 6$
$y = 2$

When $x = 12$, $12 + 3y = 15$
$3y = 3$
$y = 1$

When $x = 15$, $15 + 3y = 15$
$3y = 0$
$y = 0$

The whole number solutions are (0,5), (3,4), (6,3), (9,2), (12,1), (15,0).

48. $5n + 10d = 195$, or $0.05n + 0.1d = 1.95$;
(10,19), (0,39), (15,9). Answers may vary.

49. The value of n nickels is 0.05n.
The value of q quarters is 0.25q.

Thus, $0.05n + 0.25q = 2.35$

or $5n + 25q = 235$ Clearing decimals

When $n = 42$, $5 \cdot 42 + 25q = 235$
$210 + 25q = 235$
$25q = 25$
$q = 1$

When $n = 7$, $5 \cdot 7 + 25q = 235$
$35 + 25q = 235$
$25q = 200$
$q = 8$

When $n = 27$, $5 \cdot 27 + 25q = 235$
$135 + 25q = 235$
$25q = 100$
$q = 4$

Solutions are ordered pairs of the form (n,q). Three are (42,1), (7,8), and (27,4). Answers may vary.

50. Answers may vary. (-3,3), (2,2), (0,0)

51. $y = -2.8x + 3.5$

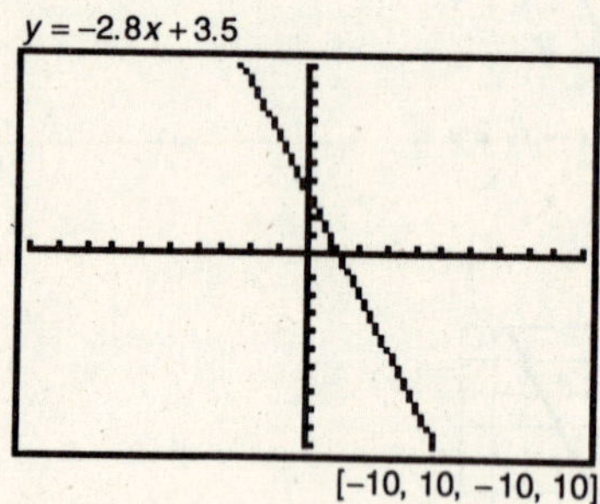

[−10, 10, −10, 10]

52. $y = 4.5x + 2.1$

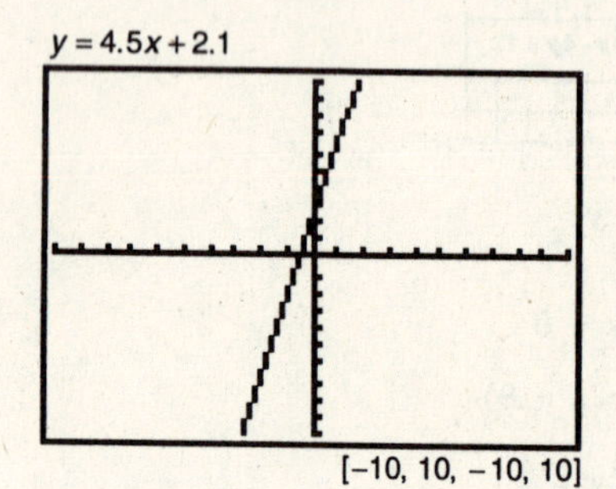

[−10, 10, −10, 10]

53. $y = \frac{2}{7}x - \frac{24}{5}$

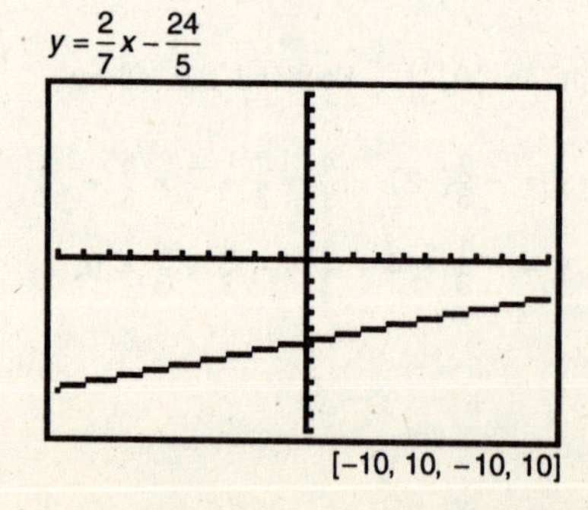

[−10, 10, −10, 10]

54. $y = -\frac{33}{8}x - \frac{45}{7}$

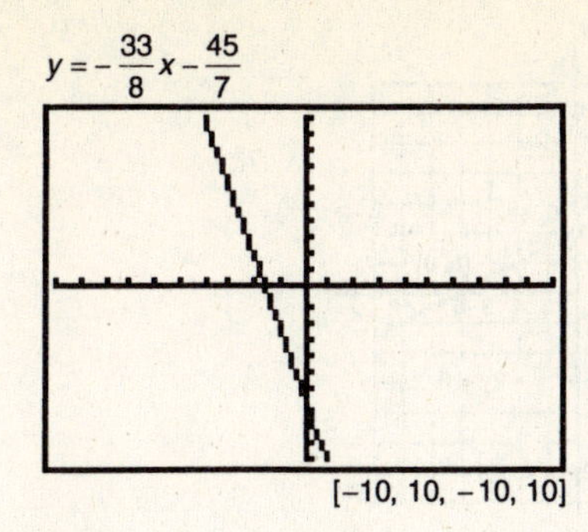

55. 

Exercise Set 3.3

1. (a) The graph crosses the y-axis at (0,3), so the y-intercept is (0,3).

   (b) The graph crosses the x-axis at (4,0), so the x-intercept is (4,0).

2. (a) (0,5); (b) (2,0)

3. (a) The graph crosses the y-axis at (0,5), so the y-intercept is (0,5).

   (b) The graph crosses the x-axis at (-3,0), so the x-intercept is (-3,0).

4. (a) (0,-4); (b) (3,0)

5. $2x + 5y = 20$

   (a) To find the y-intercept, let x = 0. This is the same as covering the x-term and then solving.

   $5y = 20$

   $y = 4$

   The y-intercept is (0,4).

   (b) To find the x-intercept, let y = 0. This is the same as covering the y-term and then solving.

   $2x = 20$

   $x = 10$

   The x-intercept is (10,0).

6. (a) (0,5); (b) (3,0)

7. $4x - 3y = 24$

   (a) To find the y-intercept, let x = 0. This is the same as covering the x-term and then solving.

   $-3y = 24$

   $y = -8$

   The y-intercept is (0,-8).

   (b) To find the x-intercept, let y = 0. This is the same as covering the y-term and then solving.

   $4x = 24$

   $x = 6$

   The x-intercept is (6,0).

8. (a) (0,-4); (b) (14,0)

9. $-6x + y = 8$

   (a) To find the y-intercept, let x = 0. This is the same as covering the x-term and then solving.

   $y = 8$

   The y-intercept is (0,8).

   (b) To find the x-intercept, let y = 0. This is the same as covering the y-term and then solving.

   $-6x = 8$

   $x = -\frac{4}{3}$

   The x-intercept is $\left(-\frac{4}{3}, 0\right)$.

10. (a) (0,10); (b) $\left(-\frac{5}{4}, 0\right)$

11. $2y - 4 = 6x$

    $-6x + 2y = 4$ Writing the equation in the form $Ax + By = C$

    (a) To find the y-intercept, let x = 0. This is the same as covering the x-term and then solving.

    $2y = 4$

    $y = 2$

    The y-intercept is (0,2).

    (b) To find the x-intercept, let y = 0. This is the same as covering the y-term and then solving.

    $-6x = 4$

    $x = -\frac{2}{3}$

    The x-intercept is $\left(-\frac{2}{3}, 0\right)$.

12. (a) (0,-2); (b) $\left(\frac{2}{3}, 0\right)$

13. $3x + 2y = 12$

    Find the y-intercept:

    $2y = 12$ Covering the x-term

    $y = 6$

    The y-intercept is (0,6).

    Find the x-intercept:

    $3x = 12$ Covering the y-term

    $x = 4$

    The x-intercept is (4,0).

To find a third point we replace x with 2 and solve for y.

$3\cdot 2 + 2y = 12$

$6 + 2y = 12$

$2y = 6$

$y = 3$

The point (2,3) appears to line up with the intercepts, so we draw the graph.

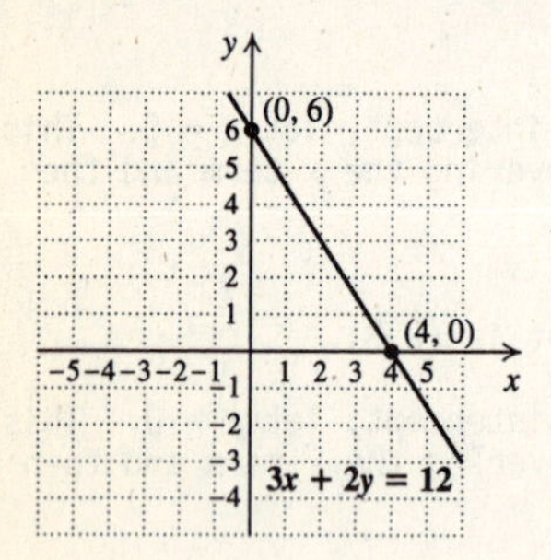

14.

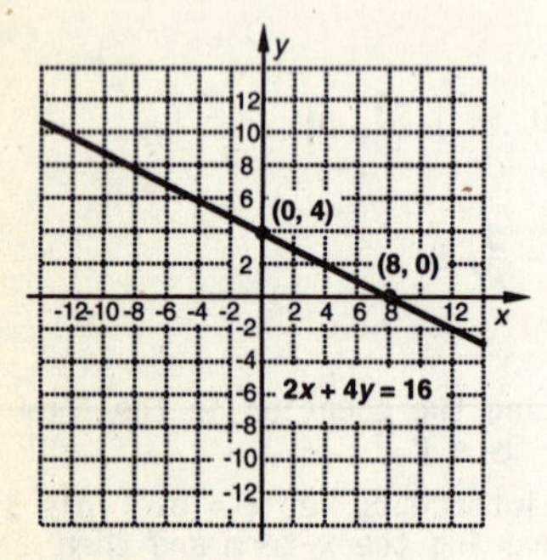

15. $x + 3y = 6$

Find the y-intercept:

$3y = 6$ Covering the x-term

$y = 2$

The y-intercept is (0,2).

Find the x-intercept:

$x = 6$ Covering the y-term

The x-intercept is (6,0).

To find a third point we replace x with 3 and solve for y.

$3 + 3y = 6$

$3y = 3$

$y = 1$

The point (3,1) appears to line up with intercepts, so we draw the graph.

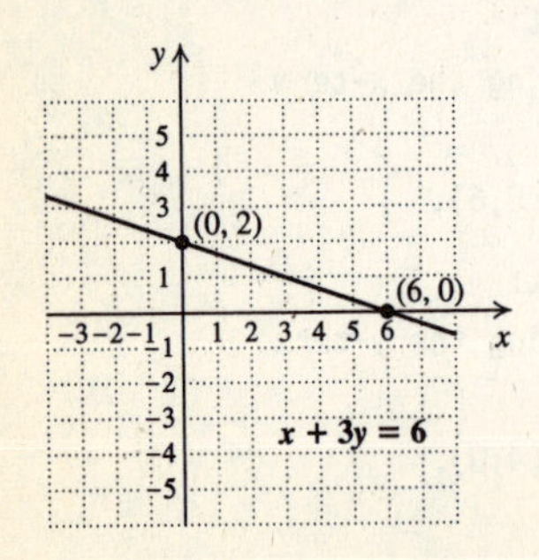

16.

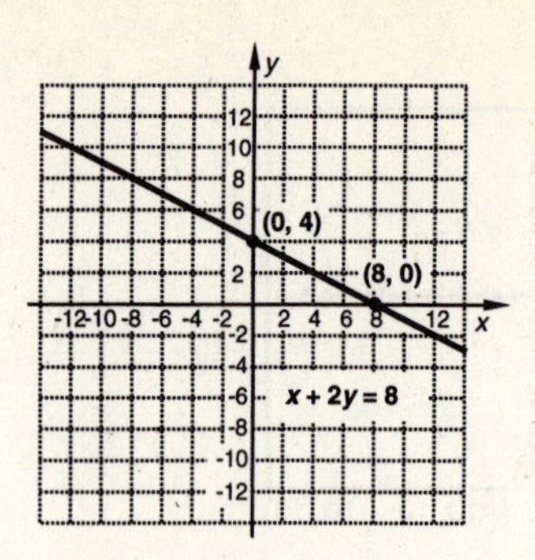

17. $-x + 2y = 4$

Find the y-intercept:

$2y = 4$ Covering the x-term

$y = 2$

The y-intercept is (0,2).

Find the x-intercept:

$-x = 4$ Covering the y-term

$x = -4$

The x-intercept is (-4,0).

To find a third point we replace x with 4 and solve for y.

$-4 + 2y = 4$

$2y = 8$

$y = 4$

The point (4,4) appears to line up with the intercepts, so we draw the graph.

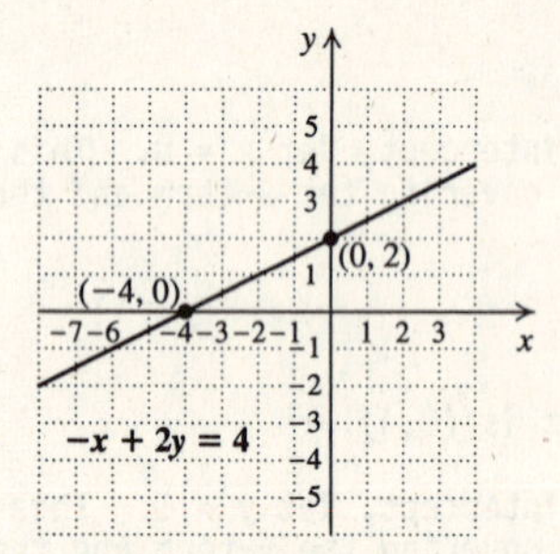

18.

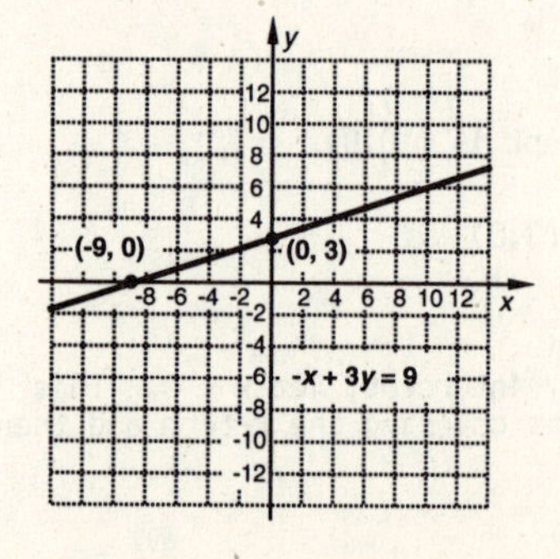

19. $3x + y = 9$

Find the y-intercept:

$y = 9$ Covering the x-term

The y-intercept is $(0,9)$.

Find the x-intercept:

$3x = 9$ Covering the y-term

$x = 3$

The x-intercept is $(3,0)$.

To find a third point we replace x with 2 and solve for y.

$3\cdot2 + y = 9$

$6 + y = 9$

$y = 3$

The point $(2,3)$ appears to line up with the intercepts, so we draw the graph.

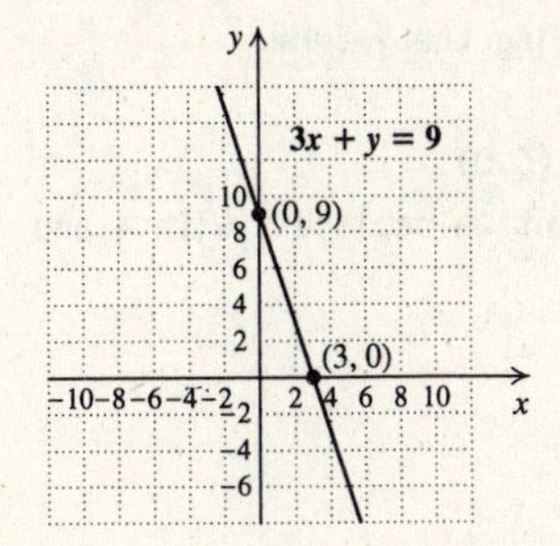

20.

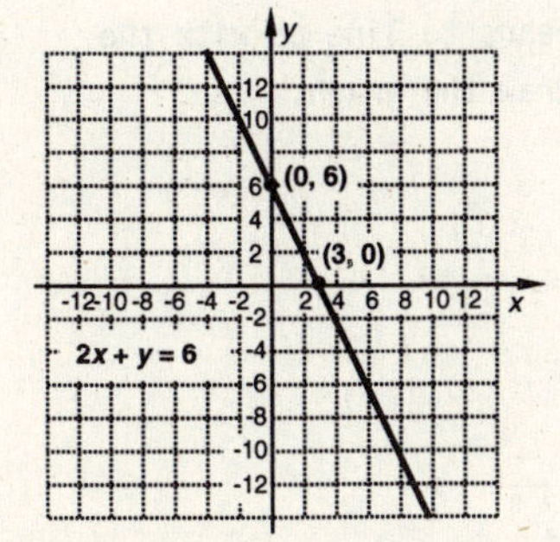

21. $2y - 2 = 6x$

We can leave the equation in the given form or rewrite it in the form $Ax + By = C$. We will use the given form.

To find the y-intercept, let $x = 0$.

$2y - 2 = 6\cdot0$

$2y - 2 = 0$

$2y = 2$

$y = 1$

The y-intercept is $(0,1)$.

Find the x-intercept:

$-2 = 6x$ Covering the y-term

$-\frac{1}{3} = x$

The x-intercept is $\left(-\frac{1}{3}, 0\right)$.

To find a third point we replace x with 1 and solve for y.

$2y - 2 = 6\cdot1$

$2y - 2 = 6$

$2y = 8$

$y = 4$

The point $(1,4)$ appears to line up with the intercepts, so we draw the graph.

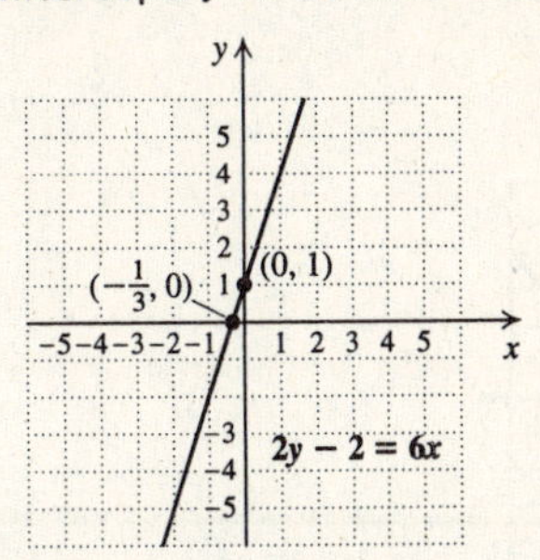

22.

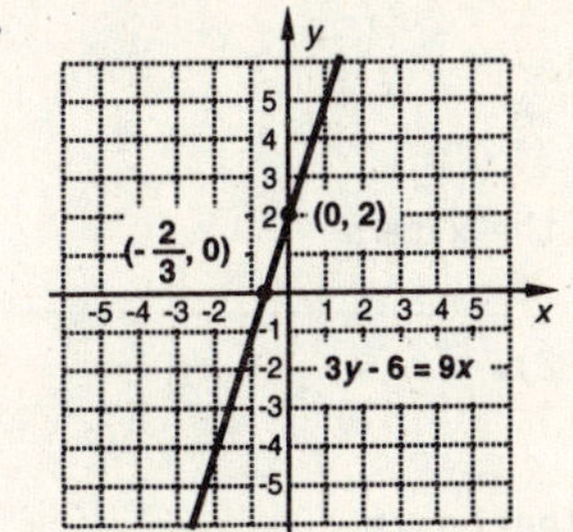

23. $3x - 9 = 3y$

We can leave the equation in the given form or rewrite it in the form $Ax + By = C$. We will use the given form.

Find the y-intercept:

$-9 = 3y$ Covering the x-term

$-3 = y$

The y-intercept is $(0,-3)$.

To find the x-intercept, let $y = 0$.

$3x - 9 = 3\cdot0$

$3x - 9 = 0$

$3x = 9$

$x = 3$

The x-intercept is $(3,0)$.

To find a third point we replace x with 1 and solve for y.

$3\cdot1 - 9 = 3y$

$3 - 9 = 3y$

$-6 = 3y$

$-2 = y$

The point $(1,-2)$ appears to line up with the intercepts, so we draw the graph.

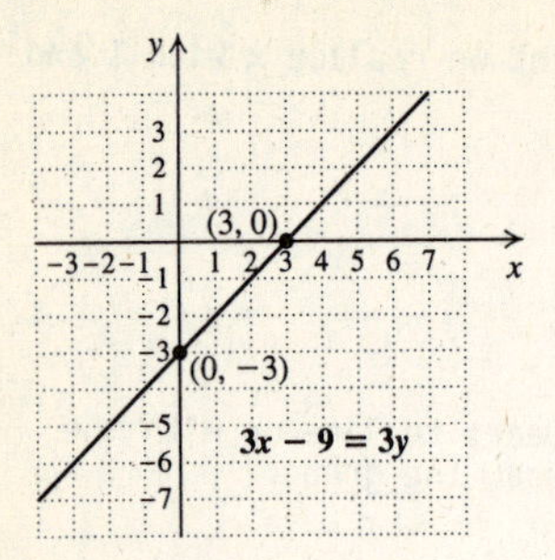

24.

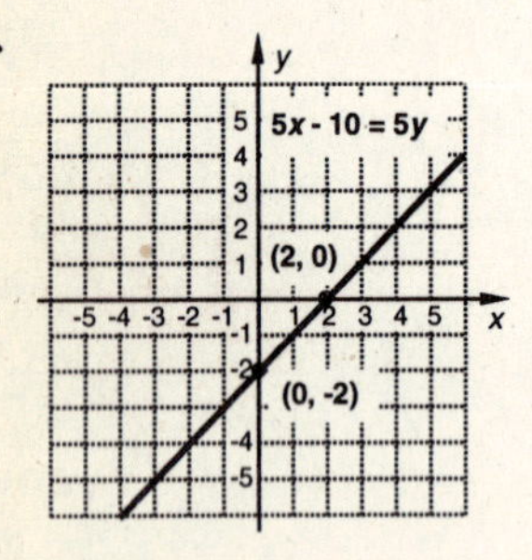

25. $2x - 3y = 6$

Find the y-intercept:

$-3y = 6$ Covering the x-term

$y = -2$

The y-intercept is (0,-2).

Find the x-intercept:

$2x = 6$ Covering the y-term

$x = 3$

The x-intercept is (3,0).

To find a third point we replace x with -3 and solve for y.

$$2(-3) - 3y = 6$$
$$-6 - 3y = 6$$
$$-3y = 12$$
$$y = -4$$

The point (-3,-4) appears to line up with the intercepts, so we draw the graph.

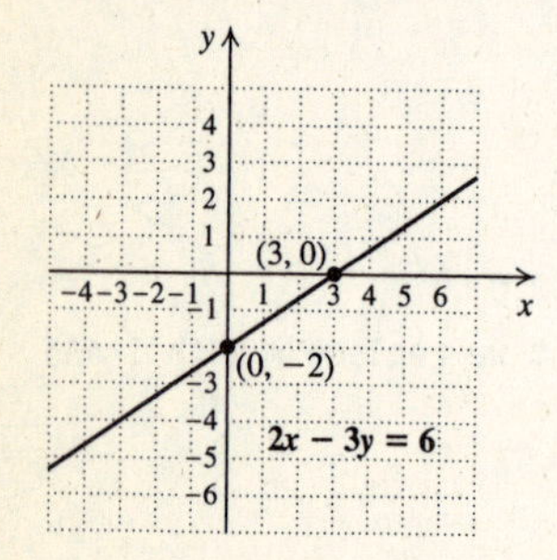

26.

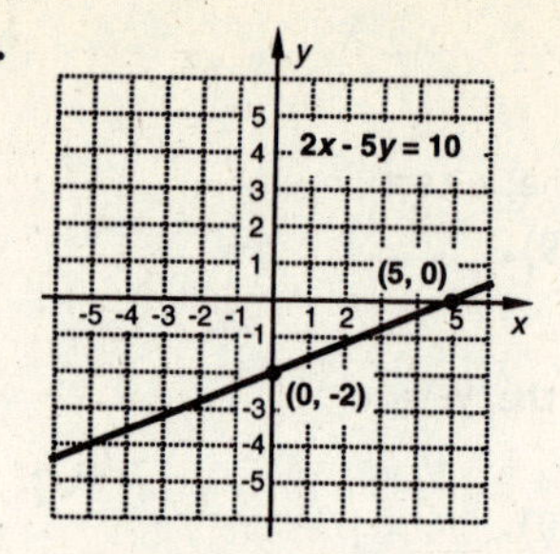

27. $4x + 5y = 20$

Find the y-intercept:

$5y = 20$ Covering the x-term

$y = 4$

The y-intercept is (0,4).

Find the x-intercept:

$4x = 20$ Covering the y-term

$x = 5$

The x-intercept is (5,0).

To find a third point we replace x with 4 and solve for y.

$$4 \cdot 4 + 5y = 20$$
$$16 + 5y = 20$$
$$5y = 4$$
$$y = \frac{4}{5}$$

The point $\left(4,\frac{4}{5}\right)$ appears to line up with the intercepts, so we draw the graph.

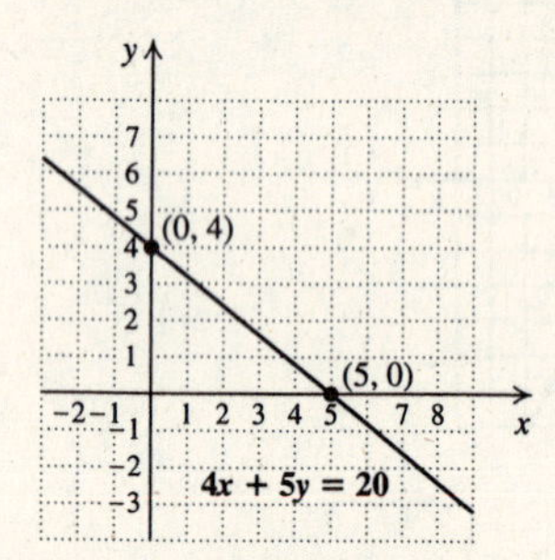

28.

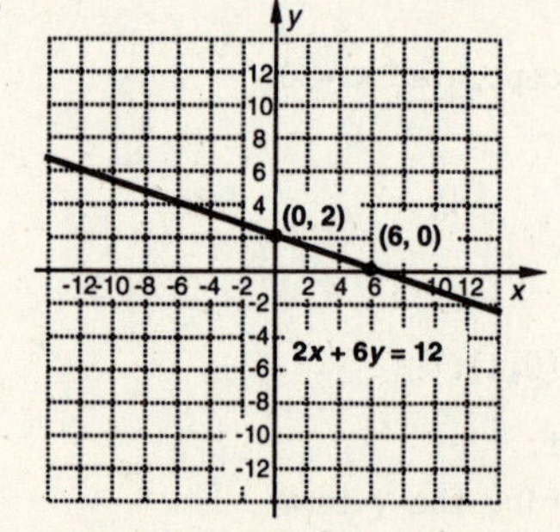

29. $2x + 3y = 8$

Find the y-intercept:

$3y = 8$ Covering the x-term

$y = \frac{8}{3}$

The y-intercept is $\left(0,\frac{8}{3}\right)$.

Find the x-intercept:

$2x = 8$ Covering the y-term

$x = 4$

The x-intercept is (4,0).

To find a third point we replace x with 1 and solve for y.

$2\cdot1 + 3y = 8$

$2 + 3y = 8$

$3y = 6$

$y = 2$

The point (1,2) appears to line up with the intercepts, so we draw the graph.

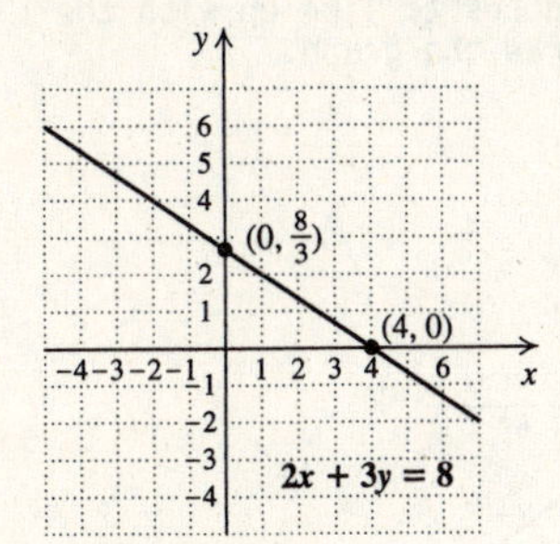

30.

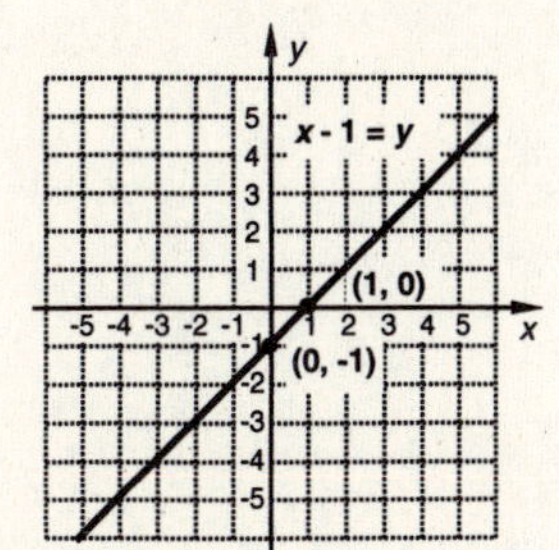

31. $x - 3 = y$

We can leave the equation in the given form or rewrite it in the form $Ax + By = C$. We will use the given form.

Find the y-intercept:

$-3 = y$ Covering the x-term

The y-intercept is (0,-3).

To find the x-intercept, let $y = 0$.

$x - 3 = 0$

$x = 3$

The x-intercept is (3,0).

To find a third point we replace x with -2 and solve for y.

$-2 - 3 = y$

$-5 = y$

The point (-2,-5) appears to line up with the intercepts, so we draw the graph.

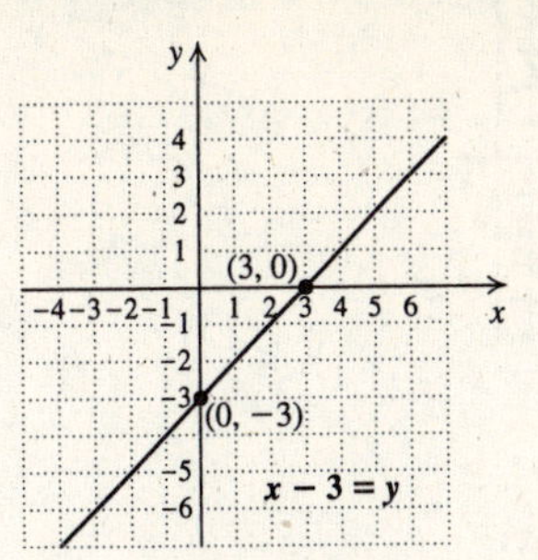

32.

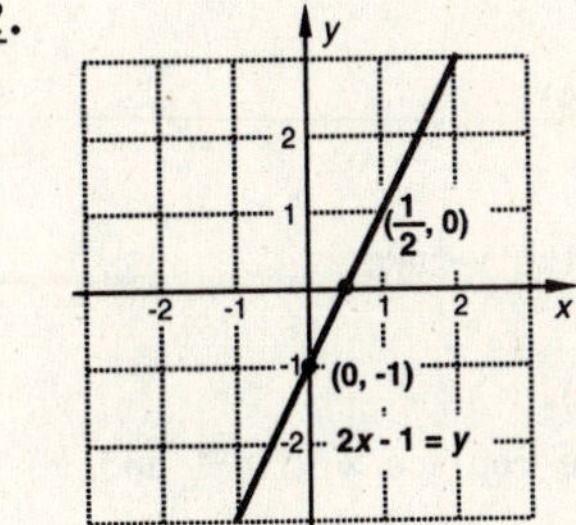

33. $3x - 2 = y$

We can leave the equation in the given form or rewrite it in the form $Ax + By = C$. We will use the given form.

Find the y-intercept:

$-2 = y$ Covering the x-term

The y-intercept is (0,-2).

To find the x-intercept, let $y = 0$.

$3x - 2 = 0$

$3x = 2$

$x = \frac{2}{3}$

The x-intercept is $\left(\frac{2}{3},0\right)$.

To find a third point we replace x with 2 and solve for y.

$3\cdot2 - 2 = y$

$6 - 2 = y$

$4 = y$

The point (2,4) appears to line up with the intercepts, so we draw the graph.

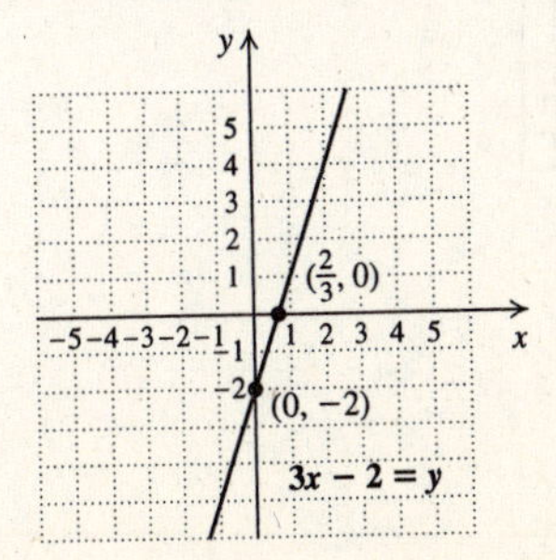

34.

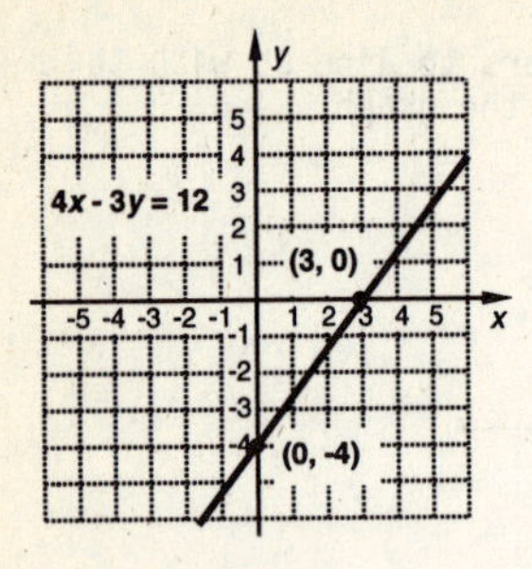

35. $6x - 2y = 18$

Find the y-intercept:

$-2y = 18$ Covering the x-term

$y = -9$

The y-intercept is (0,-9).

Find the x-intercept:

$6x = 18$ Covering the y-term

$x = 3$

The x-intercept is (3,0).

To find a third point we replace x with 1 and solve for y.

$6 \cdot 1 - 2y = 18$

$6 - 2y = 18$

$-2y = 12$

$y = -6$

The point (1,-6) appears to line up with the intercepts, so we draw the graph.

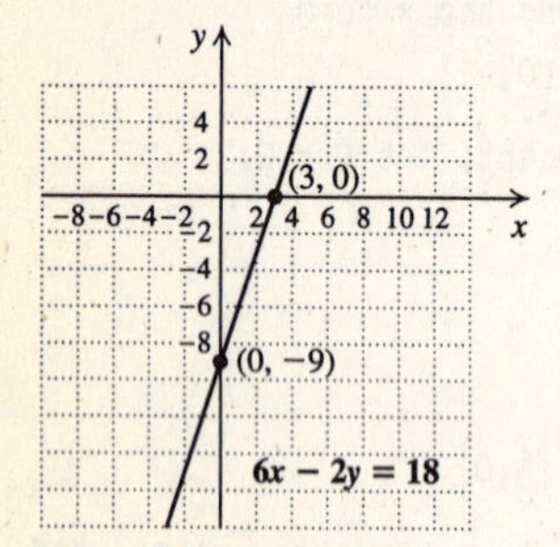

36.

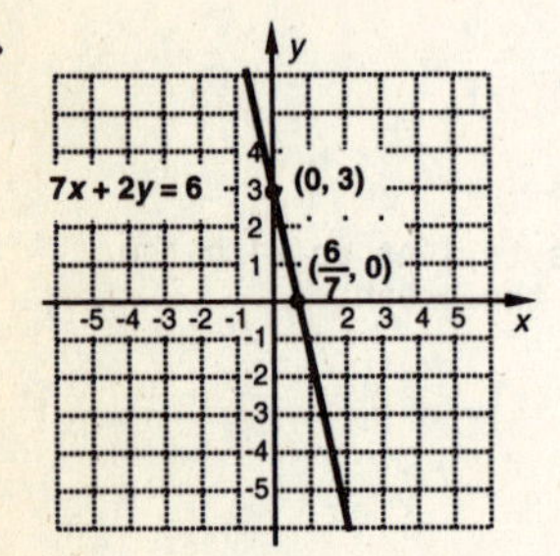

37. $3x + 4y = 5$

Find the y-intercept:

$4y = 5$ Covering the x-term

$y = \frac{5}{4}$

The y-intercept is $\left(0,\frac{5}{4}\right)$.

Find the x-intercept:

$3x = 5$ Covering the y-term

$x = \frac{5}{3}$

The x-intercept is $\left(\frac{5}{3},0\right)$.

To find a third point we replace x with 3 and solve for y.

$3 \cdot 3 + 4y = 5$

$9 + 4y = 5$

$4y = -4$

$y = -1$

The point (3,-1) appears to line up with the intercepts, so we draw the graph.

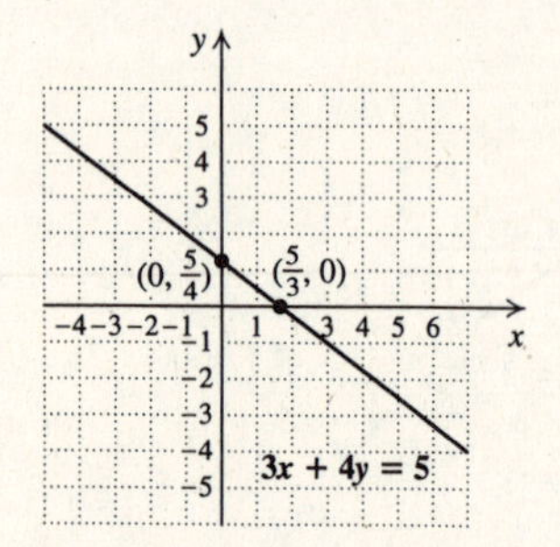

38.

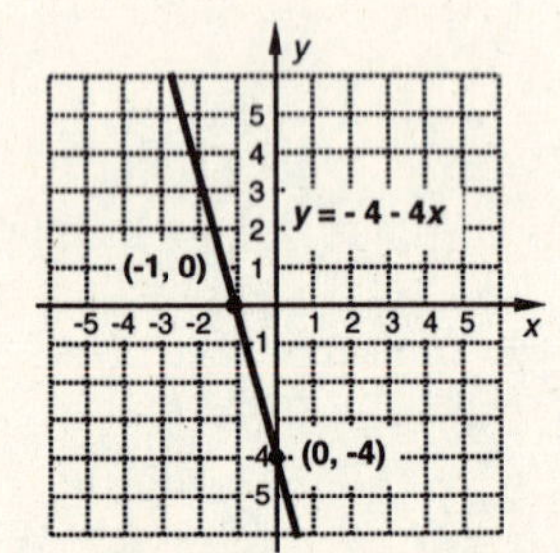

39. $y = -3 - 3x$

We can leave the equation in the given form or rewrite it in the form $Ax + By = C$. We will use the given form.

Find the y-intercept:

$y = -3$ Covering the x-term

The y-intercept is (0,-3).

To find the x-intercept, let $y = 0$.

$0 = -3 - 3x$

$3x = -3$

$x = -1$

The x-intercept is (-1,0).

To find a third point we replace x with -2 and solve for y.

$y = -3 - 3\cdot(-2)$

$y = -3 + 6$

$y = 3$

The point (-2,3) appears to line up with the intercepts, so we draw the graph.

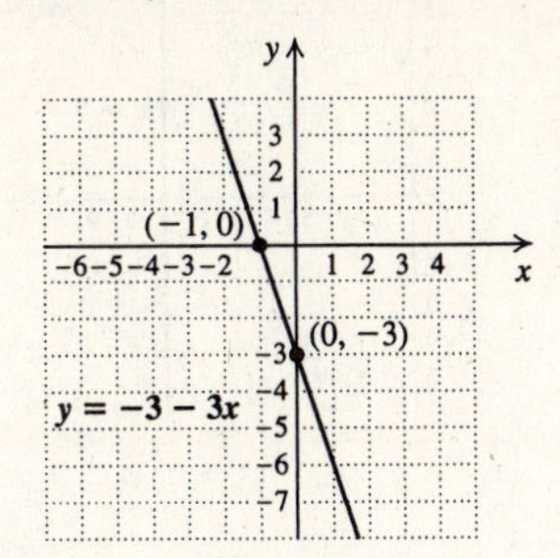

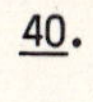

40.

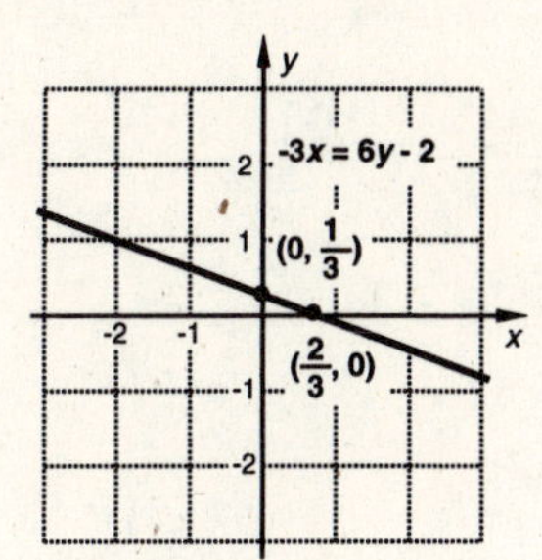

41. $-4x = 8y - 5$

We can leave the equation in the given form or rewrite it in the form Ax + By = C. We will use the given form.

To find the y-intercept, let x = 0.

$-4\cdot0 = 8y - 5$

$0 = 8y - 5$

$5 = 8y$

$\frac{5}{8} = y$

The y-intercept is $\left(0,\frac{5}{8}\right)$.

Find the x-intercept:

$-4x = -5$ Covering the y-term

$x = \frac{5}{4}$

The x-intercept is $\left(\frac{5}{4},0\right)$.

To find a third point we replace x with -5 and solve for y.

$-4(-5) = 8y - 5$

$20 = 8y - 5$

$25 = 8y$

$\frac{25}{8} = y$

The point $\left(-5,\frac{25}{8}\right)$ appears to line up with the intercepts, so we draw the graph.

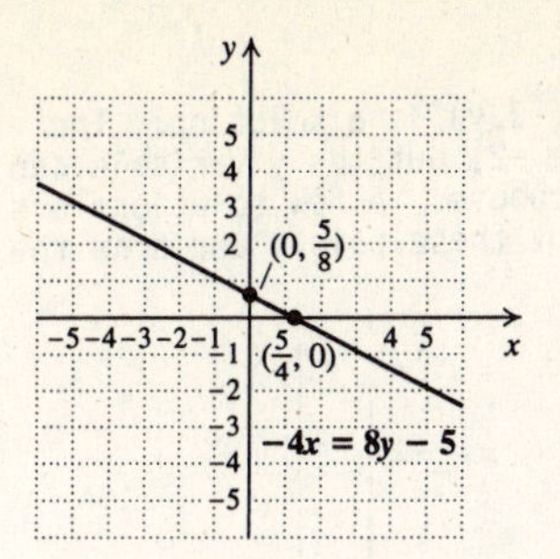

42.

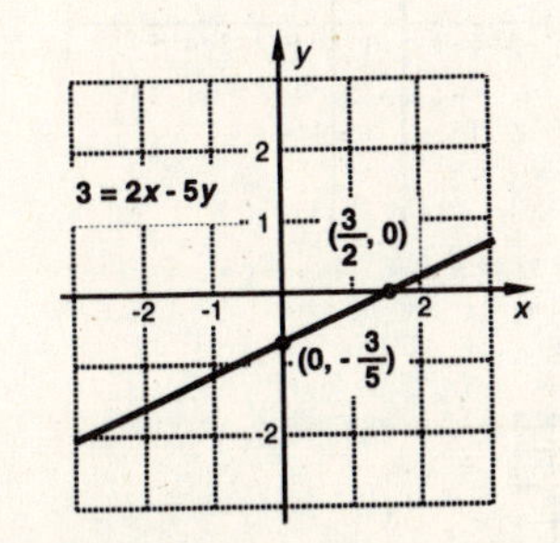

43. $y - 3x = 0$

Find the y-intercept:

$y = 0$ Covering the x-term

The y-intercept is (0,0). Note that this is also the x-intercept.

In order to graph the line, we will find a second point.

When x = 1, $y - 3\cdot1 = 0$

$y - 3 = 0$

$y = 3$

To find a third point we let x = -1 and solve for y.

$y - 3(-1) = 0$

$y + 3 = 0$

$y = -3$

The point (-1,-3) appears to line up with the other two points, so we draw the graph.

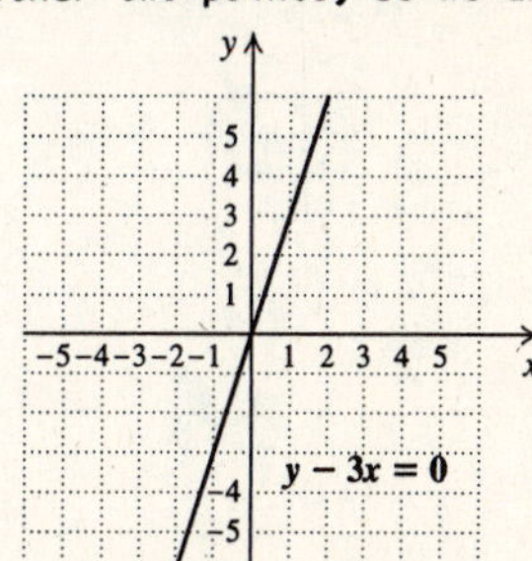

44.

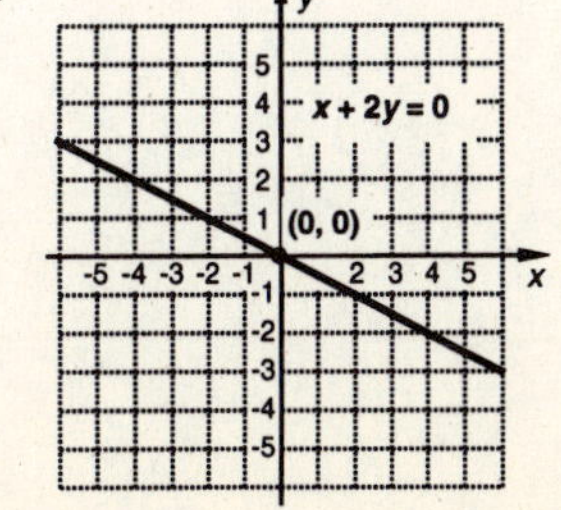

45. x = -2

Any ordered pair (-2,y) is a solution. The variable x must be -2, but the y variable can be any number we choose. A few solutions are listed below. Plot these points and draw the line.

x	y
-2	-2
-2	0
-2	4

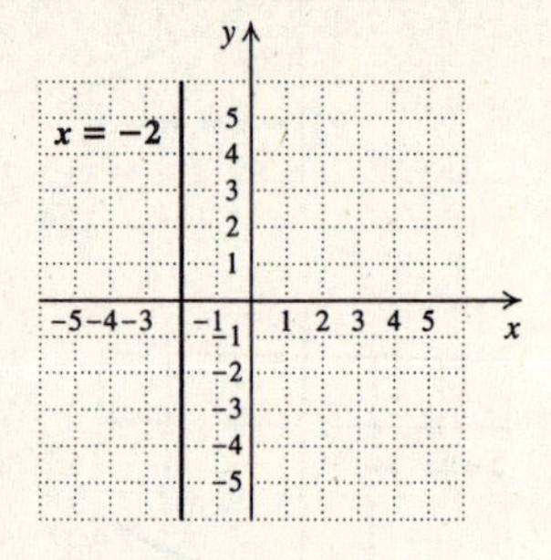

46.

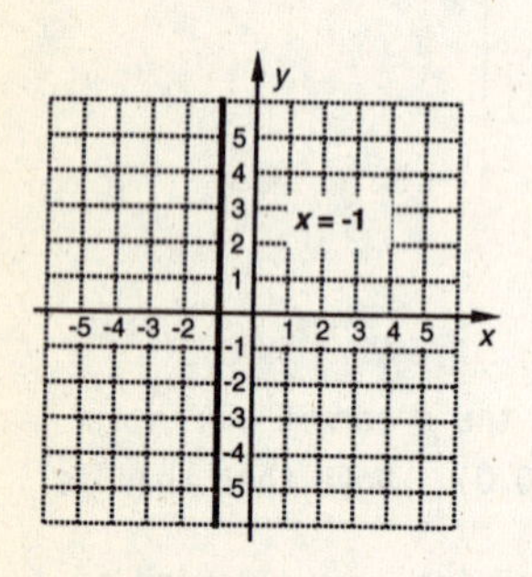

47. y = 2

Any ordered pair (x,2) is a solution. The variable y must be 2, but the x variable can be any number we choose. A few solutions are listed below. Plot these points and draw the line.

x	y
-3	2
0	2
2	2

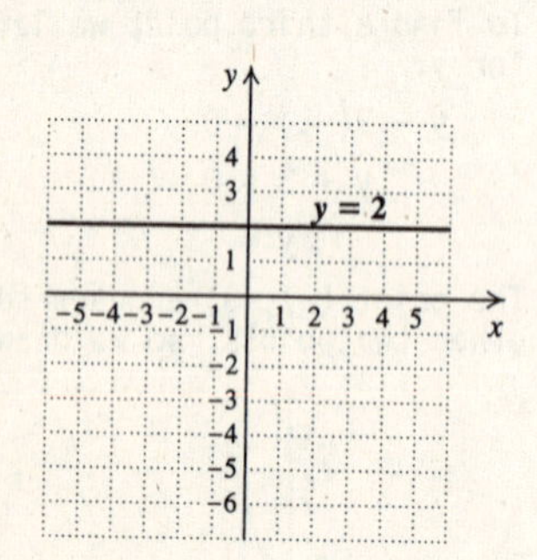

48.

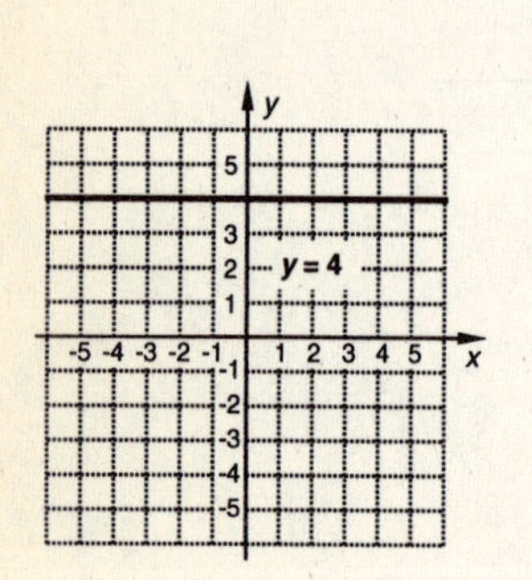

49. x = 7

Any ordered pair (7,y) is a solution. The variable x must be 7, but the y variable can be any number we choose. A few solutions are listed below. Plot these points and draw the line.

x	y
7	-1
7	4
7	5

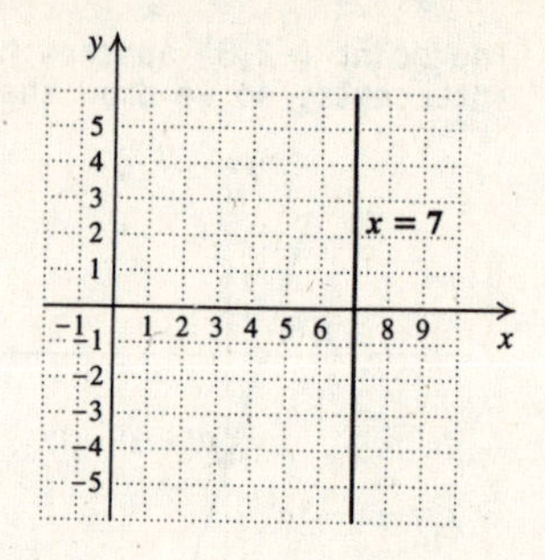

50.

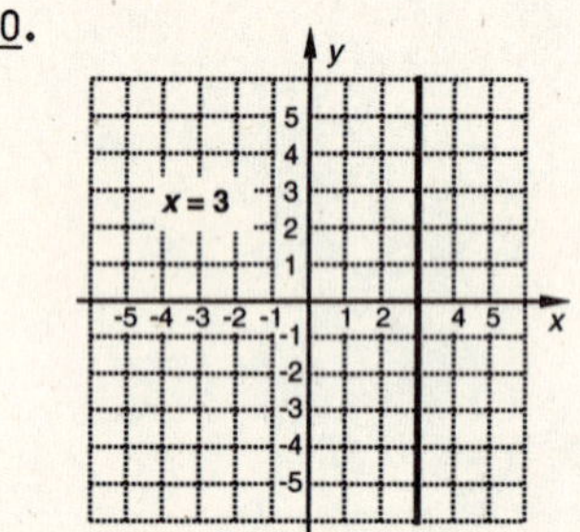

51. y = 0

Any ordered pair (x,0) is a solution. The variable y must be 0, but the x variable can be any number we choose. A few solutions are listed below. Plot these points and draw the line.

x	y
-5	0
-1	0
3	0

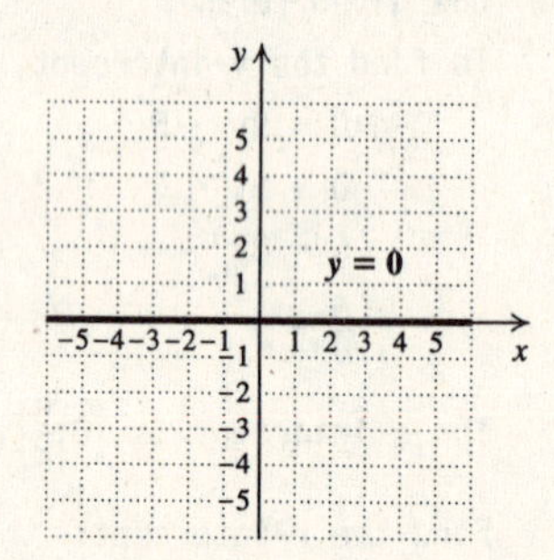

52.

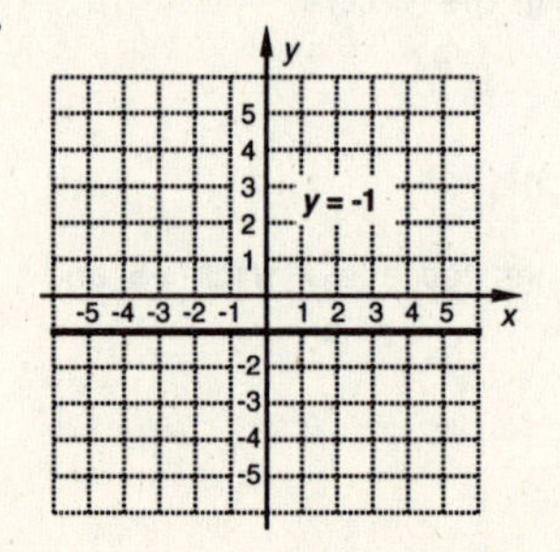

53. $x = \frac{3}{2}$

Any ordered pair $\left(\frac{3}{2},y\right)$ is a solution. The variable x must be $\frac{3}{2}$, but the y variable can be any number we choose. A few solutions are listed below. Plot these points and draw the line.

x	y
$\frac{3}{2}$	-2
$\frac{3}{2}$	0
$\frac{3}{2}$	4

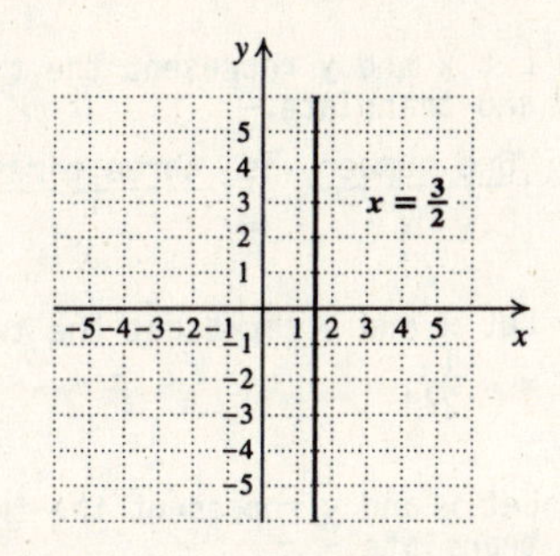

54.

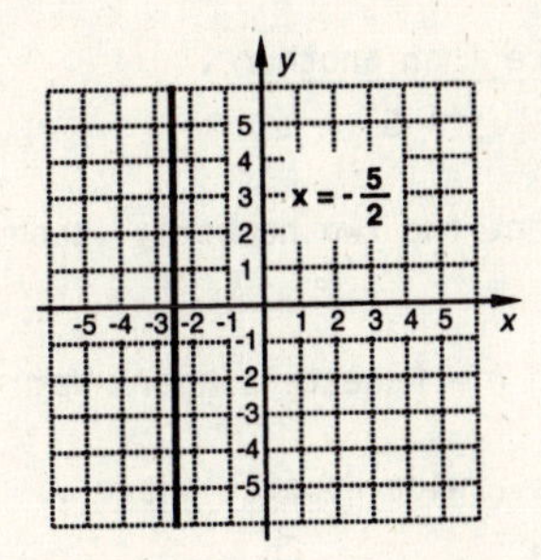

55. $3y = -5$

$y = -\frac{5}{3}$ Solving for y

Any ordered pair $\left(x,-\frac{5}{3}\right)$ is a solution. A few solutions are listed below. Plot these points and draw the line.

x	y
-3	$-\frac{5}{3}$
0	$-\frac{5}{3}$
2	$-\frac{5}{3}$

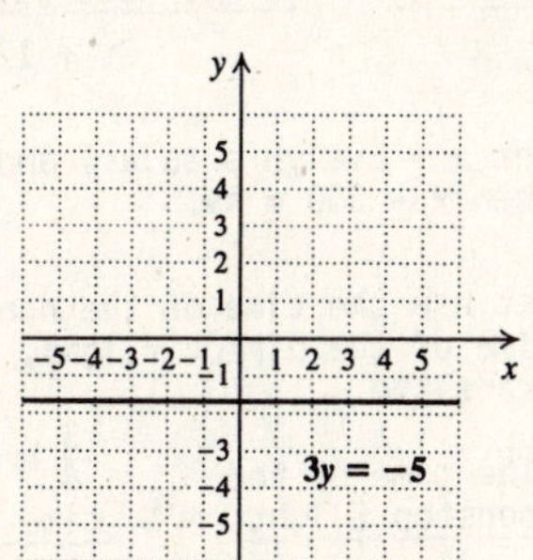

56.

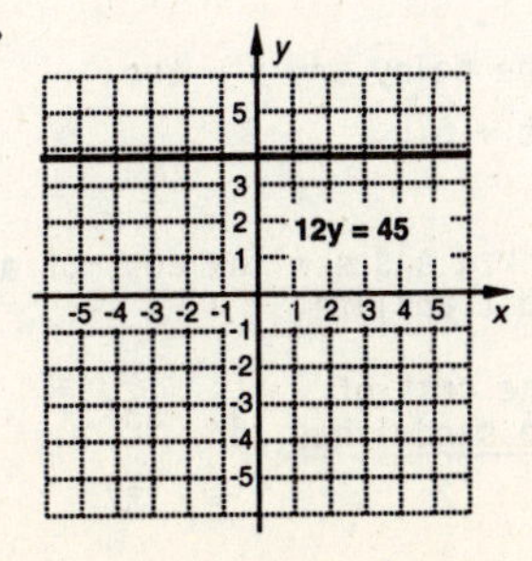

57. $4x + 3 = 0$

$4x = -3$

$x = -\frac{3}{4}$ Solving for x

Any ordered pair $\left(-\frac{3}{4},y\right)$ is a solution. A few solutions are listed below. Plot these points and draw the line.

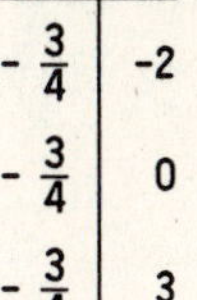

x	y
$-\frac{3}{4}$	-2
$-\frac{3}{4}$	0
$-\frac{3}{4}$	3

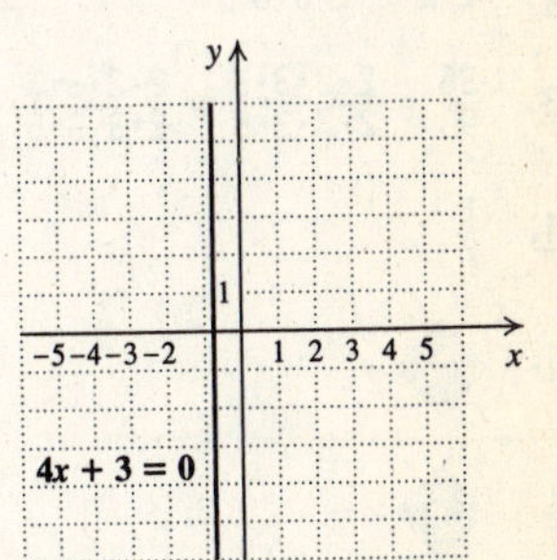

58.

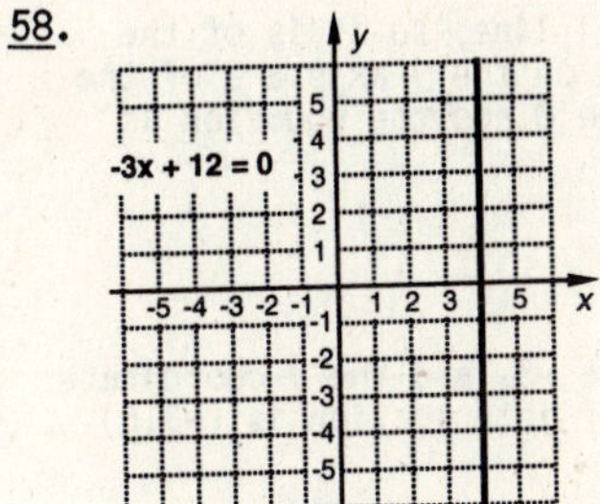

59. $18 - 3y = 0$

$-3y = -18$

$y = 6$ Solving for y

Any ordered pair (x,6) is a solution. A few solutions are listed below. Plot these points and draw the line.

x	y
-4	6
0	6
2	6

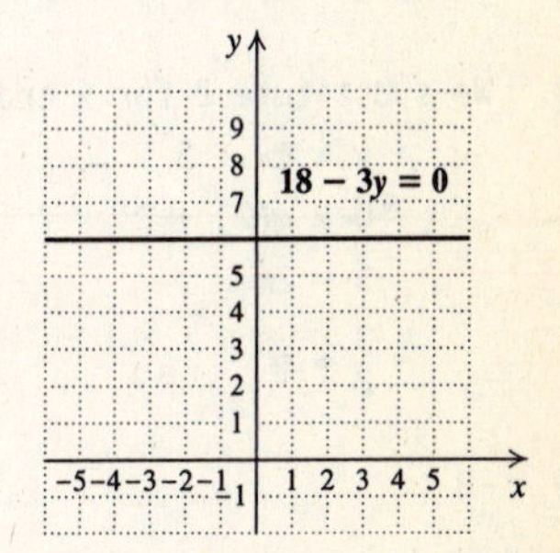

60.

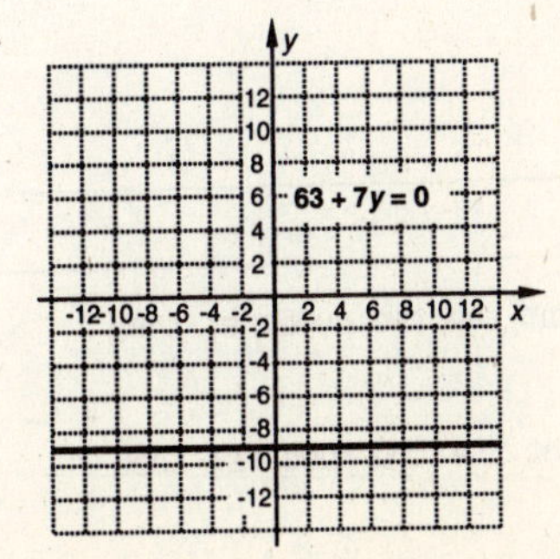

61. We begin by factoring 98 in any way we can:

$98 = 2 \cdot 49$

The factor 49 is not prime, so we factor it again:

$98 = 2 \cdot 49 = 2 \cdot 7 \cdot 7$

Both 2 and 7 are prime, so the prime factorization of 98 is $2 \cdot 7 \cdot 7$.

62. $2 \cdot 2 \cdot 2 \cdot 2 \cdot 3 \cdot 5$

63. $\frac{36}{90} = \frac{2 \cdot 2 \cdot 3 \cdot 3}{2 \cdot 3 \cdot 3 \cdot 5} = \frac{\cancel{2} \cdot 2 \cdot \cancel{3} \cdot \cancel{3}}{\cancel{2} \cdot \cancel{3} \cdot \cancel{3} \cdot 5} = \frac{2}{5}$

64. $\frac{1}{7}$

65. ◆

66. ◆

67. The y-axis is a vertical line, so it is of the form x = a. All points on the y-axis are of the form (0,y), so a must be 0 and the equation is x = 0.

68. y = 0

69. The x-coordinate must be -3, and the y-coordinate must be 6. The point of intersection is (-3,6).

70. y = -5

71. A line parallel to the y-axis has an equation of the form x = a. Since the line is 13 units to the right of the y-axis, all points on the line are of the form (13,y). Thus, a is 13, and the equation is x = 13.

72. y = 2.8

73. We substitute 2 for x and 0 for y and solve for m:

$$y = mx + 3$$
$$0 = m(2) + 3$$
$$-3 = 2m$$
$$-\frac{3}{2} = m$$

74. -8

75. ◆

Exercise Set 3.4

1. Let x and y represent the two numbers; then x + y = 27.

2. Let x and y represent the two numbers; then x + y = 53.

3. Let x and y represent the two numbers. We translate.

A number plus twice another number is 65.

$x + 2y = 65$

4. Let x and y represent the two numbers; then x + 2y = 93.

5. Let x and y represent the two numbers. We reword and translate.

One number is three times another number.

$x = 3y$

6. Let x and y represent the two numbers; then $x = \frac{1}{2}y$.

7. Let x and y represent the two numbers. We translate.

One number is 5 more than another.

$x = y + 5$

8. We let x and y represent the two numbers; then x = y - 7.

9. Let h = Hank's age and n = Nanette's age. We translate.

Hank's age plus 7 is twice Nanette's age.

$h + 7 = 2n$

10. Let a = Lisa's age and b = Lou's age; then a = 2b - 5.

11. Let x = Lois' salary and y = Roberta's salary. We reword and translate.

Lois' salary is $170 more than three times Roberta's salary.

$x = 3y + 170$

12. Let x = Evelyn's salary and y = Eric's salary; then x + 200 = 4y.

13. Let n = the time of the nonstop flight and d = the time of the direct flight. We reword and translate.

The time of the nonstop flight is $\frac{3}{4}$ hr more than half the time of the direct flight.

$n = \frac{1}{2}d + \frac{3}{4}$

14. Let d = the length of the delay and t = the flight time; then $d = \frac{1}{2}t + 5$.

15. Let p = the cost of a pizza and s = the cost of a sandwich. We reword and translate.

The cost of three pizzas plus the cost of two sandwiches is $37.

$3p + 2s = 37$

16. Let m = the cost of an entree and d = the cost of a dessert; then m = 2d.

17. a) We substitute and calculate.

1 hr: d = 55·1 = 55 mi

2 hr: d = 55·2 = 110 mi

5 hr: d = 55·5 = 275 mi

10 hr: d = 55·10 = 550 mi

b) We plot the points found in part (a) and any others that we may calculate and draw the line they determine.

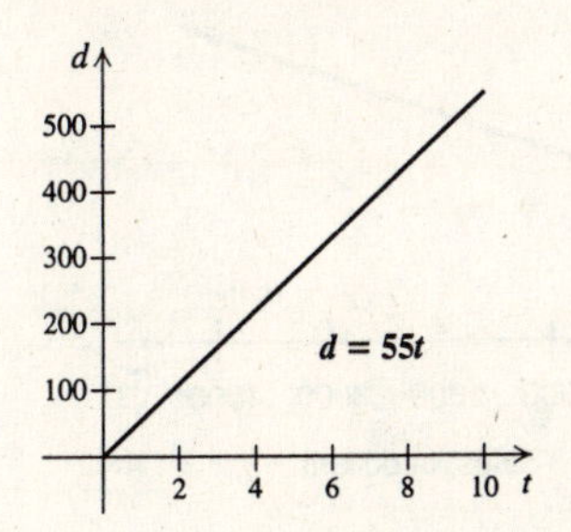

18. a) 1 in.; 2.5 in.; 4 in.; 6 in.

b)

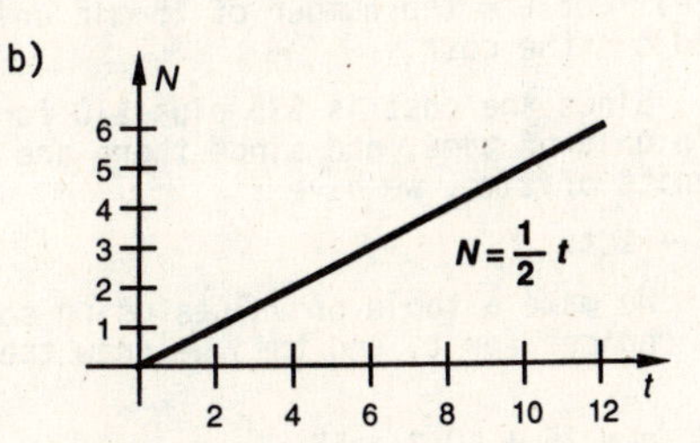

19. a) We substitute and calculate.

Size 4: y = 4 - 2 = 2

Size 5: y = 5 - 2 = 3

Size 6: y = 6 - 2 = 4

Size 7: y = 7 - 2 = 5

Size 8: y = 8 - 2 = 6

b) We plot the points found in part (a) and any others that we may calculate and draw the line they determine.

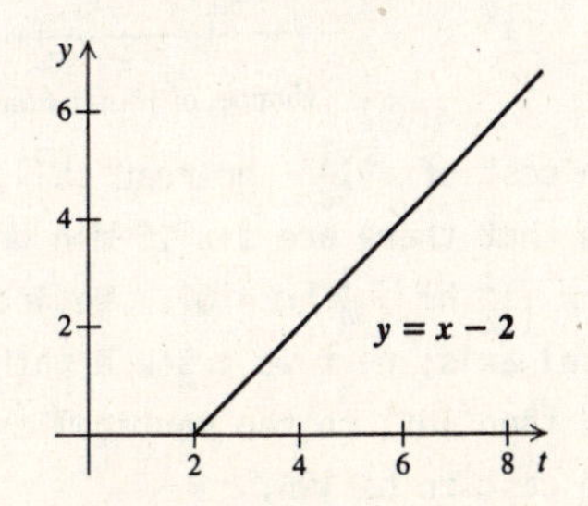

20. a) \$0.46; \$2.89; \$4.24

b)

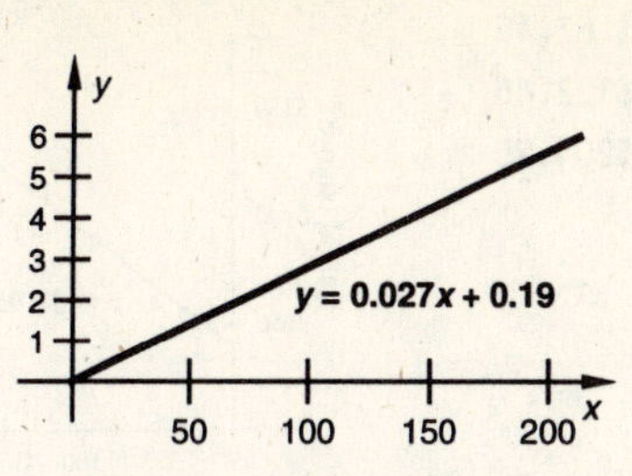

21. a) We substitute and calculate.

$$2\frac{1}{2} + w = 15$$

$$w = 12\frac{1}{2}$$

Sandy is $12\frac{1}{2}$ times more likely to die from lung cancer than Polly.

b) We use the point found in part (a) and others that we calculate to make a table and draw the graph.

When t = 1, 1 + w = 15, or w = 14.

When t = 5, 5 + w = 15, or w = 10.

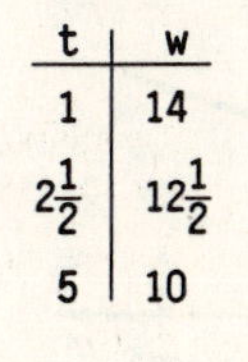

t	w
1	14
$2\frac{1}{2}$	$12\frac{1}{2}$
5	10

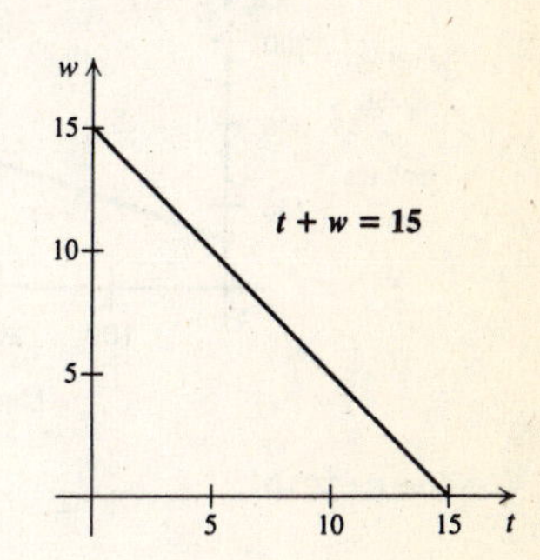

22. a) $33\frac{1}{3}$ lb

b)

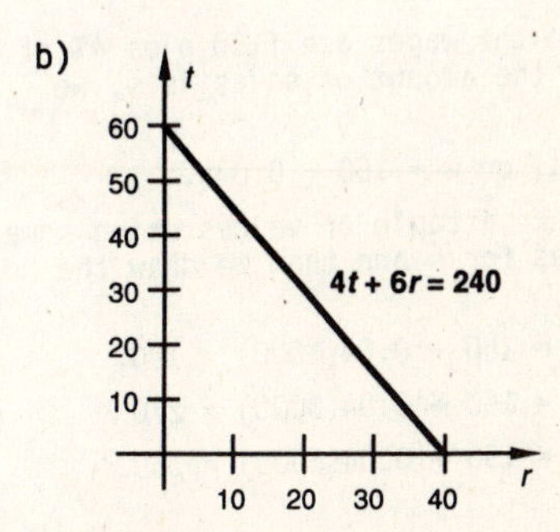

23. Familiarize. Let m = mileage and c = cost.

Translate. Since the cost of a rental is \$39.95 plus 55¢ for each mile, and since m miles are to be driven, we have

c = 39.95 + 0.55m.

Carry out. We make a table of values using some convenient choices for m, and then we draw the graph.

When m = 50, c = 39.95 + 0.55(50) = 67.45

When m = 150, c = 39.95 + 0.55(150) = 122.45.

When m = 300, c = 39.95 + 0.55(300) = 204.95.

Mileage	Cost
50	$ 67.45
150	$122.45
300	$204.95

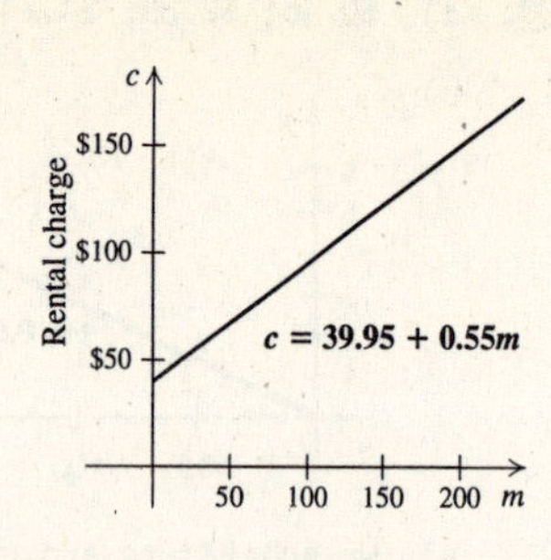

To estimate the cost of renting a 20-ft truck for one day and driving 180 mi, we locate 180 on the horizontal axis. From there we trace a path up to the line and then left to the vertical axis. We estimate the cost of the rental at $140.

Check. We could calculate the exact cost.

$c = 39.95 + 0.55(180) = \$138.95$

Our estimate is close enough to serve as a good approximation. The rental firm could use the graph for other quick cost estimates.

State. The cost is about $140.

24.

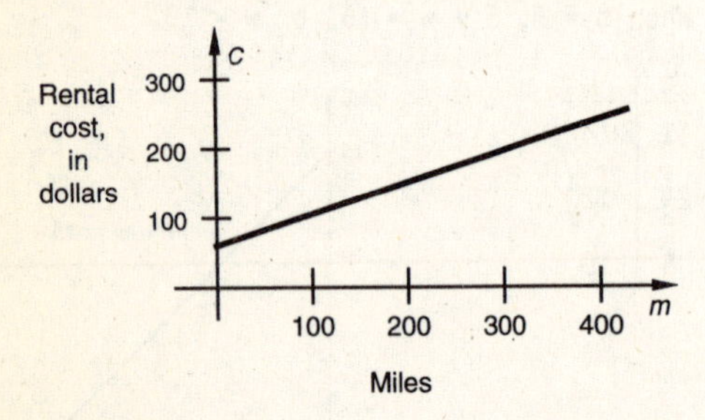

About $210

25. Familiarize. Let s = weekly sales and w = weekly wages.

Translate. Since the wages are $150 plus 4% of sales, and since the amount of sales is s, we have

$w = 150 + 4\%s$, or $w = 150 + 0.04s$.

Carry out. We make a table of values using some convenient choices for s and then we draw the graph.

When $s = 1000$, $w = 150 + 0.04(1000) = 190$.

When $s = 3000$, $w = 150 + 0.04(3000) = 270$.

When $s = 5000$, $w = 150 + 0.04(5000) = 350$.

Sales	Wages
$1000	$190
$3000	$270
$5000	$350

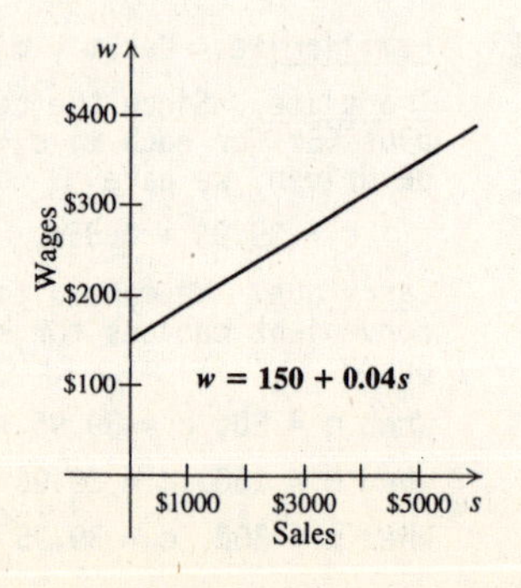

To estimate the wages paid when a salesperson sells $4500 in merchandise in one week, we locate $4500 on the horizontal axis. From there we trace a path up to the line and then left to the vertical axis. We estimate the wages to be $330.

Check. We could calculate the exact wages.

$w = 150 + 0.04(4500) = \$330$

Our estimate is accurate. The graph could be used for other quick estimates of wages.

State. The wages are $330.

26.

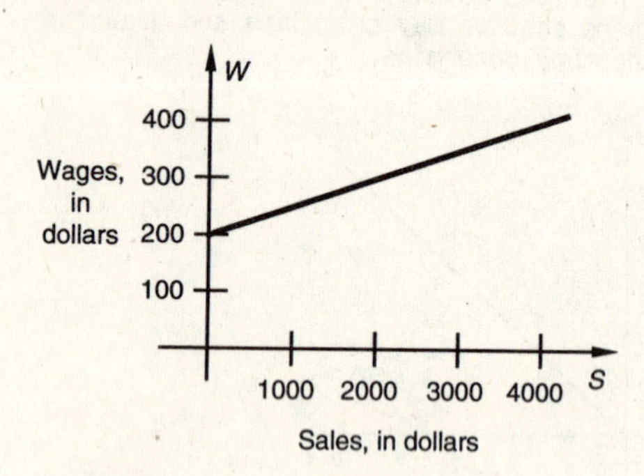

About $375

27. Familiarize. Let t = the number of 15-min units of time and c = the cost.

Translate. Since the cost is $35 plus $10 for each 15-min unit of time, and since there are t 15-min units of time, we have

$c = 35 + 10t$.

Carry out. We make a table of values using some convenient choices for t, and then we draw the graph.

When $t = 2$, $c = 35 + 10 \cdot 2 = 55$.

When $t = 5$, $c = 35 + 10 \cdot 5 = 85$.

When $t = 8$, $c = 35 + 10 \cdot 8 = 115$.

15-min Time Units	Cost
2	$ 55
5	$ 85
8	$115

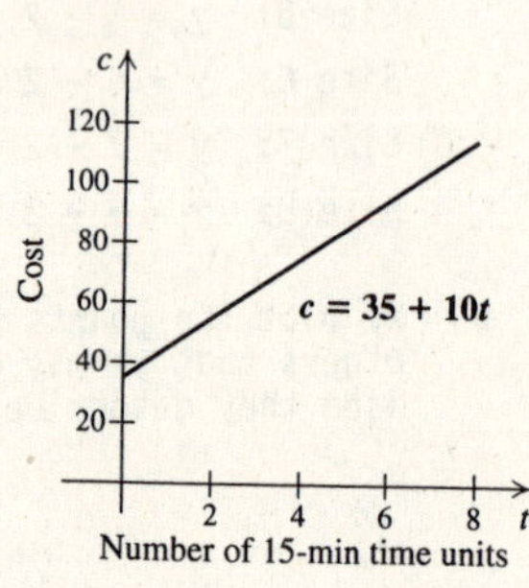

To estimate the cost of a $1\frac{1}{2}$ - hr road call, we first determine that there are six 15-min units of time in $1\frac{1}{2}$ hr $\left[1\frac{1}{2} \text{ hr} \div \frac{1}{4} \text{ hr} = 6\right]$. We locate 6 on the horizontal axis; next we trace a path up to the line and then left to the vertical axis. We estimate the cost to be $95.

Check. We could calculate the exact cost.

$c = 35 + 10 \cdot 6 = \$95$

Our estimate is accurate. The graph could be used for other quick estimates of the cost of a road call.

State. The cost of a $1\frac{1}{2}$ - hr road call is \$95.

28.

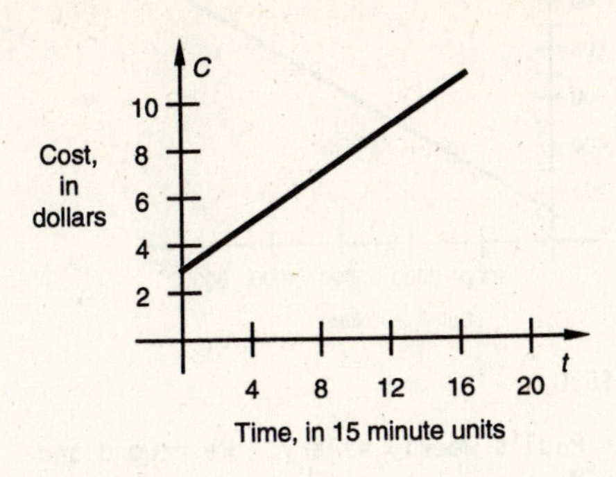

About \$10

29. Familiarize. Let n = the number of people and p = the number of pounds of cheese needed. Then n - 10 = the number of people in excess of 10.

Translate. Since the number of pounds of cheese needed is 3 lb plus $\frac{2}{9}$ lb for each person in excess of 10, and since the number of people in excess of 10 is n - 10, we have

$$p = 3 + \frac{2}{9}(n - 10), \text{ or}$$

$$p = 3 + \frac{2}{9}n - \frac{20}{9}, \text{ or}$$

$$p = \frac{2}{9}n + \frac{7}{9}, \; n \geqslant 10.$$

Carry out. We make a table of values using some convenient choices for n, and then we draw the graph.

When n = 12, $p = \frac{2}{9} \cdot 12 + \frac{7}{9} = \frac{31}{9}$, or $3\frac{4}{9}$.

When n = 18, $p = \frac{2}{9} \cdot 18 + \frac{7}{9} = \frac{43}{9}$, or $4\frac{7}{9}$.

When n = 24, $p = \frac{2}{9} \cdot 24 + \frac{7}{9} = \frac{55}{9}$, or $6\frac{1}{9}$.

Number of people	Pounds of Cheese
12	$3\frac{4}{9}$
18	$4\frac{7}{9}$
24	$6\frac{1}{9}$

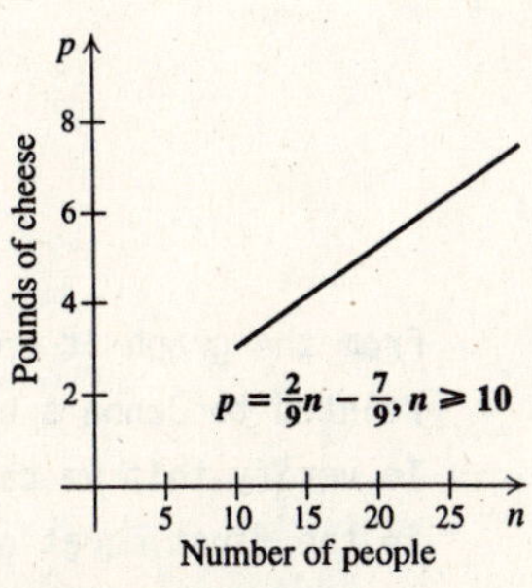

To estimate the amount of cheese needed for a party of 21, we locate 21 on the horizontal axis. Then we trace a path up to the line and left to the vertical axis. We estimate the amount to be 5 lb.

Check. We could calculate the exact amount.

$$p = \frac{2}{9} \cdot 21 + \frac{7}{9} = \frac{49}{9}, \text{ or } 5\frac{4}{9} \text{ lb.}$$

Our estimate is close enough to serve as a good approximation. The catering firm could use the graph for other quick estimates.

State. About 5 lb of cheese is needed for a party of 21.

30.

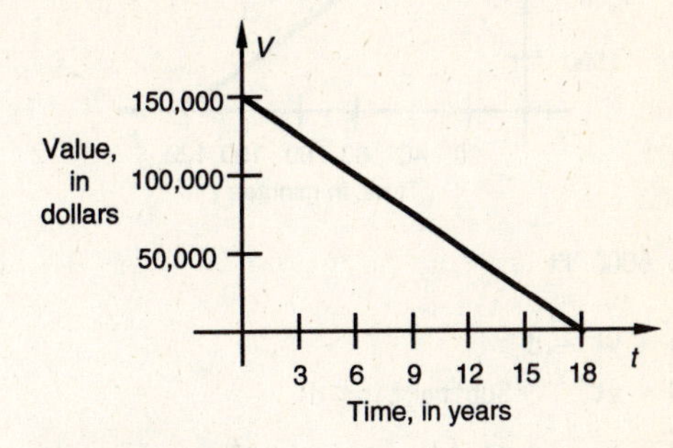

About \$80,000

31. Familiarize. Let t = the time of the descent and a = the altitude.

Translate. The altitude is 32,000 ft less 3000 ft for each minute of the descent. We have

$$a = 32,000 - 3000t.$$

Carry out. We make a table of values using some convenient choices for t, and then we draw the graph.

When t = 2, a = 32,000 - 3000·2 = 26,000.

When t = 6, a = 32,000 - 3000·6 = 14,000.

When t = 10, a = 32,000 - 3000·10 = 2000.

Time of Descent	Altitude
2	26,000
6	14,000
10	2000

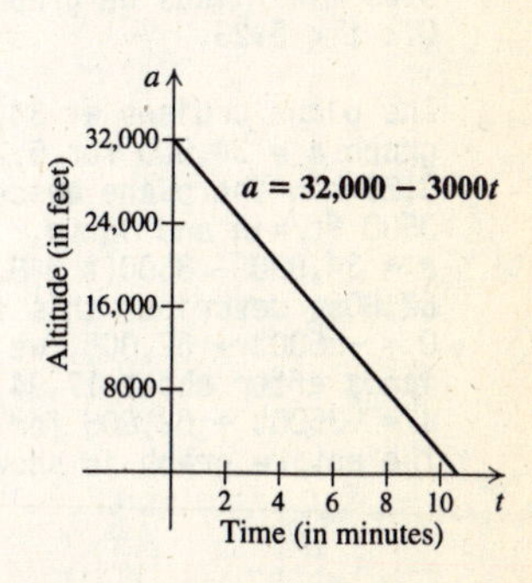

To estimate the altitude 8 min into the descent, we locate 8 on the horizontal axis. Then we trace a path up to the line and left to the vertical axis. We estimate the altitude to be 8000 ft.

Check. We could calculate the exact altitude.

$$a = 32,000 - 3000 \cdot 8 = 8000 \text{ ft.}$$

Our estimate is accurate. The graph could be used for other quick estimates of altitude.

State. The altitude 8 min into the descent is 8000 ft.

32.

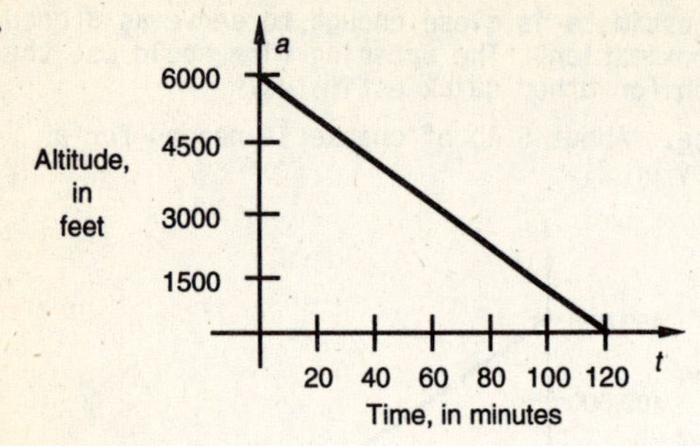

About 5000 ft

33. $s = vt + d$

$s - d = vt$ Subtracting d

$\frac{s - d}{v} = t$ Dividing by v

34. 25

35. 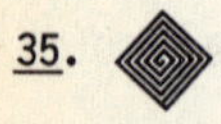

36.

37. 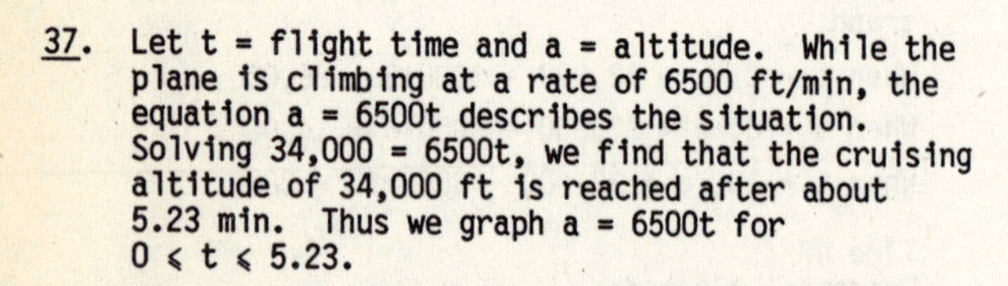
Let t = flight time and a = altitude. While the plane is climbing at a rate of 6500 ft/min, the equation a = 6500t describes the situation. Solving 34,000 = 6500t, we find that the cruising altitude of 34,000 ft is reached after about 5.23 min. Thus we graph a = 6500t for $0 \leq t \leq 5.23$.

The plane cruises at 34,000 ft for 3 min, so we graph a = 34,000 for $5.23 < t \leq 8.23$. After 8.23 min the plane descends at a rate of 3500 ft/min and lands. The equation a = 34,000 - 3500(t - 8.23), or a = -3500t + 62,805, describes this situation. Solving 0 = -3500t + 62,805, we find that the plane lands after about 17.94 min. Thus we graph a = -3500t + 62,805 for $8.23 < t \leq 17.94$. The entire graph is shown below.

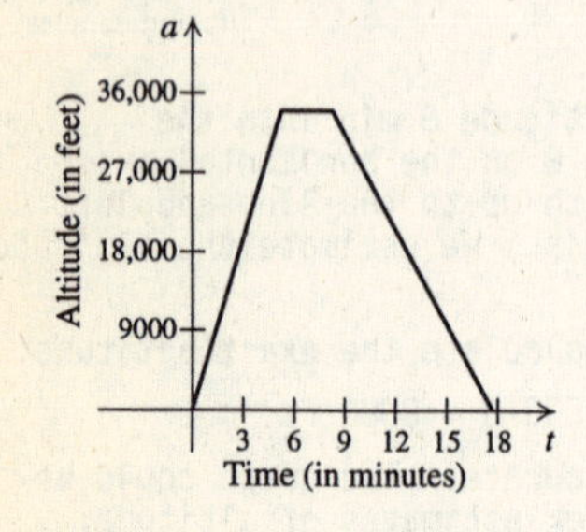

38.

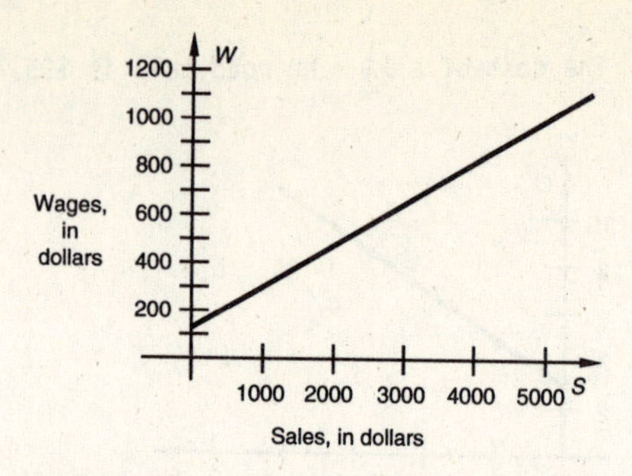

About \$550

39. Let s = Paul's weekly salary. We reword and translate.

Peggy's salary	is	twice Paul's salary	less	\$150.
p	=	2s	-	150

Paul's salary	is	\$70	more than	half of Jenna's salary.
s	=	70	+	$\frac{1}{2}j$

We solve the second equation for j.

$$s = 70 + \frac{1}{2}j$$

$$s - 70 = \frac{1}{2}j$$

$$2(s - 70) = j$$

$$2s - 140 = j$$

We make a table of values by choosing values for s and finding the corresponding values of j and p. Then we draw the graph by plotting the points (j,p) and drawing a line through them.

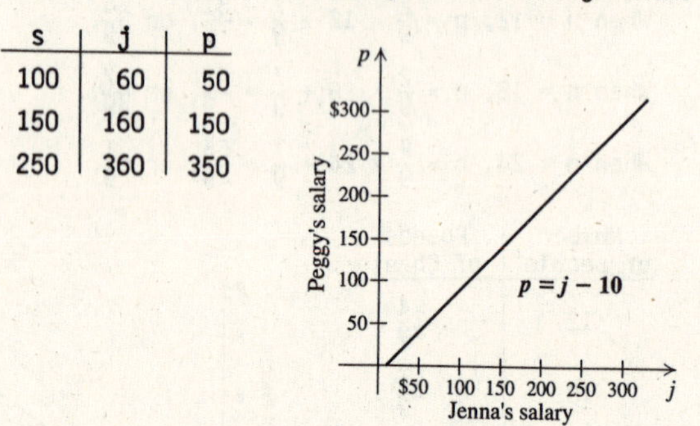

s	j	p
100	60	50
150	160	150
250	360	350

From the graph it appears that Peggy's salary is related to Jenna's by the equation p = j - 10. To verify this we can substitute $70 + \frac{1}{2}j$ for s in the first equation and simplify.

$p = 2s - 150$

$p = 2\left[70 + \frac{1}{2}j\right] - 150$ Substituting

$p = 140 + j - 150$

$p = j - 10$

40.

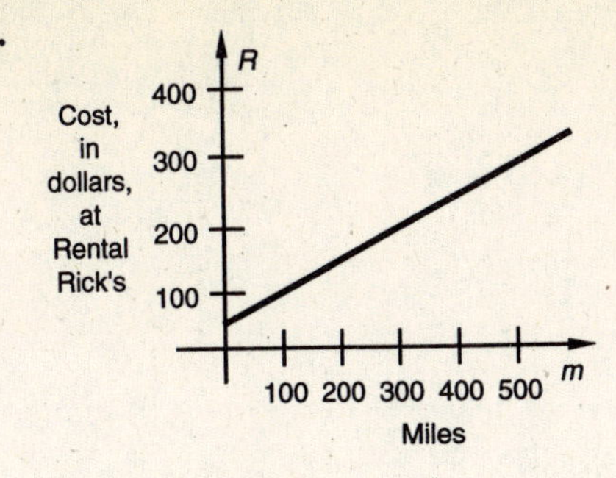

$c = 38 + 0.475m$

41. $280.13

42. $376.40

Exercise Set 4.1

1. $2^4 \cdot 2^3 = 2^{4+3} = 2^7$ (Adding exponents)

2. 3^7

3. $8^5 \cdot 8^9 = 8^{5+9} = 8^{14}$ (Adding exponents)

4. n^{23}

5. $x^4 \cdot x^3 = x^{4+3} = x^7$

6. y^{16}

7. $9^{17} \cdot 9^{21} = 9^{17+21} = 9^{38}$

8. t^{16}

9. $(3y)^4(3y)^8 = (3y)^{4+8} = (3y)^{12}$

10. $(2t)^{25}$

11. $(7y)^1(7y)^{16} = (7y)^{1+16} = (7y)^{17}$

12. $8x$

13. $(a^2b^7)(a^3b^2) = a^2b^7a^3b^2$ Using an associative law
$= a^2a^3b^7b^2$ Using a commutative law
$= a^5b^9$ Adding exponents

14. $m^{11}n^6$

15. $(xy^9)(x^3y^5) = xy^9x^3y^5$
$= xx^3y^9y^5$
$= x^4y^{14}$

16. $a^{12}b^4$

17. $r^3 \cdot r^7 \cdot r^2 = r^{3+7+2} = r^{12}$

18. s^{11}

19. $x^3(xy^4)(xy) = x^3xy^4xy$
$= x^3 \cdot x \cdot x \cdot y^4 \cdot y$
$= x^5y^5$

20. a^8b^2

21. $\frac{7^5}{7^2} = 7^{5-2} = 7^3$ (Subtracting exponents)

22. 4^4

23. $\frac{8^{12}}{8^6} = 8^{12-6} = 8^6$ (Subtracting exponents)

24. 9^{12}

25. $\frac{y^9}{y^5} = y^{9-5} = y^4$

26. x

27. $\frac{(5a)^7}{(5a)^6} = (5a)^{7-6} = (5a)^1 = 5a$

28. $3m$

29. $\frac{6^5x^8}{6^2x^3} = 6^{5-2}x^{8-3} = 6^3x^5$

30. 3^5a^2

31. $\frac{18m^5}{6m^2} = \frac{18}{6}m^{5-2} = 3m^3$

32. $5n^4$

33. $\frac{a^9b^7}{a^2b} = a^{9-2}b^{7-1} = a^7b^6$

34. r^7s^7

35. $\frac{m^9n^8}{m^0n^4} = m^{9-0}n^{8-4} = m^9n^4$

36. a^8b^9

37. When $x = -12$, $x^0 = (-12)^0 = 1$. (Any nonzero number raised to the 0 power is 1.)

38. 1

39. When $x = -4$, $5x^0 = 5(-4)^0 = 5 \cdot 1 = 5$

40. 7

41. When $n \neq 0$, $n^0 = 1$. (Any nonzero number raised to the 0 power is 1.)

42. 1

43. $10^0 = 1$

44. 1

45. $5^1 - 5^0 = 5 - 1 = 4$

46. -7

47. $(x^3)^4 = x^{3 \cdot 4} = x^{12}$ (Multiplying exponents)

48. a^{24}

49. $(2^3)^8 = 2^{3 \cdot 8} = 2^{24}$ (Multiplying exponents)

50. 5^{21}

51. $(m^7)^5 = m^{7 \cdot 5} = m^{35}$

52. n^{18}

53. $(a^{25})^3 = a^{25 \cdot 3} = a^{75}$

54. a^{75}

55. $(3x)^2 = 3^2x^2$ Raising each factor to the second power

$= 9x^2$ Simplifying

56. $25a^2$

57. $(-2a)^3 = (-2)^3a^3 = -8a^3$

58. $-27x^3$

59. $(4m^3)^2 = 4^2(m^3)^2 = 16m^6$

60. $25n^8$

61. $(3a^2b)^3 = 3^3(a^2)^3b^3 = 27a^6b^3$

62. $125x^3y^6$

63. $(a^3b^2)^5 = (a^3)^5(b^2)^5 = a^{15}b^{10}$

64. $m^{24}n^{30}$

65. $(-5x^4y^5)^2 = (-5)^2(x^4)^2(y^5)^2 = 25x^8y^{10}$

66. $81a^{20}b^{28}$

67. $\left(\frac{a}{4}\right)^3 = \frac{a^3}{4^3} = \frac{a^3}{64}$ (Raising the numerator and the denominator to the third power)

68. $\frac{81}{x^4}$

69. $\left(\frac{7}{5a}\right)^2 = \frac{7^2}{(5a)^2} = \frac{49}{5^2a^2} = \frac{49}{25a^2}$

70. $\frac{64x^3}{27}$

71. $\left(\frac{a^2}{b^3}\right)^4 = \frac{(a^2)^4}{(b^3)^4} = \frac{a^8}{b^{12}}$

72. $\frac{x^{15}}{y^{20}}$

73. $\left(\frac{y^3}{2}\right)^2 = \frac{(y^3)^2}{2^2} = \frac{y^6}{4}$

74. $\frac{a^{15}}{27}$

75. $\left(\frac{5x^2}{y^3}\right)^3 = \frac{(5x^2)^3}{(y^3)^3} = \frac{5^3(x^2)^3}{y^9} = \frac{125x^6}{y^9}$

76. $\frac{49y^{10}}{x^8}$

77. $\left(\frac{a^3}{-2b^5}\right)^4 = \frac{(a^3)^4}{(-2b^5)^4} = \frac{a^{12}}{(-2)^4(b^5)^4} = \frac{a^{12}}{16b^{20}}$

78. $\frac{x^{20}}{81y^{12}}$

79. $\left(\frac{2a^2}{3b^4}\right)^3 = \frac{(2a^2)^3}{(3b^4)^3} = \frac{2^3(a^2)^3}{3^3(b^4)^3} = \frac{8a^6}{27b^{12}}$

80. $\frac{9x^{10}}{16y^6}$

81. $\left[\frac{4x^3y^5}{3z^7}\right]^2 = \frac{(4x^3y^5)^2}{(3z^7)^2} = \frac{4^2(x^3)^2(y^5)^2}{3^2(z^7)^2} = \frac{16x^6y^{10}}{9z^{14}}$

82. $\frac{125a^{21}}{8b^{15}c^3}$

83. $3s + 3t + 24 = 3s + 3t + 3\cdot8 = 3(s + t + 8)$

84. $-7(x + 2)$

85. $9x + 2y - 4x - 2y = (9 - 4)x + (2 - 2)y = 5x$

86. 37.5%

87. ◈

88. ◈

89. $\frac{w^{50}}{w^x} = w^x$

$w^{50-x} = w^x$ Using the quotient rule

$50 - x = x$ The exponents are the same.

$50 = 2x$

$25 = x$

90. y^{5x}

91. $a^{5k} \div a^{3k} = a^{5k-3k} = a^{2k}$

92. a^{4t}

93. $\frac{\left(\frac{1}{2}\right)^4}{\left(\frac{1}{2}\right)^5} = \frac{\left(\frac{1}{2}\right)^4}{\left(\frac{1}{2}\right)^4\left(\frac{1}{2}\right)} = \frac{\left(\frac{1}{2}\right)^{4-4}}{\frac{1}{2}} = \frac{\left(\frac{1}{2}\right)^0}{\frac{1}{2}} = \frac{1}{\frac{1}{2}} =$

$1 \cdot \frac{2}{1} = 2$

94. 1

95. Since the bases are the same, the expression with the larger exponent is larger. Thus, $3^5 > 3^4$.

96. $4^2 < 4^3$

97. Since the exponents are the same, the expression with the larger base is larger. Thus, $4^3 < 5^3$.

98. $4^3 < 3^4$

99. Choose any number except 0. For example, let $x = 1$.

$3x^2 = 3\cdot1^2 = 3\cdot1 = 3$, but

$(3x)^2 = (3\cdot1)^2 = 3^2 = 9$.

100. Choose any number except 0.

101. Choose any number except 2. For example, let $x = 0$.

$\frac{x+2}{2} = \frac{0+2}{2} = \frac{2}{2} = 1$, but $x = 0$.

102. Choose any number except 0 or 1.

103. $A = P(1 + r)^t$

We substitute 10,400 for P, 8.5% for r, and 5 for t.

$A = 10,400(1 + 8.5\%)^5$

$A = 10,400(1 + 0.085)^5$

$A = 10,400(1.085)^5$

$A \approx 15,638.03$ Using a calculator to do the computations

There is \$15,638.03 in the account at the end of 5 years.

104. \$27,087.01

105. $A = s^2 = (5x)^2 = 5^2x^2 = 25x^2$

106. $343a^3$

Exercise Set 4.2

1. Three monomials are added, so $x^2 - 10x + 25$ is a trinomial.

2. Monomial

3. The polynomial $x^3 - 7x^2 + 2x - 4$ is none of these because it is composed of four monomials.

4. Binomial

5. Two monomials are added, so $4x^2 - 25$ is a binomial.

6. None of these

7. The polynomial $40x$ is a monomial because it is the product of a constant and a variable raised to a whole number power.

8. Trinomial

9. $2 - 3x + x^2 = 2 + (-3x) + x^2$

The terms are 2, $-3x$, and x^2.

10. $2x^2$, $3x$, -4

11. $5x^3 + 6x^2 - 3x^2$

Like terms: $6x^2$ and $-3x^2$ Same exponent and variable

12. $3x^2$ and $-2x^2$

13. $2x^4 + 5x - 7x - 3x^4$

Like terms: $2x^4$ and $-3x^4$ Same exponent
Like terms: $5x$ and $-7x$ and variable

14. $-3t$ and $-2t$, t^3 and $-5t^3$

15. $-3x + 6$

The coefficient of $-3x$, the first term, is -3.

The coefficient of 6, the second term, is 6.

16. 2, -4

17. $5x^2 + 3x + 3$

The coefficient of $5x^2$, the first term, is 5.

The coefficient of $3x$, the second term, is 3.

The coefficient of 3, the third term, is 3.

18. 3, -5, 2

19. $-7x^3 + 6x^2 + 3x + 7$

The coefficient of $-7x^3$, the first term, is -7.

The coefficient of $6x^2$, the second term, is 6.

The coefficient of $3x$, the third term, is 3.

The coefficient of 7, the fourth term, is 7.

20. 5, 1, -1, 2

21. $-5x^4 + 6x^3 - 3x^2 + 8x - 2$

The coefficient of $-5x^4$, the first term, is -5.

The coefficient of $6x^3$, the second term, is 6.

The coefficient of $-3x^2$, the third term, is -3.

The coefficient of $8x$, the fourth term, is 8.

The coefficient of -2, the fifth term, is -2.

22. 7, -4, -4, 5

23. $2x - 4 = 2x^1 - 4x^0$

The degree of $2x$ is 1.

The degree of -4 is 0.

The degree of the polynomial is 1, the largest exponent.

24. 0, 1; 1

25. $3x^2 - 5x + 2 = 3x^2 - 5x^1 + 2x^0$

The degree of $3x^2$ is 2.

The degree of $-5x$ is 1.

The degree of 2 is 0.

The degree of the polynomial is 2, the largest exponent.

26. 3, 2, 0; 3

27. $-7x^3 + 6x^2 + 3x + 7 = -7x^3 + 6x^2 + 3x^1 + 7x^0$
The degree of $-7x^3$ is 3.
The degree of $6x^2$ is 2.
The degree of $3x$ is 1.
The degree of 7 is 0.
The degree of the polynomial is 3, the largest exponent.

28. 4, 2, 1, 0; 4

29. $x^2 - 3x + x^6 - 9x^4 = x^2 - 3x^1 + x^6 - 9x^4$
The degree of x^2 is 2.
The degree of $-3x$ is 1.
The degree of x^6 is 6.
The degree of $-9x^4$ is 4.
The degree of the polynomial is 6, the largest exponent.

30. 1, 2, 0, 3; 3

31. See the answer section in the text.

32. $3x^2 + 8x^5 - 46x^3 + 6x - 2.4 - \frac{1}{2}x^4$

Term	Coefficient	Degree of Term	Degree of Polynomial
$8x^5$	8	5	5
$-\frac{1}{2}x^4$	$-\frac{1}{2}$	4	
$-46x^3$	-46	3	
$3x^2$	3	2	
$6x$	6	1	
-2.4	-2.4	0	

33. $2x - 5x = (2 - 5)x = -3x$

34. $10x^2$

35. $x - 9x = 1x - 9x = (1 - 9)x = -8x$

36. $-4x$

37. $5x^3 + 6x^3 + 4 = (5 + 6)x^3 + 4 = 11x^3 + 4$

38. $4x^4 + 5$

39. $5x^3 + 6x - 4x^3 - 7x = (5 - 4)x^3 + (6 - 7)x = 1x^3 + (-1)x = x^3 - x$

40. $4a^4$

41. $6b^5 + 3b^2 - 2b^5 - 3b^2 = (6 - 2)b^5 + (3 - 3)b^2 = 4b^5 + 0b^2 = 4b^5$

42. $6x^2 - 3x$

43. $\frac{1}{4}x^5 - 5 + \frac{1}{2}x^5 - 2x - 37 =$
$\left(\frac{1}{4} + \frac{1}{2}\right)x^5 - 2x + (-5 - 37) = \frac{3}{4}x^5 - 2x - 42$

44. $\frac{1}{6}x^3 + 2x - 12$

45. $6x^2 + 2x^4 - 2x^2 - x^4 - 4x^2 =$
$6x^2 + 2x^4 - 2x^2 - 1x^4 - 4x^2 =$
$(6 - 2 - 4)x^2 + (2 - 1)x^4 = 0x^2 + 1x^4 = 0 + x^4 = x^4$

46. $-x^3$

47. $\frac{1}{4}x^3 - x^2 - \frac{1}{6}x^2 + \frac{3}{8}x^3 + \frac{5}{16}x^3 =$
$\frac{1}{4}x^3 - 1x^2 - \frac{1}{6}x^2 + \frac{3}{8}x^3 + \frac{5}{16}x^3 =$
$\left(\frac{1}{4} + \frac{3}{8} + \frac{5}{16}\right)x^3 + \left(-1 - \frac{1}{6}\right)x^2 =$
$\left(\frac{4}{16} + \frac{6}{16} + \frac{5}{16}\right)x^3 + \left(-\frac{6}{6} - \frac{1}{6}\right)x^2 = \frac{15}{16}x^3 - \frac{7}{6}x^2$

48. 0

49. $3x^4 - 5x^6 - 2x^4 + 6x^6 = x^4 + x^6 = x^6 + x^4$

50. $x^4 - 2x^3 + 1$

51. $-2x + 4x^3 - 7x + 9x^3 + 8 = -9x + 13x^3 + 8 = 13x^3 - 9x + 8$

52. $x^2 - 4x + 1$

53. $3x + 3x + 3x - x^2 - 4x^2 = 9x - 5x^2 = -5x^2 + 9x$

54. $-4x^3 - 6x$

55. $-x + \frac{3}{4} + 15x^4 - x - \frac{1}{2} - 3x^4 = -2x + \frac{1}{4} + 12x^4 =$
$12x^4 - 2x + \frac{1}{4}$

56. $4x^3 + x - \frac{1}{2}$

57. $-5x + 2 = -5\cdot 4 + 2 = -20 + 2 = -18$

58. -11

59. $2x^2 - 5x + 7 = 2\cdot 4^2 - 5\cdot 4 + 7 = 2\cdot 16 - 20 + 7 = 32 - 20 + 7 = 19$

60. 59

61. $x^3 - 5x^2 + x = 4^3 - 5\cdot 4^2 + 4 = 64 - 5\cdot 16 + 4 = 64 - 80 + 4 = -12$

62. 51

63. $3x + 5 = 3(-1) + 5 = -3 + 5 = 2$

64. 8

65. $x^2 - 2x + 1 = (-1)^2 - 2(-1) + 1 = 1 + 2 + 1 = 4$

66. -10

67. $-3x^3 + 7x^2 - 3x - 2 =$
$-3(-1)^3 + 7(-1)^2 - 3(-1) - 2 =$
$-3(-1) + 7\cdot 1 + 3 - 2 = 3 + 7 + 3 - 2 = 11$

68. -4

69. $0.4r^2 - 40r + 1039 = 0.4(18)^2 - 40(18) + 1039 =$
$0.4(324) - 720 + 1039 = 129.6 - 720 + 1039 =$
448.6

There are approximately 449 accidents daily involving an 18-year-old driver.

70. 399

71. $11.12t^2 = 11.12(10)^2 = 11.12(100) = 1112$

A skydiver has fallen approximately 1112 ft 10 seconds after jumping from a plane.

72. 3091 ft

73. Evaluate the polynomial for $x = 75$:

$280x - 0.4x^2 = 280\cdot 75 - 0.4(75)^2 =$
$21{,}000 - 0.4(5625) = 21{,}000 - 2250 = 18{,}750$

The total revenue is $18,750.

74. $24,000

75. Evaluate the polynomial for $x = 500$:

$5000 + 0.6(500)^2 = 5000 + 0.6(250{,}000) =$
$5000 + 150{,}000 = 155{,}000$

The total cost is $155,000.

76. $258,500

77. $2\pi r = 2(3.14)(10)$ Substituting 3.14 for π and 10 for r
$= 62.8$

The circumference is 62.8 cm.

78. 31.4 ft

79. $\pi r^2 = 3.14(5)^2$ Substituting 3.14 for π and 5 for r
$= 3.14(25)$
$= 78.5$

The area is 78.5 m^2.

80. 314 in^2

81. When $t = 1$, $-t^2 + 6t - 4 = -1^2 + 6\cdot 1 - 4 =$
$-1 + 6 - 4 = 1$.
When $t = 2$, $-t^2 + 6t - 4 = -2^2 + 6\cdot 2 - 4 =$
$-4 + 12 - 4 = 4$.
When $t = 3$, $-t^2 + 6t - 4 = -3^2 + 6\cdot 3 - 4 =$
$-9 + 18 - 4 = 5$.
When $t = 4$, $-t^2 + 6t - 4 = -4^2 + 6\cdot 4 - 4 =$
$-16 + 24 - 4 = 4$.
When $t = 5$, $-t^2 + 6t - 4 = -5^2 + 6\cdot 5 - 4 =$
$-25 + 30 - 4 = 1$.

We complete the table. Then we plot the points and connect them with a smooth curve.

t	$-t^2 + 6t - 4$
1	1
2	4
3	5
4	4
5	1

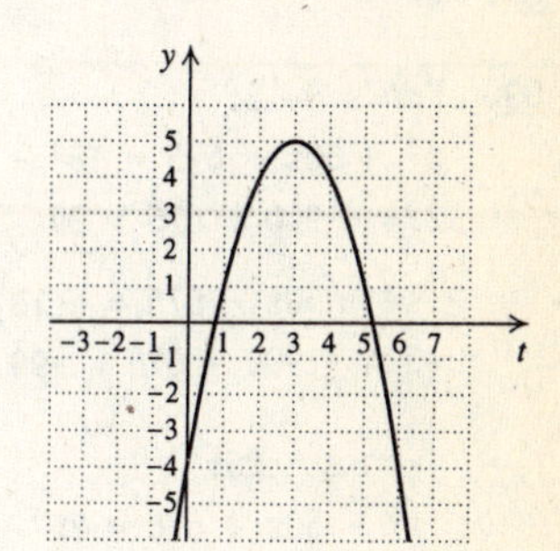

82.

t	$-t^2 + 10t - 18$
3	3
4	6
5	7
6	6
7	3

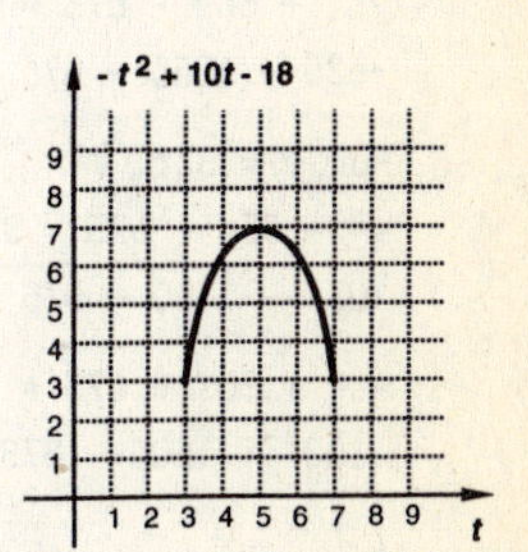

83. Familiarize. The page numbers on facing pages are consecutive integers. If x = the smaller number, then x + 1 = the larger number.

Translate. We reword and translate.

First integer plus second integer is 549.
$x \quad + \quad (x + 1) \quad = \quad 549$

Carry out. We solve the equation.

$x + (x + 1) = 549$
$2x + 1 = 549$
$2x = 548$
$x = 274$

If x is 274, then x + 1 is 274 + 1, or 275.

Check. 274 and 275 are consecutive integers, and their sum is 549. The numbers check.

State. The page numbers are 274 and 275.

84. $0.125, or 12.5¢

85. ◈

86. ◈

87. $\frac{9}{2}x^8 + \frac{1}{9}x^2 + \frac{1}{2}x^9 + \frac{9}{2}x + \frac{9}{2}x^9 + \frac{8}{9}x^2 + \frac{1}{2}x - \frac{1}{2}x^8$

$= \left(\frac{1}{2} + \frac{9}{2}\right)x^9 + \left(\frac{9}{2} - \frac{1}{2}\right)x^8 + \left(\frac{1}{9} + \frac{8}{9}\right)x^2 + \left(\frac{9}{2} + \frac{1}{2}\right)x$

$= \frac{10}{2}x^9 + \frac{8}{2}x^8 + \frac{9}{9}x^2 + \frac{10}{2}x$

$= 5x^9 + 4x^8 + x^2 + 5x$

88. $3x^6$

89. For $s = 18$:

$s^2 - 50s + 675 = 18^2 - 50(18) + 675 =$
$324 - 900 + 675 = 99$

$-s^2 + 50s - 675 = -(18)^2 + 50(18) - 675 =$
$-324 + 900 - 675 = -99$

For $s = 25$:

$s^2 - 50s + 675 = 25^2 - 50(25) + 675 =$
$625 - 1250 + 675 = 50$

$-s^2 + 50s - 675 = -(25)^2 + 50(25) - 675 =$
$-625 + 1250 - 675 = -50$

For $s = 32$:

$s^2 - 50s + 675 = 32^2 - 50(32) + 675 =$
$1024 - 1600 + 675 = 99$

$-s^2 + 50s - 675 = -(32)^2 + 50(32) - 675 =$
$-1024 + 1600 - 675 = -99$

90. 50

91. Answers may vary. Use any ax^5-term, where a is an integer, and 3 other terms with different degrees each less than degree 5, and integer coefficients. Three answers are $-6x^5 + 14x^4 - x^2 + 11$, $x^5 - 8x^3 + 3x + 1$, and $23x^5 + 2x^4 - x^2 + 5x$.

92. Answers may vary.

$0.2y^4 - y + \frac{5}{2}$, $-\frac{8}{7}y^4 + 5.5y^3 - 2y^2$,

$2.9y^4 - 4y^2 - \frac{11}{3}$

93. $(5m^5)^2 = 25m^{10}$

The degree is 10.

94. Answers may vary.

$9y^4$, $-\frac{3}{2}y^4$, $4.2y^4$

95. When $d = 0$, $-0.0064d^2 + 0.8d + 2 =$
$-0.0064(0)^2 + 0.8(0) + 2 = 0 + 0 + 2 = 2.$

When $d = 30$, $-0.0064(30)^2 + 0.8(30) + 2 =$
$-5.76 + 24 + 2 = 20.24.$

When $d = 60$, $-0.0064(60)^2 + 0.8(60) + 2 =$
$-23.04 + 48 + 2 = 26.96.$

When $d = 90$, $-0.0064(90)^2 + 0.8(90) + 2 =$
$-51.84 + 72 + 2 = 22.16.$

When $d = 120$, $-0.0064(120)^2 + 0.8(120) + 2 =$
$-92.16 + 96 + 2 = 5.84.$

We complete the table. Then we plot the points and connect them with a smooth curve.

d	$-0.0064d^2 + 0.8d + 2$
0	2
30	20.24
60	26.96
90	22.16
120	5.84

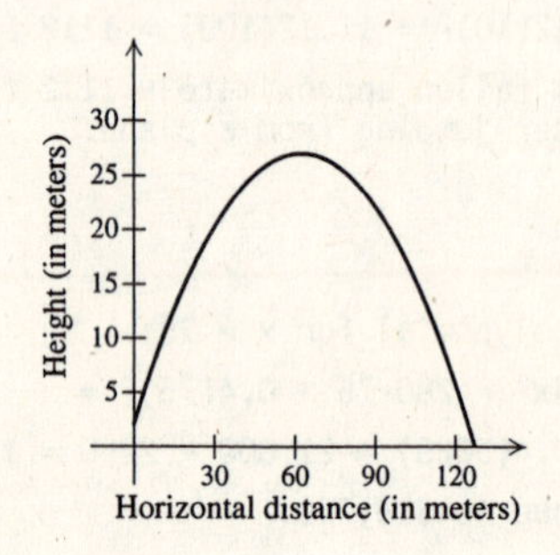

96. $x^3 - 2x^2 - 6x + 3$

Exercise Set 4.3

1. $(3x + 2) + (-4x + 3) = (3 - 4)x + (2 + 3) = -x + 5$

2. $-x + 3$

3. $(-6x + 2) + (x^2 + x - 3) =$
$x^2 + (-6 + 1)x + (2 - 3) = x^2 - 5x - 1$

4. $x^2 + 3x - 5$

5. $(x^2 - 9) + (x^2 + 9) = (1 + 1)x^2 + (-9 + 9) = 2x^2$

6. $3x^3 - 4x^2$

7. $(3x^2 - 5x + 10) + (2x^2 + 8x - 40) =$
$(3 + 2)x^2 + (-5 + 8)x + (10 - 40) = 5x^2 + 3x - 30$

8. $6x^4 + 3x^3 + 4x^2 - 3x + 2$

9. $(1.2x^3 + 4.5x^2 - 3.8x) + (-3.4x^3 - 4.7x^2 + 23) =$
$(1.2 - 3.4)x^3 + (4.5 - 4.7)x^2 - 3.8x + 23 =$
$-2.2x^3 - 0.2x^2 - 3.8x + 23$

10. $2.8x^4 - 0.6x^2 + 1.8x - 3.2$

11. $(1 + 4x + 6x^2 + 7x^3) + (5 - 4x + 6x^2 - 7x^3) =$
$(1 + 5) + (4 - 4)x + (6 + 6)x^2 + (7 - 7)x^3 =$
$6 + 0x + 12x^2 + 0x^3 = 6 + 12x^2$, or $12x^2 + 6$

12. $3x^4 - 4x^3 + x^2 + x + 4$

13. $(9x^8 - 7x^4 + 2x^2 + 5) + (8x^7 + 4x^4 - 2x) =$
$9x^8 + 8x^7 + (-7 + 4)x^4 + 2x^2 - 2x + 5 =$
$9x^8 + 8x^7 - 3x^4 + 2x^2 - 2x + 5$

14. $4x^5 + 9x^2 + 1$

15. $\left[\frac{1}{4}x^4 + \frac{2}{3}x^3 + \frac{5}{8}x^2 + 7\right] + \left[-\frac{3}{4}x^4 + \frac{3}{8}x^2 - 7\right] =$
$\left[\frac{1}{4} - \frac{3}{4}\right]x^4 + \frac{2}{3}x^3 + \left[\frac{5}{8} + \frac{3}{8}\right]x^2 + (7 - 7) =$
$-\frac{2}{4}x^4 + \frac{2}{3}x^3 + \frac{8}{8}x^2 + 0 =$
$-\frac{1}{2}x^4 + \frac{2}{3}x^3 + x^2$

16. $\frac{2}{15}x^9 - \frac{2}{5}x^5 + \frac{1}{4}x^4 + \frac{1}{4}x^2 + \frac{15}{2}$

17. $(0.02x^5-0.2x^3+x+0.08) + (-0.01x^5+x^4-0.8x-0.02) =$
$(0.02-0.01)x^5+x^4-0.2x^3+(1-0.8)x+(0.08-0.02) =$
$0.01x^5 + x^4 - 0.2x^3 + 0.2x + 0.06$

18. $0.10x^6 + 0.02x^3 + 0.22x + 0.55$

19.
$$\begin{array}{rrrr} -3x^4 & + 6x^2 & + 2x & - 1 \\ & - 3x^2 & + 2x & + 1 \\ \hline -3x^4 & + 3x^2 & + 4x & + 0 \end{array}$$
$-3x^4 + 3x^2 + 4x$

20. $-4x^3 + 4x^2 + 6x$

21. Rewrite the problem so the coefficients of like terms have the same number of decimal places.
$$\begin{array}{rrrrr} 0.15x^4 & + 0.10x^3 & - \ 0.90x^2 & & \\ & -0.01x^3 & + \ 0.01x^2 & + x & \\ 1.25x^4 & & + \ 0.11x^2 & & + 0.01 \\ & 0.27x^3 & & & + 0.99 \\ -0.35x^4 & & + 15.00x^2 & & - 0.03 \\ \hline 1.05x^4 & + 0.36x^3 & + 14.22x^2 & + x & + 0.97 \end{array}$$

22. $1.3x^4 + 0.35x^3 + 9.53x^2 + 2x + 0.96$

23. Two equivalent expressions for the opposite of $-5x$ are

a) $-(-5x)$ and

b) $5x$. (Changing the sign)

24. $-(x^2 - 3x)$, $-x^2 + 3x$

25. Two equivalent expressions for the opposite of $-x^2 + 10x - 2$ are

a) $-(-x^2 + 10x - 2)$ and

b) $x^2 - 10x + 2$. (Changing the sign of every term)

26. $-(-4x^3 - x^2 - x)$, $4x^3 + x^2 + x$

27. Two equivalent expressions for the opposite of $12x^4 - 3x^3 + 3$ are

a) $-(12x^4 - 3x^3 + 3)$ and

b) $-12x^4 + 3x^3 - 3$. (Changing the sign of every term)

28. $-(4x^3 - 6x^2 - 8x + 1)$, $-4x^3 + 6x^2 + 8x - 1$

29. We change the sign of every term inside parentheses.
$-(3x - 7) = -3x + 7$

30. $2x - 4$

31. We change the sign of every term inside parentheses.
$-(4x^2 - 3x + 2) = -4x^2 + 3x - 2$

32. $6a^3 - 2a^2 + 9a - 1$

33. We change the sign of every term inside parentheses.
$-\left[-4x^4 + 6x^2 + \frac{3}{4}x - 8\right] = 4x^4 - 6x^2 - \frac{3}{4}x + 8$

34. $5x^4 - 4x^3 + x^2 - 0.9$

35. $(3x + 2) - (-4x + 3) = 3x + 2 + 4x - 3$
Changing the sign of every term inside parentheses
$= 7x - 1$

36. $13x - 1$

37. $(-6x + 2) - (x^2 + x - 3) = -6x + 2 - x^2 - x + 3$
$= -x^2 - 7x + 5$

38. $x^2 - 13x + 13$

39. $(x^2 - 9) - (x^2 + 9) = x^2 - 9 - x^2 - 9 = -18$

40. $-x^3 + 6x^2$

41. $(6x^4 + 3x^3 - 1) - (4x^2 - 3x + 3)$
$= 6x^4 + 3x^3 - 1 - 4x^2 + 3x - 3$
$= 6x^4 + 3x^3 - 4x^2 + 3x - 4$

42. $-3x^3 + x^2 + 2x - 3$

43. $(1.2x^3 + 4.5x^2 - 3.8x) - (-3.4x^3 - 4.7x^2 + 23)$
$= 1.2x^3 + 4.5x^2 - 3.8x + 3.4x^3 + 4.7x^2 - 23$
$= 4.6x^3 + 9.2x^2 - 3.8x - 23$

44. $-1.8x^4 - 0.6x^2 - 1.8x + 4.6$

45. $(5x^2 + 6) - (3x^2 - 8) = 5x^2 + 6 - 3x^2 + 8$
$= 2x^2 + 14$

46. $7x^3 - 9x^2 - 2x + 10$

47. $(6x^5 - 3x^4 + x + 1) - (8x^5 + 3x^4 - 1)$
$= 6x^5 - 3x^4 + x + 1 - 8x^5 - 3x^4 + 1$
$= -2x^5 - 6x^4 + x + 2$

48. $-x^2 - 2x + 4$

49. $(6x^2 + 2x) - (-3x^2 - 7x + 8)$
$= 6x^2 + 2x + 3x^2 + 7x - 8$
$= 9x^2 + 9x - 8$

50. $7x^3 + 3x^2 + 2x - 1$

51. $\frac{5}{8}x^3 - \frac{1}{4}x - \frac{1}{3} - \left(-\frac{1}{8}x^3 + \frac{1}{4}x - \frac{1}{3}\right)$
$= \frac{5}{8}x^3 - \frac{1}{4}x - \frac{1}{3} + \frac{1}{8}x^3 - \frac{1}{4}x + \frac{1}{3}$
$= \frac{6}{8}x^3 - \frac{2}{4}x$
$= \frac{3}{4}x^3 - \frac{1}{2}x$

52. $\frac{3}{5}x^3 - 0.11$

53. $(0.08x^3 - 0.02x^2 + 0.01x) - (0.02x^3 + 0.03x^2 - 1)$
$= 0.08x^3 - 0.02x^2 + 0.01x - 0.02x^3 - 0.03x^2 + 1$
$= 0.06x^3 - 0.05x^2 + 0.01x + 1$

54. $0.1x^4 - 0.9$

55. $x^2 + 5x + 6$
$\underline{-(x^2 + 2x\quad)}$

$x^2 + 5x + 6$
$\underline{-x^2 - 2x}$ Changing signs and removing parentheses
$3x + 6$ Adding

56. $-x^2 + 1$

57. $5x^4 + 6x^3 - 9x^2$
$\underline{-(-6x^4 - 6x^3 \quad + 8x + 9)}$

$5x^4 + 6x^3 - 9x^2$
$\underline{6x^4 + 6x^3 \quad - 8x - 9}$ Changing signs and removing parentheses
$11x^4 + 12x^3 - 9x^2 - 8x - 9$ Adding

58. $5x^4 - 6x^3 - x^2 + 5x + 15$

59. $3x^4 + 6x^2 + 8x - 1$
$\underline{-(4x^5 - 6x^4 \quad - 8x - 7)}$

$3x^4 + 6x^2 + 8x - 1$
$\underline{-4x^5 + 6x^4 \quad + 8x + 7}$ Changing signs and removing parentheses
$-4x^5 + 9x^4 + 6x^2 + 16x + 6$ Adding

60. $-4x^5 - 6x^3 + 8x^2 - 5x - 2$

61. $x^5 \quad - 1$
$\underline{-(x^5 - x^4 + x^3 - x^2 + x - 1)}$

$x^5 \quad - 1$
$\underline{-x^5 + x^4 - x^3 + x^2 - x + 1}$ Changing sings and removing parentheses
$x^4 - x^3 + x^2 - x$ Adding

62. $2x^4 - 2x^3 + 2x^2$

63. a)

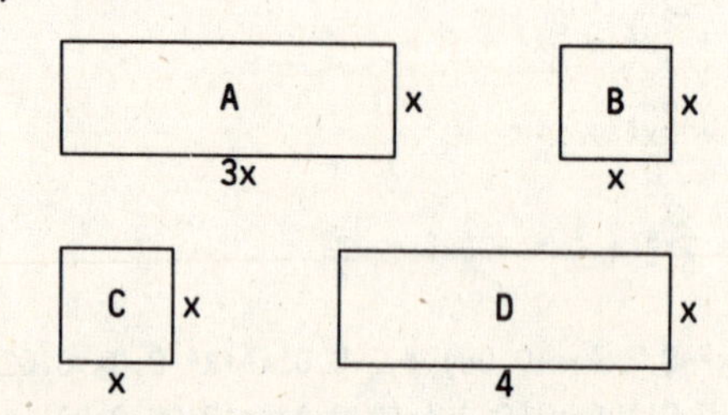

Familiarize. The area of a rectangle is the product of the length and the width.

Translate. The sum of the areas is found as follows:

Area of A + Area of B + Area of C + Area of D

$3x\cdot x + x\cdot x + x\cdot x + 4\cdot x$

Carry out. We collect like terms.

$3x^2 + x^2 + x^2 + 4x = 5x^2 + 4x$

Check. We can go over our calculations. We can also assign some value to x, say 2, and carry out the computation of the area in two ways.

Sum of areas: $3\cdot 2\cdot 2 + 2\cdot 2 + 2\cdot 2 + 4\cdot 2 =$
$12 + 4 + 4 + 8 = 28$

Substituting in the polynomial: $5(2)^2 + 4\cdot 2 =$
$20 + 8 = 28$

Since the results are the same, our solution is probably correct.

State. A polynomial for the sum of the areas is $5x^2 + 4x$.

b) For $x = 3$: $5x^2 + 4x = 5\cdot 3^2 + 4\cdot 3 =$
$5\cdot 9 + 4\cdot 3 = 45 + 12 = 57$

When $x = 3$, the sum of the areas is 57 square units.

For $x = 8$: $5x^2 + 4x = 5\cdot 8^2 + 4\cdot 8 =$
$5\cdot 64 + 4\cdot 8 = 320 + 32 = 352$

When $x = 8$, the sum of the areas is 352 square units.

64. a) $r^2\pi + 13\pi$

b) 38π; 140.69π

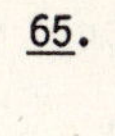

65.

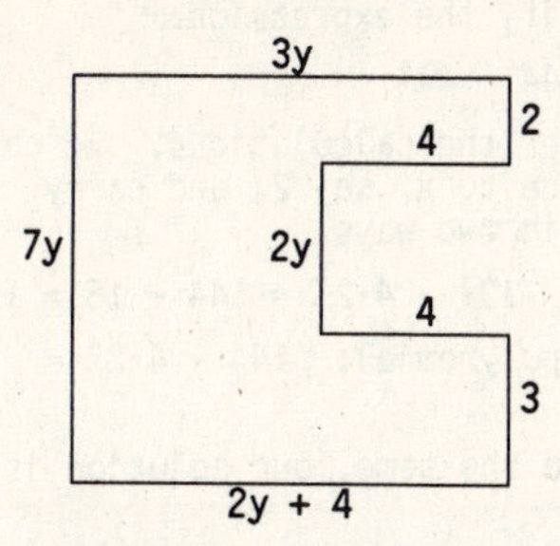

Familiarize. The perimeter is the sum of the lengths of the sides.

Translate. The sum of the lengths is found as follows:

$3y + 7y + (2y + 4) + 3 + 4 + 2y + 4 + 2$

Carry out. We collect like terms.

$(3 + 7 + 2 + 2)y + (4 + 3 + 4 + 4 + 2) =$

$14y + 17$

Check. We can go over our calculations. We can also assign some value to y, say 3, and carry out the computation of the perimeter in two ways.

Sum of lengths: $3\cdot3 + 7\cdot3 + (2\cdot3 + 4) + 3 + 4 + 2\cdot3 + 4 + 2 = 9 + 21 + 10 + 3 + 4 + 6 + 4 + 2 = 59$

Substituting in the polynomial: $14\cdot3 + 17 = 42 + 17 = 59$

Since the results are the same, our solution is probably correct.

State. A polynomial for the perimeter of the figure is $14y + 17$.

66. $11\frac{1}{2}a + 10$

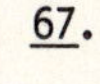

67.

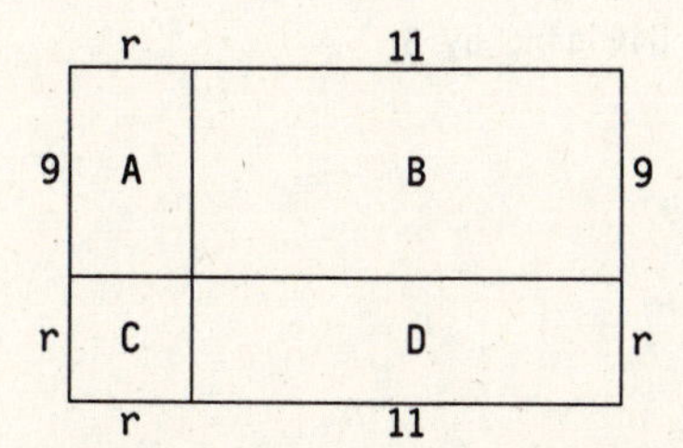

The area of the figure can be found by finding the sum of the areas of the four rectangles A, B, C, and D. The area of a rectangle is the product of the length and the width.

$$\text{Area of A} + \text{Area of B} + \text{Area of C} + \text{Area of D}$$

$$= 9\cdot r + 11\cdot 9 + r\cdot r + 11\cdot r$$

$$= 9r + 99 + r^2 + 11r$$

An algebraic expression for the area of the figure is $9r + 99 + r^2 + 11r$.

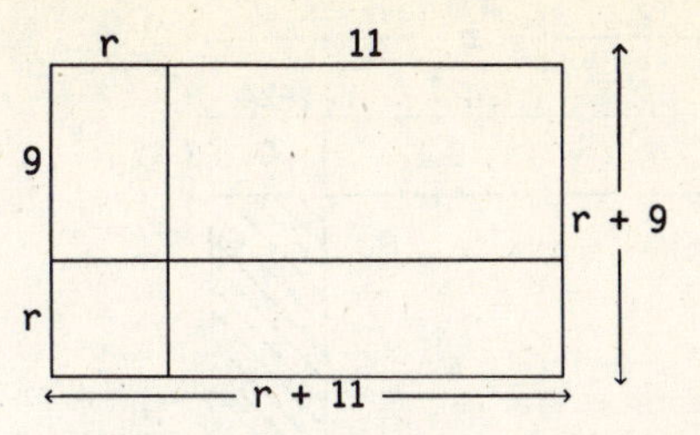

The length and width of the figure can be expressed as $r + 11$ and $r + 9$, respectively. The area of this figure (a rectangle) is the product of the length and width. An algebraic expression for the area is $(r + 11)\cdot(r + 9)$.

The algebraic expressions $9r + 99 + r^2 + 11r$ and $(r + 11)\cdot(r + 9)$ represent the same area.

68. $(x + 3)^2$, $x^2 + 3x + 3x + 9$

69.

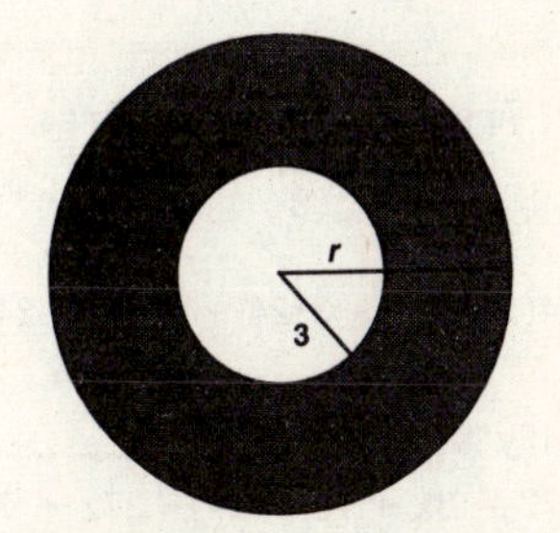

Familiarize. Recall that the area of a circle is the product of π and the square of the radius, r^2. $A = \pi r^2$

Translate.

$$\text{Area of circle with radius } r - \text{Area of circle with radius } 3 = \text{Shaded area}$$

$$\pi\cdot r^2 - \pi\cdot 3^2 = \text{Shaded area}$$

Carry out. We simplify the expression.

$$\pi\cdot r^2 - \pi\cdot 3^2 = \pi r^2 - 9\pi$$

Check. We can go over our calculations. We can also assign some value to r, say 5, and carry out the computation in two ways.

Difference of areas: $\pi\cdot5^2 - \pi\cdot3^2 = 25\pi - 9\pi = 16\pi$

Substituting in the polynomial: $\pi\cdot5^2 - 9\pi = 25\pi - 9\pi = 16\pi$

Since the results are the same, our solution is probably correct.

State. A polynomial for the shaded area is $\pi r^2 - 9\pi$.

70. $m^2 - 28$

71.

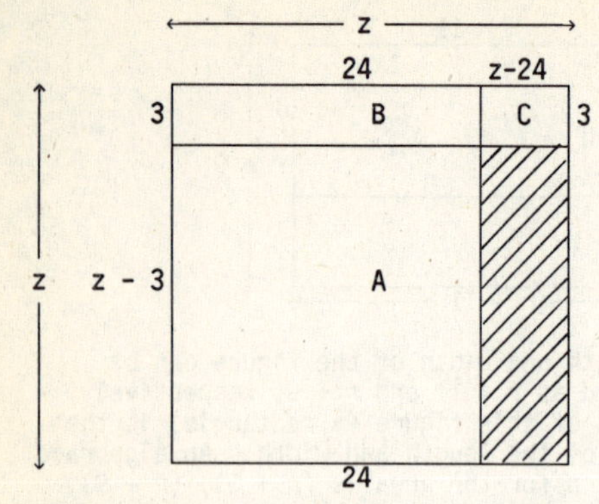

Familiarize. We label the sides of A, B, and C with additional information. The area of the square is $z\cdot z$, or z^2. The area of the shaded section is z^2 minus the areas of sections A, B, and C.

Translate.

Area of shaded section	=	Area of square	-	Area of A	-	Area of B	-	Area of C

Area of shaded section $= z\cdot z - 24(z - 3) - 3\cdot 24 - 3(z - 24)$

Carry out. We simplify the expression.

$$z^2 - 24z + 72 - 72 - 3z + 72 = z^2 - 27z + 72$$

Check. We can go over our calculations. We can also assign some value to z, say 30, and carry out the computation in two ways.

Difference of areas: $30\cdot 30 - 24\cdot 27 - 3\cdot 24 - 3\cdot 6 = 900 - 648 - 72 - 18 = 162$

Substituting in the polynomial: $30^2 - 27\cdot 30 + 72 = 900 - 810 + 72 = 162$

Since the results are the same, our solution is probably correct.

State. A polynomial for the shaded area is $z^2 - 27z + 72$.

72. $\pi x^2 - 2x^2$, or $(\pi - 2)x^2$

73.

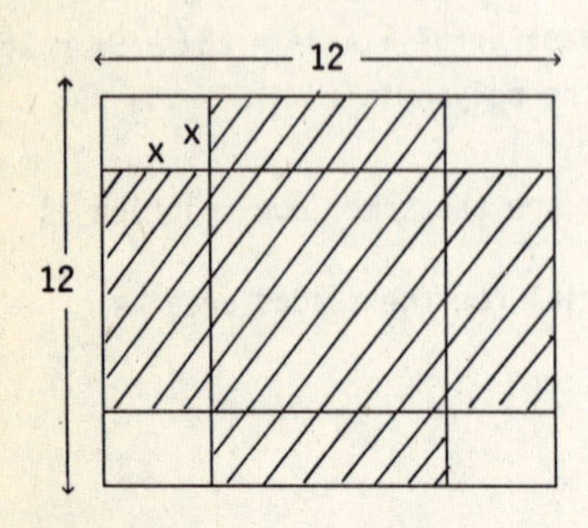

Familiarize. The area of each corner that is not shaded is $x\cdot x$, or x^2.

Area of shaded section	=	Area of square	-	Area of four corners

Area of shaded section $= 12^2 - 4\cdot x^2$

Carry out. We simplify the expression.

$$12\cdot 12 - 4\cdot x^2 = 144 - 4x^2$$

Check. We can go over the calculations. We can also assign some value to x, say 2, and carry out the computation in two ways.

Difference of areas: $12^2 - 4\cdot 2^2 = 144 - 16 = 128$

Substituting in the polynomial: $144 - 4\cdot 2^2 = 144 - 16 = 128$

Since the results are the same, our solution is probably correct.

State. A polynomial for the shaded area is $144 - 4x^2$.

74. $y^2 - 4y + 4$

75.

$1.5x - 2.7x = 23 - 5.6x$

$10(1.5x - 2.7x) = 10(23 - 5.6x)$ Clearing decimals

$15x - 27x = 230 - 56x$

$-12x = 230 - 56x$ Collecting like terms

$44x = 230$ Adding 56x

$x = \frac{230}{44}$ Dividing by 44

$x = \frac{115}{22}$ Simplifying

76. 1

77. $8(x - 2) = 16$

$8x - 16 = 16$ Multiplying to remove parentheses

$8x = 32$ Adding 16

$x = 4$ Dividing by 8

78. $-\frac{76}{3}$

79. ◈

80. ◈

81. $(4a^2 - 3a) + (7a^2 - 9a - 13) - (6a - 9)$

$= 4a^2 - 3a + 7a^2 - 9a - 13 - 6a + 9$

$= 11a^2 - 18a - 4$

82. $5x^2 - 9x - 1$

83. $(-8y^2 - 4) - (3y + 6) - (2y^2 - y)$

$= -8y^2 - 4 - 3y - 6 - 2y^2 + y$

$= -10y^2 - 2y - 10$

84. $4x^3 - 5x^2 + 6$

85. $(-y^4 - 7y^3 + y^2) + (-2y^4 + 5y - 2) - (-6y^3 + y^2)$
$= -y^4 - 7y^3 + y^2 - 2y^4 + 5y - 2 + 6y^3 - y^2$
$= -3y^4 - y^3 + 5y - 2$

86. $2 + x + 2x^2 + 4x^3$

87. $(345.099x^3 - 6.178x) - (-224.508x^3 + 8.99x)$
$= 345.099x^3 - 6.178x + 224.508x^3 - 8.99x$
$= 569.607x^3 - 15.168x$

88. $28x + 2x^2$

89. Familiarize. The surface area is $2\ell w + 2\ell h + 2wh$, where ℓ = length, w = width, and h = height of the rectangular solid. Here we have $\ell = 9$, $w = a$, and $h = 5$.

Translate. We substitute in the formula above.

$2\cdot9\cdot a + 2\cdot9\cdot5 + 2\cdot a\cdot5$

Carry out. We simplify the expression.

$2\cdot9\cdot a + 2\cdot9\cdot5 + 2\cdot a\cdot5$
$= 18a + 90 + 10a$
$= 28a + 90$

Check. We can go over the calculations. We can also assign some value to a, say 7, and carry out the computation in two ways.

Using the formula: $2\cdot9\cdot7 + 2\cdot9\cdot5 + 2\cdot7\cdot5 =$
$126 + 90 + 70 = 286$

Substituting in the polynomial: $28\cdot7 + 90 =$
$196 + 90 = 286$

Since the results are the same, our solution is probably correct.

State. A polynomial for the surface area is $28a + 90$.

90. 

91. a) $R - C = 280x - 0.4x^2 - (5000 + 0.6x^2)$
$= 280x - 0.4x^2 - 5000 - 0.6x^2$
$= -x^2 + 280x - 5000$

A polynomial for total profit is $-x^2 + 280x - 5000$.

b) Evaluate the polynomial for x = 75:
$-x^2 + 280x - 5000 = -(75)^2 + 280(75) - 5000$
$= -5625 + 21,000 - 5000$
$= 10,375$

The total profit on the production and sale of 75 stereos is \$10,375.

c) Evaluate the polynomial for x = 100:
$-x^2 + 280x - 5000 = -(100)^2 + 280(100) - 5000$
$= -10,000 + 28,000 - 5000$
$= 13,000$

The total profit on the production and sale of 100 stereos is \$13,000.

Exercise Set 4.4

1. $(6x^2)(7) = (6\cdot7)x^2 = 42x^2$

2. $-10x^2$

3. $(-x^3)(-x) = (-1x^3)(-1x) = (-1)(-1)(x^3\cdot x) = x^4$

4. $-x^6$

5. $(-x^5)(x^3) = (-1x^5)(1x^3) = (-1)(1)(x^5\cdot x^3) = -x^8$

6. x^8

7. $(3x^4)(2x^2) = (3\cdot2)(x^4\cdot x^2) = 6x^6$

8. $20x^8$

9. $(7t^5)(4t^3) = (7\cdot4)(t^5\cdot t^3) = 28t^8$

10. $30a^4$

11. $(-0.1x^6)(0.2x^4) = (-0.1)(0.2)(x^6\cdot x^4) = -0.02x^{10}$

12. $-0.12x^9$

13. $\left(-\frac{1}{5}x^3\right)\left(-\frac{1}{3}x\right) = \left(-\frac{1}{5}\right)\left(-\frac{1}{3}\right)(x^3\cdot x) = \frac{1}{15}x^4$

14. $-\frac{1}{20}x^{12}$

15. $(-4x^2)(0) = 0$ Any number multiplied by 0 is 0.

16. $4m^5$

17. $(3x^2)(-4x^3)(2x^6) = (3)(-4)(2)(x^2\cdot x^3\cdot x^6) = -24x^{11}$

18. $60y^{12}$

19. $3x(-x + 5) = 3x(-x) + 3x(5)$
$= -3x^2 + 15x$

20. $8x^2 - 12x$

21. $4x(x + 1) = 4x(x) + 4x(1)$
$= 4x^2 + 4x$

22. $3x^2 + 6x$

23. $(x + 7)5x = x \cdot 5x + 7 \cdot 5x$
$= 5x^2 + 35x$

24. $3x^2 - 18x$

25. $x^2(x^3 + 1) = x^2(x^3) + x^2(1)$
$= x^5 + x^2$

26. $-2x^5 + 2x^3$

27. $3x(2x^2 - 6x + 1) = 3x(2x^2) + 3x(-6x) + 3x(1)$
$= 6x^3 - 18x^2 + 3x$

28. $-8x^4 + 24x^3 + 20x^2 - 4x$

29. $4x^2(3x + 6) = 4x^2(3x) + 4x^2(6)$
$= 12x^3 + 24x^2$

30. $-10x^3 + 5x^2$

31. $-6x^2(x^2 + x) = -6x^2(x^2) - 6x^2(x)$
$= -6x^4 - 6x^3$

32. $-4x^4 + 4x^3$

33. $3y^2(6y^4 + 8y^3) = 3y^2(6y^4) + 3y^2(8y^3)$
$= 18y^6 + 24y^5$

34. $4y^7 - 24y^6$

35. $3x^4(14x^{50} + 20x^{11} + 6x^{57} + 60x^{15})$
$= 3x^4(14x^{50}) + 3x^4(20x^{11}) + 3x^4(6x^{57}) + 3x^4(60x^{15})$
$= 42x^{54} + 60x^{15} + 18x^{61} + 180x^{19}$

36. $20x^{38} - 50x^{25} + 25x^{14}$

37. $(x + 6)(x + 3) = (x + 6)x + (x + 6)3$
$= x \cdot x + 6 \cdot x + x \cdot 3 + 6 \cdot 3$
$= x^2 + 6x + 3x + 18$
$= x^2 + 9x + 18$

38. $x^2 + 7x + 10$

39. $(x + 5)(x - 2) = (x + 5)x + (x + 5)(-2)$
$= x \cdot x + 5 \cdot x + x(-2) + 5(-2)$
$= x^2 + 5x - 2x - 10$
$= x^2 + 3x - 10$

40. $x^2 + 4x - 12$

41. $(x - 4)(x - 3) = (x - 4)x + (x - 4)(-3)$
$= x \cdot x - 4 \cdot x + x(-3) - 4(-3)$
$= x^2 - 4x - 3x + 12$
$= x^2 - 7x + 12$

42. $x^2 - 10x + 21$

43. $(x + 3)(x - 3) = (x + 3)x + (x + 3)(-3)$
$= x \cdot x + 3 \cdot x + x(-3) + 3(-3)$
$= x^2 + 3x - 3x - 9$
$= x^2 - 9$

44. $x^2 - 36$

45. $(5 - x)(5 - 2x) = (5 - x)5 + (5 - x)(-2x)$
$= 5 \cdot 5 - x \cdot 5 + 5(-2x) - x(-2x)$
$= 25 - 5x - 10x + 2x^2$
$= 25 - 15x + 2x^2$

46. $18 + 12x + 2x^2$

47. $(2x + 5)(2x + 5) = (2x + 5)2x + (2x + 5)5$
$= 2x \cdot 2x + 5 \cdot 2x + 2x \cdot 5 + 5 \cdot 5$
$= 4x^2 + 10x + 10x + 25$
$= 4x^2 + 20x + 25$

48. $9x^2 - 24x + 16$

49. $(3y - 4)(3y + 4) = (3y - 4)3y + (3y - 4)4$
$= 3y \cdot 3y - 4 \cdot 3y + 3y \cdot 4 - 4 \cdot 4$
$= 9y^2 - 12y + 12y - 16$
$= 9y^2 - 16$

50. $4y^2 - 1$

51. $\left(x - \frac{5}{2}\right)\left(x + \frac{2}{5}\right) = \left(x - \frac{5}{2}\right)x + \left(x - \frac{5}{2}\right)\frac{2}{5}$
$= x \cdot x - \frac{5}{2} \cdot x + x \cdot \frac{2}{5} - \frac{5}{2} \cdot \frac{2}{5}$
$= x^2 - \frac{5}{2}x + \frac{2}{5}x - 1$
$= x^2 - \frac{25}{10}x + \frac{4}{10}x - 1$
$= x^2 - \frac{21}{10}x - 1$

52. $x^2 + \frac{17}{6}x + 2$

53. $(x^2 + x + 1)(x - 1)$
$= (x^2 + x + 1)x + (x^2 + x + 1)(-1)$
$= x^2 \cdot x + x \cdot x + 1 \cdot x + x^2(-1) + x(-1) + 1(-1)$
$= x^3 + x^2 + x - x^2 - x - 1$
$= x^3 - 1$

54. $x^3 + x^2 + 4$

55. $(2x + 1)(2x^2 + 6x + 1)$
$= 2x(2x^2 + 6x + 1) + 1(2x^2 + 6x + 1)$
$= 2x \cdot 2x^2 + 2x \cdot 6x + 2x \cdot 1 + 1 \cdot 2x^2 + 1 \cdot 6x + 1 \cdot 1$
$= 4x^3 + 12x^2 + 2x + 2x^2 + 6x + 1$
$= 4x^3 + 14x^2 + 8x + 1$

56. $12x^3 - 10x^2 - x + 1$

57. $(y^2 - 3)(3y^2 - 6y + 2)$
$= y^2(3y^2 - 6y + 2) - 3(3y^2 - 6y + 2)$
$= y^2 \cdot 3y^2 + y^2(-6y) + y^2 \cdot 2 - 3 \cdot 3y^2 - 3(-6y) - 3 \cdot 2$
$= 3y^4 - 6y^3 + 2y^2 - 9y^2 + 18y - 6$
$= 3y^4 - 6y^3 - 7y^2 + 18y - 6$

58. $3y^4 + 18y^3 - 18y - 3$

59. $(x^3 + x^2)(x^3 + x^2 - x)$
$= x^3(x^3 + x^2 - x) + x^2(x^3 + x^2 - x)$
$= x^3 \cdot x^3 + x^3 \cdot x^2 + x^3(-x) + x^2 \cdot x^3 + x^2 \cdot x^2 + x^2(-x)$
$= x^6 + x^5 - x^4 + x^5 + x^4 - x^3$
$= x^6 + 2x^5 - x^3$

60. $x^6 - 2x^5 + 2x^4 - x^3$

61. $(-5x^3 - 7x^2 + 1)(2x^2 - x)$
$= (-5x^3 - 7x^2 + 1)2x^2 + (-5x^3 - 7x^2 + 1)(-x)$
$= -5x^3 \cdot 2x^2 - 7x^2 \cdot 2x^2 + 1 \cdot 2x^2 - 5x^3(-x) - 7x^2(-x) + 1(-x)$
$= -10x^5 - 14x^4 + 2x^2 + 5x^4 + 7x^3 - x$
$= -10x^5 - 9x^4 + 7x^3 + 2x^2 - x$

62. $-20x^5 + 25x^4 - 4x^3 - 5x^2 - 2$

63.
$$\begin{array}{rl}
1 + x + x^2 & \\
-1 - x + x^2 & \\
\hline
x^2 + x^3 + x^4 & \text{Multiplying the top row by } x^2 \\
-x - x^2 - x^3 & \text{Multiplying by } -x \\
-1 - x - x^2 & \text{Multiplying by } -1 \\
\hline
-1 - 2x - x^2 \qquad + x^4 & \text{Collecting like terms}
\end{array}$$
This result can also be written in descending order: $x^4 - x^2 - 2x - 1$

64. $x^4 - 2x^3 + 3x^2 - 2x + 1$

65.
$$\begin{array}{rl}
2x^2 + 3x - 4 & \\
2x^2 + x - 2 & \\
\hline
-4x^2 - 6x + 8 & \text{Multiplying by } -2 \\
2x^3 + 3x^2 - 4x & \text{Multiplying by } x \\
4x^4 + 6x^3 - 8x^2 & \text{Multiplying by } 2x^2 \\
\hline
4x^4 + 8x^3 - 9x^2 - 10x + 8 & \text{Collecting like terms}
\end{array}$$

66. $4x^4 - 12x^3 - 5x^2 + 17x + 6$

67. We will multiply horizontally while still aligning like terms.
$(x + 1)(x^3 + 7x^2 + 5x + 4)$
$= x^4 + 7x^3 + 5x^2 + 4x$ Multiplying by x
$+ x^3 + 7x^2 + 5x + 4$ Multiplying by 1
$= x^4 + 8x^3 + 12x^2 + 9x + 4$

68. $x^4 + 7x^3 + 19x^2 + 21x + 6$

69.
$$\begin{array}{rl}
2x^2 + x - 2 & \\
-2x^2 + 4x - 5 & \\
\hline
-10x^2 - 5x + 10 & \text{Multiplying by } -5 \\
8x^3 + 4x^2 - 8x & \text{Multiplying by } 4x \\
-4x^4 - 2x^3 + 4x^2 & \text{Multiplying by } -2x^2 \\
\hline
-4x^4 + 6x^3 - 2x^2 - 13x + 10 &
\end{array}$$

70. $-6x^4 + 4x^3 + 36x^2 - 20x + 2$

71. We will multiply horizontally, while still aligning like terms.
$(2x + 1)(x^3 - 4x^2 + 3x - 2)$
$= 2x^4 - 8x^3 + 6x^2 - 4x$
$+ x^3 - 4x^2 + 3x - 2$
$= 2x^4 - 7x^3 + 2x^2 - x - 2$

72. $4x^4 - 5x^3 + 14x^2 + 11x - 3$

73.
$$\begin{array}{r}
x^3 + x^2 + x + 1 \\
x - 1 \\
\hline
-x^3 - x^2 - x - 1 \\
x^4 + x^3 + x^2 + x \\
\hline
x^4 \qquad\qquad - 1
\end{array}$$

74. $x^4 - 3x^3 + 3x^2 - 4x + 4$

75.
$$\begin{array}{r}
x^3 + x^2 - x - 3 \\
x - 3 \\
\hline
-3x^3 - 3x^2 + 3x + 9 \\
x^4 + x^3 - x^2 - 3x \\
\hline
x^4 - 2x^3 - 4x^2 \qquad + 9
\end{array}$$

76. $x^4 + 3x^3 - 5x^2 + 16$

77. $-\frac{1}{4} - \frac{1}{2} = -\frac{1}{4} - \frac{1}{2} \cdot \frac{2}{2} = -\frac{1}{4} - \frac{2}{4} = -\frac{3}{4}$

78. $4(4x - 6y + 9)$

79. 

80.

81. 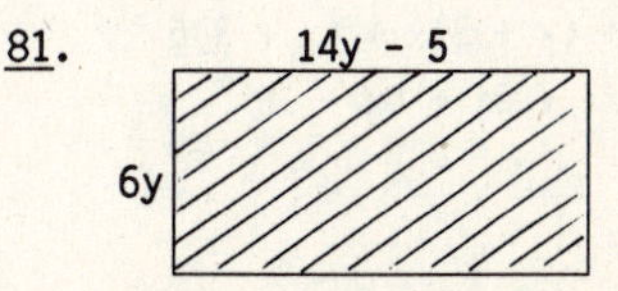

The shaded area is the product of the length and width of the rectangle:
$6y(14y - 5) = 6y \cdot 14y + 6y(-5)$
$= 84y^2 - 30y$

82. $78t^2 + 40t$

83.

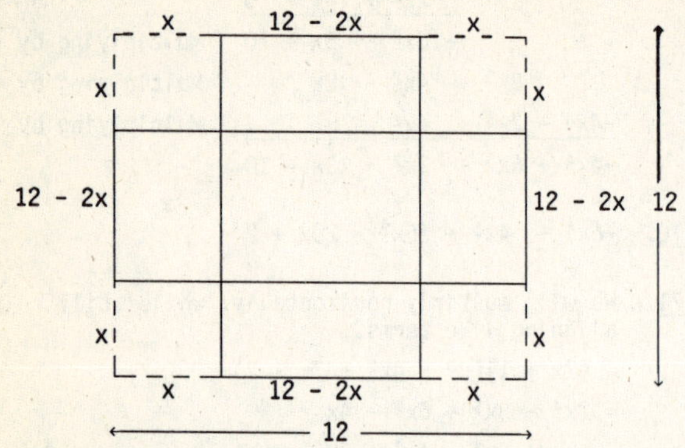

The dimensions of the box are 12 - 2x by 12 - 2x by x. The volume is the product of the dimensions (volume = length × width × height):

$$\text{Volume} = (12 - 2x)(12 - 2x)x$$
$$= (144 - 48x + 4x^2)x$$
$$= 144x - 48x^2 + 4x^3, \text{ or } 4x^3 - 48x^2 + 144x$$

The outside surface area is the sum of the area of the bottom and the areas of the four sides. The dimensions of the bottom are 12 - 2x by 12 - 2x, and the dimensions of each side are x by 12 - 2x.

$$\text{Surface area} = \text{Area of bottom} + 4\cdot\text{Area of each side}$$
$$= (12 - 2x)(12 - 2x) + 4\cdot x(12 - 2x)$$
$$= 144 - 48x + 4x^2 + 48x - 8x^2$$
$$= 144 - 4x^2, \text{ or } -4x^2 + 144$$

84. $x^3 - 5x^2 + 8x - 4$

85. Let b = the length of the base. Then b + 4 = the height. Let A represent the area.

$$\text{Area} = \frac{1}{2} \times \text{base} \times \text{height}$$
$$A = \frac{1}{2} \cdot b \cdot (b + 4)$$
$$A = \frac{1}{2}b(b + 4)$$
$$A = \frac{1}{2}b^2 + 2b$$

86. 8 ft by 16 ft

87. $(x + 3)(x + 6) + (x + 3)(x + 6)$
$= (x + 3)x + (x + 3)6 + (x + 3)x + (x + 3)6$
$= x^2 + 3x + 6x + 18 + x^2 + 3x + 6x + 18$
$= 2x^2 + 18x + 36$

88. $2x^2 - 18x + 28$

89. $(x + 5)^2 - (x - 3)^2$
$= (x + 5)(x + 5) - (x - 3)(x - 3)$
$= (x + 5)x + (x + 5)5 - [(x - 3)x - (x - 3)3]$
$= x^2 + 5x + 5x + 25 - (x^2 - 3x - 3x + 9)$
$= x^2 + 10x + 25 - (x^2 - 6x + 9)$
$= x^2 + 10x + 25 - x^2 + 6x - 9$
$= 16x + 16$

90. $2x^2 - 20x + 52$

Exercise Set 4.5

1. $(x + 1)(x^2 + 3)$
F O I L
$= x\cdot x^2 + x\cdot 3 + 1\cdot x^2 + 1\cdot 3$
$= x^3 + 3x + x^2 + 3$, or $x^3 + x^2 + 3x + 3$

2. $x^3 - x^2 - 3x + 3$

3. $(x^3 + 2)(x + 1)$
F O I L
$= x^3\cdot x + x^3\cdot 1 + 2\cdot x + 2\cdot 1$
$= x^4 + x^3 + 2x + 2$

4. $x^5 + 12x^4 + 2x + 24$

5. $(y + 2)(y - 3)$
F O I L
$= y\cdot y + y\cdot(-3) + 2\cdot y + 2\cdot(-3)$
$= y^2 - 3y + 2y - 6$
$= y^2 - y - 6$

6. $a^2 + 4a + 4$

7. $(3x + 2)(3x + 3)$
F O I L
$= 3x\cdot 3x + 3x\cdot 3 + 2\cdot 3x + 2\cdot 3$
$= 9x^2 + 9x + 6x + 6$
$= 9x^2 + 15x + 6$

8. $8x^2 + 10x + 2$

9. $(5x - 6)(x + 2)$
F O I L
$= 5x\cdot x + 5x\cdot 2 + (-6)\cdot x + (-6)\cdot 2$
$= 5x^2 + 10x - 6x - 12$
$= 5x^2 + 4x - 12$

10. $x^2 - 64$

11. $(3t - 1)(3t + 1)$
F O I L
$= 3t\cdot 3t + 3t\cdot 1 + (-1)\cdot 3t + (-1)\cdot 1$
$= 9t^2 + 3t - 3t - 1$
$= 9t^2 - 1$

12. $4m^2 + 12m + 9$

13. $(4x - 2)(x - 1)$

F O I L

$= 4x\cdot x + 4x\cdot(-1) + (-2)\cdot x + (-2)\cdot(-1)$

$= 4x^2 - 4x - 2x + 2$

$= 4x^2 - 6x + 2$

14. $6x^2 - x - 1$

15. $\left(p - \frac{1}{4}\right)\left(p + \frac{1}{4}\right)$

F O I L

$= p\cdot p + p\cdot\frac{1}{4} + \left(-\frac{1}{4}\right)\cdot p + \left(-\frac{1}{4}\right)\cdot\frac{1}{4}$

$= p^2 + \frac{1}{4}p - \frac{1}{4}p - \frac{1}{16}$

$= p^2 - \frac{1}{16}$

16. $q^2 + \frac{3}{2}q + \frac{9}{16}$

17. $(x - 0.1)(x + 0.1)$

F O I L

$= x\cdot x + x\cdot(0.1) + (-0.1)\cdot x + (-0.1)(0.1)$

$= x^2 + 0.1x - 0.1x - 0.01$

$= x^2 - 0.01$

18. $x^2 - 0.1x - 0.12$

19. $(2x^2 + 6)(x + 1)$

F O I L

$= 2x^3 + 2x^2 + 6x + 6$

20. $4x^3 - 2x^2 + 6x - 3$

21. $(-2x + 1)(x + 6)$

F O I L

$= -2x^2 - 12x + x + 6$

$= -2x^2 - 11x + 6$

22. $6x^2 - 4x - 16$

23. $(a + 7)(a + 7)$

F O I L

$= a^2 + 7a + 7a + 49$

$= a^2 + 14a + 49$

24. $4y^2 + 20y + 25$

25. $(1 + 2x)(1 - 3x)$

F O I L

$= 1 - 3x + 2x - 6x^2$

$= 1 - x - 6x^2$

26. $-3x^2 - 5x - 2$

27. $(x^2 + 3)(x^3 - 1)$

F O I L

$= x^5 - x^2 + 3x^3 - 3$, or $x^5 + 3x^3 - x^2 - 3$

28. $2x^5 + x^4 - 6x - 3$

29. $(3x^2 - 2)(x^4 - 2)$

F O I L

$= 3x^6 - 6x^2 - 2x^4 + 4$, or $3x^6 - 2x^4 - 6x^2 + 4$

30. $x^{20} - 9$

31. $(3x^5 + 2)(2x^2 + 6)$

F O I L

$= 6x^7 + 18x^5 + 4x^2 + 12$

32. $1 + 3x^2 - 2x - 6x^3$, or $-6x^3 + 3x^2 - 2x + 1$

33. $(8x^3 + 1)(x^3 + 8)$

$= 8x^6 + 64x^3 + x^3 + 8$

$= 8x^6 + 65x^3 + 8$

34. $20 - 8x^2 - 10x + 4x^3$, or $4x^3 - 8x^2 - 10x + 20$

35. $(4x^2 + 3)(x - 3)$

F O I L

$= 4x^3 - 12x^2 + 3x - 9$

36. $14x^2 - 53x + 14$

37. $(4y^4 + y^2)(y^2 + y)$

F O I L

$= 4y^6 + 4y^5 + y^4 + y^3$

38. $10y^{12} + 16y^9 + 6y^6$

39. $(x + 4)(x - 4) = x^2 - 4^2$ Product of sum and difference of the same two terms

$= x^2 - 16$

40. $x^2 - 1$

41. $(2x + 1)(2x - 1) = (2x)^2 - 1^2$ Product of sum and difference of the same two terms

$= 4x^2 - 1$

42. $x^4 - 1$

43. $(5m - 2)(5m + 2) = (5m)^2 - 2^2$ Product of sum and difference of the same two terms

$= 25m^2 - 4$

44. $9x^8 - 4$

45. $(2x^2 + 3)(2x^2 - 3) = (2x^2)^2 - 3^2$ Product of sum and difference of the same two terms
$= 4x^4 - 9$

46. $36x^{10} - 25$

47. $(3x^4 - 4)(3x^4 + 4) = (3x^4)^2 - 4^2$
$= 9x^8 - 16$

48. $t^4 - 0.04$

49. $(x^6 - x^2)(x^6 + x^2) = (x^6)^2 - (x^2)^2$
$= x^{12} - x^4$

50. $4x^6 - 0.09$

51. $(x^4 + 3x)(x^4 - 3x) = (x^4)^2 - (3x)^2$
$= x^8 - 9x^2$

52. $\frac{9}{16} - 4x^6$

53. $(x^{12} - 3)(x^{12} + 3) = (x^{12})^2 - 3^2$
$= x^{24} - 9$

54. $144 - 9x^4$

55. $(2y^8 + 3)(2y^8 - 3) = (2y^8)^2 - 3^2$
$= 4y^{16} - 9$

56. $m^2 - \frac{4}{9}$

57. $(x + 2)^2 = x^2 + 2\cdot x\cdot 2 + 2^2$ Square of a binomial
$= x^2 + 4x + 4$

58. $4x^2 - 4x + 1$

59. $(3x^2 + 1) = (3x^2)^2 + 2\cdot 3x^2\cdot 1 + 1^2$ Square of a binomial
$= 9x^4 + 6x^2 + 1$

60. $9x^2 + \frac{9}{2}x + \frac{9}{16}$

61. $\left(a - \frac{1}{2}\right)^2 = a^2 - 2\cdot a \cdot \frac{1}{2} + \left(\frac{1}{2}\right)^2$ Square of a binomial
$= a^2 - a + \frac{1}{4}$

62. $4a^2 - \frac{4}{5}a + \frac{1}{25}$

63. $(3 + x)^2 = 3^2 + 2\cdot 3\cdot x + x^2$
$= 9 + 6x + x^2$

64. $x^6 - 2x^3 + 1$

65. $(x^2 + 1)^2 = (x^2)^2 + 2\cdot x^2\cdot 1 + 1^2$
$= x^4 + 2x^2 + 1$

66. $64x^2 - 16x^3 + x^4$

67. $(2 - 3x^4)^2 = 2^2 - 2\cdot 2\cdot 3x^4 + (3x^4)^2$
$= 4 - 12x^4 + 9x^8$

68. $36x^6 - 24x^3 + 4$

69. $(5 + 6t^2)^2 = 5^2 + 2\cdot 5\cdot 6t^2 + (6t^2)^2$
$= 25 + 60t^2 + 36t^4$

70. $9p^4 - 6p^3 + p^2$

71. $(7x - 0.3)^2 = (7x)^2 - 2(7x)(0.3) + (0.3)^2$
$= 49x^2 - 4.2x + 0.09$

72. $16a^2 - 4.8a + 0.36$

73. $5a^3(2a^2 - 1)$
$= 5a^3\cdot 2a^2 - 5a^3\cdot 1$ Multiplying each term of the binomial by the monomial
$= 10a^5 - 5a^3$

74. $a^3 - a^2 - 10a + 12$

75. $(x^2 - 5)(x^2 + x - 1)$
$= x^4 + x^3 - x^2$ Multiplying horizontally and aligning like terms
$\quad - 5x^2 - 5x + 5$
$= x^4 + x^3 - 6x^2 - 5x + 5$

76. $27x^6 - 9x^5$

77. $(3 - 2x^3)^2 = 3^2 - 2\cdot 3\cdot 2x^3 + (2x^3)^2$ Squaring a binomial
$= 9 - 12x^3 + 4x^6$

78. $x^2 - 8x^4 + 16x^6$

79. $4x(x^2 + 6x - 3)$
$= 4x\cdot x^2 + 4x\cdot 6x + 4x(-3)$ Multiplying each term of the trinomial by the monomial
$= 4x^3 + 24x^2 - 12x$

80. $-8x^6 + 48x^3 + 72x$

81. $\left(2x^2 - \frac{1}{2}\right)\left(2x^2 - \frac{1}{2}\right)$ Squaring a binomial
$= (2x^2)^2 - 2\cdot 2x^2 \cdot \frac{1}{2} + \left(\frac{1}{2}\right)^2$
$= 4x^4 - 2x^2 + \frac{1}{4}$

82. $x^4 - 2x^2 + 1$

83. $(-1 + 3p)(1 + 3p)$
$= (3p - 1)(3p + 1)$ Product of the sum and difference of the same two terms
$= (3p)^2 - 1^2$
$= 9p^2 - 1$, or $-1 + 9p^2$

84. $-9q^2 + 4$

85. $3t^2(5t^3 - t^2 + t)$

$= 3t^2 \cdot 5t^3 + 3t^2(-t^2) + 3t^2 \cdot t$ Multiplying each term of the trinomial by the monomial

$= 15t^5 - 3t^4 + 3t^3$

86. $-6x^5 - 48x^3 + 54x^2$

87. $(6x^4 + 4)^2$ Squaring a binomial

$= (6x^4)^2 + 2 \cdot 6x^4 \cdot 4 + 4^2$

$= 36x^8 + 48x^4 + 16$

88. $64a^2 + 80a + 25$

89. $(3x + 2)(4x^2 + 5)$ Product of two binomials; use FOIL

$= 3x \cdot 4x^2 + 3x \cdot 5 + 2 \cdot 4x^2 + 2 \cdot 5$

$= 12x^3 + 15x + 8x^2 + 10$, or

$12x^3 + 8x^2 + 15x + 10$

90. $6x^4 - 3x^2 - 63$

91. $(8 - 6x^4)^2$ Squaring a binomial

$= 8^2 - 2 \cdot 8 \cdot 6x^4 + (6x^4)^2$

$= 64 - 96x^4 + 36x^8$

92. $\frac{3}{25}x^4 + 4x^2 - 63$

93. Using columns to multiply a binomial and a trinomial

$$\begin{array}{r} t^2 + t + 1 \\ t - 1 \\ \hline -t^2 - t - 1 \\ t^3 + t^2 + t \quad\quad \\ \hline t^3 \quad\quad\quad\quad - 1 \end{array}$$

94. $y^3 + 125$

95. $3^2 + 4^2 = 9 + 16 = 25$

$(3 + 4)^2 = 7^2 = 49$

96. 85; 169

97. $9^2 - 5^2 = 81 - 25 = 56$

$(9 - 5)^2 = 4^2 = 16$

98. 105; 49

99.

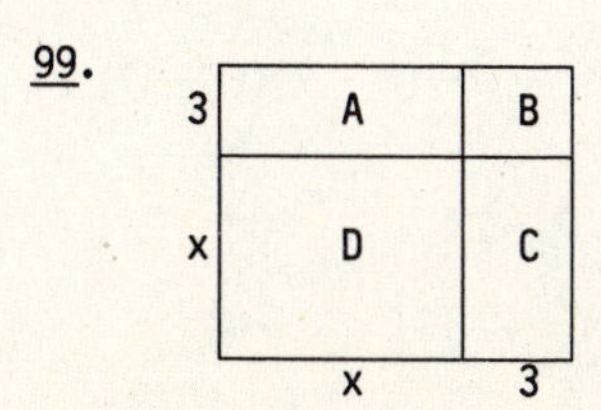

We can find the shaded area in two ways.

Method 1: The figure is a square with side $x + 3$, so the area is $(x + 3)^2 = x^2 + 6x + 9$.

Method 2: We add the areas of A, B, C, and D.

$3 \cdot x + 3 \cdot 3 + 3 \cdot x + x \cdot x = 3x + 9 + 3x + x^2 =$

$x^2 + 6x + 9$.

Either way we find that the total shaded area is $x^2 + 6x + 9$.

100. $a^2 + 2a + 1$

101.

3 A B

t D C

t 4

We can find the shaded area in two ways.

Method 1: The figure is a rectangle with dimensions $t + 4$ by $t + 3$, so the area is $(t + 4)(t + 3) = t^2 + 3t + 4t + 12 = t^2 + 7t + 12$.

Method 2: We add the areas of A, B, C, and D. $3 \cdot t + 3 \cdot 4 + 4 \cdot t + t \cdot t = 3t + 12 + 4t + t^2 = t^2 + 7t + 12$.

Either way, we find that the area is $t^2 + 7t + 12$.

102. $x^2 + 7x + 10$

103.

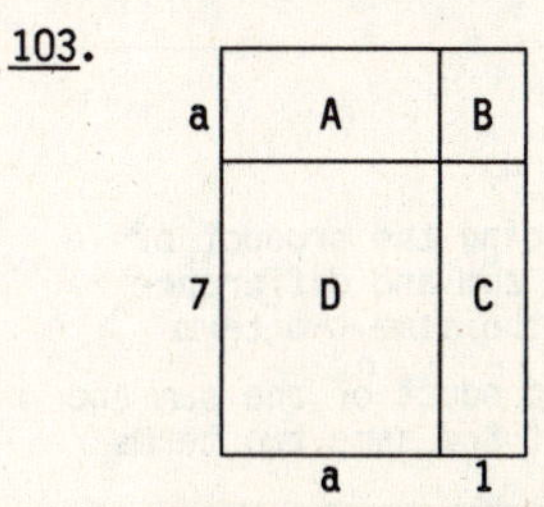

We can find the shaded area in two ways.

Method 1: The figure is a rectangle with dimensions $a + 7$ by $a + 1$, so the area is $(a + 7)(a + 1) = a^2 + a + 7a + 7 = a^2 + 8a + 7$.

Method 2: We add the areas of A, B, C, and D. $a \cdot a + a \cdot 1 + 7 \cdot 1 + 7 \cdot a = a^2 + a + 7 + 7a = a^2 + 8a + 7$

Either way, we find that the area is $a^2 + 8a + 7$.

104. $t^2 + 13t + 36$

105. Familiarize. Let t = the number of watts used by the television set. Then $10t$ = the number of watts used by the lamps, and $40t$ = the number of watts used by the air conditioner.

Translate.

Lamp watts	+	Air conditioner watts	+	Television watts	=	Total watts
$10t$	+	$40t$	+	t	=	2550

<u>Carry out</u>. We solve the equation.

$$10t + 40t + t = 2550$$
$$51t = 2550$$
$$t = 50$$

The possible solution is:

Television, t: 50 watts

Lamps, 10t: 10·50, or 500 watts

Air conditioner, 40t: 40·50, or 2000 watts

<u>Check</u>. The number of watts used by the lamps, 500, is 10 times 50, the number used by the television. The number of watts used by the air conditioner, 2000, is 40 times 50, the number used by the television. Also, 50 + 500 + 2000 = 2550, the total wattage used.

<u>State</u>. The television uses 50 watts, the lamps use 500 watts, and the air conditioner uses 2000 watts.

<u>106</u>. $\frac{28}{27}$

<u>107</u>. 

<u>108</u>.

<u>109</u>. $4y(y + 5)(2y + 8)$

$= 4y(2y^2 + 8y + 10y + 40)$

$= 4y(2y^2 + 18y + 40)$

$= 8y^3 + 72y^2 + 160y$

<u>110</u>. $80x^3 + 24x^2 - 216x$

<u>111</u>. $[(3x - 2)(3x + 2)](9x^2 + 4)$

$= (9x^2 - 4)(9x^2 + 4)$ Finding the product of the sum and difference of the same two terms

$= 81x^4 - 16$ Finding the product of the sum and difference of the same two terms again

<u>112</u>. $16x^4 - 1$

<u>113</u>. $(5t^3 - 3)^2(5t^3 + 3)^2$

$= [(5t^3 - 3)(5t^3 + 3)][(5t^3 - 3)(5t^3 + 3)]$

$= (25t^6 - 9)(25t^6 - 9)$

$= (25t^6)^2 - 2\cdot 25t^6\cdot 9 + 9^2$

$= 625t^{12} - 450t^6 + 81$

<u>114</u>. $5a^2 + 12a - 9$

<u>115</u>. $(67.58x + 3.225)^2$

$= (67.58x)^2 + 2(67.58x)(3.225) + (3.225)^2$

$= 4567.0564x^2 + 435.891x + 10.400625$

<u>116</u>. $x^2 - 8x + 16$

<u>117</u>. $18 \times 22 = (20 - 2)(20 + 2) = 400 - 4 = 396$

<u>118</u>. 9951

<u>119</u>. $(x + 2)(x - 5) = (x + 1)(x - 3)$

$x^2 - 5x + 2x - 10 = x^2 - 3x + x - 3$

$x^2 - 3x - 10 = x^2 - 2x - 3$

$-3x - 10 = -2x - 3$ Adding $-x^2$

$-3x + 2x = 10 - 3$ Adding 2x and 10

$-x = 7$

$x = -7$

<u>120</u>. 0

<u>121</u>. If w = the width, then w + 1 = the length, and (w + 1) + 1, or w + 2 = the height.

Volume = length × width × height

$= (w + 1) \cdot w \cdot (w + 2)$

$= (w^2 + w)(w + 2)$

$= w^3 + 2w^2 + w^2 + 2w$

$= w^3 + 3w^2 + 2w$

<u>122</u>. $\ell^3 - \ell$

<u>123</u>. If h = the height, then h - 1 = the length and (h - 1) - 1, or h - 2 = the width.

Volume = length × width × height

$= (h - 1) \cdot (h - 2) \cdot h$

$= (h^2 - 2h - h + 2)h$

$= (h^2 - 3h + 2)h$

$= h^3 - 3h^2 + 2h$

<u>124</u>. $Q(Q - 14) - 5(Q - 14)$, or $(Q - 5)(Q - 14)$

<u>125</u>.

F
7
F
F - 7
17
F - 17

The area of the entire figure is F·F, or F^2.

The area of the section not shaded is (F - 7)(F - 17)

Area of shaded region = Area of entire figure - Area of section not shaded

Area of shaded region = $F^2 - (F - 7)(F - 17)$

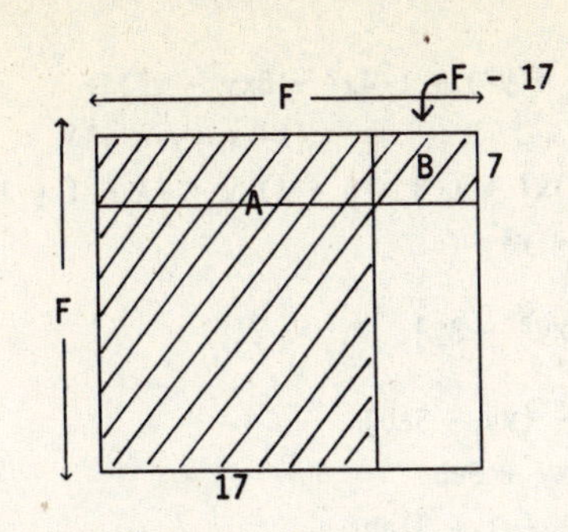

Area of shaded region = Area of A + Area of B

Area of shaded region = 17F + 7(F - 17)

= 17F + 7F - 119

= 24F - 119

126. (y + 1)(y - 1), y(y + 1) - y - 1

127. a)

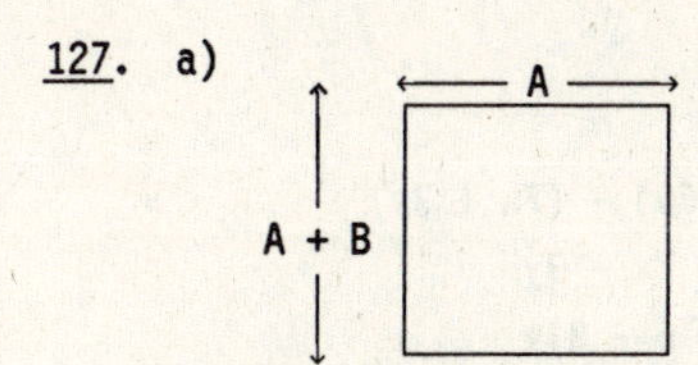

The area of the entire rectangle is $A(A + B)$, or $A^2 + AB$.

b)

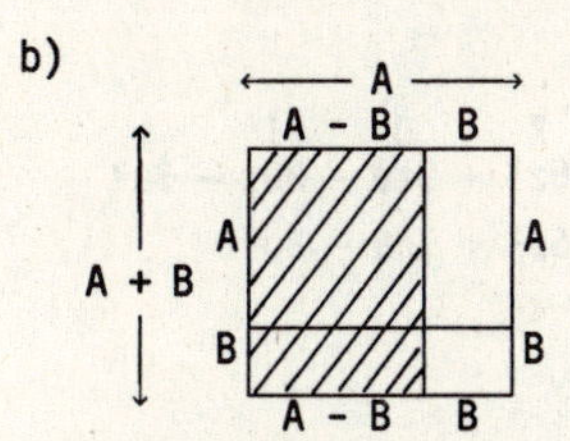

The sum of the areas of the two unshaded rectangles is $A \cdot B + B \cdot B$, or $AB + B^2$.

c) Area in part (a) - area in part (b)

$= A^2 + AB - (AB + B^2)$

$= A^2 + AB - AB - B^2$

$= A^2 - B^2$

d) The area of the shaded region is $(A + B)(A - B) = A^2 - B^2$. This is the same as the polynomial found in part (c).

128. 10, 11, and 12

129. $(10x + 5)^2 = (10x)^2 + 2 \cdot 10x \cdot 5 + 5^2$

$= 100x^2 + 100x + 25$

$= 100(x^2 + x) + 25$

To square any two-digit number ending in 5 mentally, add the first digit to its square, multiply by 100, and add 25.

Exercise Set 4.6

1. We replace x by 3 and y by -2.

$x^2 - y^2 + xy = 3^2 - (-2)^2 + 3(-2) = 9 - 4 - 6 = -1$

2. 19

3. We replace x by 2, y by -3, and z by -1.

$xyz^2 + z = 2(-3)(-1)^2 + (-1) = -6 - 1 = -7$

4. -1

5. Evaluate the polynomial for P = 10,000 and i = 0.08.

$A = P(1 + i)^2 = 10{,}000(1 + 0.08)^2$

$= 10{,}000(1.08)^2$

$= 10{,}000(1.1664)$

$= 11{,}664$

At 8% interest for 2 years, $10,000 will grow to $11,664.

6. $11,449

7. Evaluate the polynomial for P = 10,000 and i = 0.08.

$A = P(1 + i)^3 = 10{,}000(1 + 0.08)^3$

$= 10{,}000(1.08)^3$

$= 10{,}000(1.259712)$

$= 12{,}597.12$

At 8% interest for 3 years, $10,000 will grow to $12,597.12.

8. $12,250.43

9. Evaluate the polynomial for h = 4, r = 3/4, and $\pi \approx 3.14$.

$2\pi rh + 3\pi r^2 \approx 2(3.14)\left(\frac{3}{4}\right)(4) + 3(3.14)\left(\frac{3}{4}\right)^2$

$= 2(3.14)\left(\frac{3}{4}\right)(4) + 3(3.14)\left(\frac{9}{16}\right)$

$= 18.84 + 5.29875$

$= 24.13875$

The surface area is about 24.13875 in^2

10. 73.59375 in^2

11. $x^3y - 2xy + 3x^2 - 5$

Term	Coefficient	Degree	
x^3y	1	4	(Think: $x^3y = x^3y^1$)
$-2xy$	-2	2	(Think: $-2xy = -2x^1y^1$)
$3x^2$	3	2	
-5	-5	0	(Think: $-5 = -5x^0$)

The degree of a polynomial is the degree of the term of highest degree.

The term of highest degree is x^3y. Its degree is 4. The degree of the polynomial is 4.

12. Coefficients: 5, -1, 15, 1
Degrees: 3, 2, 1, 0; 3

13. $17x^2y^3 - 3x^3yz - 7$

Term	Coefficient	Degree
$17x^2y^3$	17	5
$-3x^3yz$	-3	5 (Think: $-3x^3yz = 3x^3y^1z^1$)
-7	-7	0 (Think: $-7 = -7x^0$)

The terms of highest degree are $17x^2y^3$ and $-3x^3yz$. Each has degree 5. The degree of the polynomial is 5.

14. Coefficients: 6, -1, 8, -1
Degrees: 0, 2, 4, 5; 5

15. $a + b - 2a - 3b = (1 - 2)a + (1 - 3)b = -a - 2b$

16. $y - 7$

17. $3x^2y - 2xy^2 + x^2$
There are no like terms, so none of the terms can be collected.

18. $m^3 + 2m^2n - 3m^2 + 3mn^2$

19. $2u^2v - 3uv^2 + 6u^2v - 2uv^2$
$= (2 + 6)u^2v + (-3 - 2)uv^2$
$= 8u^2v - 5uv^2$

20. $-2x^2 - 4xy - 2y^2$

21. $6au + 3av + 14au + 7av$
$= (6 + 14)au + (3 + 7)av$
$= 20au + 10av$

22. $3x^2y + 3z^2y + 3xy^2$

23. $(2x^2 - xy + y^2) + (-x^2 - 3xy + 2y^2)$
$= (2 - 1)x^2 + (-1 - 3)xy + (1 + 2)y^2$
$= x^2 - 4xy + 3y^2$

24. $6 - z$

25. $(r^3 + 3rs - 5s^2) - (5r^3 + rs + 4s^2)$
$= r^3 + 3rs - 5s^2 - 5r^3 - rs - 4s^2$ Adding the opposite
$= (1 - 5)r^3 + (3 - 1)rs + (-5 - 4)s^2$
$= -4r^3 + 2rs - 9s^2$

26. $-2a^4 - 8ab + 7ab^2$

27. $(r - 2s + 3) + (2r + 3s - 7)$
$= (1 + 2)r + (-2 + 3)s + (3 - 7)$
$= 3r + s - 4$

28. $-3b^3a^2 - b^2a^3 + 5ba + 3$

29. $(2x^2 - 3xy + y^2) + (-4x^2 - 6xy - y^2) + (x^2 + xy - y^2)$
$= (2 - 4 + 1)x^2 + (-3 - 6 + 1)xy + (1 - 1 - 1)y^2$
$= -x^2 - 8xy - y^2$

30. $3x^3 - x^2y + xy^2 - 3y^3$

31. $(xy - ab) - (xy - 3ab)$
$= xy - ab - xy + 3ab$
$= (1 - 1)xy + (-1 + 3)ab$
$= 0xy + 2ab$
$= 2ab$

32. $x^4y^2 + y + 2x$

33. $(-2a + 7b - c) + (-3b + 4c - 8d)$
$= -2a + (7 - 3)b + (-1 + 4)c - 8d$
$= -2a + 4b + 3c - 8d$

34. $15a^2b - 4ab$

35. $(4x + 5y) + (-5x + 6y) - (7x + 3y)$
$= 4x + 5y - 5x + 6y - 7x - 3y$
$= (4 - 5 - 7)x + (5 + 6 - 3)y$
$= -8x + 8y$

36. $-5b$

37. $(3z - u)(2z + 3u) = \overset{F}{6z^2} + \overset{O}{9zu} - \overset{I}{2uz} - \overset{L}{3u^2}$
$= 6z^2 + 7zu - 3u^2$

38. $a^3 + a^2b - ab^2 - b^3$

39. $(a^2b - 2)(a^2b - 5) = \overset{F}{a^4b^2} - \overset{O}{5a^2b} - \overset{I}{2a^2b} + \overset{L}{10}$
$= a^4b^2 - 7a^2b + 10$

40. $x^2y^2 + 3xy - 28$

41. $(a^3 + bc)(a^3 - bc) = (a^3) - (bc)^2$
$[(A + B)(A - B) = A^2 - B^2]$
$= a^6 - b^2c^2$

42. $m^4 + m^2n^2 + n^4$

43.
$$\begin{array}{r} y^4x + \phantom{y^4x + {}} y^2 + 1 \\ \underline{y^2 + 1} \\ y^4x + \phantom{y^4x + {}} y^2 + 1 \\ \underline{y^6x + y^4 \phantom{{}+ y^4x} + y^2 \phantom{{}+ 1}} \\ y^6x + y^4 + y^4x + 2y^2 + 1 \end{array}$$

44. $a^3 - b^3$

45. $(3xy - 1)(4xy + 2)$

F O I L

$= 12x^2y^2 + 6xy - 4xy - 2$

$= 12x^2y^2 + 2xy - 2$

46. $m^6n^2 + 2m^3n - 48$

47. $(3 - c^2d^2)(4 + c^2d^2)$

F O I L

$= 12 + 3c^2d^2 - 4c^2d^2 - c^4d^4$

$= 12 - c^2d^2 - c^4d^4$

48. $30x^2 - 28xy + 6y^2$

49. $(m^2 - n^2)(m + n)$

F O I L

$= m^3 + m^2n - mn^2 - n^3$

50. $0.4p^2q^2 - 0.02pq - 0.02$

51. $(xy + x^5y^5)(x^4y^4 - xy)$

F O I L

$= x^5y^5 - x^2y^2 + x^9y^9 - x^6y^6$

$= x^9y^9 - x^6y^6 + x^5y^5 - x^2y^2$

52. $x^2 + xy^3 - 2y^6$

53. $(x + h)^2$

$= x^2 + 2xh + h^2$ $[(A + B)^2 = A^2 + 2AB + B^2]$

54. $9a^2 + 12ab + 4b^2$

55. $(r^3t^2 - 4)^2$

$= (r^3t^2)^2 - 2\cdot r^3t^2\cdot 4 + 4^2$

$[(A - B)^2 = A^2 - 2AB + B^2]$

$= r^6t^4 - 8r^3t^2 + 16$

56. $9a^4b^2 - 6a^2b^3 + b^4$

57. $(p^4 + m^2n^2)^2$

$= (p^4)^2 + 2\cdot p^4\cdot m^2n^2 + (m^2n^2)^2$

$[(A + B)^2 = A^2 + 2AB + B^2]$

$= p^8 + 2p^4m^2n^2 + m^4n^4$

58. $a^2b^2 + 2abcd + c^2d^2$

59. $(2a - b)(2a + b) = (2a)^2 - b^2 = 4a^2 - b^2$

60. $x^2 - y^2$

61. $(c^2 - d)(c^2 + d) = (c^2)^2 - d^2$

$= c^4 - d^2$

62. $p^6 - 25q^2$

63. $(ab + cd^2)(ab - cd^2) = (ab)^2 - (cd^2)^2$

$= a^2b^2 - c^2d^4$

64. $x^2y^2 - p^2q^2$

65. $(x + y - 3)(x + y + 3)$

$= [(x + y) - 3][(x + y) + 3]$

$= (x + y)^2 - 3^2$

$= x^2 + 2xy + y^2 - 9$

66. $p^2 + 2pq + q^2 - 16$

67. $[x + y + z][x - (y + z)]$

$= [x + (y + z)][x - (y + z)]$

$= x^2 - (y + z)^2$

$= x^2 - (y^2 + 2yz + z^2)$

$= x^2 - y^2 - 2yz - z^2$

68. $a^2 - b^2 - 2bc - c^2$

69. $(a + b + c)(a - b - c)$

$= [a + (b + c)][a - (b + c)]$

$= a^2 - (b + c)^2$

$= a^2 - (b^2 + 2bc + c^2)$

$= a^2 - b^2 - 2bc - c^2$

70. $9x^2 + 12x + 4 - 25y^2$

71. The figure is a square with side $x + y$. Thus the area is $(x + y)^2 = x^2 + 2xy + y^2$.

72. $a^2 + ac + ab + bc$

73. The figure is a parallelogram with base $x + z$ and height $x - z$. Thus the area is $(x + z)(x - z) = x^2 - z^2$.

74. $\frac{1}{2}a^2b^2 - 2$

75. Locate December, 1989, on the horizontal scale. Then move up to the line representing white office paper and left to the vertical scale to read the information being sought. In December, 1989, the price being paid for white office paper was $60 per ton.

76. December, 1988

77. Locate the highest point on the line representing newsprint. Then move down to the horizontal scale to read the information being sought. The value of newsprint peaked in December, 1987.

78. December, 1990

79.

80.

81. It is helpful to add additional labels to the figure.

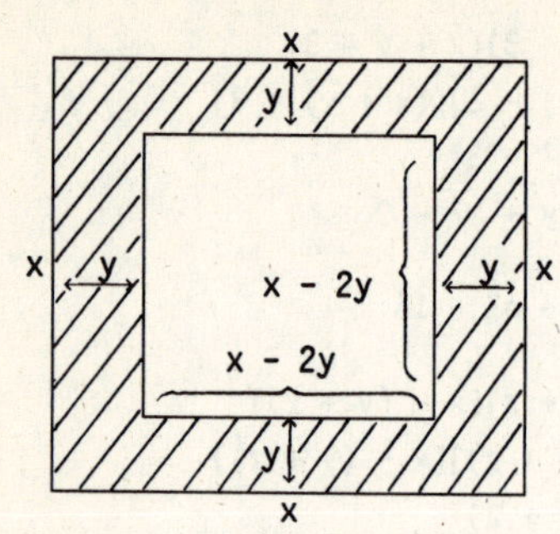

The area of the large square is $x \cdot x$, or x^2. The area of the small square is $(x - 2y)(x - 2y)$, or $(x - 2y)^2$.

$$\underline{\text{Area of shaded region}} = \underline{\text{Area of large square}} - \underline{\text{Area of small square}}$$

$$\text{Area of shaded region} = x^2 - (x - 2y)^2$$

$$= x^2 - (x^2 - 4xy + 4y^2)$$

$$= x^2 - x^2 + 4xy - 4y^2$$

$$= 4xy - 4y^2$$

82. $2\pi ab - \pi b^2$

83. It is helpful to add additional labels to the figure.

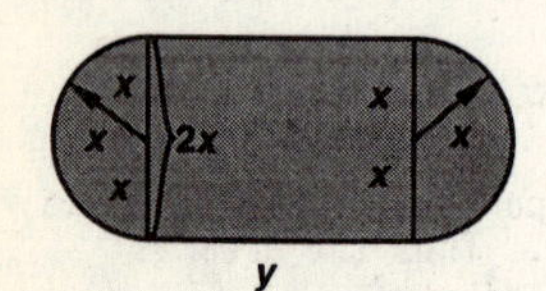

The two semicircles make a circle with radius x. The area of that circle is πx^2. The area of the rectangle is $2x \cdot y$. The sum of the two areas, $\pi x^2 + 2xy$, is the area of the shaded region.

84. $a^2 - 4b^2$

85. Evaluate the polynomial for h = 165 and A = 30:

$$0.041h - 0.018A - 2.69$$

$$= 0.041(165) - 0.018(30) - 2.69$$

$$= 6.765 - 0.54 - 2.69$$

$$= 3.535$$

The lung capacity is 3.535 liters.

86. 33

87. ◈

Exercise Set 4.7

1. $\dfrac{24x^4 - 4x^3}{8} = \dfrac{24x^4}{8} - \dfrac{4x^3}{8}$

$= \dfrac{24}{8}x^4 - \dfrac{4}{8}x^3$ Dividing coefficients

$= 3x^4 - \dfrac{1}{2}x^3$

To check, we multiply the quotient by 8:

$$\left(3x^4 - \frac{1}{2}x^3\right)8 = 24x^4 - 4x^3$$

The answer checks.

2. $2a^4 - \frac{1}{2}a^2$

3. $\dfrac{u - 2u^2 - u^5}{u}$

$= \dfrac{u}{u} - \dfrac{2u^2}{u} - \dfrac{u^5}{u}$

$= 1 - 2u - u^4$

Check: We multiply.

$$\begin{array}{r} 1 - 2u - u^4 \\ u \\ \hline u - 2u^2 - u^5 \end{array}$$

4. $50x^4 - 7x^3 + x$

5. $(15t^3 + 24t^2 - 6t) \div 3t$

$= \dfrac{15t^3 + 24t^2 - 6t}{3t}$

$= \dfrac{15t^3}{3t} + \dfrac{24t^2}{3t} - \dfrac{6t}{3t}$

$= 5t^2 + 8t - 2$

Check: We multiply.

$$\begin{array}{r} 5t^2 + 8t - 2 \\ 3t \\ \hline 15t^3 + 24t^2 - 6t \end{array}$$

6. $5t^2 + 3t - 6$

7. $(20x^6 - 20x^4 - 5x^2) \div (-5x^2)$

$= \dfrac{20x^6 - 20x^4 - 5x^2}{-5x^2}$

$= \dfrac{20x^6}{-5x^2} - \dfrac{20x^4}{-5x^2} - \dfrac{5x^2}{-5x^2}$

$= -4x^4 + 4x^2 + 1$

Check: We multiply.

$$\begin{array}{r} -4x^4 + 4x^2 + 1 \\ -5x^2 \\ \hline 20x^6 - 20x^4 - 5x^2 \end{array}$$

8. $-3x^4 - 4x^3 + 1$

9. $(24x^5 - 40x^4 + 6x^3) \div (4x^3)$

$= \dfrac{24x^5 - 40x^4 + 6x^3}{4x^3}$

$= \dfrac{24x^5}{4x^3} - \dfrac{40x^4}{4x^3} + \dfrac{6x^3}{4x^3}$

$= 6x^2 - 10x + \dfrac{3}{2}$

Check: We multiply.

$$\begin{array}{r} 6x^2 - 10x + \frac{3}{2} \\ 4x^3 \\ \hline 24x^5 - 40x^4 + 6x^3 \end{array}$$

10. $2x^3 - 3x^2 - \frac{1}{3}$

11. $\frac{8x^2 - 3x + 1}{2}$

$= \frac{8x^2}{2} - \frac{3x}{2} + \frac{1}{2}$

$= 4x^2 - \frac{3}{2}x + \frac{1}{2}$

Check: We multiply.

$$\begin{array}{r} 4x^2 - \frac{3}{2}x + \frac{1}{2} \\ 2 \\ \hline 8x^2 - 3x + 1 \end{array}$$

12. $2x^2 + x - \frac{2}{3}$

13. $\frac{2x^3 + 6x^2 + 4x}{2x}$

$= \frac{2x^3}{2x} + \frac{6x^2}{2x} + \frac{4x}{2x}$

$= x^2 + 3x + 2$

Check: We multiply.

$$\begin{array}{r} x^2 + 3x + 2 \\ 2x \\ \hline 2x^3 + 6x^2 + 4x \end{array}$$

14. $2x^2 - 3x + 5$

15. $\frac{9r^2s^2 + 3r^2s - 6rs^2}{-3rs}$

$= \frac{9r^2s^2}{-3rs} + \frac{3r^2s}{-3rs} - \frac{6rs^2}{-3rs}$

$= -3rs - r + 2s$

Check: We multiply.

$$\begin{array}{r} -3rs - r + 2s \\ -3rs \\ \hline 9r^2s^2 + 3r^2s - 6rs^2 \end{array}$$

16. $1 - 2x^2y + 3x^4y^5$

17. $x + 2\overline{)x^2 + 4x + 4}$ with quotient $x + 2$

$x^2 + 2x$

$2x + 4 \leftarrow (x^2 + 4x) - (x^2 + 2x) = 2x$

$2x + 4$

$0 \leftarrow (2x + 4) - (2x + 4)$

The answer is $x + 2$.

18. $x - 3$

19. $x - 5\overline{)x^2 - 10x - 25}$ with quotient $x - 5$

$x^2 - 5x$

$-5x - 25 \leftarrow (x^2-10x) - (x^2-5x) = -5x$

$-5x + 25$

$-50 \leftarrow (-5x - 25) - (-5x + 25)$

The answer is $x - 5 + \frac{-50}{x - 5}$, or $x - 5 - \frac{50}{x - 5}$.

20. $x + 4 - \frac{32}{x + 4}$

21. $x + 6\overline{)x^2 + 4x - 14}$ with quotient $x - 2$

$x^2 + 6x$

$-2x - 14 \leftarrow (x^2 + 4x) - (x^2 + 6x) = -2x$

$-2x - 12$

$-2 \leftarrow (-2x - 14) - (-2x - 12)$

The answer is $x - 2 + \frac{-2}{x + 6}$, or $x - 2 - \frac{2}{x + 6}$.

22. $x + 7 + \frac{5}{x - 2}$

23. $x + 3\overline{)x^2 + 0x - 9}$ with quotient $x - 3$ ← Filling in the missing term

$x^2 + 3x$

$-3x - 9 \leftarrow x^2 - (x^2 + 3x) = -3x$

$-3x - 9$

$0 \leftarrow (-3x - 9) - (-3x - 9)$

The answer is $x - 3$.

24. $x - 5$

25. $x + 1\overline{)x^5 + 0x^4 + 0x^3 + 0x^2 + 0x + 1}$ with quotient $x^4 - x^3 + x^2 - x + 1$ ← Filling in missing terms

$x^5 + x^4$

$-x^4 \leftarrow x^5 - (x^5 + x^4)$

$-x^4 - x^3$

$x^3 \leftarrow -x^4 - (-x^4 - x^3)$

$x^3 + x^2$

$-x^2 \leftarrow x^3 - (x^3 + x^2)$

$-x^2 - x$

$x + 1 \leftarrow -x^2-(-x^2-x)$

$x + 1$

$0 \leftarrow (x+1)-(x+1)$

The answer is $x^4 - x^3 + x^2 - x + 1$.

26. $x^4 + x^3 + x^2 + x + 1$

27. $4x + 3\overline{)8x^3 - 22x^2 - 5x + 12}$ with quotient $2x^2 - 7x + 4$

$8x^3 + 6x^2$

$-28x^2 - 5x \leftarrow (8x^3-22x^2)-(8x^3+6x^2) = -28x^2$

$-28x^2 - 21x$

$16x + 12 \leftarrow (-28x^2 - 5x) - (-28x^2 - 21x) = 16x$

$16x + 12$

$0 \leftarrow (16x+12)-(16x+12)$

The answer is $2x^2 - 7x + 4$.

28. $x^2 - 3x + 1$

29. $x^3 - 7\overline{)x^6 - 13x^3 + 42}$ with quotient $x^3 - 6$

$x^6 - 7x^3$

$-6x^3 + 42 \leftarrow (x^6-13x^3)-(x^6-7x^3)=-6x^3$

$-6x^3 + 42$

$0 \leftarrow (-6x^3 + 42) - (-6x^3 + 42)$

The answer is $x^3 - 6$.

30. $x^3 + 8$

31. $x - 2\overline{)x^4 + 0x^3 + 0x^2 + 0x - 16}$ Quotient: $x^3 + 2x^2 + 4x + 8$

$\underline{x^4 - 2x^3}$

$2x^3 \quad \leftarrow x^4 - (x^4 - 2x^3) = 2x^3$

$\underline{2x^3 - 4x^2}$

$4x^2 \quad \leftarrow 2x^3 - (2x^3 - 4x^2) = 4x^2$

$\underline{4x^2 - 8x}$

$8x - 16 \leftarrow 4x^2 - (4x^2 - 8x) = 8x$

$\underline{8x - 16}$

$0 \leftarrow (8x - 16) - (8x - 16)$

The answer is $x^3 + 2x^2 + 4x + 8$.

32. $x^3 + 3x^2 + 9x + 27$

33. $t - 1\overline{)t^3 - t^2 + t - 1}$ Quotient: $t^2 + 1$

$\underline{t^3 - t^2}$

$0 + t - 1 \quad \leftarrow (t^3 - t^2) - (t^3 - t^2) = 0$

$\underline{t - 1}$

$0 \quad \leftarrow (t - 1) - (t - 1)$

The answer is $t^2 + 1$.

34. $t^2 - 2t + 3 - \dfrac{4}{t + 1}$

35. Familiarize. Let w = the width. Then w + 15 = the length. We draw a picture.

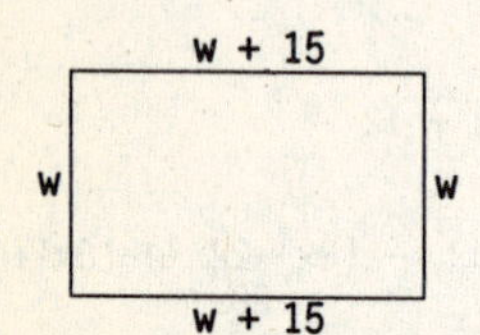

We will use the fact that the perimeter is 640 ft to find w (the width). Then we can find w + 15 (the length) and multiply the length and the width to find the area.

Translate.

Width + Width + Length + Length = Perimeter

$w + w + (w + 15) + (w + 15) = 640$

Carry out.

$$w + w + (w + 15) + (w + 15) = 640$$
$$4w + 30 = 640$$
$$4w = 610$$
$$w = 152.5$$

If the width is 152.5, then the length is 152.5 + 15, or 167.5 The area is (167.5)(152.5), or 25,543.75 ft^2.

Check. The length, 167.5 ft, is 15 ft greater than the width, 152.5 ft. The perimeter is 152.5 + 152.5 + 167.5 + 167.5, or 640 ft. We should also recheck the computation we used to find the area. The answer checks.

State. The area is 25,543.75 ft^2.

36. $\left\{x \middle| x < -\frac{12}{5}\right\}$

37. To plot (4,-1) we start at the origin and move 4 units to the right and then down 1 unit. To plot (0,5) we start at the origin and move 0 units horizontally and up 5 units. To plot (-2,3) we start at the origin and move 2 units to the left and then up 3 units. To plot (-3,0) we start at the origin and move 3 units to the left and 0 units vertically.

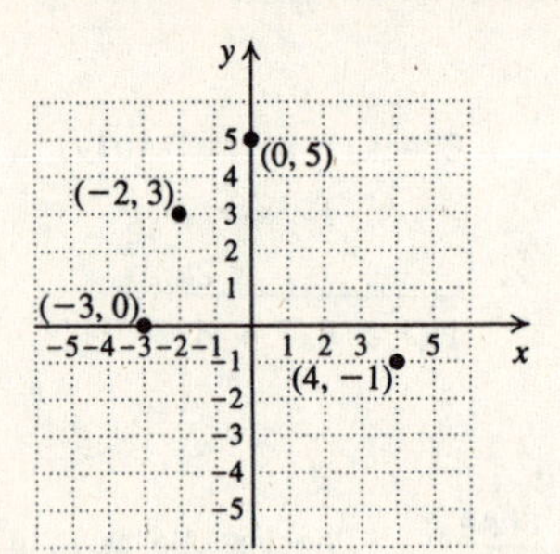

38. III

39.

40.

41. $x^2 + 4\overline{)x^4 + 9x^2 + 20}$ Quotient: $x^2 + 5$

$\underline{x^4 + 4x^2}$

$5x^2 + 20$

$\underline{5x^2 + 20}$

0

The answer is $x^2 + 5$.

42. $y^3 - ay^2 + a^2y - a^3 + \dfrac{a^2 + a^4}{y + a}$

43. $5a^2 - 7a - 2\overline{)5a^3 + 8a^2 - 23a - 1}$ Quotient: $a + 3$

$\underline{5a^3 - 7a^2 - 2a}$

$15a^2 - 21a - 1$

$\underline{15a^2 - 21a - 6}$

5

The answer is $a + 3 + \dfrac{5}{5a^2 - 7a - 2}$.

44. $5y + 2 + \dfrac{-10y + 11}{3y^2 - 5y - 2}$

45. $(4x^5 - 14x^3 - x^2 + 3) + (2x^5 + 3x^4 + x^3 - 3x^2 + 5x)$

$= 6x^5 + 3x^4 - 13x^3 - 4x^2 + 5x + 3$

$$
\begin{array}{r}
2x^2 + x - 3 \\
3x^3 - 2x - 1\overline{)\,6x^5 + 3x^4 - 13x^3 - 4x^2 + 5x + 3} \\
\underline{6x^5 \qquad - 4x^3 - 2x^2 \qquad\qquad} \\
3x^4 - 9x^3 - 2x^2 + 5x \qquad \\
\underline{3x^4 \qquad\quad - 2x^2 - x \qquad} \\
-9x^3 \qquad\quad + 6x + 3 \\
\underline{-9x^3 \qquad\quad + 6x + 3} \\
0
\end{array}
$$

The answer is $2x^2 + x - 3$.

46. $5x^5 + 5x^4 - 8x^2 - 8x + 2$

47.
$$
\begin{array}{r}
3a^{2h} + 2a^h - 5 \\
2a^h + 3\overline{)\,6a^{3h} + 13a^{2h} - 4a^h - 15} \\
\underline{6a^{3h} + 9a^{2h} \qquad\qquad\qquad} \\
4a^{2h} - 4a^h \qquad\quad \\
\underline{4a^{2h} + 6a^h \qquad\quad} \\
-10a^h - 15 \\
\underline{-10a^h - 15} \\
0
\end{array}
$$

The answer is $3a^{2h} + 2a^h - 5$.

48. -5

49.
$$
\begin{array}{r}
2x + (3c + 2) \\
x - 1\overline{)\,2x^2 + 3cx - 8} \\
\underline{2x^2 - 2x \qquad\quad} \\
(3c + 2)x - 8 \\
\underline{(3c + 2)x - (3c + 2)} \\
-8 + (3c + 2)
\end{array}
$$

We set the remainder equal to 0:

$-8 + 3c + 2 = 0$

$3c - 6 = 0$

$3c = 6$

$c = 2$

Thus, c must be 2.

50. 1

Exercise Set 4.8

1. $3^{-2} = \frac{1}{3^2} = \frac{1}{9}$

2. $\frac{1}{2^3} = \frac{1}{8}$

3. $10^{-4} = \frac{1}{10^4} = \frac{1}{10,000}$

4. $\frac{1}{5^6} = \frac{1}{15,625}$

5. $7^{-3} = \frac{1}{7^3} = \frac{1}{343}$

6. $\frac{1}{5^2} = \frac{1}{25}$

7. $a^{-3} = \frac{1}{a^3}$

8. $\frac{1}{x^2}$

9. $\frac{1}{y^{-4}} = y^4$

10. t^7

11. $\frac{1}{z^{-n}} = z^n$

12. h^m

13. $2^{-1} = \frac{1}{2^1} = \frac{1}{2}$

14. $\frac{3}{2}$

15. $\left(\frac{1}{4}\right)^{-2} = \frac{1}{\left(\frac{1}{4}\right)^2} = \frac{1}{\frac{1}{16}} = 1 \cdot \frac{16}{1} = 16$

16. $\frac{1}{\left(\frac{4}{5}\right)^2} = \frac{25}{16}$

17. $\frac{1}{4^3} = 4^{-3}$

18. 5^{-2}

19. $\frac{1}{x^3} = x^{-3}$

20. y^{-2}

21. $\frac{1}{a^4} = a^{-4}$

22. t^{-5}

23. $\frac{1}{p^n} = p^{-n}$

24. m^{-n}

25. $\frac{1}{5} = \frac{1}{5^1} = 5^{-1}$

26. 8^{-1}

27. $\frac{1}{t} = \frac{1}{t^1} = t^{-1}$

28. m^{-1}

29. $3^{-5} \cdot 3^8 = 3^{-5+8} = 3^3$

30. 5

31. $x^{-2} \cdot x = x^{-2+1} = x^{-1}$, or $\frac{1}{x}$

32. 1

33. $x^{-7} \cdot x^{-6} = x^{-13}$, or $\frac{1}{x^{13}}$

34. y^{-13}, or $\frac{1}{y^{13}}$

35. $\frac{m^6}{m^{12}} = m^{6-12} = m^{-6}$, or $\frac{1}{m^6}$

36. p^{-1}, or $\frac{1}{p}$

37. $\frac{(8x)^6}{(8x)^{10}} = (8x)^{6-10} = (8x)^{-4}$, or $\frac{1}{(8x)^4}$

38. $(9t)^{-7}$, or $\frac{1}{(9t)^7}$

39. $\frac{18^9}{18^9} = 18^{9-9} = 18^0 = 1$

40. 1

41. $(a^{-3}b^{-5})(a^{-4}b^{-6}) = a^{-3+(-4)}b^{-5+(-6)} =$

$a^{-7}b^{-11}$, or $\frac{1}{a^7b^{11}}$

42. $x^{-5}y^{-9}$, or $\frac{1}{x^5y^9}$

43. $\frac{x^7}{x^{-2}} = x^{7-(-2)} = x^9$

44. t^{11}

45. $\frac{z^{-6}}{z^{-2}} = z^{-6-(-2)} = z^{-4}$, or $\frac{1}{z^4}$

46. y^{-4}, or $\frac{1}{y^4}$

47. $\frac{x^{-5}}{x^{-8}} = x^{-5-(-8)} = x^3$

48. y^5

49. $\frac{x}{x^{-1}} = x^{1-(-1)} = x^2$

50. x^5

51. $(a^{-3})^5 = a^{-3 \cdot 5} = a^{-15}$, or $\frac{1}{a^{15}}$

52. x^{-30}, or $\frac{1}{x^{30}}$

53. $(5^2)^{-3} = 5^{2(-3)} = 5^{-6}$, or $\frac{1}{5^6}$

54. 9^{-12}, or $\frac{1}{9^{12}}$

55. $(x^{-3})^{-4} = x^{(-3)(-4)} = x^{12}$

56. a^{30}

57. $(m^{-3})^7 = m^{-3 \cdot 7} = m^{-21}$, or $\frac{1}{m^{21}}$

58. n^{-16}, or $\frac{1}{n^{16}}$

59. $(ab)^{-3} = a^{-3}b^{-3}$, or $\frac{1}{a^3b^3}$

60. $m^{-5}n^{-5}$, or $\frac{1}{m^5n^5}$

61. $(5ab)^{-2} = 5^{-2}a^{-2}b^{-2}$, or $\frac{1}{5^2a^2b^2}$, or $\frac{1}{25a^2b^2}$

62. $4^{-2}x^{-2}y^{-2}$, or $\frac{1}{16x^2y^2}$

63. $(6x^{-5})^2 = 6^2x^{-10} = 36x^{-10}$, or $\frac{36}{x^{10}}$

64. $81a^{-16}$, or $\frac{81}{a^{16}}$

65. $(x^4y^5)^{-3} = (x^4)^{-3}(y^5)^{-3} = x^{4(-3)}y^{5(-3)} =$

$x^{-12}y^{-15}$, or $\frac{1}{x^{12}y^{15}}$

66. $t^{-20}x^{-12}$, or $\frac{1}{t^{20}x^{12}}$

67. $(x^{-6}y^{-2})^{-4} = (x^{-6})^{-4}(y^{-2})^{-4} =$

$= x^{(-6)(-4)}y^{(-2)(-4)} = x^{24}y^8$

68. $x^{10}y^{35}$

69. $(3x^3y^{-8}z^{-3})^2 = 3^2(x^3)^2(y^{-8})^2(z^{-3})^2 =$

$9x^6y^{-16}z^{-6}$, or $\frac{9x^6}{y^{16}z^6}$

70. $8a^6y^{-12}z^{-15}$, or $\frac{8a^6}{y^{12}z^{15}}$

71. $(x^3y^{-4}z^{-5})(x^{-4}y^{-2}z^9) = x^{3+(-4)}y^{-4+(-2)}z^{-5+9} =$

$x^{-1}y^{-6}z^4$, or $\frac{z^4}{xy^6}$

72. $a^{-8}b^5c^4$, or $\frac{b^5c^4}{a^8}$

73. $(m^{-4}n^7p^3)(m^9n^{-2}p^{-10}) = m^{-4+9}n^{7+(-2)}p^{3+(-10)} =$

$m^5n^5p^{-7}$, or $\frac{m^5n^5}{p^7}$

74. $t^{-14}p^3m^6$, or $\frac{p^3m^6}{t^{14}}$

75. $\left[\frac{y^2}{2}\right]^{-3} = \frac{(y^2)^{-3}}{2^{-3}} = \frac{y^{-6}}{2^{-3}}$, or $y^{-6} \cdot \frac{1}{2^{-3}} = \frac{1}{y^6} \cdot 2^3 =$

$\frac{8}{y^6}$

76. $\frac{9}{a^8}$

77. $\left[\frac{3}{a^2}\right]^3 = \frac{3^3}{(a^2)^3} = \frac{27}{a^6}$

78. $\dfrac{49}{x^{14}}$

79. $\left(\dfrac{x^2y}{z}\right)^3 = \dfrac{(x^2)^3y^3}{z^3} = \dfrac{x^6y^3}{z^3}$

80. $\dfrac{m^3}{n^{12}p^3}$

81. $\left(\dfrac{a^2b}{cd^3}\right)^{-2} = \dfrac{(a^2)^{-2}b^{-2}}{c^{-2}(d^3)^{-2}} = \dfrac{a^{-4}b^{-2}}{c^{-2}d^{-6}}$, or

$a^{-4}b^{-2} \cdot \dfrac{1}{c^{-2}} \cdot \dfrac{1}{d^{-6}} = \dfrac{1}{a^4} \cdot \dfrac{1}{b^2} \cdot c^2d^6 = \dfrac{c^2d^6}{a^4b^2}$

82. $\dfrac{27b^{12}}{8a^6}$

83. 2.14×10^3

Since the exponent is positive, the decimal point will move to the right.

2.140. The decimal point moves right 3 places.

$2.14 \times 10^3 = 2140$

84. 892

85. 6.92×10^{-3}

Since the exponent is negative, the decimal point will move to the left.

.006.92 The decimal point moves left 3 places.

$6.92 \times 10^{-3} = 0.00692$

86. 0.000726

87. 7.84×10^8

Since the exponent is positive, the decimal point will move to the right.

7.84000000. 8 places

$7.84 \times 10^8 = 784,000,000$

88. 13,500,000

89. 8.764×10^{-10}

Since the exponent is negative, the decimal point will move to the left.

0.0000000008.764 10 places

$8.764 \times 10^{-10} = 0.0000000008764$

90. 0.009043

91. $10^8 = 1 \times 10^8$

Since the exponent is positive, the decimal point will move to the right.

1.00000000. 8 places

$10^8 = 100,000,000$

92. 10,000

93. $10^{-4} = 1 \times 10^{-4}$

Since the exponent is negative, the decimal point will move to the left.

0.0001. 4 places

$10^{-4} = 0.0001$

94. 0.0000001

95. $25,000 = 2.5 \times 10^n$

To write 2.5 as 25,000 we move the decimal point 4 places to the right. Thus, n is 4 and

$25,000 = 2.5 \times 10^4$.

96. 7.15×10^4

97. $0.00371 = 3.71 \times 10^n$

To write 3.71 as 0.00371 we move the decimal point 3 places to the left. Thus, n is -3 and

$0.00371 = 3.71 \times 10^{-3}$.

98. 8.14×10^{-2}

99. $78,000,000,000 = 7.8 \times 10^n$

To write 7.8 as 78,000,000,000 we move the decimal point 10 places to the right. Thus, n is 10 and

$78,000,000,000 = 7.8 \times 10^{10}$.

100. 3.7×10^{12}

101. $907,000,000,000,000,000 = 9.07 \times 10^n$

To write 9.07 as 907,000,000,000,000,000 we move the decimal point 17 places to the right. Thus, n is 17 and

$907,000,000,000,000,000 = 9.07 \times 10^{17}$.

102. 1.68×10^{14}

103. $0.00000374 = 3.74 \times 10^n$

To write 3.74 as 0.00000374 we move the decimal point 6 places to the left. Thus n is -6 and

$0.00000374 = 3.74 \times 10^{-6}$.

104. 2.75×10^{-10}

105. $0.000000018 = 1.8 \times 10^n$

To write 1.8 as 0.000000018 we move the decimal point 8 places to the left. Thus, n is -8 and

$0.000000018 = 1.8 \times 10^{-8}$.

106. 2×10^{-11}

107. $10{,}000{,}000 = 1 \times 10^n$, or 10^n

To write 1 as 10,000,000 we move the decimal point 7 places to the right. Thus, n is 7 and

$10{,}000{,}000 = 10^7$.

108. 10^{11}

109. $0.000000001 = 1 \times 10^n$, or 10^n

To write 1 as 0.000000001 we move the decimal point 9 places to the left. Thus, n is -9 and

$0.000000001 = 10^{-9}$.

110. 10^{-7}

111. $(3 \times 10^4)(2 \times 10^5) = (3 \cdot 2) \times (10^4 \cdot 10^5)$
$= 6 \times 10^{4+5}$ Adding exponents
$= 6 \times 10^9$

112. 6.46×10^5

113. $(5.2 \times 10^5)(6.5 \times 10^{-2}) = (5.2 \cdot 6.5) \times (10^5 \cdot 10^{-2})$
$= 33.8 \times 10^3$

The answer is not yet in scientific notation since 33.8 is not a number between 1 and 10. We convert to scientific notation:

$33.8 \times 10^3 = (3.38 \times 10) \times 10^3 = 3.38 \times 10^4$

114. 6.106×10^{-11}

115. $(9.9 \times 10^{-6})(8.23 \times 10^{-8})$
$= (9.9 \cdot 8.23) \times (10^{-6} \cdot 10^{-8})$
$= 81.477 \times 10^{-14}$

The answer is not yet in scientific notation because 81.477 is not between 1 and 10. We convert to scientific notation:

$81.477 \times 10^{-14} = (8.1477 \times 10) \times 10^{-14} =$
8.1477×10^{-13}

116. 1.123×10^{-5}

117. $\frac{8.5 \times 10^8}{3.4 \times 10^{-5}} = \frac{8.5}{3.4} \times \frac{10^8}{10^{-5}}$
$= 2.5 \times 10^{8-(-5)}$
$= 2.5 \times 10^{13}$

118. 2.24×10^{-7}

119. $(3.0 \times 10^6) \div (6.0 \times 10^9) = \frac{3.0 \times 10^6}{6.0 \times 10^9}$
$= \frac{3.0}{6.0} \times \frac{10^6}{10^9}$
$= 0.5 \times 10^{-3}$

The answer is not yet in scientific notation because 0.5 is not between 1 and 10. We convert to scientific notation:

$0.5 \times 10^{-3} = (5 \times 10^{-1}) \times 10^{-3} = 5 \times 10^{-4}$

120. 9.375×10^2

121. $\frac{7.5 \times 10^{-9}}{2.5 \times 10^{12}} = \frac{7.5}{2.5} \times \frac{10^{-9}}{10^{12}}$
$= 3 \times 10^{-21}$

122. 5×10^{-24}

123. Familiarize. We express 3064 and 249 million in scientific notation

$3064 = 3.064 \times 10^n$

To write 3.064 as 3064 we move the decimal point 3 places to the right, so n is 3 and $3064 = 3.064 \times 10^3$.

$249{,}000{,}000 = 2.49 \times 10^n$

To write 2.49 as 249,000,000 we move the decimal point 8 places to the right, so n is 8 and $249{,}000{,}000 = 2.49 \times 10^8$.

Let p = the part of the population that are members of the Professional Bowlers Association.

Translate. We reword the problem.

What is	number of members	divided by	population of the U.S.?
p =	(3.064×10^3)	$\div$	(2.49×10^8)

Carry out. We do the computation.

$p = (3.064 \times 10^3) \div (2.49 \times 10^8)$
$p = (3.064 \div 2.49) \times (10^3 \div 10^8)$
$p \approx 1.231 \times 10^{3-8}$
$p \approx 1.231 \times 10^{-5}$

Check. We review our computation. Also, the answer seems reasonable since it is smaller than either of the original numbers.

State. Approximately 1.231×10^{-5} of the population are members of the Professional Bowlers Association.

124. 3.3×10^{-2}

125. Familiarize. There are 365 days in one year. Express 6.5 million and 365 in scientific notation.

$6.5 \text{ million} = 6{,}500{,}000 = 6.5 \times 10^n$

To write 6.5 as 6,500,000 we move the decimal point 6 places to the right, so n is 6 and $6{,}500{,}000 = 6.5 \times 10^6$.

$365 = 3.65 \times 10^n$

To write 3.65 as 365 we move the decimal point 2 places to the right, so n is 2 and $365 = 3.65 \times 10^2$.

Let p = the amount of popcorn Americans eat in one year.

Translate. We reword the problem.

What is	daily consumption	times	number of days in a year?
p =	(6.5×10^6)	$\times$	(3.65×10^2)

Carry out. We do the computation.

$p = (6.5 \times 10^6) \times (3.65 \times 10^2)$

$p = (6.5 \times 3.65) \times (10^6 \times 10^2)$

$p = 23.725 \times 10^8$

$p = (2.3725 \times 10) \times 10^8$

$p = 2.3725 \times 10^9$

Check. We review the computation. Also, the answer seems reasonable since it is larger than 6.5 million.

State. Americans eat 2.3725×10^9 gal of popcorn each year.

126. 1.095×10^9 gal

127. Familiarize. There are 60 seconds in one minute and 60 minutes in one hour, so there are 60(60), or 3600 seconds in one hour. There are 24 hours in one day and 365 days in one year, so there are 3600(24)(365), or 31,536,000 seconds in one year.

We express 3600, 31,536,000 and 4,200,000 in scientific notation:

$3600 = 3.6 \times 10^n$

To write 3.6 as 3600 we move the decimal point 3 places to the right, so n is 3 and $3600 = 3.6 \times 10^3$.

$31{,}536{,}000 = 3.1536 \times 10^n$

To write 3.1536 as 31,536,000 we move the decimal point 7 places to the right, so n is 7 and $31{,}536{,}000 = 3.1536 \times 10^7$.

$4{,}200{,}000 = 4.2 \times 10^n$

To write 4.2 as 4,200,000 we move the decimal point 6 places to the right, so n is 6 and $4{,}200{,}000 = 4.2 \times 10^6$.

Let h = the discharge in one hour and y = the discharge in one year.

Translate. We reword and write two equations. To find the discharge in one hour:

What is	number of seconds in one hour	times	discharge per second?
h =	(3.6×10^3)	$\times$	(4.2×10^6)

To find the discharge in one year:

What is	number of seconds in one year	times	discharge per second?
y =	(3.1536×10^7)	$\times$	(4.2×10^6)

Carry out. We do the computations.

$h = (3.6 \times 10^3) \times (4.2 \times 10^6)$

$h = (3.6 \times 4.2) \times (10^3 \times 10^6)$

$h = 15.12 \times 10^9 = (1.512 \times 10) \times 10^9$

$h = 1.512 \times 10^{10}$

$y = (3.1536 \times 10^7) \times (4.2 \times 10^6)$

$y = (3.1536 \times 4.2) \times (10^7 \times 10^6)$

$y = 13.24512 \times 10^{13} = (1.324512 \times 10) \times 10^{13}$

$y = 1.324512 \times 10^{14}$

Check. We can review the computations. Also, the answers seem reasonable since they are both larger than the numbers we started with.

State. In one hour 1.512×10^{10} cu ft of water is discharged. In one year 1.324512×10^{14} cu ft of water is discharged.

128. 6.7×10^{-2}

129. $-9a + 17a = (-9 + 17)a = 8a$

130. $-17x$

131. To plot (-4,1) we start at the origin and move 4 units to the left and then up 1 unit. To plot (-3,-2) we start at the origin and move 3 units to the left and then down 2 units. To plot (5,2) we start at the origin and move 5 units to the right and then up 2 units. To plot (-1,4) we start at the origin and move 1 unit to the left and then up 4 units.

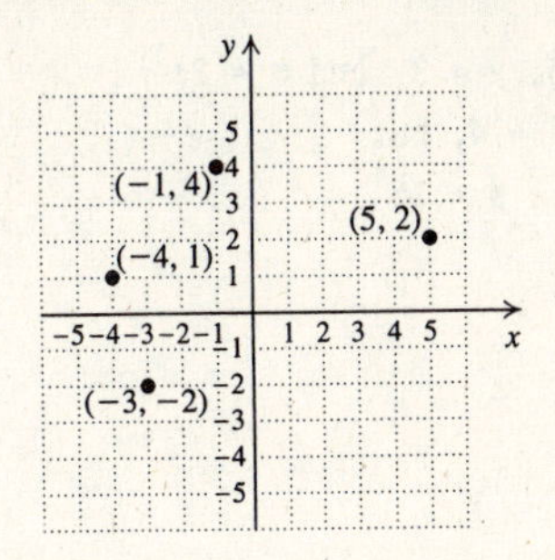

132. I and IV

133. $\dfrac{(5.2 \times 10^6)(6.1 \times 10^{-11})}{1.28 \times 10^{-3}}$

$= \dfrac{(5.2)(10.6)}{1.28} \times \dfrac{10^6 \cdot 10^{-11}}{10^{-3}}$

$= 24.78125 \times 10^{6+(-11)-(-3)}$

$= 24.78125 \times 10^{-2}$

$= (2.478125 \times 10) \times 10^{-2}$

$= 2.478125 \times 10^{-1}$

134. 1.5234375×10^7

135. $\{2.1 \times 10^6[(2.5 \times 10^{-3}) \div (5.0 \times 10^{-5})]\} \div (3.0 \times 10^{17})$

$= \{2.1 \times 10^6[0.5 \times 10^2]\} \div (3.0 \times 10^{17})$ Dividing inside the brackets first

$= \{1.05 \times 10^8\} \div (3.0 \times 10^{17})$ Multiplying inside the braces

$= 0.35 \times 10^{-9}$ Dividing

$= (3.5 \times 10^{-1}) \times 10^{-9}$ Writing 0.35 in scientific notation

$= 3.5 \times 10^{-10}$ Simplifying

136. a) 1.6×10^2

b) 2.5×10^{-11}

137. $4^{-3}\cdot 8\cdot 16 = (2^2)^{-3}\cdot 2^3\cdot 2^4 = 2^{-6}\cdot 2^3\cdot 2^4 = 2$

138. 4

139. $(5^{-12})^2 5^{25} = 5^{-24}5^{25} = 5$

140. 7

141. $\left(\frac{1}{a}\right)^{-n} = \frac{1^{-n}}{a^{-n}} = \frac{1}{a^{-n}} = a^n$

142. 2.5

143. False; let x = 2, y = 3, m = 4, and n = 2:
$2^4\cdot 3^2 = 16\cdot 9 = 144$, but
$(2\cdot 3)^{4\cdot 2} = 6^8 = 1{,}679{,}616$

144. False

145. False; let x = 5, y = 3, and m = 2:
$(5 - 3)^2 = 2^2 = 4$, but
$5^2 - 3^2 = 25 - 9 = 16$

Exercise Set 5.1

1. Answers may vary. $6x^3 = (6x)(x^2) = (3x^2)(2x) = (2x^2)(3x)$

2. Answers may vary. $(3x^2)(3x^2)$, $(9x)(x^3)$, $(3x)(3x^3)$

3. Answers may vary. $-9x^5 = (-3x^2)(3x^3) = (-x)(9x^4) = (3x^2)(-3x^3)$

4. Answers may vary. $(-4x)(3x^5)$, $(-6x^2)(2x^4)$, $(12x^3)(-x^3)$

5. Answers may vary. $24x^4 = (6x)(4x^3) = (-3x^2)(-8x^2) = (2x^3)(12x)$

6. Answers may vary. $(3x)(5x^4)$, $(x^3)(15x^2)$, $(3x^2)(5x^3)$

7. $x^2 - 4x = x \cdot x - x \cdot 4$
$= x(x - 4)$

8. $x(x + 8)$

9. $2x^2 + 6x = 2x \cdot x + 2x \cdot 3$
$= 2x(x + 3)$

10. $3x(x - 1)$

11. $x^3 + 6x^2 = x^2 \cdot x + x^2 \cdot 6$
$= x^2(x + 6)$

12. $x^2(4x^2 + 1)$

13. $8x^4 - 24x^2 = 8x^2 \cdot x^2 - 8x^2 \cdot 3$
$= 8x^2(x^2 - 3)$

14. $5x^3(x^2 + 2)$

15. $2x^2 + 2x - 8 = 2 \cdot x^2 + 2 \cdot x - 2 \cdot 4$
$= 2(x^2 + x - 4)$

16. $3(2x^2 + x - 5)$

17. $17x^5y^3 + 34x^3y^2 + 51xy$
$= 17xy \cdot x^4y^2 + 17xy \cdot 2x^2y + 17xy \cdot 3$
$= 17xy(x^4y^2 + 2x^2y + 3)$

18. $16xy^2(x^5y^2 - 2x^4y - 3)$

19. $6x^4 - 10x^3 + 3x^2 = x^2 \cdot 6x^2 - x^2 \cdot 10x + x^2 \cdot 3$
$= x^2(6x^2 - 10x + 3)$

20. $x(5x^4 + 10x - 8)$

21. $x^5y^5 + x^4y^3 + x^3y^3 - x^2y^2$
$= x^2y^2 \cdot x^3y^3 + x^2y^2 \cdot x^2y + x^2y^2 \cdot xy + x^2y^2(-1)$
$= x^2y^2(x^3y^3 + x^2y + xy - 1)$

22. $x^3y^3(x^6y^3 - x^4y^2 + xy + 1)$

23. $2x^7 - 2x^6 - 64x^5 + 4x^3$
$= 2x^3 \cdot x^4 - 2x^3 \cdot x^3 - 2x^3 \cdot 32x^2 + 2x^3 \cdot 2$
$= 2x^3(x^4 - x^3 - 32x^2 + 2)$

24. $5(2x^3 + 5x^2 + 3x - 4)$

25. $1.6x^4 - 2.4x^3 + 3.2x^2 + 6.4x$
$= 0.8x(2x^3) - 0.8x(3x^2) + 0.8x(4x) + 0.8x(8)$
$= 0.8x(2x^3 - 3x^2 + 4x + 8)$

26. $0.5x^2(5x^4 - x^2 + 10x + 20)$

27. $\frac{5}{3}x^6 + \frac{4}{3}x^5 + \frac{1}{3}x^4 + \frac{1}{3}x^3$
$= \frac{1}{3}x^3(5x^3) + \frac{1}{3}x^3(4x^2) + \frac{1}{3}x^3(x) + \frac{1}{3}x^3(1)$
$= \frac{1}{3}x^3(5x^3 + 4x^2 + x + 1)$

28. $\frac{1}{7}x(5x^6 + 3x^4 - 6x^2 - 1)$

29. $y(y + 3) + 4(y + 3)$
$= (y + 3)(y + 4)$ Factoring out the common binomial factor $y + 3$

30. $(b - 5)(b - 3)$

31. $x^2(x + 3) + 2(x + 3)$
$= (x + 3)(x^2 + 2)$ Factoring out the common binomial factor $x + 3$

32. $(2z + 1)(3z^2 + 1)$

33. $y^2(y + 8) + (y + 8) = y^2(y + 8) + 1(y + 8)$
$= (y + 8)(y^2 + 1)$ Factoring out the common factor

34. $(x - 7)(x^2 - 3)$

35. $x^3 + 3x^2 + 2x + 6 = x^3 + 3x^2 + 2x + 6$
$= x^2(x + 3) + 2(x + 3)$ Factoring each binomial
$= (x + 3)(x^2 + 2)$ Factoring out the common factor

36. $(2z + 1)(3z^2 + 1)$

37. $2x^3 + 6x^2 + x + 3 = \underbrace{2x^3 + 6x^2} + \underbrace{x + 3}$
$= 2x^2(x + 3) + 1(x + 3)$ Factoring each binomial
$= (x + 3)(2x^2 + 1)$

38. $(3x + 2)(x^2 + 1)$

39. $8x^3 - 12x^2 + 6x - 9 = 4x^2(2x - 3) + 3(2x - 3)$
$= (2x - 3)(4x^2 + 3)$

40. $(2x - 5)(5x^2 + 2)$

41. $12x^3 - 16x^2 + 3x - 4 = 4x^2(3x - 4) + 1(3x - 4)$
Factoring 1 out of the second binomial
$= (3x - 4)(4x^2 + 1)$

42. $(6x - 7)(3x^2 + 5)$

43. $x^3 + 8x^2 - 3x - 24 = x^2(x + 8) - 3(x + 8)$
$= (x + 8)(x^2 - 3)$

44. $(x + 6)(2x^2 - 5)$

45. $w^3 - 7w^2 + 4w - 28 = w^2(w - 7) + 4(w - 7)$
$= (w - 7)(w^2 + 4)$

46. $(y + 8)(y^2 - 2)$

47. $x^3 - x^2 - 2x + 5 = x^2(x - 1) - 1(2x - 5)$
This polynomial is not factorable using factoring by grouping.

48. Not factorable by grouping

49. $2x^3 - 8x^2 - 9x + 36 = 2x^2(x - 4) - 9(x - 4)$
$= (x - 4)(2x^2 - 9)$

50. $(5g - 1)(4g^2 - 5)$

51. Graph: $y = x - 6$
The equation is in the form $y = mx + b$, so we know the y-intercept is (0,-6). We find two other pairs.
When $x = 5$, $y = 5 - 6 = -1$.
When $x = 2$, $y = 2 - 6 = -4$.

x	y
0	-6
5	-1
2	-4

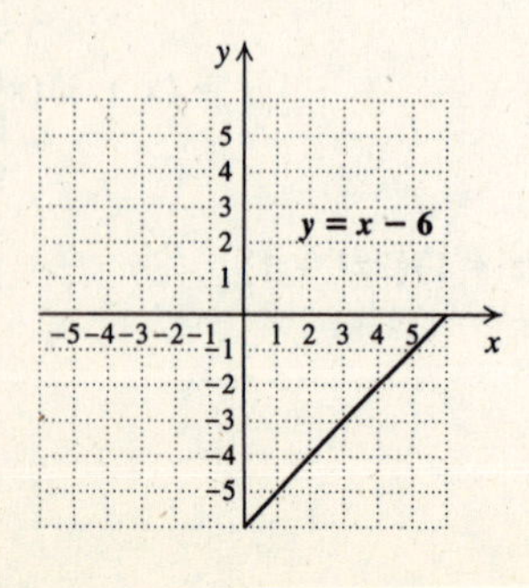

52. $\{x | x \leq \frac{14}{5}\}$

53. $-13 - (-25) = -13 + 25$ Adding an opposite
$= 12$

54. $p = 2A - q$

55. $(y + 5)(y + 7) = y^2 + 7y + 5y + 35$ Using FOIL
$= y^2 + 12y + 35$

56. $y^2 + 14y + 49$

57. $(y + 7)(y - 7) = y^2 - 7^2 = y^2 - 49$
$[(A + B)(A - B) = A^2 - B^2]$

58. $y^2 - 14y + 49$

59.

60.

61. $4x^5 + 6x^3 + 6x^2 + 9 = 2x^3(2x^2 + 3) + 3(2x^2 + 3)$
$= (2x^2 + 3)(2x^3 + 3)$

62. $(x^2 + 1)(x^4 + 1)$

63. $x^{12} + x^7 + x^5 + 1 = x^7(x^5 + 1) + 1(x^5 + 1) =$
$(x^5 + 1)(x^7 + 1)$

64. Not factorable by grouping

65. $p^3 - p^2 + 3p + 3 = p^2(p - 1) + 3(p + 1)$
This polynomial is not factorable using factoring by grouping.

66. $(x^2 + 2x + 3)(a + 1)$

67.

68.

Exercise Set 5.2

1. $x^2 + 8x + 15$
Since the constant term and coefficient of the middle term are both positive, we look for a factorization of 15 in which both factors are positive. Their sum must be 8.

Pairs of factors of 15	Sums of factors
1, 15	16
3, 5	8

The numbers we want are 3 and 5.
$x^2 + 8x + 15 = (x + 3)(x + 5)$

2. $(x + 2)(x + 3)$

3. $x^2 + 7x + 12$

Since the constant term is positive and the coefficient of the middle term is positive, we look for a factorization of 12 in which both factors are positive. Their sum must be 7.

Pairs of factors of 12	Sums of factors
1, 12	13
2, 6	8
3, 4	7

The numbers we want are 3 and 4.

$x^2 + 7x + 12 = (x + 3)(x + 4)$.

4. $(x + 1)(x + 8)$

5. $x^2 - 6x + 9$

Since the constant term is positive and the coefficient of the middle term is negative, we look for a factorization of 9 in which both factors are negative. Their sum must be -6.

Pairs of factors of 9	Sums of factors
-1, -9	-10
-3, -3	-6

The numbers we want are -3 and -3.

$x^2 - 6x + 9 = (x - 3)(x - 3)$, or $(x - 3)^2$.

6. $(y + 4)(y + 7)$

7. $x^2 + 9x + 14$

Since the constant term is positive and the coefficient of the middle term is positive, we look for a factorization of 14 in which both factors are positive. Their sum must be 9.

Pairs of factors of 14	Sums of factors
1, 14	15
2, 7	9

The numbers we want are 2 and 7.

$x^2 + 9x + 14 = (x + 2)(x + 7)$.

8. $(a + 5)(a + 6)$

9. $b^2 + 5b + 4$

Since the constant term is positive and the coefficient of the middle term is positive, we look for a factorization of 4 in which both factors are positive. Their sum must be 5.

Pairs of factors of 4	Sums of factors
1, 4	5
2, 2	4

The numbers we want are 1 and 4.

$b^2 + 5b + 4 = (b + 1)(b + 4)$

10. $\left(x - \frac{1}{5}\right)^2$

11. $x^2 + \frac{2}{3}x + \frac{1}{9}$

Since the constant term is positive and the coefficient of the middle term is positive, we look for a factorization of $\frac{1}{9}$ in which both factors are positive. Their sum must be $\frac{2}{3}$.

Pairs of factors of $\frac{1}{9}$	Sums of factors
$1, \frac{1}{9}$	$\frac{10}{9}$
$\frac{1}{3}, \frac{1}{3}$	$\frac{2}{3}$

The numbers we want are $\frac{1}{3}$ and $\frac{1}{3}$.

$x^2 + \frac{2}{3}x + \frac{1}{9} = \left(x + \frac{1}{3}\right)\left(x + \frac{1}{3}\right)$, or $\left(x + \frac{1}{3}\right)^2$.

12. $(z - 1)(z - 7)$

13. $d^2 - 7d + 10$

Since the constant term is positive and the coefficient of the middle term is negative, we look for a factorization of 10 in which both factors are negative. Their sum must be -7.

Pairs of factors of 10	Sums of factors
-1, -10	-11
-2, -5	-7

The numbers we want are -2 and -5.

$d^2 - 7d + 10 = (d - 2)(d - 5)$.

14. $(x - 3)(x - 5)$

15. $y^2 - 11y + 10$

Since the constant term is positive and the coefficient of the middle term is negative, we look for a factorization of 10 in which both factors are negative. Their sum must be -11.

Pairs of factors of 10	Sums of factors
-1, -10	-11
-2, -5	-7

The numbers we want are -1 and -10.

$y^2 - 11y + 10 = (y - 1)(y - 10)$.

16. $(x + 3)(x - 5)$

17. $x^2 + x - 42$

Since the constant term is negative, we look for a factorization of -42 in which one factor is positive and one factor is negative. Their sum must be 1, the coefficient of the middle term, so the positive factor must have the larger absolute value. Thus we consider only pairs of factors in which the positive factor has the larger absolute value.

Pairs of factors of -42	Sums of factors
-1, 42	41
-2, 21	19
-3, 14	11
-6, 7	1

The numbers we want are -6 and 7.

$x^2 + x - 42 = (x - 6)(x + 7)$.

18. $(x - 3)(x + 5)$

19. $2x^2 - 14x - 36 = 2(x^2 - 7x - 18)$

After factoring out the common factor, 2, we consider $x^2 - 7x - 18$. Since the constant term is negative, we look for a factorization of -18 in which one factor is positive and one factor is negative. Their sum must be -7, the coefficient of the middle term, so the negative factor must have the larger absolute value. Thus we consider only pairs of factors in which the negative factor has the larger absolute value.

Pairs of factors of -18	Sums of factors
1, -18	-17
2, -9	-7
3, -6	-3

The numbers we want are 2 and -9. The factorization of $x^2 - 7x - 18$ is $(x + 2)(x - 9)$. We must not forget the common factor, 2. The factorization of $2x^2 - 14x - 36$ is $2(x + 2)(x - 9)$.

20. $3(y + 4)(y - 7)$

21. $x^3 - 6x^2 - 16x = x(x^2 - 6x - 16)$

After factoring out the common factor, x, we consider $x^2 - 6x - 16$. Since the constant term is negative, we look for a factorization of -16 in which one factor is positive and one factor is negative. Their sum must be -6, the coefficient of the middle term, so the negative factor must have the larger absolute value. Thus we consider only pairs of factors in which the negative factor has the larger absolute value.

Pairs of factors of -16	Sums of factors
1, -16	-15
2, -8	-6

The numbers we want are 2 and -8.

Then $x^2 - 6x - 16 = (x + 2)(x - 8)$, so $x^3 - 6x^2 - 16x = x(x + 2)(x - 8)$.

22. $x(x + 6)(x - 7)$

23. $y^2 - 4y - 45$

Since the constant term is negative, we look for a factorization of -45 in which one factor is positive and one factor is negative. Their sum must be -4, the coefficient of the middle term, so the negative factor must have the larger absolute value. Thus we consider only pairs of factors in which the negative factor has the larger absolute value.

Pairs of factors of -45	Sums of factors
1, -45	-44
3, -15	-12
5, -9	-4

The numbers we want are 5 and -9.

$y^2 - 4y - 45 = (y + 5)(y - 9)$.

24. $(x + 5)(x - 12)$

25. $-2x - 99 + x^2 = x^2 - 2x - 99$

Since the constant term is negative, we look for a factorization of -99 in which one factor is positive and one factor is negative. Their sum must be -2, the coefficient of the middle term, so the negative factor must have the larger absolute value. Thus we consider only pairs of factors in which the negative factor has the larger absolute value.

Pairs of factors of -99	Sums of factors
1, -99	-98
3, -33	-30
9, -11	-2

The numbers we want are 9 and -11.

$-2x - 99 + x^2 = (x + 9)(x - 11)$.

26. $(x - 6)(x + 12)$

27. $c^4 + c^3 - 56c^2 = c^2(c^2 + c - 56)$

After factoring out the common factor, c^2, we consider $c^2 + c - 56$. Since the constant term is negative, we look for a factorization of -56 in which one factor is positive and one factor is negative. Their sum must be 1, so the positive factor must have the larger absolute value. Thus we consider only pairs of factors in which the positive factor has the larger absolute value.

Pairs of factors of -56	Sums of factors
-1, 56	55
-2, 28	26
-4, 14	12
-7, 8	1

The numbers we want are -7 and 8. The factorization of $c^2 + c - 56$ is $(c - 7)(c + 8)$, so $c^4 + c^3 - 56c^2 = c^2(c - 7)(c + 8)$.

28. $5(b - 3)(b + 8)$

29. $2a^2 + 4a - 70 = 2(a^2 + 2a - 35)$

After factoring out the common factor, 2, we consider $a^2 + 2a - 35$. Since the constant term is negative, we look for a factorization of -35 in which one factor is positive and one factor is negative. Their sum must be 2, so the positive factor must have the larger absolute value. Thus we consider only pairs of factors in which the positive factor has the larger absolute value.

Pairs of factors of -35	Sums of factors
-1, 35	34
-5, 7	2

The numbers we want are -5 and 7. The factorization of $a^2 + 2a - 35$ is $(a - 5)(a + 7)$, so $2a^2 + 4a - 70 = 2(a - 5)(a + 7)$.

30. $x^3(x + 2)(x - 1)$

31. $x^2 + x + 1$

Since the constant term and the coefficient of the middle term are both positive, we look for a factorization of 1 in which both factors are positive. The sum must be 1. The only possible pair of factors is 1 and 1, but their sum is not 1. Thus, this polynomial is not factorable into polynomials with integer coefficients.

32. Not factorable

33. $7 - 2p + p^2 = p^2 - 2p + 7$

Since the constant term is positive and the coefficient of the middle term is negative, we look for a factorization of 7 in which both factors are negative. Their sum must be -2. The only possible pair of factors is -1 and -7, but their sum is not -2. Thus, this polynomial is not factorable into polynomials with integer coefficients.

34. Not factorable

35. $x^2 + 20x + 100$

We look for two factors, both positive, whose product is 100 and whose sum is 20.

They are 10 and 10. $10 \cdot 10 = 100$ and $10 + 10 = 20$.

$x^2 + 20x + 100 = (x + 10)(x + 10)$, or $(x + 10)^2$.

36. $(x + 9)(x + 11)$

37. $3x^3 - 63x^2 - 300x = 3x(x^2 - 21x - 100)$

After factoring out the common factor, 3x, we consider $x^2 - 21x - 100$. We look for two factors, one positive and one negative, whose product is -100 and whose sum is -21.

They are 4 and -25. $4 \cdot (-25) = -100$ and $4 + (-25) = -21$.

$x^2 - 21x - 100 = (x + 4)(x - 25)$, so
$3x^3 - 63x^2 - 300x = 3x(x + 4)(x - 25)$.

38. $2x(x - 8)(x - 12)$

39. $x^2 - 21x - 72$

We look for two factors, one positive and one negative, whose product is -72 and whose sum is -21. They are 3 and -24.

$x^2 - 21x - 72 = (x + 3)(x - 24)$

40. $4(x + 5)^2$

41. $x^2 - 25x + 144$

We look for two factors, both negative, whose product is 144 and whose sum is -25. They are -9 and -16.

$x^2 - 25x + 144 = (x - 9)(x - 16)$

42. $(y - 9)(y - 12)$

43. $a^4 + a^3 - 132a^2 = a^2(a^2 + a - 132)$

After factoring out the common factor, a^2, we consider $a^2 + a - 132$. We look for two factors, one positive and one negative, whose product is -132 and whose sum is 1. They are -11 and 12.

$a^2 + a - 132 = (a - 11)(a + 12)$, so
$a^4 + a^3 - 132a^2 = a^2(a - 11)(a + 12)$.

44. $a^4(a - 6)(a + 15)$

45. $120 - 23x + x^2 = x^2 - 23x + 120$

We look for two factors, both negative, whose product is 120 and whose sum is -23. They are -8 and -15.

$x^2 - 23x + 120 = (x - 8)(x - 15)$

46. $(d + 6)(d + 16)$

47. First write the polynomial in descending order and factor out -1.

$108 - 3x - x^2 = -x^2 - 3x + 108 = -1(x^2 + 3x - 108)$

Now we factor the polynomial $x^2 + 3x - 108$. We look for two factors, one positive and one negative, whose product is -108 and whose sum is 3. They are -9 and 12.

$x^2 + 3x - 108 = (x - 9)(x + 12)$

The final answer must include the -1 which was factored out above.

$-x^2 - 3x + 108 = -1(x - 9)(x + 12)$.

Using the distributive law to find $-1(x - 9)$, we see that $-1(x - 9)(x + 12)$ can also be expressed as $(-x + 9)(x + 12)$, or $(9 - x)(12 + x)$.

48. $-1(y - 16)(y + 7)$, or $(16 - y)(7 + y)$

49. $y^2 - 0.2y - 0.08$

We look for two factors, one positive and one negative, whose product is -0.08 and whose sum is -0.2. They are -0.4 and 0.2.

$y^2 - 0.2y - 0.08 = (y - 0.4)(y + 0.2)$

50. $(t - 0.5)(t + 0.2)$

51. $p^2 + 3pq - 10q^2 = p^2 + 3qp - 10q^2$

Think of 3q as a "coefficient" of p. Then we look for factors of $-10q^2$ whose sum is 3q. They are 5q and -2q.

$p^2 + 3pq - 10q^2 = (p + 5q)(p - 2q)$.

52. $(a - 3b)(a + b)$

53. $m^2 + 5mn + 5n^2 = m^2 + 5nm + 5n^2$

We look for factors of $5n^2$ whose sum is 5n. The only reasonable possibilities are shown below.

Pairs of factors of $5n^2$	Sums of factors
5n, n	6n
-5n, -n	-6n

There are no factors whose sum is 5n. Thus, the polynomial is not factorable into polynomials with integer coefficients.

54. $(x - 8y)(x - 3y)$

55. $s^2 - 2st - 15t^2 = s^2 - 2ts - 15t^2$

We look for factors of $-15t^2$ whose sum is -2t. They are -5t and 3t.

$s^2 - 2st - 15t^2 = (s - 5t)(s + 3t)$

56. $(b + 10c)(b - 2c)$

57. $2x^3 - 10x^2 + 12x = 2x(x^2 - 5x + 6)$

After factoring out the common factor, 2x, we consider $x^2 - 5x + 6$. We look for two factors, both negative, whose product is 6 and whose sum is -5. They are -3 and -2.

$x^2 - 5x + 6 = (x - 3)(x - 2)$, so
$2x^3 - 10x^2 + 12x = 2x(x - 3)(x - 2)$.

58. $3a^4(a - 6)(a - 2)$

59. $7a^9 - 28a^8 - 35a^7 = 7a^7(a^2 - 4a - 5)$

After factoring out the common factor, $7a^7$, we consider $a^2 - 4a - 5$. We look for two factors, one positive and one negative, whose product is -5 and whose sum is -4. They are -5 and 1.

$a^2 - 4a - 5 = (a - 5)(a + 1)$, so
$7a^9 - 28a^8 - 35a^7 = 7a^7(a - 5)(a + 1)$.

60. $6x^8(x - 7)(x + 2)$

61. $(x + 6)(3x + 4) = 3x^2 + 4x + 18x + 24$ Using FOIL
$= 3x^2 + 22x + 24$

62. $49w^2 + 84w + 36$

63. Familiarize. Let n = the number of people arrested the year before.

Translate. We reword the problem.

Number arrested last year	less 1.2% of	that number	is 29,090.
n	- 1.2%	· n	= 29,090

Carry out. We solve the equation.

$n - 1.2\%\cdot n = 29,090$
$1\cdot n - 0.012n = 29,090$
$0.988n = 29,090$
$n \approx 29,443$ Rounding

Check. 1.2% of 29,443 is $0.012(29,443) \approx 353$ and $29,443 - 353 = 29,090$. The answer checks.

State. Approximately 29,443 people were arrested last year.

64. 100°, 25°, 55°

65.

66.

67. $y^2 + my + 50$

We look for pairs of factors whose product is 50. The sum of each pair is represented by m.

Pairs of factors whose product is 50	Sums of factors
1, 50	51
-1, -50	-51
2, 25	27
-2, -25	-27
5, 10	15
-5, -10	-15

The polynomial $y^2 + my + 50$ can be factored if m is 51, -51, 27, -27, 15, or -15.

68. 49, -49, 23, -23, 5, -5

69. $x^2 - \frac{1}{2}x - \frac{3}{16}$

We look for two factors, one positive and one negative, whose product is $-\frac{3}{16}$ and whose sum is $-\frac{1}{2}$.

They are $-\frac{3}{4}$ and $\frac{1}{4}$.

$-\frac{3}{4}\cdot\frac{1}{4} = -\frac{3}{16}$ and $-\frac{3}{4} + \frac{1}{4} = -\frac{2}{4} = -\frac{1}{2}$.

$x^2 - \frac{1}{2}x - \frac{3}{16} = \left(x - \frac{3}{4}\right)\left(x + \frac{1}{4}\right)$

70. $\left(x - \frac{1}{2}\right)\left(x + \frac{1}{4}\right)$

71. $x^2 + \frac{30}{7}x - \frac{25}{7}$

We look for two factors, one positive and one negative, whose product is $-\frac{25}{7}$ and whose sum is $\frac{30}{7}$.

They are 5 and $-\frac{5}{7}$.

$5 \cdot \left(-\frac{5}{7}\right) = -\frac{25}{7}$ and $5 + \left(-\frac{5}{7}\right) = \frac{35}{7} + \left(-\frac{5}{7}\right) = \frac{30}{7}$.

$x^2 + \frac{30}{7}x - \frac{25}{7} = (x + 5)\left(x - \frac{5}{7}\right)$

72. $\frac{1}{3}x(x + 3)(x - 2)$

73. $b^{2n} + 7b^n + 10$

Consider this trinomial as $(b^n)^2 + 7b^n + 10$. We look for numbers p and q such that $b^{2n} + 7b^n + 10 = (b^n + p)(b^n + q)$. We find two factors, both positive, whose product is 10 and whose sum is 7. They are 5 and 2.

$b^{2n} + 7b^n + 10 = (b^n + 5)(b^n + 2)$

74. $(a^m - 7)(a^m - 4)$

75. $(x + 1)a^2 + (x + 1)3a + (x + 1)2$

$= (x + 1)(a^2 + 3a + 2)$

After factoring out the common factor x + 1, we consider $a^2 + 3a + 2$. We look for two factors, both positive, whose product is 2 and whose sum is 3. They are 1 and 2.

$a^2 + 3a + 2 = (a + 1)(a + 2)$, so

$(x + 1)a^2 + (x + 1)3a + (x + 1)2 = (x + 1)(a + 1)(a + 2)$.

76. $(a - 5)(x + 9)(x - 1)$

77. We first label the drawing with additional information.

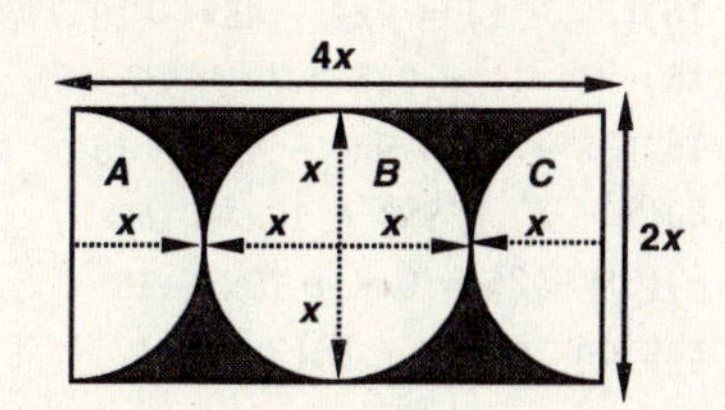

4x represents the length of the rectangle and 2x the width. The area of the rectangle is $4x \cdot 2x$, or $8x^2$.

The area of semicircle A is $\frac{1}{2}\pi x^2$.

The area of circle B is πx^2.

The area of semicircle C is $\frac{1}{2}\pi x^2$.

Area of shaded region = Area of rectangle − Area of A − Area of B − Area of C

Area of shaded region = $8x^2 - \frac{1}{2}\pi x^2 - \pi x^2 - \frac{1}{2}\pi x^2$

$= 8x^2 - 2\pi x^2$

$= 2x^2(4 - \pi)$

The shaded area can be represented by $2x^2(4 - \pi)$.

78. $x^2(\pi - 1)$

Exercise Set 5.3

1. $2x^2 - 7x - 4$

(1) Look for a common factor. There is none (other than 1 or -1).

(2) Because $2x^2$ can be factored as $2x \cdot x$, we have this possibility:

$(2x + \quad)(x + \quad)$

(3) Their are 3 pairs of factors of -4 and they can be listed two ways:

-4,1   4,-1   2,-2

and 1,-4   -1,4   -2,2

(4) Look for Outer and Inner products resulting from steps (2) and (3) for which the sum is the middle term, -7x. We try some possibilities:

$(2x - 4)(x + 1) = 2x^2 - 2x - 4$

$(2x + 4)(x - 1) = 2x^2 + 2x - 4$

$(2x + 2)(x - 2) = 2x^2 - 2x - 4$

$(2x + 1)(x - 4) = 2x^2 - 7x - 4$

The factorization is $(2x + 1)(x - 4)$.

2. $(3x - 4)(x + 1)$

3. $5x^2 + x - 18$

(1) There is no common factor (other than 1 or -1).

(2) Because $5x^2$ can be factored as $5x \cdot x$, we have this possibility:

$(5x + \quad)(x + \quad)$

(3) There are 6 pairs of factors of -18 and they can be listed two ways:

-18,1   18,-1   -9,2   9,-2   -6,3   6,-3

and 1,-18   -1,18   2,-9   -2,9   3,-6   -3,6

(4) Look for Outer and Inner products resulting from steps (2) and (3) for which the sum of is x. We try some possibilities:

$(5x - 18)(x + 1) = 5x^2 - 13x - 18$

$(5x + 18)(x - 1) = 5x^2 + 13x - 18$

$(5x - 9)(x + 2) = 5x^2 + x - 18$

The factorization is $(5x - 9)(x + 2)$.

4. $(3x + 5)(x - 3)$

5. $6x^2 + 23x + 7$

(1) There is no common factor (other than 1 or -1).

(2) Because $6x^2$ can be factored as $6x \cdot x$ or $3x \cdot 2x$, we have these possibilities:

$(6x + \quad)(x + \quad)$ and $(3x + \quad)(2x + \quad)$

(3) There are 2 pairs of factors of 7 and they can be listed two ways:

7,1   -7,-1

and 1,7   -1,-7

(4) Look for Outer and Inner products resulting from steps (2) and (3) for which the sum is 23x. Since all signs are positive, we need consider only plus signs. We try some possibilities:

$(6x + 7)(x + 1) = 6x^2 + 13x + 7$

$(3x + 7)(2x + 1) = 6x^2 + 17x + 7$

$(6x + 1)(x + 7) = 6x^2 + 43x + 7$

$(3x + 1)(2x + 7) = 6x^2 + 23x + 7$

The factorization is $(3x + 1)(2x + 7)$.

6. $(2x + 3)(3x + 2)$

7. $3x^2 + 4x + 1$

(1) There is no common factor (other than 1 or -1).

(2) Because $3x^2$ can be factored as $3x \cdot x$, we have this possibility:

$(3x + \quad)(x + \quad)$

(3) There are 2 pairs of factors of 1. In this case they can be listed only one way:

1,1 -1,-1

(4) Look for Outer and Inner products resulting from steps (2) and (3) for which the sum is 4x. Since all signs are positive, we need consider only plus signs. There is only one such possibility:

$(3x + 1)(x + 1) = 3x^2 + 4x + 1$

The factorization is $(3x + 1)(x + 1)$.

8. $(7x + 1)(x + 2)$

9. $4x^2 + 4x - 15$

(1) There is no common factor (other than 1 or -1).

(2) Because $4x^2$ can be factored as $4x \cdot x$ or $2x \cdot 2x$, we have these possibilities:

$(4x + \quad)(x + \quad)$ and $(2x + \quad)(2x + \quad)$

(3) There are 4 pairs of factors of -15 and they can be listed two ways:

	15,-1	-15,1	5,-3	-5,3
and	-1,15	1,-15	-3,5	3,-5

(4) We try some possibilities:

$(4x + 15)(x - 1) = 4x^2 + 11x - 15$

$(2x + 15)(2x - 1) = 4x^2 + 28x - 15$

$(4x - 15)(x + 1) = 4x^2 - 11x - 15$

$(2x - 15)(2x + 1) = 4x^2 - 28x - 15$

$(4x + 5)(x - 3) = 4x^2 - 7x - 15$

$(2x + 5)(2x - 3) = 4x^2 + 4x - 15$

The factorization is $(2x + 5)(2x - 3)$.

10. $(3x - 2)(3x + 4)$

11. $2x^2 - x - 1$

(1) There is no common factor (other than 1 or -1).

(2) Because $2x^2$ can be factored as $2x \cdot x$, we have this possibility:

$(2x + \quad)(x + \quad)$

(3) There is 1 pair of factors of -1 and it can be listed two ways:

-1,1 and 1,-1

(4) We try some possibilities:

$(2x - 1)(x + 1) = 2x^2 + x - 1$

$(2x + 1)(x - 1) = 2x^2 - x - 1$

The factorization is $(2x + 1)(x - 1)$.

12. $(3x - 5)(5x + 2)$

13. $9x^2 + 18x - 16$

(1) There is no common factor (other than 1 or -1).

(2) Because $9x^2$ can be factored as $9x \cdot x$ or $3x \cdot 3x$, we have these possibilities:

$(9x + \quad)(x + \quad)$ and $(3x + \quad)(3x + \quad)$

(3) There are 5 pairs of factors of -16 and they can be listed two ways:

	16,-1	-16,1	8,-2	-8,2	4,-4
and	-1,16	1,-16	-2,8	2,-8	-4,4

(4) We try some possibilities:

$(9x + 16)(x - 1) = 9x^2 + 7x - 16$

$(3x + 16)(3x - 1) = 9x^2 + 45x - 16$

$(9x - 16)(x + 1) = 9x^2 - 7x - 16$

$(3x - 16)(3x + 1) = 9x^2 - 45x - 16$

$(9x + 8)(x - 2) = 9x^2 - 10x - 16$

$(3x + 8)(3x - 2) = 9x^2 + 18x - 16$

The factorization is $(3x + 8)(3x - 2)$.

14. $(2x + 1)(x + 2)$

15. $3x^2 - 5x - 2$

(1) There is no common factor (other than 1 or -1).

(2) Because $3x^2$ can be factored as $3x \cdot x$, we have this possibility:

$(3x + \quad)(x + \quad)$

(3) There are 2 pairs of factors of -2 and they can be listed two ways:

	2,-1	-2,1
and	-1,2	1,-2

(4) We try some possibilities:

$(3x + 2)(x - 1) = 3x^2 - x - 2$

$(3x - 2)(x + 1) = 3x^2 + x - 2$

$(3x - 1)(x + 2) = 3x^2 + 5x - 2$

$(3x + 1)(x - 2) = 3x^2 - 5x - 2$

The factorization is $(3x + 1)(x - 2)$.

16. $(6x - 5)(3x + 2)$

17. $12x^2 + 31x + 20$

(1) There is no common factor (other than 1 or -1).

(2) Because $12x^2$ can be factored as $12x \cdot x$, $6x \cdot 2x$, or $4x \cdot 3x$, we have these possibilities:

$(12x + \ \ )(x + \ \ )$ and $(6x + \ \ )(2x + \ \ )$ and $(4x + \ \ )(3x + \ \ )$

(3) Since all signs are positive, we need consider only positive pairs of factors of 20. There are 3 such pairs and they can be listed two ways:

20,1 10,2 5,4

and 1,20 2,10 4,5

(4) We can immediately reject all possibilities in which either factor has a common factor, such as $(12x + 20)$ or $(6x + 4)$, because we determined at the outset that there are no common factors. We try some of the remaining possibilities:

$(12x + 1)(x + 20) = 12x^2 + 241x + 20$

$(12x + 5)(x + 4) = 12x^2 + 53x + 20$

$(6x + 1)(2x + 20) = 12x^2 + 122x + 20$

$(4x + 5)(3x + 4) = 12x^2 + 31x + 20$

The factorization is $(4x + 5)(3x + 4)$.

18. $(3x + 5)(5x - 2)$

19. $14x^2 + 19x - 3$

(1) There is no common factor (other than 1 or -1).

(2) Because $14x^2$ can be factored as $14x \cdot x$, or $7x \cdot 2x$, we have these possibilities:

$(14x + \ \ )(x + \ \ )$ and $(7x + \ \ )(2x + \ \ )$

(3) There are 2 pairs of factors of -3 and they can be listed two ways:

-1,3 -3,1

and 3,-1 1,-3

(4) We try some possibilities:

$(14x - 1)(x + 3) = 14x^2 + 41x - 3$

$(7x - 1)(2x + 3) = 14x^2 + 19x - 3$

The factorization is $(7x - 1)(2x + 3)$.

20. $(7x + 4)(5x + 2)$

21. $9x^2 + 18x + 8$

(1) There is no common factor (other than 1 or -1).

(2) Because $9x^2$ can be factored as $9x \cdot x$ or $3x \cdot 3x$, we have these possibilities:

$(9x + \ \ )(x + \ \ )$ and $(3x + \ \ )(3x + \ \ )$

(3) Since all signs are positive, we need consider only positive pairs of factors of 8. There are 2 such pairs and they can be listed two ways:

8,1 4,2

and 1,8 2,4

(4) We try some possibilities:

$(9x + 8)(x + 1) = 9x^2 + 17x + 8$

$(3x + 8)(3x + 1) = 9x^2 + 27x + 8$

$(9x + 4)(x + 2) = 9x^2 + 22x + 8$

$(3x + 4)(3x + 2) = 9x^2 + 18x + 8$

The factorization is $(3x + 4)(3x + 2)$.

22. Prime

23. $49 - 42x + 9x^2 = 9x^2 - 42x + 49$

(1) There is no common factor (other than 1 or -1).

(2) Because $9x^2$ can be factored as $9x \cdot x$ or $3x \cdot 3x$, we have these possibilities:

$(9x + \ \ )(x + \ \ )$ and $(3x + \ \ )(3x + \ \ )$

(3) Since 49 is positive and the middle term is negative, we need consider only negative pairs of factors of 49. There are 2 such pairs and one pair can be listed two ways:

-49,-1 -7,-7

and -1,-49

(4) We try some possibilities:

$(9x - 49)(x - 1) = 9x^2 - 58x + 49$

$(3x - 49)(3x - 1) = 9x^2 - 150x + 49$

$(9x - 7)(x - 7) = 9x^2 - 70x + 49$

$(3x - 7)(3x - 7) = 9x^2 - 42x + 49$

The factorization is $(3x - 7)(3x - 7)$, or $(3x - 7)^2$. This can also be expressed as follows:

$(3x - 7)^2 = (-1)^2(3x - 7)^2 = [-1 \cdot (3x - 7)]^2 = (-3x + 7)^2$, or $(7 - 3x)^2$

24. $(5x + 4)^2$

25. $24x^2 + 47x - 2$

(1) There is no common factor (other than 1 or -1).

(2) Because $24x^2$ can be factored as $24x \cdot x$, $12x \cdot 2x$, $6x \cdot 4x$, or $3x \cdot 8x$, we have these possibilities:

$(24x + \ \ )(x + \ \ )$ and $(12x + \ \ )(2x + \ \ )$ and $(6x + \ \ )(4x + \ \ )$ and $(3x + \ \ )(8x + \ \ )$

(3) There are 2 pairs of factors of -2 and they can be listed two ways:

2,-1 -2,1

and -1, 2 1,-2

(4) We can immediately reject all possibilities in which either factor has a common factor, such as $(24x + 2)$ or $(12x - 2)$, because we determined at the outset that there are no common factors. We try some of the remaining possibilities:

$(24x - 1)(x + 2) = 24x^2 + 47x - 2$

The factorization is $(24x - 1)(x + 2)$.

26. $(8a + 3)(2a + 9)$

27. $35x^2 - 57x - 44$

(1) There is no common factor (other than 1 or -1).

(2) Because $35x^2$ can be factored as $35x \cdot x$ or $7x \cdot 5x$, we have these possibilities:

$(35x + \ )(x + \ )$ and $(7x + \ )(5x + \ )$

(3) There are 6 pairs of factors of -44 and they can be listed two ways:

	1,-44	-1,44	2,-22	-2,22	4,-11	-4,11
and	-44,1	44,-1	-22,2	22,-2	-11,4	11,-4

(4) We try some possibilities:

$(35x + 1)(x - 44) = 35x^2 - 1539x - 44$

$(7x + 1)(5x - 44) = 35x^2 - 303x - 44$

$(35x + 2)(x - 22) = 35x^2 - 768x - 44$

$(7x + 2)(5x - 22) = 35x^2 - 144x - 44$

$(35x + 4)(x - 11) = 35x^2 - 381x - 44$

$(7x + 4)(5x - 11) = 35x^2 - 57x - 44$

The factorization is $(7x + 4)(5x - 11)$.

28. $(3a - 1)(3a + 5)$

29. $2x^2 - 6x - 19$

(1) There is no common factor (other than 1 or -1).

(2) Because $2x^2$ can be factored as $2x \cdot x$, we have this possibility:

$(2x + \ )(x + \ )$

(3) There are 2 pairs of factors of -19 and they can be listed two ways:

	19,-1	-19,1
and	-1,19	1,-19

(4) We try some possibilities:

$(2x + 19)(x - 1) = 2x^2 + 17x - 19$

$(2x - 1)(x + 19) = 2x^2 + 37x - 19$

The other two possibilities will only change the sign of the middle terms in these trials.

We must conclude that $2x^2 - 6x - 19$ is prime.

30. $(2x + 5)(x - 3)$

31. $12x^2 + 28x - 24$

(1) We factor out the common factor, 4:

$4(3x^2 + 7x - 6)$

Then we factor the trinomial $3x^2 + 7x - 6$.

(2) Because $3x^2$ can be factored as $3x \cdot x$, we have this possibility:

$(3x + \ )(x + \ )$

(3) There are 4 pairs of factors of -6 and they can be listed two ways:

	6,-1	-6,1	3,-2	-3,2
and	-1,6	1,-6	-2,3	2,-3

(4) We can immediately reject all possibilities in which either factor has a common factor, such as $(3x + 6)$ or $(3x - 3)$, because we factored out the largest common factor at the outset. We try some of the remaining possibilities:

$(3x - 1)(x + 6) = 3x^2 + 17x - 6$

$(3x - 2)(x + 3) = 3x^2 + 7x - 6$

The factorization of $3x^2 + 7x - 6$ is $(3x - 2)(x + 3)$. We must include the common factor in order to get a factorization of the original trinomial.

$12x^2 + 28x - 24 = 4(3x - 2)(x + 3)$

32. $3(2x + 1)(x + 5)$

33. $30x^2 - 24x - 54$

(1) Factor out the common factor, 6:

$6(5x^2 - 4x - 9)$

Then we factor the trinomial $5x^2 - 4x - 9$.

(2) Because $5x^2$ can be factored as $5x \cdot x$, we have this possibility:

$(5x + \ )(x + \ )$

(3) There are 3 pairs of factors of -9 and they can be listed two ways:

	9,-1	-9,1	3,-3
and	-1,9	1,-9	-3,3

(4) We try some possibilities:

$(5x + 9)(x - 1) = 5x^2 + 4x - 9$

$(5x - 9)(x + 1) = 5x^2 - 4x - 9$

The factorization of $5x^2 - 4x - 9$ is $(5x - 9)(x + 1)$. We must include the common factor in order to get a factorization of the original trinomial.

$30x^2 - 24x - 54 = 6(5x - 9)(x + 1)$

34. $5(4x - 1)(x - 1)$

35. $4x + 6x^2 - 10 = 6x^2 + 4x - 10$

(1) Factor out the common factor, 2:

$2(3x^2 + 2x - 5)$

Then we factor the trinomial $3x^2 + 2x - 5$.

(2) Because $3x^2$ can be factored as $3x \cdot x$, we have this possibility:

$(3x + \ )(x + \ )$

(3) There are 2 pairs of factors of -5 and they can be listed two ways:

	5,-1	-5,1
and	-1,5	1,-5

(4) We try some possibilities:

$(3x + 5)(x - 1) = 3x^2 + 2x - 5$

Then $3x^2 + 2x - 5 = (3x + 5)(x - 1)$, so $6x^2 + 4x - 10 = 2(3x + 5)(x - 1)$.

36. $3(2x - 3)(3x + 1)$

37. $3x^2 - 4x + 1$

(1) There is no common factor (other than 1 or -1).

(2) Because $3x^2$ can be factored as $3x \cdot x$, we have this possibility:

$(3x + \quad)(x + \quad)$

(3) Since 1 is positive and the middle term is negative, we need consider only negative factor pairs of 1. The only such pair is -1,-1.

(4) There is only one possibility:

$(3x - 1)(x - 1) = 3x^2 - 4x + 1$

The factorization is $(3x - 1)(x - 1)$.

38. $(2x - 3)(3x - 2)$

39. $12x^2 - 28x - 24$

(1) Factor out the common factor, 4:

$4(3x^2 - 7x - 6)$

Then we factor the trinomial $3x^2 - 7x - 6$.

(2) Because $3x^2$ can be factored as $3x \cdot x$, we have this possibility:

$(3x + \quad)(x + \quad)$

(3) There are 4 pairs of factors of -6 and they can be listed two ways:

	6,-1	-6,1	3,-2	-3,2
and	-1,6	1,-6	-2,3	2,-3

(4) We can immediately reject all possibilities in which either factor has a common factor, such as $(3x - 6)$ or $(3x + 3)$, because we factored out the largest common factor at the outset. We try some of the remaining possibilities:

$(3x - 1)(x + 6) = 3x^2 + 17x - 6$

$(3x - 2)(x + 3) = 3x^2 + 7x - 6$

$(3x + 2)(x - 3) = 3x^2 - 7x - 6$

Then $3x^2 - 7x - 6 = (3x + 2)(x - 3)$, so $12x^2 - 28x - 24 = 4(3x + 2)(x - 3)$.

40. $3(2x - 1)(x - 5)$

41. $-1 + 2x^2 - x = 2x^2 - x - 1$

(1) There is no common factor (other than 1 or -1).

(2) Because $2x^2$ can be factored as $2x \cdot x$, we have this possibility:

$(2x + \quad)(x + \quad)$

(3) There is 1 pair of factors of -1 and it can be listed two ways:

1,-1 and -1,1

(4) We try some possibilities:

$(2x + 1)(x - 1) = 2x^2 - x - 1$

The factorization is $(2x + 1)(x - 1)$.

42. $(5x - 3)(3x - 2)$

43. $9x^2 - 18x - 16$

(1) There is no common factor (other than 1 or -1).

(2) Because $9x^2$ can be factored as $9x \cdot x$, or $3x \cdot 3x$. We have these possibilities:

$(9x + \quad)(x + \quad)$ and $(3x + \quad)(3x + \quad)$

(3) There are 5 pairs of factors of -16 and they can be listed two ways:

	16,-1	-16,1	8,-2	-8,2	4,-4
and	-1,16	1,-16	-2,8	2,-8	-4,4

(4) We try some possibilities:

$(9x + 16)(x - 1) = 9x^2 + 7x - 16$

$(3x + 16)(3x - 1) = 9x^2 + 45x - 16$

$(9x + 8)(x - 2) = 9x^2 - 10x - 16$

$(3x + 8)(3x - 2) = 9x^2 + 18x - 16$

$(3x - 8)(3x + 2) = 9x^2 - 18x - 16$

The factorization is $(3x - 8)(3x + 2)$.

44. $7(2x + 1)(x + 2)$

45. $15x^2 - 25x - 10$

(1) Factor out the common factor, 5:

$5(3x^2 - 5x - 2)$

Then we factor the trinomial $3x^2 - 5x - 2$. This was done in Exercise 15. We know that $3x^2 - 5x - 2 = (3x + 1)(x - 2)$, so $15x^2 - 25x - 10 = 5(3x + 1)(x - 2)$.

46. $(6x + 5)(3x - 2)$

47. $12x^3 + 31x^2 + 20x$

(1) We factor out the common factor, x:

$x(12x^2 + 31x + 20)$

Then we factor the trinomial $12x^2 + 31x + 20$. This was done in Exercise 17. We know that $12x^2 + 31x + 20 = (3x + 4)(4x + 5)$, so $12x^3 + 31x^2 + 20x = x(3x + 4)(4x + 5)$.

48. $x(5x - 2)(3x + 5)$

49. $14x^4 + 19x^3 - 3x^2$

(1) Factor out the common factor, x^2:

$x^2(14x^2 + 19x - 3)$

Then we factor the trinomial $14x^2 + 19x - 3$. This was done in Exercise 19. We know that $14x^2 + 19x - 3 = (7x - 1)(2x + 3)$, so $14x^4 + 19x^3 - 3x^2 = x^2(7x - 1)(2x + 3)$.

50. $2x^2(5x + 2)(7x + 4)$

51. $168x^3 - 45x^2 + 3x$

(1) Factor out the common factor, 3x:

$3x(56x^2 - 15x + 1)$

Then we factor the trinomial $56x^2 - 15x + 1$.

(2) Because $56x^2$ can be factored as $56x \cdot x$, $28x \cdot 2x$, $14x \cdot 4x$, or $7x \cdot 8x$, we have these possibilities:

$(56x + \quad)(x + \quad)$ and $(28x + \quad)(2x + \quad)$ and

$(14x + \quad)(4x + \quad)$ and $(7x + \quad)(8x + \quad)$

(3) Since 1 is positive and the middle term is negative we need consider only the negative factor pair -1, -1.

(4) We try some possibilities:

$(56x - 1)(x - 1) = 56x^2 - 57x + 1$

$(28x - 1)(2x - 1) = 56x^2 - 30x + 1$

$(14x - 1)(4x - 1) = 56x^2 - 18x + 1$

$(7x - 1)(8x - 1) = 56x^2 - 15x + 1$

Then $56x^2 - 15x + 1 = (7x - 1)(8x - 1)$, so $168x^3 - 45x^2 + 3x = 3x(7x - 1)(8x - 1)$.

52. $24x^3(3x + 2)(2x + 1)$

53. $15x^2 - 19x + 6$

(1) There is no common factor (other than 1 or -1).

(2) Because $15x^2$ can be factored as $15x \cdot x$ or $5x \cdot 3x$, we have these possibilities:

$(15x + \quad)(x + \quad)$ and $(5x + \quad)(3x + \quad)$

(3) Since 6 is positive and the middle term is negative, we need consider only negative factor pairs. There are 2 such pairs and they can be listed two ways:

	-6,-1	-3,-2
and	-1,-6	-2,-3

(4) We can immediately reject all possibilities in which either factor has a common factor, such as $(15x - 6)$ or $(3x - 3)$, because we determined at the outset that there is no common factor. We try some of the remaining possibilities:

$(15x - 1)(x - 6) = 15x^2 - 91x + 6$

$(15x - 2)(x - 3) = 15x^2 - 47x + 6$

$(5x - 6)(3x - 1) = 15x^2 - 23x + 6$

$(5x - 3)(3x - 2) = 15x^2 - 19x + 6$

The factorization is $(5x - 3)(3x - 2)$.

54. $(3x + 2)(3x + 4)$

55. $25t^2 + 80t + 64$

(1) There is no common factor (other than 1 or -1).

(2) Because $25t^2$ can be factored as $25t \cdot t$ or $5t \cdot 5t$, we have these possibilities:

$(25t + \quad)(t + \quad)$ and $(5t + \quad)(5t + \quad)$

(3) Since all signs are positive, we need consider only positive pairs of factors. There are 4 such pairs and 3 of them can be listed two ways:

	64,1	32,2	16,4	8,8
and	1,64	2,32	4,16	

(4) We try some possibilities:

$(25t + 64)(t + 1) = 25t^2 + 89t + 64$

$(5t + 32)(5t + 2) = 25t^2 + 170t + 64$

$(25t + 16)(t + 4) = 25t^2 + 116t + 64$

$(5t + 8)(5t + 8) = 25t^2 + 80t + 64$

The factorization is $(5t + 8)(5t + 8)$, or $(5t + 8)^2$.

56. $(3x - 7)^2$

57. $6x^3 + 4x^2 - 10x$

(1) Factor out the common factor, $2x$:

$2x(3x^2 + 2x - 5)$

Then we factor the trinomial $3x^2 + 2x - 5$. We did this in Exercise 35 (after we factored 2 out of the original trinomial). We know that $3x^2 + 2x - 5 = (3x + 5)(x - 1)$, so $6x^3 + 4x^2 - 10x = 2x(3x + 5)(x - 1)$.

58. $3x(3x + 1)(2x - 3)$

59. $25x^2 + 89x + 64$

We follow the same procedure as in Exercise 55. The factorization is the first possibility we tried in step (4): $(25x + 64)(x + 1)$

60. Prime

61. $x^2 + 3x - 7$

(1) There is no common factor (other than 1 or -1).

(2) Because x^2 can be factored as $x \cdot x$, we have this possibility:

$(x + \quad)(x + \quad)$

(3) There are 2 pairs of factors of -7. Since the coefficient of x^2 is 1, we only list them one way:

7,-1   -7,1

(4) We try the possibilities:

$(x + 7)(x - 1) = x^2 + 6x - 7$

$(x - 7)(x + 1) = x^2 - 6x - 7$

Neither possibility works. Thus, $x^2 + 3x - 7$ is prime. (Note that we could have also used the method of Section 5.2 since the leading coefficient of this trinomial is 1.)

62. Prime

63. $12m^2 + mn - 20n^2$

(1) There is no common factor (other than 1 or -1).

(2) Because $12m^2$ can be factored as $12m \cdot m$, $6m \cdot 2m$, or $3m \cdot 4m$, we have these possibilities:

$(12m + \quad)(m + \quad)$ and $(6m + \quad)(2m + \quad)$ and

$(3m + \quad)(4m + \quad)$

(3) There are 6 pairs of factors of $-20n^2$ and they can be listed two ways:

$20n,-n$ $-20n,n$ $10n,-2n$ $-10n,2n$

$-n,20n$ $n,-20n$ $-2n,10n$ $2n,-10n$

$5n,-4n$ $-5n,4n$

$-4n,5n$ $4n,-5n$

(4) We can immediately reject all possibilities in which either factor has a common factor, such as $(12m + 20n)$ or $(4m - 2n)$, because we determined at the outset that there is no common factor. We try some of the remaining possibilities:

$(12m - n)(m + 20n) = 12m^2 + 239mn - 20n^2$

$(12m + 5n)(m - 4n) = 12m^2 - 43mn - 20n^2$

$(3m - 20n)(4m + n) = 12m^2 - 77mn - 20n^2$

$(3m - 4n)(4m + 5n) = 12m^2 - mn - 20n^2$

$(3m + 4n)(4m - 5n) = 12m^2 + mn - 20n^2$

The factorization is $(3m + 4n)(4m - 5n)$.

64. $(4a + 3b)(3a + 2b)$

65. $6a^2 - ab - 15b^2$

(1) There is no common factor (other than 1 or -1).

(2) Because $6a^2$ can be factored as $6a \cdot a$ or $3a \cdot 2a$. We have these possibilities:

$(6a + \quad)(a + \quad)$ and $(3a + \quad)(2a + \quad)$

(3) There are 4 pairs of factors of $-15b^2$ and they can be listed two ways:

$15b,-b$ $-15b,b$ $5b,-3b$ $-5b,3b$

and $-b,15b$ $b,-15b$ $-3b,5b$ $3b,-5b$

(4) We can immediately reject all possibilities in which either factor has a common factor, such as $(6a + 15b)$ or $(3a - 3b)$, because we determined at the outset that there is no common factor. We try some of the remaining possibilities:

$(6a - b)(a + 15b) = 6a^2 + 89ab - 15b^2$

$(3a - b)(2a + 15b) = 6a^2 + 43ab - 15b^2$

$(6a + 5b)(a - 3b) = 6a^2 - 13ab - 15b^2$

$(3a + 5b)(2a - 3b) = 6a^2 + ab - 15b^2$

$(3a - 5b)(2a + 3b) = 6a^2 - ab - 15b^2$

The factorization is $(3a - 5b)(2a + 3b)$.

66. $(3p + 2q)(p - 6q)$

67. $9a^2 + 18ab + 8b^2$

(1) There is no common factor (other than 1 or -1).

(2) Because $9a^2$ can be factored as $9a \cdot a$ or $3a \cdot 3a$, we have these possibilities:

$(9a + \quad)(a + \quad)$ and $(3a + \quad)(3a + \quad)$

(3) Since all signs are positive, we need consider only pairs of factors of $8b^2$ with positive coefficients. There are 2 such pairs and they can be listed two ways:

$8b,b$ $4b,2b$

and $b,8b$ $2b,4b$

(4) We try some possibilities:

$(9a + 8b)(a + b) = 9a^2 + 17ab + 8b^2$

$(3a + 8b)(3a + b) = 9a^2 + 27ab + 8b^2$

$(9a + 4b)(a + 2b) = 9a^2 + 22ab + 8b^2$

$(3a + 4b)(3a + 2b) = 9a^2 + 18ab + 8b^2$

The factorization is $(3a + 4b)(3a + 2b)$.

68. $2(5s - 3t)(s + t)$

69. $35p^2 + 34pq + 8q^2$

(1) There is no common factor (other than 1 or -1).

(2) Because $35p^2$ can be factored as $35p \cdot p$ or $7p \cdot 5p$, we have these possibilities:

$(35p + \quad)(p + \quad)$ and $(7p + \quad)(5p + \quad)$

(3) Since all signs are positive, we need consider only pairs of factors of $8q^2$ with positive coefficients. There are 2 such pairs and they can be listed two ways:

$8q,q$ $4q,2q$

and $q,8q$ $2q,4q$

(4) We try some possibilities:

$(35p + 8q)(p + q) = 35p^2 + 43pq + 8q^2$

$(7p + 8q)(5p + q) = 35p^2 + 47pq + 8q^2$

$(35p + 4q)(p + 2q) = 35p^2 + 74pq + 8q^2$

$(7p + 4q)(5p + 2q) = 35p^2 + 34pq + 8q^2$

The factorization is $(7p + 4q)(5p + 2q)$.

70. $3(2a + 5b)(5a + 2b)$

71. $18x^2 - 6xy - 24y^2$

(1) Factor out the common factor, 6:

$6(3x^2 - xy - 4y^2)$

Then we factor the trinomial $3x^2 - xy - 4y^2$.

(2) Because $3x^2$ can be factored as $3x \cdot x$, we have this possibility:

$(3x + \quad)(x + \quad)$

(3) There are 3 pairs of factors of $-4y^2$ and they can be listed two ways:

$4y,-y$ $-4y,y$ $2y,-2y$

and $-y,4y$ $y,-4y$ $-2y,2y$

(4) We try some possibilities:

$(3x + 4y)(x - y) = 3x^2 + xy - 4y^2$

$(3x - 4y)(x + y) = 3x^2 - xy - 4y^2$

Then $3x^2 - xy - 4y^2 = (3x - 4y)(x + y)$, so $18x^2 - 6xy - 24y^2 = 6(3x - 4y)(x + y)$.

72. $5(3a - 4b)(a + b)$

73. $y^2 + 4y + y + 4 = y(y + 4) + 1(y + 4)$
$= (y + 4)(y + 1)$

74. $(x + 5)(x + 2)$

75. $x^2 - 4x - x + 4 = x(x - 4) - 1(x - 4)$
$= (x - 4)(x - 1)$

76. $(a + 5)(a - 2)$

77. $6x^2 + 4x + 9x + 6 = 2x(3x + 2) + 3(3x + 2)$
$= (3x + 2)(2x + 3)$

78. $(3x - 2)(x + 1)$

79. $3x^2 - 4x - 12x + 16 = x(3x - 4) - 4(3x - 4)$
$= (3x - 4)(x - 4)$

80. $(4 - 3y)(6 - 5y)$

81. $35x^2 - 40x + 21x - 24 = 5x(7x - 8) + 3(7x - 8)$
$= (7x - 8)(5x + 3)$

82. $(4x - 3)(2x - 7)$

83. $4x^2 + 6x - 6x - 9 = 2x(2x + 3) - 3(2x + 3)$
$= (2x + 3)(2x - 3)$

84. $(x^2 - 3)(2x^2 - 5)$

85. $2x^2 - 7x - 4$

a) First look for a common factor. There is none (other than 1).

b) Multiply the leading coefficient and the constant, 2 and -4: $2(-4) = -8$.

c) Try to factor -8 so that the sum of the factors is -7.

Pairs of factors of -8	Sums of factors
-1, 8	7
1, -8	-7
-2, 4	2
2, -4	-2

d) Split the middle term: $-7x = 1x - 8x$

e) Factor by grouping:

$2x^2 - 7x - 4 = 2x^2 + x - 8x - 4$
$= x(2x + 1) - 4(2x + 1)$
$= (2x + 1)(x - 4)$

86. $(x + 1)(3x - 4)$

87. $5x^2 + x - 18$

a) First look for a common factor. There is none (other than 1).

b) Multiply the leading coefficient and the constant, 5 and -18: $5(-18) = -90$.

c) Try to factor -90 so that the sum of the factors is 1.

Pairs of factors of -90	Sums of factors
-1, 90	89
1, -90	-89
-2, 45	43
2, -45	-43
-3, 30	27
3, -30	-27
-5, 18	13
5, - 18	-13
-6, 15	9
6, -15	-9
-9, 10	1
9, -10	-1

d) Split the middle term: $x = -9x + 10x$

e) Factor by grouping:

$5x^2 + x - 18 = 5x^2 - 9x + 10x - 18$
$= x(5x - 9) + 2(5x - 9)$
$= (5x - 9)(x + 2)$

88. $(x - 3)(3x + 5)$

89. $6x^2 + 23x + 7$

a) First look for a common factor. There is none (other than 1).

b) Multiply the leading coefficient and the constant, 6 and 7: $6 \cdot 7 = 42$.

c) Try to factor 42 so that the sum of the factors is 23. We only need to consider positive factors.

Pairs of factors of 42	Sums of factors
1, 42	43
2, 21	23
3, 14	17
6, 7	13

d) Split the middle term: $23x = 2x + 21x$

e) Factor by grouping:

$6x^2 + 23x + 7 = 6x^2 + 2x + 21x + 7$
$= 2x(3x + 1) + 7(3x + 1)$
$= (3x + 1)(2x + 7)$

90. $(2x + 3)(3x + 2)$

91. $3x^2 + 4x + 1$

a) First look for a common factor. There is none (other than 1).

b) Multiply the leading coefficient and the constant, 3 and 1: $3 \cdot 1 = 3$.

c) Try to factor 3 so that the sum of the factors is 4. The numbers we want are 1 and 3: $1 \cdot 3 = 3$ and $1 + 3 = 4$.

d) Split the middle term: $4x = 1x + 3x$

e) Factor by grouping:

$3x^2 + 4x + 1 = 3x^2 + x + 3x + 1$
$= x(3x + 1) + 1(3x + 1)$
$= (3x + 1)(x + 1)$

92. $(x + 2)(7x + 1)$

93. $4x^2 + 4x - 15$

a) First look for a common factor. There is none (other than 1).

b) Multiply the leading coefficient and the constant, 4 and -15: $4(-15) = -60$.

c) Try to factor -60 so that the sum of the factors is 4.

Pairs of factors of -60	Sums of factors
-1, 60	59
1, -60	-59
-2, 30	28
2, -30	-28
-3, 20	17
3, -20	-17
-4, 15	11
4, -15	-11
-5, 12	7
5, -12	-7
-6, 10	4
6, -10	-4

d) Split the middle term: $4x = -6x + 10x$

e) Factor by grouping:

$4x^2 + 4x - 15 = 4x^2 - 6x + 10x - 15$
$= 2x(2x - 3) + 5(2x - 3)$
$= (2x - 3)(2x + 5)$

94. $(3x + 4)(3x - 2)$

95. $2x^2 - x - 1$

a) First look for a common factor. There is none (other than 1).

b) Multiply the leading coefficient and the constant, 2 and -1: $2(-1) = -2$.

c) Try to factor -2 so that the sum of the factors is -1. The numbers we want are -2 and 1: $-2\cdot1 = -2$ and $-2 + 1 = -1$.

d) Split the middle term: $-x = -2x + 1x$

e) Factor by grouping:

$2x^2 - x - 1 = 2x^2 - 2x + x - 1$
$= 2x(x - 1) + 1(x - 1)$
$= (x - 1)(2x + 1)$

96. $(3x - 5)(5x + 2)$

97. $9x^2 + 18x - 16$

a) First look for a common factor. There is none (other than 1).

b) Multiply the leading coefficient and the constant, 9 and -16: $9(-16) = -144$.

c) Try to factor -144 so that the sum of the factors is 18.

Pairs of factors of -144	Sums of factors
-1, 144	143
1, -144	-143
-2, 72	70
2, -72	-70
-3, 48	45
3, -48	-45
-4, 36	32
4, -36	-32
-6, 24	18
6, -24	-18
-8, 18	10
8, -18	-10
-9, 16	7
9, -16	-7
-12, 12	0

d) Split the middle term: $18x = -6x + 24x$

e) Factor by grouping:

$9x^2 + 18x - 16 = 9x^2 - 6x + 24x - 16$
$= 3x(3x - 2) + 8(3x - 2)$
$= (3x - 2)(3x + 8)$

98. $(x + 2)(2x + 1)$

99. $3x^2 - 5x - 2$

a) First look for a common factor. There is none (other than 1).

b) Multiply the leading coefficient and the constant, 3 and -2: $3(-2) = -6$.

c) Try to factor -6 so that the sum of the factors is -5. The numbers we want are 1 and -6: $1(-6) = -6$ and $1 + (-6) = -5$.

d) Split the middle term: $-5x = 1x - 6x$

e) Factor by grouping:

$3x^2 - 5x - 2 = 3x^2 + x - 6x - 2$
$= x(3x + 1) - 2(3x + 1)$
$= (3x + 1)(x - 2)$

100. $(3x + 2)(6x - 5)$

101. $12x^2 + 31x + 20$

a) First look for a common factor. There is none (other than 1).

b) Multiply the leading coefficient and the constant, 12 and 20: $12\cdot20 = 240$.

c) Try to factor 240 so that the sum of the factors is 31. We only need to consider positive factors.

Pairs of factors of 240	Sums of factors
1, 240	241
2, 120	122
3, 80	83
4, 60	64
5, 48	53
6, 40	46
8, 30	38
10, 24	34
12, 20	32
15, 16	31

d) Split the middle term: $31x = 15x + 16x$

e) Factor by grouping:

$$12x^2 + 31x + 20 = 12x^2 + 15x + 16x + 20$$
$$= 3x(4x + 5) + 4(4x + 5)$$
$$= (4x + 5)(3x + 4)$$

102. $(3x + 5)(5x - 2)$

103. $14x^2 + 19x - 3$

a) First look for a common factor. There is none (other than 1).

b) Multiply the leading coefficient and the constant, 14 and -3: $14(-3) = -42$.

c) Try to factor -42 so that the sum of the factors is 19.

Pairs of factors of -42	Sums of factors
-1, 42	41
1, -42	-41
-2, 21	19
2, -21	-19
-3, 14	11
3, -14	-11
-6, 7	1
6, -7	-1

d) Split the middle term: $19x = -2x + 21x$

e) Factor by grouping:

$$14x^2 + 19x - 3 = 14x^2 - 2x + 21x - 3$$
$$= 2x(7x - 1) + 3(7x - 1)$$
$$= (7x - 1)(2x + 3)$$

104. $(7x + 4)(5x + 2)$

105. $9x^2 + 18x + 8$

a) First look for a common factor. There is none (other than 1).

b) Multiply the leading coefficient and the constant, 9 and 8: $9\cdot8 = 72$.

c) Try to factor 72 so that the sum of the factors is 18. We only need to consider positive factors.

Pairs of factors of 72	Sums of factors
1, 72	73
2, 36	38
3, 24	27
4, 18	22
6, 12	18
8, 9	17

d) Split the middle term: $18x = 6x + 12x$.

e) Factor by grouping:

$$9x^2 + 18x + 8 = 9x^2 + 6x + 12x + 8$$
$$= 3x(3x + 2) + 4(3x + 2)$$
$$= (3x + 2)(3x + 4)$$

106. $(2 - 3x)(3 - 2x)$, or $(3x - 2)(2x - 3)$

107. $49 - 42x + 9x^2 = 9x^2 - 42x + 49$

a) First look for a common factor. There is none (other than 1).

b) Multiply the leading coefficient and the constant, 9 and 49: $9\cdot49 = 441$

c) Try to factor 441 so that the sum of the factors is -42. We only need to consider negative factors.

Pairs of factors of 441	Sums of factors
-1, -441	-442
-3, -147	-150
-7, -63	-70
-9, -49	-58
-21, -21	-42

d) Split the middle term: $-42x = -21x - 21x$

e) Factor by grouping:

$$9x^2 - 42x + 49 = 9x^2 - 21x - 21x + 49$$
$$= 3x(3x - 7) - 7(3x - 7)$$
$$= (3x - 7)(3x - 7), \text{ or}$$
$$(3x - 7)^2$$

108. $(5x + 4)^2$

109. Familiarize. We will use the formula $C = 2\pi r$, where C is circumference and r is radius, to find the radius in kilometers. Then we will multiply that number by 0.62 to find the radius in miles.

Translate.

$$\text{Circumference} = 2 \cdot \pi \cdot \text{radius}$$
$$40{,}000 \approx 2(3.14)r$$

Carry out. First we solve the equation.

$$40{,}000 \approx 2(3.14)r$$
$$40{,}000 \approx 6.28r$$
$$6369 \approx r$$

Then we multiply to find the radius in miles:

$6369(0.62) \approx 3949$

Check. If $r = 6369$, then $2\pi r = 2(3.14)(6369) \approx 40{,}000$. We should also recheck the multiplication we did to find the radius in miles. Both values check.

State. The radius of the earth is about 6369 km or 3949 mi. (These values may differ slightly if a different approximation is used for π.)

110. 40°

111. Graph: $y = \frac{2}{5}x - 1$

Because the equation is in the form $y = mx + b$, we know the y-intercept is $(0,-1)$. We find two other points on the line, substituting multiples of 5 for x to avoid fractions.

When $x = -5$, $y = \frac{2}{5}(-5) - 1 = -2 - 1 = -3$.

When $x = 5$, $y = \frac{2}{5}(5) - 1 = 2 - 1 = 1$.

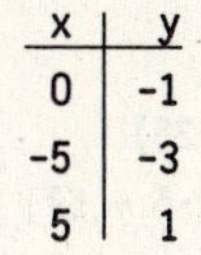

x	y
0	-1
-5	-3
5	1

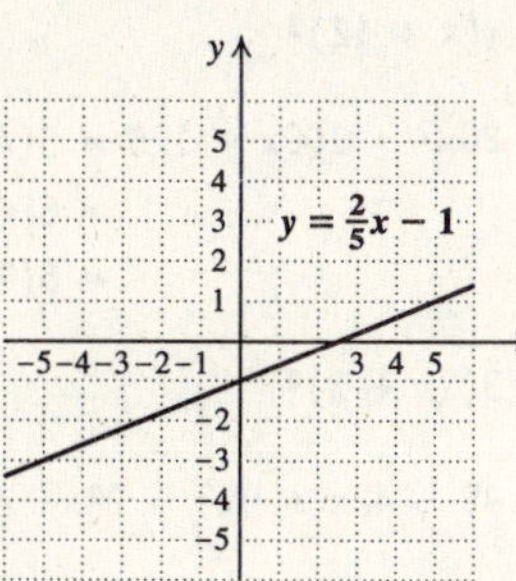

112. y^8

113.

114.

115. $9x^{10} - 12x^5 + 4 = 9(x^5)^2 - 12x^5 + 4$

(1) There is no common factor (other than 1 or -1).

(2) Because $9x^{10}$, or $9(x^5)^2$, can be factored as $9x^5 \cdot x^5$ or $3x^5 \cdot 3x^5$ we have these possibilities:

$(9x^5 + \quad)(x^5 + \quad)$ and $(3x^5 + \quad)(3x^5 + \quad)$

(3) Since the middle term is negative and the constant is positive, we need consider only negative pairs of factors of 4. There are 2 such pairs and one of them can be listed two ways:

-1,-4   -2,-2

and   -4,-1

(4) We try some possibilities:

$(9x^5 - 1)(x^5 - 4) = 9x^{10} - 37x^5 + 4$

$(3x^5 - 2)(3x^5 - 2) = 9x^{10} - 12x^5 + 4$

The factorization of $9x^{10} - 12x^5 + 4$ is $(3x^5 - 2)(3x^5 - 2)$, or $(3x^5 - 2)^2$.

116. $(4x^5 + 1)^2$

117. $20x^{2n} + 16x^n + 3 = 20(x^n)^2 + 16x^n + 3$

(1) There is no common factor (other than 1 or -1).

(2) Because $20x^{2n}$ can be factored as $20x^n \cdot x^n$, $10x^n \cdot 2x^n$, or $5x^n \cdot 4x^n$, we have these possibilities:

$(20x^n + \quad)(x^n + \quad)$ and $(10x^n + \quad)(2x^n + \quad)$ and $(5x^n + \quad)(4x^n + \quad)$

(3) Since all signs are positive, we need consider only the positive factor pair 3, 1 when factoring 3. This pair can also be listed as 1, 3.

(4) We try some possibilities:

$(20x^n + 3)(x^n + 1) = 20x^{2n} + 23x^n + 3$

$(10x^n + 3)(2x^n + 1) = 20x^{2n} + 16x^n + 3$

The factorization is $(10x^n + 3)(2x^n + 1)$.

118. $-(3x^m - 4)(5x^m - 2)$

119. $3x^{6a} - 2x^{3a} - 1 = 3(x^{3a})^2 - 2x^{3a} - 1$

(1) There is no common factor (other than 1 or -1).

(2) Because $3x^{6a}$ can be factored as $3x^{3a} \cdot x^{3a}$, we have this possibility:

$(3x^{3a} + \quad)(x^{3a} + \quad)$

(3) There is 1 pair of factors of -1 and it can be listed two ways:

-1,1   and   1,-1

(4) We try these possibilities:

$(3x^{3a} - 1)(x^{3a} + 1) = 3x^{6a} + 2x^{3a} - 1$

$(3x^{3a} + 1)(x^{3a} - 1) = 3x^{6a} - 2x^{3a} - 1$

The factorization is $(3x^{3a} + 1)(x^{3a} - 1)$.

120. $x(x^n - 1)^2$

121. $3(a + 1)^{n+1}(a + 3)^2 - 5(a + 1)^n(a + 3)^3$

$= (a + 1)^n(a + 3)^2[3(a + 1) - 5(a + 3)]$ Removing the common factors

$= (a + 1)^n(a + 3)^2[3a + 3 - 5a - 15]$ Simplifying

$= (a + 1)^n(a + 3)^2(-2a - 12)$ inside the brackets

$= (a + 1)^n(a + 3)^2(-2)(a + 6)$ Removing the common factor

$= -2(a + 1)^n(a + 3)^2(a + 6)$ Rearranging

Exercise Set 5.4

1. $x^2 - 14x + 49$

a) We know that x^2 and 49 are squares.

b) There is no minus sign before either x^2 or 49.

c) If we multiply the square roots, x and 7, and double the product, we get $2 \cdot x \cdot 7 = 14x$. This is the opposite of the remaining term, $-14x$.

Thus, $x^2 - 14x + 49$ is a trinomial square.

2. Yes

3. $x^2 + 16x - 64$

Both x^2 and 64 are squares, but there is a minus sign before 64. Thus, $x^2 + 16x - 64$ is not a trinomial square.

4. No

5. $x^2 - 3x + 9$

a) Both x^2 and 9 are squares.

b) There is no minus sign before either x^2 or 9.

c) If we multiply the square roots, x and 3, and double the product, we get $2 \cdot x \cdot 3 = 6x$. This is neither the remaining term nor its opposite.

Thus, $x^2 - 3x + 9$ is not a trinomial square.

6. No

7. $8x^2 + 40x + 25$

Only one term, 25, is a square. Thus, $8x^2 + 40x + 25$ is not a trinomial square.

8. No

9. $x^2 - 14x + 49 = x^2 - 2 \cdot x \cdot 7 + 7^2 = (x - 7)^2$

$$\begin{array}{ccccc} \uparrow & \uparrow\uparrow\uparrow & \uparrow & & \\ A^2 - & 2\ A\ B & + B^2 & = & (A - B)^2 \end{array}$$

10. $(x - 8)^2$

11. $x^2 + 16x + 64 = x^2 + 2 \cdot x \cdot 8 + 8^2 = (x + 8)^2$

$$\begin{array}{ccccc} \uparrow & \uparrow\uparrow\uparrow & \uparrow & & \\ A^2 + & 2\ A\ B & + B^2 & = & (A + B)^2 \end{array}$$

12. $(x + 7)^2$

13. $x^2 - 2x + 1 = x^2 - 2 \cdot x \cdot 1 + 1^2 = (x - 1)^2$

14. $(x + 1)^2$

15. $4 + 4x + x^2 = x^2 + 4x + 4$ Changing the order

$= x^2 + 2 \cdot x \cdot 2 + 2^2$

$= (x + 2)^2$

16. $(x - 2)^2$

17. $9x^2 + 6x + 1 = (3x)^2 + 2 \cdot 3x \cdot 1 + 1^2$

$= (3x + 1)^2$

18. $(5x - 1)^2$

19. $49 - 56y + 16y^2 = 16y^2 - 56y + 49$

$= (4y)^2 - 2 \cdot 4y \cdot 7 + 7^2$

$= (4y - 7)^2$

We could also factor as follows:

$49 - 56y + 16y^2 = 7^2 - 2 \cdot 7 \cdot 4y + (4y)^2$

$= (7 - 4y)^2$

20. $3(4m + 5)^2$

21. $2x^2 - 4x + 2 = 2(x^2 - 2x + 1)$

$= 2(x^2 - 2 \cdot x \cdot 1 + 1^2)$

$= 2(x - 1)^2$

22. $2(x - 10)^2$

23. $x^3 - 18x^2 + 81x = x(x^2 - 18x + 81)$

$= x(x^2 - 2 \cdot x \cdot 9 + 9^2)$

$= x(x - 9)^2$

24. $x(x + 12)^2$

25. $20x^2 + 100x + 125 = 5(4x^2 + 20x + 25)$

$= 5[(2x)^2 + 2 \cdot 2x \cdot 5 + 5^2]$

$= 5(2x + 5)^2$

26. $3(2x + 3)^2$

27. $49 - 42x + 9x^2 = 7^2 - 2 \cdot 7 \cdot 3x + (3x)^2 = (7 - 3x)^2$

28. $(8 - 7x)^2$, or $(7x - 8)^2$

29. $5y^2 + 10y + 5 = 5(y^2 + 2y + 1)$

$= 5(y^2 + 2 \cdot y \cdot 1 + 1^2)$

$= 5(y + 1)^2$

30. $2(a + 7)^2$

31. $2 + 20x + 50x^2 = 2(1 + 10x + 25x^2)$

$= 2[1^2 + 2 \cdot 1 \cdot 5x + (5x)^2]$

$= 2(1 + 5x)^2$

32. $7(1 - a)^2$, or $7(a - 1)^2$

33. $4p^2 + 12pq + 9q^2 = (2p)^2 + 2 \cdot 2p \cdot 3q + (3q)^2$

$= (2p + 3q)^2$

34. $(5m + 2n)^2$

35. $a^2 - 14ab + 49b^2 = a^2 - 2 \cdot a \cdot 7b + (7b)^2$

$= (a - 7b)^2$

36. $(x - 3y)^2$

37. $64m^2 + 16mn + n^2 = (8m)^2 + 2 \cdot 8m \cdot n + n^2$

$= (8m + n)^2$

38. $(9p - q)^2$

39. $16s^2 - 40st + 25t^2 = (4s)^2 - 2\cdot 4s\cdot 5t + (5t)^2$
$= (4s - 5t)^2$

40. $4(3a + 4b)^2$

41. $x^2 - 4$
a) The first expression is a square: x^2
The second expression is a square: $4 = 2^2$
b) The terms have different signs.
$x^2 - 4$ is a difference of squares.

42. Yes

43. $x^2 + 36$
The terms do not have different signs.
$x^2 + 36$ is not a difference of squares.

44. No

45. $x^2 - 35$
The second expression, 35, is not a square.
$x^2 - 35$ is not a difference of squares.

46. No

47. $16x^2 - 25$
a) The first expression is a square:
$16x^2 = (4x)^2$
The second expression is a square: $25 = 5^2$
b) The terms have different signs.
$16x^2 - 25$ is a difference of squares.

48. Yes

49. $y^2 - 4 = y^2 - 2^2 = (y + 2)(y - 2)$

50. $(x + 6)(x - 6)$

51. $p^2 - 9 = p^2 - 3^2 = (p + 3)(p - 3)$

52. $(q + 1)(q - 1)$

53. $-49 + t^2 = t^2 - 49 = t^2 - 7^2 = (t + 7)(t - 7)$

54. $(m + 8)(m - 8)$

55. $a^2 - b^2 = (a + b)(a - b)$

56. $(p + q)(p - q)$

57. $25t^2 - m^2 = (5t)^2 - m^2 = (5t + m)(5t - m)$

58. $(w + 7z)(w - 7z)$

59. $100 - k^2 = 10^2 - k^2 = (10 + k)(10 - k)$

60. $(9 + w)(9 - w)$

61. $16a^2 - 9 = (4a)^2 - 3^2 = (4a + 3)(4a - 3)$

62. $(5x + 2)(5x - 2)$

63. $4x^2 - 25y^2 = (2x)^2 - (5y)^2 = (2x + 5y)(2x - 5y)$

64. $(3a + 4b)(3a - 4b)$

65. $8x^2 - 98 = 2(4x^2 - 49) = 2[(2x)^2 - 7^2] =$
$2(2x + 7)(2x - 7)$

66. $6(2x + 3)(2x - 3)$

67. $36x - 49x^3 = x(36 - 49x^2) = x[6^2 - (7x)^2] =$
$x(6 + 7x)(6 - 7x)$

68. $x(4 + 9x)(4 - 9x)$

69. $49a^4 - 81 = (7a^2)^2 - 9^2 = (7a^2 + 9)(7a^2 - 9)$

70. $(5a^2 + 3)(5a^2 - 3)$

71. $x^4 - 1 = (x^2)^2 - 1^2$
$= (x^2 + 1)(x^2 - 1)$
$= (x^2 + 1)(x + 1)(x - 1)$ Factoring further; $x^2 - 1$ is a difference of squares

72. $(x^2 + 4)(x + 2)(x - 2)$

73. $4x^4 - 64 = 4(x^4 - 16) = 4[(x^2)^2 - 4^2]$
$= 4(x^2 + 4)(x^2 - 4)$
$= 4(x^2 + 4)(x + 2)(x - 2)$ Factoring further; $x^2 - 4$ is a difference of squares

74. $5(x^2 + 4)(x + 2)(x - 2)$

75. $1 - y^8 = 1^2 - (y^4)^2$
$= (1 + y^4)(1 - y^4)$
$= (1 + y^4)(1 + y^2)(1 - y^2)$ Factoring $1 - y^4$
$= (1 + y^4)(1 + y^2)(1 + y)(1 - y)$
Factoring $1 - y^2$

76. $(x^4 + 1)(x^2 + 1)(x + 1)(x - 1)$

77. $3x^3 - 24x^2 + 48x = 3x(x^2 - 8x + 16)$
$= 3x(x^2 - 2\cdot x\cdot 4 + 4^2)$
$= 3x(x - 4)^2$

78. $2a^2(a - 9)^2$

79. $x^{12} - 16 = (x^6)^2 - 4^2$
$= (x^6 + 4)(x^6 - 4)$
$= (x^6 + 4)(x^3 + 2)(x^3 - 2)$ Factoring $x^6 - 4$

80. $(x^4 + 9)(x^2 + 3)(x^2 - 3)$

81. $y^2 - \frac{1}{16} = y^2 - \left(\frac{1}{4}\right)^2$
$= \left(y + \frac{1}{4}\right)\left(y - \frac{1}{4}\right)$

82. $\left(\frac{1}{5} + x\right)\left(\frac{1}{5} - x\right)$

83. $a^8 - 2a^7 + a^6 = a^6(a^2 - 2a + 1)$
$= a^6(a^2 - 2 \cdot a \cdot 1 + 1^2)$
$= a^6(a - 1)^2$

84. $x^6(x - 4)^2$

85. $25 - \frac{1}{49}x^2 = 5^2 - \left(\frac{1}{7}x\right)^2$
$= \left(5 + \frac{1}{7}x\right)\left(5 - \frac{1}{7}x\right)$

86. $\left(2 + \frac{1}{3}y\right)\left(2 - \frac{1}{3}y\right)$

87. $16m^4 - t^4 = (4m^2)^2 - (t^2)^2$
$= (4m^2 + t^2)(4m^2 - t^2)$
$= (4m^2 + t^2)(2m + t)(2m - t)$ Factoring $4m^2 - t^2$

88. $(1 + a^2b^2)(1 + ab)(1 - ab)$

89. Familiarize. Let s = the score on the fourth test.

Translate.

The average score is at least 90.

$\frac{96 + 98 + 89 + s}{4} \geqslant 90$

Carry out. We solve the inequality.

$\frac{96 + 98 + 89 + s}{4} \geqslant 90$

$96 + 98 + 89 + s \geqslant 360$ Multiplying by 4

$283 + s \geqslant 360$

$s \geqslant 77$

Check. We can obtain a partial check by substituting one number greater than or equal to 77 and another number less than 77 in the inequality. This is left to the student.

State. A score of 77 or better on the last test will earn Bonnie an A in the course.

90. 3.125 L

91. $(x^3y^5)(x^9y^7) = x^{3+9}y^{5+7} = x^{12}y^{12}$

92. $25a^4b^6$

93.

94.

95. $49x^2 - 216$

There is no common factor (other than 1 or -1). Also, $49x^2$ is a square, but 216 is not so this expression is not the difference of squares. It cannot be factored, so it is prime.

96. Prime

97. $18x^3 + 12x^2 + 2x = 2x(9x^2 + 6x + 1)$
$= 2x[(3x)^2 + 2 \cdot 3x \cdot 1 + 1^2]$
$= 2x(3x + 1)^2$

98. $2(81x^2 - 41)$

99. $x^8 - 2^8 = (x^4 + 2^4)(x^4 - 2^4) =$
$(x^4 + 2^4)(x^2 + 2^2)(x^2 - 2^2) =$
$(x^4 + 2^4)(x^2 + 2^2)(x + 2)(x - 2)$, or
$(x^4 + 16)(x^2 + 4)(x + 2)(x - 2)$

100. $4x^2(x + 1)(x - 1)$

101. $3x^5 - 12x^3 = 3x^3(x^2 - 4) = 3x^3(x + 2)(x - 2)$

102. $3\left(x + \frac{1}{3}\right)\left(x - \frac{1}{3}\right)$, or $\frac{1}{3}(3x + 1)(3x - 1)$

103. $18x^3 - \frac{8}{25}x = 2x\left(9x^2 - \frac{4}{25}\right) = 2x\left(3x + \frac{2}{5}\right)\left(3x - \frac{2}{5}\right)$

104. $(x + 1.5)(x - 1.5)$

105. $0.49p - p^3 = p(0.49 - p^2) = p(0.7 + p)(0.7 - p)$

106. $(0.8x + 1.1)(0.8x - 1.1)$

107. $(x + 3)^2 - 9 = [(x + 3) + 3][(x + 3) - 3] =$
$x(x + 6)$

108. $(y - 5 + 6q)(y - 5 - 6q)$

109. $x^2 - \left(\frac{1}{x}\right)^2 = \left(x + \frac{1}{x}\right)\left(x - \frac{1}{x}\right)$

110. $(a^n + 7b^n)(a^n - 7b^n)$

111. $81 - b^{4k} = 9^2 - (b^{2k}) = (9 + b^{2k})(9 - b^{2k}) =$
$(9 + b^{2k})[3^2 - (b^k)^2] =$
$(9 + b^{2k})(3 + b^k)(3 - b^k)$

112. $(x + 3)(x - 3)(x^2 + 1)$

113. $9b^{2n} + 12b^n + 4 = (3b^n)^2 + 2 \cdot 3b^n \cdot 2 + 2^2 = (3b^n + 2)^2$

114. $16(x^2 - 3)^2$

115. $(y + 3)^2 + 2(y + 3) + 1$
$= (y + 3)^2 + 2 \cdot (y + 3) \cdot 1 + 1^2$
$= [(y + 3) + 1]^2$
$= (y + 4)^2$

116. $(7x + 4)^2$

117. $27x^3 - 63x^2 - 147x + 343$
$= 9x^2(3x - 7) - 49(3x - 7)$
$= (3x - 7)(9x^2 - 49)$
$= (3x - 7)(3x + 7)(3x - 7)$, or
$(3x - 7)^2(3x + 7)$

118. $(x - 1)(2x^2 + x + 1)$

119. $a^2 + 2a + 1 - 9 = (a + 1)^2 - 9$
$= [(a + 1) + 3][(a + 1) - 3)]$
$= (a + 4)(a - 2)$

120. $(y + x + 7)(y - x - 1)$

121. If $cy^2 + 6y + 1$ is the square of a binomial, then $2 \cdot a \cdot 1 = 6$ where $a^2 = c$. Then $a = 3$, so $c = a^2 = 3^2 = 9$. (The polynomial is $9y^2 + 6y + 1$.)

122. 16

123. See the answer section in the text.

124. 0, 2

Exercise Set 5.5

1. $2x^2 - 128 = 2(x^2 - 64)$ 2 is a common factor
$= 2(x^2 - 8^2)$ Difference of squares
$= 2(x + 8)(x - 8)$

2. $3(t + 3)(t - 3)$

3. $a^2 + 25 - 10a = a^2 - 10a + 25$
$= a^2 - 2 \cdot a \cdot 5 + 5^2$ Trinomial square
$= (a - 5)^2$

4. $(y + 7)^2$

5. $2x^2 - 11x + 12$

There is no common factor (other than 1). This polynomial has three terms, but it is not a trinomial square. Multiply the leading coefficient and the constant, 2 and 12: $2 \cdot 12 = 24$. Try to factor 24 so that the sum of the factors is -11. The numbers we want are -3 and -8: $-3(-8) = 24$ and $-3 + (-8) = -11$. Split the middle term and factor by grouping.

$2x^2 - 11x + 12 = 2x^2 - 3x - 8x + 12$
$= x(2x - 3) - 4(2x - 3)$
$= (2x - 3)(x - 4)$

6. $(2y - 5)(4y + 1)$

7. $x^3 + 24x^2 + 144x = x(x^2 + 24x + 144)$ x is a common factor
$= x(x^2 + 2 \cdot x \cdot 12 + 12^2)$ Trinomial square
$= x(x + 12)^2$

8. $x(x - 9)^2$

9. $x^3 + 3x^2 - 4x - 12$
$= x^2(x + 3) - 4(x + 3)$ Factoring by grouping
$= (x + 3)(x^2 - 4)$
$= (x + 3)(x + 2)(x - 2)$ Factoring the difference of squares

10. $(x + 5)(x - 5)^2$

11. $24x^2 - 54 = 6(4x^2 - 9)$ 6 is a common factor
$= 6[(2x)^2 - 3^2]$ Difference of squares
$= 6(2x + 3)(2x - 3)$

12. $2(2x + 7)(2x - 7)$

13. $20x^3 - 4x^2 - 72x = 4x(5x^2 - x - 18)$ 4x is a common factor
$= 4x(5x + 9)(x - 2)$ Factoring the trinomial using trial and error

14. $3x(x + 3)(3x - 5)$

15. $x^2 + 4$ is a sum of squares. It is prime.

16. Prime

17. $x^4 + 7x^2 - 3x^3 - 21x = x(x^3 + 7x - 3x^2 - 21)$
$= x[x(x^2 + 7) - 3(x^2 + 7)]$
$= x[(x^2 + 7)(x - 3)]$
$= x(x^2 + 7)(x - 3)$

18. $m(m + 8)(m^2 + 8)$

19. $x^5 - 14x^4 + 49x^3 = x^3(x^2 - 14x + 49)$ x^3 is a common factor
$= x^3(x^2 - 2 \cdot x \cdot 7 + 7^2)$ Trinomial square
$= x^3(x - 7)^2$

20. $2x^4(x + 2)^2$

21. $20 - 6x - 2x^2 = -2(-10 + 3x + x^2)$ -2 is a common factor
$= -2(x^2 + 3x - 10)$ Writing in descending order
$= -2(x + 5)(x - 2)$ Using trial and error

22. $-3(2x - 5)(x + 3)$, or $3(5 - 2x)(3 + x)$

23. $x^2 + 3x + 1$
There is no common factor (other than 1). This is not a trinomial square, because $2 \cdot x \cdot 1 \neq 3x$. We try factoring by trial and error. We look for two factors whose product is 1 and whose sum is 3. There are none. The polynomial cannot be factored. It is prime.

24. Prime

25. $4x^4 - 64 = 4(x^4 - 16)$ 4 is a common factor
$= 4[(x^2)^2 - 4^2]$ Difference of squares
$= 4(x^2 + 4)(x^2 - 4)$ Difference of squares
$= 4(x^2 + 4)(x + 2)(x - 2)$

26. $5x(x^2 + 4)(x + 2)(x - 2)$

27. $t^8 - 1$ Difference of squares
$= (t^4 + 1)(t^4 - 1)$ Difference of squares
$= (t^4 + 1)(t^2 + 1)(t^2 - 1)$ Difference of squares
$= (t^4 + 1)(t^2 + 1)(t + 1)(t - 1)$

28. $(1 + n^4)(1 + n^2)(1 + n)(1 - n)$

29. $x^5 - 4x^4 + 3x^3 = x^3(x^2 - 4x + 3)$ x^3 is a common factor
$= x^3(x - 3)(x - 1)$ Factoring the trinomial using trial and error

30. $x^4(x^2 - 2x + 7)$

31. $x^2 - y^2 = (x + y)(x - y)$ Difference of squares

32. $(pq + r)(pq - r)$

33. $12n^2 + 24n^3 = 12n^2(1 + 2n)$

34. $a(x^2 + y^2)$

35. $9x^2y^2 - 36xy = 9xy(xy - 4)$

36. $xy(x - y)$

37. $2\pi rh + 2\pi r^2 = 2\pi r(h + r)$

38. $5p^2q^2(2p^2q^2 + 7pq + 2)$

39. $(a + b)(x - 3) + (a + b)(x + 4)$
$= (a + b)[(x - 3) + (x + 4)]$ $(a + b)$ is a common factor
$= (a + b)(2x + 1)$

40. $(a^3 + b)(5c - 1)$

41. $(x - 1)(x + 1) - y(x + 1) = (x + 1)(x - 1 - y)$ $(x + 1)$ is a common factor

42. $(x + 1)(x + y)$

43. $n^2 + 2n + np + 2p = n(n + 2) + p(n + 2)$ Factoring by grouping
$= (n + 2)(n + p)$

44. $(a - 3)(a + y)$

45. $2x^2 - 4x + xz - 2z$
$= 2x(x - 2) + z(x - 2)$ Factoring by grouping
$= (x - 2)(2x + z)$

46. $(2y - 1)(3y + p)$

47. $x^2 + y^2 - 2xy = x^2 - 2xy + y^2$ Trinomial square
$= (x - y)^2$

48. $(a - 2b)^2$, or $(2b - a)^2$

49. $9c^2 + 6cd + d^2 = (3c)^2 + 2 \cdot 3c \cdot d + d^2$ Trinomial square
$= (3c + d)^2$

50. $(4x + 3y)^2$

51. $7p^4 - 7q^4$
$= 7(p^4 - q^4)$ 7 is a common factor
$= 7(p^2 + q^2)(p^2 - q^2)$ Factoring a difference of squares
$= 7(p^2 + q^2)(p + q)(p - q)$ Factoring a difference of squares

52. $(2xy + 3z)^2$

53. $25z^2 + 10zy + y^2$
$= (5z)^2 + 2 \cdot 5z \cdot y + y^2$ Trinomial square
$= (5z + y)^2$

54. $(a^2b^2 + 4)(ab + 2)(ab - 2)$

55. $a^5 + 4a^4b - 5a^3b^2$
$= a^3(a^2 + 4ab - 5b^2)$ a^3 is a common factor
$= a^3(a + 5b)(a - b)$ Factoring the trinomial

56. $p(2p + q)^2$

57. $a^2 - ab - 2b^2 = (a - 2b)(a + b)$ Using trial and error

58. $(3b + a)(b - 6a)$

59. $2mn - 360n^2 + m^2 = m^2 + 2mn - 360n^2$ Rewriting
$= (m + 20n)(m - 18n)$ Using trial and error

60. $(xy + 5)(xy + 3)$

61. $m^2n^2 - 4mn - 32 = (mn - 8)(mn + 4)$ Using trial and error

62. $(pq + 6)(pq + 1)$

63. $a^5b^2 + 3a^4b - 10a^3$
$= a^3(a^2b^2 + 3ab - 10)$ a^3 is a common factor
$= a^3(ab + 5)(ab - 2)$ Factoring the trinomial

64. $n^4(mn + 8)(mn - 4)$

65. $49m^2 - 112mn + 64n^2$
$= (7m)^2 - 2 \cdot 7m \cdot 8n + (8n)^2$ Trinomial square
$= (7m - 8n)^2$

66. $2t^2(s^3 + 3t)(s^3 + 2t)$

67. $x^6 + x^5y - 2x^4y^2$
$= x^4(x^2 + xy - 2y)$ x^4 is a common factor
$= x^4(x + 2y)(x - y)$ Factoring the trinomial

68. $a^2(1 + bc)^2$

69. $36a^2 - 15a + \frac{25}{16} = (6a)^2 - 2 \cdot 6a \cdot \frac{5}{4} + \left(\frac{5}{4}\right)^2$
$= \left(6a - \frac{5}{4}\right)^2$

70. $\left(\frac{1}{9}x - \frac{4}{3}\right)^2$, or $\frac{1}{9}\left(\frac{1}{3}x - 4\right)^2$

71. $\frac{1}{4}a^2 + \frac{1}{3}ab + \frac{1}{9}b^2 = \left(\frac{1}{2}a\right)^2 + 2 \cdot \frac{1}{2}a \cdot \frac{1}{3}b + \left(\frac{1}{3}b\right)^2$
$= \left(\frac{1}{2}a + \frac{1}{3}b\right)^2$

72. $(0.1x - 0.5y)^2$, or $0.01(x - 5y)^2$

73. $81a^4 - b^4 = (9a^2)^2 - (b^2)^2$ Difference of squares
$= (9a^2 + b^2)(9a^2 - b^2)$ Difference of squares
$= (9a^2 + b^2)(3a + b)(3a - b)$

74. $(1 + 4x^6y^6)(1 + 2x^3y^3)(1 - 2x^3y^3)$

75. $w^3 - 7w^2 - 4w + 28$
$= w^2(w - 7) - 4(w - 7)$ Factoring by grouping
$= (w - 7)(w^2 - 4)$
$= (w - 7)(w + 2)(w - 2)$ Factoring a difference of squares

76. $(y + 8)(y + 1)(y - 1)$

77.

$y = -4x + 7$	
11 ?	$-4(-1) + 7$
	$4 + 7$
11	11 TRUE

Since 11 = 11 is true, (-1,11) is a solution.

$y = -4x + 7$	
7 ?	$-4 \cdot 0 + 7$
	$0 + 7$
7	7 TRUE

Since 7 = 7 is true, (0,7) is a solution.

$y = -4x + 7$	
-5 ?	$-4 \cdot 3 + 7$
	$-12 + 7$
-5	-5 TRUE

Since -5 = -5 is true, (3,-5) is a solution.

78.

79. $A = aX + bX - 7$
$A + 7 = aX + bX$
$A + 7 = X(a + b)$
$\frac{A + 7}{a + b} = X$

80. $\{x | x < 32\}$

81.

82.

83. $6x^2 - xy - 15y^2$
$= 6 \cdot 1^2 - 1 \cdot 1 - 15 \cdot 1^2$
$= 6 - 1 - 15$
$= -10$

$(2x + 3y)(3x - 5y)$
$= (2 \cdot 1 + 3 \cdot 1)(3 \cdot 1 - 5 \cdot 1)$
$= 5(-2)$
$= -10$

Since the value of both expressions is -10, the factorization is probably correct.

84. 49

85. $18 + y^3 - 9y - 2y^2$
$= y^3 - 2y^2 - 9y + 18$
$= y^2(y - 2) - 9(y - 2)$
$= (y - 2)(y^2 - 9)$
$= (y - 2)(y + 3)(y - 3)$

86. $-(x^2 + 2)(x + 3)(x - 3)$

87. $a^3 + 4a^2 + a + 4 = a^2(a + 4) + 1(a + 4)$
$= (a + 4)(a^2 + 1)$

88. $(x + 1)(x + 2)(x - 2)$

89. $x^4 - 7x^2 - 18 = (x^2 - 9)(x^2 + 2)$
$= (x + 3)(x - 3)(x^2 + 2)$

90. $3(x + 2)(x - 2)(x + 1)(x - 1)$

91. $x^3 - x^2 - 4x + 4 = x^2(x - 1) - 4(x - 1)$
$= (x - 1)(x^2 - 4)$
$= (x - 1)(x + 2)(x - 2)$

92. $(y + 1)(y - 7)(y + 3)$

93. $y^2(y - 1) - 2y(y - 1) + (y - 1)$
$= (y - 1)(y^2 - 2y + 1)$
$= (y - 1)(y - 1)^2$
$= (y - 1)^3$

94. $(2x + 3y - 2)(3x - y - 3)$

95. $(y + 4)^2 + 2x(y + 4) + x^2$
$= (y + 4)^2 + 2\cdot(y + 4)\cdot x + x^2$ Trinomial square
$= [(y + 4) + x]^2$

96. $(2a + b + 4)(a - b + 5)$

97. $x^{2k} - 2^{2k} = x^{2\cdot 4} - 2^{2\cdot 4}$ Substituting 4 for k
$= x^8 - 2^8$
$= x^8 - 256$
$= (x^4 + 16)(x^4 - 16)$
$= (x^4 + 16)(x^2 + 4)(x^2 - 4)$
$= (x^4 + 16)(x^2 + 4)(x + 2)(x - 2)$

98. 

Exercise Set 5.6

1. $(x + 8)(x + 6) = 0$

$x + 8 = 0$ or $x + 6 = 0$
$x = -8$ or $x = -6$

Check:
For -8:
$(x + 8)(x + 6) = 0$
$(-8 + 8)(-8 + 6) \; ? \; 0$
$0\cdot(-2)$ |
0 | 0 TRUE

For -6:
$(x + 8)(x + 6) = 0$
$(-6 + 8)(-6 + 6) \; ? \; 0$
$2\cdot 0$ |
0 | 0 TRUE

The solutions are -8 and -6.

2. -3, -2

3. $(x - 3)(x + 5) = 0$
$x - 3 = 0$ or $x + 5 = 0$
$x = 3$ or $x = -5$

Check:
For 3:
$(x - 3)(x + 5) = 0$
$(3 - 3)(3 + 5) \; ? \; 0$
$0\cdot 8$ |
0 | 0 TRUE

For -5:
$(x - 3)(x + 5) = 0$
$(-5 - 3)(-5 + 5) \; ? \; 0$
$-8\cdot 0$ |
0 | 0 TRUE

The solutions are 3 and -5.

4. -9, 3

5. $(x + 12)(x - 11) = 0$

$x + 12 = 0$ or $x - 11 = 0$
$x = -12$ or $x = 11$
The solutions are -12 and 11.

6. 13, -53

7. $x(x + 5) = 0$

$x = 0$ or $x + 5 = 0$
$x = 0$ or $x = -5$
The solutions are 0 and -5.

8. 0, -7

9. $0 = y(y + 10)$

$y = 0$ or $y + 10 = 0$
$y = 0$ or $y = -10$
The solutions are 0 and -10.

10. 0, 21

11. $(2x + 5)(x + 4) = 0$

$2x + 5 = 0$ or $x + 4 = 0$
$2x = -5$ or $x = -4$
$x = -\frac{5}{2}$ or $x = -4$

The solutions are $-\frac{5}{2}$ and -4.

12. $-\frac{9}{2}$, -8

13. $(5x + 1)(4x - 12) = 0$

$5x + 1 = 0$ or $4x - 12 = 0$
$5x = -1$ or $4x = 12$
$x = -\frac{1}{5}$ or $x = 3$

The solutions are $-\frac{1}{5}$ and 3.

14. $-\frac{9}{4}$, $\frac{1}{2}$

15. $(7x - 28)(28x - 7) = 0$

$7x - 28 = 0$ or $28x - 7 = 0$
$7x = 28$ or $28x = 7$
$x = 4$ or $x = \frac{7}{28} = \frac{1}{4}$

The solutions are 4 and $\frac{1}{4}$.

16. $\frac{11}{12}$, $\frac{5}{8}$

17. $2x(3x - 2) = 0$

$2x = 0$ or $3x - 2 = 0$
$x = 0$ or $3x = 2$
$x = 0$ or $x = \frac{2}{3}$

The solutions are 0 and $\frac{2}{3}$.

18. 0, $\frac{9}{8}$

19. $\frac{1}{2}x\left(\frac{2}{3}x - 12\right) = 0$

$\frac{1}{2}x = 0$ or $\frac{2}{3}x - 12 = 0$
$x = 0$ or $\frac{2}{3}x = 12$
$x = 0$ or $x = \frac{3}{2} \cdot 12 = 18$

The solutions are 0 and 18.

20. 0, 8

21. $\left(\frac{1}{5} + 2x\right)\left(\frac{1}{9} - 3x\right) = 0$

$\frac{1}{5} + 2x = 0$ or $\frac{1}{9} - 3x = 0$
$2x = -\frac{1}{5}$ or $-3x = -\frac{1}{9}$
$x = \frac{1}{2}\left(-\frac{1}{5}\right)$ or $x = -\frac{1}{3}\left(-\frac{1}{9}\right)$
$x = -\frac{1}{10}$ or $x = \frac{1}{27}$

The solutions are $-\frac{1}{10}$ and $\frac{1}{27}$.

22. $\frac{1}{21}$, $\frac{18}{11}$

23. $(0.3x - 0.1)(0.05x - 1) = 0$

$0.3x - 0.1 = 0$ or $0.05x - 1 = 0$
$0.3x = 0.1$ or $0.05x = 1$
$x = \frac{0.1}{0.3}$ or $x = \frac{1}{0.05}$
$x = \frac{1}{3}$ or $x = 20$

The solutions are $\frac{1}{3}$ and 20.

24. 3, 50

25. $9x(3x - 2)(2x - 1) = 0$

$9x = 0$ or $3x - 2 = 0$ or $2x - 1 = 0$
$x = 0$ or $3x = 2$ or $2x = 1$
$x = 0$ or $x = \frac{2}{3}$ or $x = \frac{1}{2}$

The solutions are 0, $\frac{2}{3}$, and $\frac{1}{2}$.

26. 5, -55, $\frac{1}{5}$

27. $x^2 + 6x + 5 = 0$
$(x + 5)(x + 1) = 0$ Factoring
$x + 5 = 0$ or $x + 1 = 0$ Using the principle of zero product
$x = -5$ or $x = -1$
The solutions are -5 and -1.

28. -6, -1

29. $x^2 + 7x - 18 = 0$
$(x + 9)(x - 2) = 0$ Factoring
$x + 9 = 0$ or $x - 2 = 0$ Using the principal of zero products
$x = -9$ or $x = 2$
The solutions are -9 and 2.

30. -7, 3

31. $x^2 - 8x + 15 = 0$
$(x - 5)(x - 3) = 0$
$x - 5 = 0$ or $x - 3 = 0$
$x = 5$ or $x = 3$
The solutions are 5 and 3.

32. 7, 2

33. $x^2 - 8x = 0$
$x(x - 8) = 0$
$x = 0$ or $x - 8 = 0$
$x = 0$ or $x = 8$
The solutions are 0 and 8.

34. 0, 3

35. $x^2 + 19x = 0$
$x(x + 19) = 0$
$x = 0$ or $x + 19 = 0$
$x = 0$ or $x = -19$
The solutions are 0 and -19.

36. 0, -12

37. $x^2 = 16$
$x^2 - 16 = 0$ Adding -16
$(x - 4)(x + 4) = 0$
$x - 4 = 0$ or $x + 4 = 0$
$x = 4$ or $x = -4$
The solutions are 4 and -4.

38. -10, 10

39. $9x^2 - 4 = 0$
$(3x - 2)(3x + 2) = 0$
$3x - 2 = 0$ or $3x + 2 = 0$
$3x = 2$ or $3x = -2$
$x = \frac{2}{3}$ or $x = -\frac{2}{3}$
The solutions are $\frac{2}{3}$ and $-\frac{2}{3}$.

40. $-\frac{3}{2}, \frac{3}{2}$

41. $0 = 6x + x^2 + 9$
$0 = x^2 + 6x + 9$ Writing in descending order
$0 = (x + 3)(x + 3)$
$x + 3 = 0$ or $x + 3 = 0$
$x = -3$ or $x = -3$
There is only one solution, -3.

42. -5

43. $x^2 + 16 = 8x$
$x^2 - 8x + 16 = 0$ Adding -8x
$(x - 4)(x - 4) = 0$
$x - 4 = 0$ or $x - 4 = 0$
$x = 4$ or $x = 4$
There is only one solution, 4.

44. 1

45. $5x^2 = 6x$
$5x^2 - 6x = 0$
$x(5x - 6) = 0$
$x = 0$ or $5x - 6 = 0$
$x = 0$ or $5x = 6$
$x = 0$ or $x = \frac{6}{5}$
The solutions are 0 and $\frac{6}{5}$.

46. $0, \frac{8}{7}$

47. $6x^2 - 4x = 10$
$6x^2 - 4x - 10 = 0$
$2(3x^2 - 2x - 5) = 0$
$2(3x - 5)(x + 1) = 0$
$3x - 5 = 0$ or $x + 1 = 0$
$3x = 5$ or $x = -1$
$x = \frac{5}{3}$ or $x = -1$
The solutions are $\frac{5}{3}$ and -1.

48. $-\frac{5}{3}, 4$

49. $12y^2 - 5y = 2$
$12y^2 - 5y - 2 = 0$
$(4y + 1)(3y - 2) = 0$
$4y + 1 = 0$ or $3y - 2 = 0$
$4y = -1$ or $3y = 2$
$y = -\frac{1}{4}$ or $y = \frac{2}{3}$
The solutions are $-\frac{1}{4}$ and $\frac{2}{3}$.

50. -5, -1

51. $x(x - 5) = 14$
$x^2 - 5x = 14$ Multiplying on the left side
$x^2 - 5x - 14 = 0$ Adding -14
$(x - 7)(x + 2) = 0$
$x - 7 = 0$ or $x + 2 = 0$
$x = 7$ or $x = -2$
The solutions are 7 and -2.

52. $\frac{2}{3}$, -1

53.
$$64m^2 - 25 = 56$$
$$64m^2 - 81 = 0$$
$$(8m - 9)(8m + 9) = 0$$
$$8m - 9 = 0 \quad \text{or} \quad 8m + 9 = 0$$
$$8m = 9 \quad \text{or} \quad 8m = -9$$
$$m = \frac{9}{8} \quad \text{or} \quad m = -\frac{9}{8}$$
The solutions are $\frac{9}{8}$ and $-\frac{9}{8}$.

54. $-\frac{7}{10}$, $\frac{7}{10}$

55.
$$3x^2 + 8x = 9 + 2x$$
$$3x^2 + 8x - 2x - 9 = 0 \quad \text{Adding -2x and -9}$$
$$3x^2 + 6x - 9 = 0 \quad \text{Collecting like terms}$$
$$3(x^2 + 2x - 3) = 0$$
$$3(x + 3)(x - 1) = 0$$
$$x + 3 = 0 \quad \text{or} \quad x - 1 = 0$$
$$x = -3 \quad \text{or} \quad x = 1$$
The solutions are -3 and 1.

56. 9, -2

57.
$$(3x + 5)(x + 3) = 7$$
$$3x^2 + 14x + 15 = 7 \quad \text{Multiplying on the left}$$
$$3x^2 + 14x + 8 = 0$$
$$(3x + 2)(x + 4) = 0$$
$$3x + 2 = 0 \quad \text{or} \quad x + 4 = 0$$
$$3x = -2 \quad \text{or} \quad x = -4$$
$$x = -\frac{2}{3} \quad \text{or} \quad x = -4$$
The solutions are $-\frac{2}{3}$ and -4.

58. $\frac{6}{5}$, -1

59. We let y = 0 and solve for x.
$$0 = x^2 - x - 6$$
$$0 = (x - 3)(x + 2)$$
$$x - 3 = 0 \quad \text{or} \quad x + 2 = 0$$
$$x = 3 \quad \text{or} \quad x = -2$$
The x-intercepts are (3,0) and (-2,0).

60. (-4,0), (1,0)

61. We let y = 0 and solve for x.
$$0 = x^2 + 2x - 8$$
$$0 = (x + 4)(x - 2)$$
$$x + 4 = 0 \quad \text{or} \quad x - 2 = 0$$
$$x = -4 \quad \text{or} \quad x = 2$$
The x-intercepts are (-4,0) and (2,0).

62. (5,0), (-3,0)

63. We let y = 0 and solve for x.
$$0 = 2x^2 + 3x - 9$$
$$0 = (2x - 3)(x + 3)$$
$$2x - 3 = 0 \quad \text{or} \quad x + 3 = 0$$
$$2x = 3 \quad \text{or} \quad x = -3$$
$$x = \frac{3}{2} \quad \text{or} \quad x = -3$$
The x-intercepts are $\left(\frac{3}{2},0\right)$ and (-3,0).

64. $\left(-\frac{5}{2},0\right)$, (2,0)

65. $(a + b)^2$

66. $a^2 + b^2$

67. $2x + 5 < 19$

68. $\frac{1}{2}x - 7 > 24$

69.

70.

71.

72.

73. a)
$$x = -3 \quad \text{or} \quad x = 4$$
$$x + 3 = 0 \quad \text{or} \quad x - 4 = 0$$
$$(x + 3)(x - 4) = 0 \quad \text{Principle of zero products}$$
$$x^2 - x - 12 = 0 \quad \text{Multiplying}$$

b)
$$x = -3 \quad \text{or} \quad x = -4$$
$$x + 3 = 0 \quad \text{or} \quad x + 4 = 0$$
$$(x + 3)(x + 4) = 0$$
$$x^2 + 7x + 12 = 0$$

c)
$$x = \frac{1}{2} \quad \text{or} \quad x = \frac{1}{2}$$
$$x - \frac{1}{2} = 0 \quad \text{or} \quad x - \frac{1}{2} = 0$$
$$\left(x - \frac{1}{2}\right)\left(x - \frac{1}{2}\right) = 0$$
$$x^2 - x + \frac{1}{4} = 0, \quad \text{or}$$
$$4x^2 - 4x + 1 = 0 \quad \text{Multiplying by 4}$$

d)
$$x = 5 \quad \text{or} \quad x = -5$$
$$x - 5 = 0 \quad \text{or} \quad x + 5 = 0$$
$$(x - 5)(x + 5) = 0$$
$$x^2 - 25 = 0$$

e) $x = 0$ or $x = 0.1$ or $x = \frac{1}{4}$

$x = 0$ or $x - 0.1 = 0$ or $x - \frac{1}{4} = 0$

$x = 0$ or $x - \frac{1}{10} = 0$ or $x - \frac{1}{4} = 0$

$x\left(x - \frac{1}{10}\right)\left(x - \frac{1}{4}\right) = 0$

$x\left(x^2 - \frac{7}{20}x + \frac{1}{40}\right) = 0$

$x^3 - \frac{7}{20}x^2 + \frac{1}{40}x = 0$, or

$40x^3 - 14x^2 + x = 0$ Multiplying by 40

74. -5, 4

75. $y(y + 8) = 16(y - 1)$

$y^2 + 8y = 16y - 16$

$y^2 - 8y + 16 = 0$

$(y - 4)(y - 4) = 0$

$y - 4 = 0$ or $y - 4 = 0$

$y = 4$ or $y = 4$

The solution is 4.

76. 5, 3

77. $x^2 - \frac{1}{64} = 0$

$\left(x - \frac{1}{8}\right)\left(x + \frac{1}{8}\right) = 0$

$x - \frac{1}{8} = 0$ or $x + \frac{1}{8} = 0$

$x = \frac{1}{8}$ or $x = -\frac{1}{8}$

The solutions are $\frac{1}{8}$ and $-\frac{1}{8}$.

78. $-\frac{5}{6}, \frac{5}{6}$

79. $\frac{5}{16}x^2 = 5$

$\frac{5}{16}x^2 - 5 = 0$

$5\left(\frac{1}{16}x^2 - 1\right) = 0$

$5\left(\frac{1}{4}x - 1\right)\left(\frac{1}{4}x + 1\right) = 0$

$\frac{1}{4}x - 1 = 0$ or $\frac{1}{4}x + 1 = 0$

$\frac{1}{4}x = 1$ or $\frac{1}{4}x = -1$

$x = 4$ or $x = -4$

The solutions are 4 and -4.

80. $-\frac{5}{9}, \frac{5}{9}$

81. a) $3(3x^2 - 4x + 8) = 3 \cdot 0$ Multiplying (a) by 3

$9x^2 - 12x + 24 = 0$

(a) and $9x^2 - 12x + 24 = 0$ are equivalent.

b) $(x - 6)(x + 3) = x^2 - 3x - 18$

(b) and $x^2 - 3x - 18 = 0$ are equivalent.

c) $4(x^2 + 2x + 9) = 4 \cdot 0$ Multiplying (c) by 4

$4x^2 + 8x + 36 = 0$

(c) and $4x^2 + 8x + 36 = 0$ are equivalent.

d) $2(2x + 5)(x + 4) = 2 \cdot 0$ Multiplying (d) by 2

$2(x + 4)(2x + 5) = 0$

$(2x + 8)(2x + 5) = 0$

(d) and $(2x + 8)(2x - 5) = 0$ are equivalent.

e) $5x^2 - 5 = 5(x^2 - 1) = 5(x + 1)(x - 1) =$

$(x + 1)(5)(x - 1) = (x + 1)(5x - 5)$

(e) and $(x + 1)(5x - 5) = 0$ are equivalent.

f) $2(x^2 + 10x - 2) = 2 \cdot 0$ Multiplying (f) by 2

$2x^2 + 20x - 4 = 0$

(f) and $2x^2 + 20x - 4 = 0$ are equivalent.

82. ◈

83. ◈

84. 3.45, -1.65

85. -2.33, -6.77

86. -0.25, 0.88

87. -4.59, -9.15

88. 4.55, -3.23

89. -3.25, -6.75

Exercise Set 5.7

1. Familiarize. Let x = the number (or numbers).

Translate. We reword the problem.

Four times the square of a number	minus	the number,	is	3.
$4x^2$	$-$	x	$=$	3

Carry out. We solve the equation.

$4x^2 - x = 3$

$4x^2 - x - 3 = 0$

$(4x + 3)(x - 1) = 0$

$4x + 3 = 0$ or $x - 1 = 0$

$4x = -3$ or $x = 1$

$x = -\frac{3}{4}$ or $x = 1$

Check. For $-\frac{3}{4}$: Four times the square of $-\frac{3}{4}$ is $4\left(-\frac{3}{4}\right)^2 = 4\left(\frac{9}{16}\right) = \frac{9}{4}$. If we subtract $-\frac{3}{4}$ from $\frac{9}{4}$ we get $\frac{9}{4} - \left(-\frac{3}{4}\right) = \frac{9}{4} + \frac{3}{4} = \frac{12}{4} = 3$.

For 1: Four times the square of 1 is $4(1)^2 = 4$. If we subtract 1 from 4 we get $4 - 1 = 3$. Both numbers check.

State. There are two such numbers, $-\frac{3}{4}$ and 1.

2. 5, -5

3. Familiarize. Let x = the number (or numbers).

Translate. We reword the problem.

The square of a number	plus	8	is	six times the number.
x^2	+	8	=	$6x$

Carry out. We solve the equation.

$x^2 + 8 = 6x$

$x^2 - 6x + 8 = 0$

$(x - 4)(x - 2) = 0$

$x - 4 = 0$ or $x - 2 = 0$

$x = 4$ or $x = 2$

Check. The square of 4 is 16, and six times the number 4 is 24. Since 16 + 8 = 24, the number 4 checks. The square of 2 is 4, and six times the number 2 is 12. Since 4 + 8 = 12, the number 2 checks.

State. There are two such numbers, 4 and 2.

4. 3, 5

5. Familiarize. The page numbers on facing pages are consecutive integers. Let x = the smaller integer. Then x + 1 = the larger integer.

Translate. We reword the problem.

Smaller integer	times	larger integer	is	210.
x	$\cdot$	$(x + 1)$	=	210

Carry out.

$x(x + 1) = 210$

$x^2 + x = 210$

$x^2 + x - 210 = 0$

$(x + 15)(x - 14) = 0$

$x + 15 = 0$ or $x - 14 = 0$

$x = -15$ or $x = 14$

Check. The solutions of the equation are -15 and 14. Since a page number cannot be negative, -15 cannot be a solution of the original problem. We only need to check 14. When x = 14, then x + 1 = 15, and $14 \cdot 15 = 210$. This checks.

State. The page numbers are 14 and 15.

6. 10 and 11

7. Familiarize. Let x = the smaller even integer. Then x + 2 = the larger even integer.

Translate. We reword the problem.

Smaller even integer	times	larger even integer	is	168.
x	$\cdot$	$(x + 2)$	=	168

Carry out.

$x(x + 2) = 168$

$x^2 + 2x = 168$

$x^2 + 2x - 168 = 0$

$(x + 14)(x - 12) = 0$

$x + 14 = 0$ or $x - 12 = 0$

$x = -14$ or $x = 12$

Check. The solutions of the equation are -14 and 12. When x is -14, then x + 2 is -12 and $-14(-12) = 168$. The numbers -14 and -12 are consecutive even integers which are solutions to the problem. When x is 12, then x + 2 is 14 and $12 \cdot 14 = 168$. The numbers 12 and 14 are also consecutive even integers which are solutions to the problem.

State. We have two solutions each of which consists of a pair of numbers: -14 and -12, and 12 and 14.

8. 14 and 16, -16 and -14

9. Familiarize. Let x = the smaller odd integer. Then x + 2 = the larger odd integer.

Translate. We reword the problem.

Smaller odd integer	times	larger odd integer	is	255.
x	$\cdot$	$x + 2$	=	255

Carry out.

$x(x + 2) = 255$

$x^2 + 2x = 255$

$x^2 + 2x - 255 = 0$

$(x - 15)(x + 17) = 0$

$x - 15 = 0$ or $x + 17 = 0$

$x = 15$ or $x = -17$

Check. The solutions of the equation are 15 and -17. When x is 15, then x + 2 is 17 and $15 \cdot 17 = 255$. The numbers 15 and 17 are consecutive odd integers which are solutions to the problem. When x is -17, then x + 2 is -15 and $-17(-15) = 255$. The numbers -17 and -15 are also consecutive odd integers which are solutions to the problem.

State. We have two solutions each of which consists of a pair of numbers: 15 and 17, and -17 and -15.

10. 11 and 13, -13 and -11

11. Familiarize. Using the labels shown on the drawing in the text, we let w = the width of the rectangle and $w + 4$ = the length. Recall that the area of a rectangle is length times width.

Translate. We reword the problem.

Length times width is area.

$(w + 4) \cdot w = 96$

Carry out.

$$(w + 4)\cdot w = 96$$
$$w^2 + 4w = 96$$
$$w^2 + 4w - 96 = 0$$
$$(w + 12)(w - 8) = 0$$
$$w + 12 = 0 \quad \text{or} \quad w - 8 = 0$$
$$w = -12 \quad \text{or} \quad w = 8$$

Check. The solutions of the equation are -12 and 8. The width of a rectangle cannot have a negative measure, so -12 cannot be a solution. Suppose the width is 8 m. The length is 4 m greater than the width, so the length is 12 m and the area is 12·8, or 96 m^2. The numbers check in the original problem.

State. The length is 12 m, and the width is 8 m.

12. Length: 12 cm, width: 7 cm

13. Familiarize. First draw a picture. Let x = the length of a side of the square.

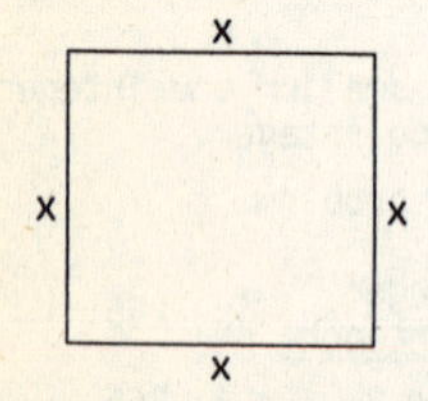

The area of the square is $x\cdot x$, or x^2. The perimeter of the square is $x + x + x + x$, or $4x$.

Translate.

Area of bookcase is 5 more than perimeter of bookcase.

$x^2 = 5 + 4x$

Carry out.

$$x^2 = 5 + 4x$$
$$x^2 - 4x - 5 = 0$$
$$(x - 5)(x + 1) = 0$$
$$x - 5 = 0 \quad \text{or} \quad x + 1 = 0$$
$$x = 5 \quad \text{or} \quad x = -1$$

Check. The solutions of the equation are 5 and -1. The length of a side cannot be negative, so we only check 5. The area is 5·5, or 25. The perimeter is 5 + 5 + 5 + 5, or 20. The area, 25, is 5 more than the perimeter, 20. This checks.

State. The length of a side is 5.

14. 3 or 1

15. Familiarize. Using the labels shown on the drawing in the text, we let h = the height and $h + 10$ = the base. Recall that the formula for the area of a triangle is $\frac{1}{2}\cdot$ (base)·(height).

Translate.

Area is $\frac{1}{2}$ times the base times the height.

$28 = \frac{1}{2} \cdot (h + 10) \cdot h$

Carry out.

$$28 = \frac{1}{2}h(h + 10)$$
$$56 = h(h + 10)$$
$$56 = h^2 + 10h$$
$$0 = h^2 + 10h - 56$$
$$0 = (h + 14)(h - 4)$$
$$h + 14 = 0 \quad \text{or} \quad h - 4 = 0$$
$$h = -14 \quad \text{or} \quad h = 4$$

Check. The solutions of the equation are -14 and 4. The height of a triangle cannot have a negative length, so -14 cannot be a solution. Suppose the height is 4 cm. The base is 10 cm greater than the height, so the base is 14 cm and the area is $\frac{1}{2}\cdot 14\cdot 4$, or 28 cm^2. These numbers check.

State. The height is 4 cm and the base is 14 cm.

16. Height: 2 m, base: 10 m

17. Familiarize. We make a drawing. Let x = the length of a side of the original square. Then $x + 3$ = the length of a side of the enlarged square.

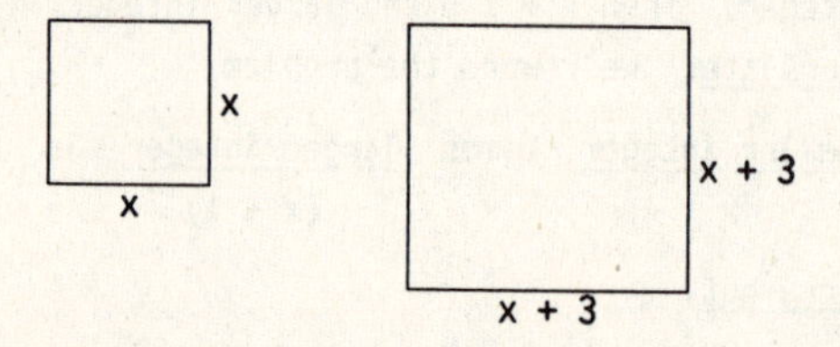

Recall that the area of a square is found by squaring the length of a side.

Translate.

Area of enlarged square is the square of the lengthened side.

$81 = (x + 3)^2$

Carry out.

$81 = (x + 3)^2$

$81 = x^2 + 6x + 9$

$0 = x^2 + 6x - 72$

$0 = (x + 12)(x - 6)$

$x + 12 = 0$ or $x - 6 = 0$

$x = -12$ or $x = 6$

Check. The solutions of the equation are -12 and 6. The length of a side cannot be negative, so -12 cannot be a solution. Suppose the length of a side of the original square is 6 m. Then the length of a side of the new square is 6 + 3, or 9 m. Its area is 9^2, or 81 m^2. The numbers check.

State. The length of a side of the original square is 6 m.

18. 4 km

19. Familiarize. Let x = the smaller odd whole number. Then x + 2 = the larger odd whole number.

Translate.

$$\underbrace{\text{Square of the smaller odd whole number}}_{x^2} + \underbrace{\text{Square of the larger odd whole number}}_{(x+2)^2} \text{ is } 74 = 74$$

Carry out.

$x^2 + (x + 2)^2 = 74$

$x^2 + x^2 + 4x + 4 = 74$

$2x^2 + 4x - 70 = 0$

$2(x^2 + 2x - 35) = 0$

$2(x + 7)(x - 5) = 0$

$x + 7 = 0$ or $x - 5 = 0$

$x = -7$ or $x = 5$

Check. The solutions of the equation are -7 and 5. The problem asks for odd whole numbers, so -7 cannot be a solution. When x is 5, x + 2 is 7. The numbers 5 and 7 are consecutive odd whole numbers. The sum of their squares, 25 + 49, is 74. The numbers check.

State. The numbers are 5 and 7.

20. 7 and 9

21. Familiarize. We will use the formula $n^2 - n = N$.

Translate. Substitute 23 for n.

$23^2 - 23 = N$

Carry out. We do the computation on the left.

$23^2 - 23 = N$

$529 - 23 = N$

$506 = N$

Check. We can recheck the computation or we can solve $n^2 - n = 506$. The answer checks.

State. 506 games will be played.

22. 182

23. Familiarize. We will use the formula $n^2 - n = N$.

Translate. Substitute 132 for N.

$n^2 - n = 132$

Carry out.

$n^2 - n = 132$

$n^2 - n - 132 = 0$

$(n - 12)(n + 11) = 0$

$n - 12 = 0$ or $n + 11 = 0$

$n = 12$ or $n = -11$

Check. The solutions of the equation are 12 and -11. Since the number of teams cannot be negative, -11 cannot be a solution. But 12 checks since $12^2 - 12 = 144 - 12 = 132$.

State. There are 12 teams in the league.

24. 10

25. Familiarize. We will use the formula $N = \frac{1}{2}(n^2 - n)$.

Translate. Substitute 40 for n.

$N = \frac{1}{2}(40^2 - 40)$

Carry out. We do the computation on the right.

$N = \frac{1}{2}(40^2 - 40)$

$N = \frac{1}{2}(1600 - 40)$

$N = \frac{1}{2}(1560)$

$N = 780$

Check. We can recheck the computation, or we can solve the equation $780 = \frac{1}{2}(n^2 - n)$. The answer checks.

State. 780 handshakes are possible.

26. 4950

27. Familiarize. We will use the formula $N = \frac{1}{2}(n^2 - n)$, since "clicks" can be substituted for handshakes.

Translate. Substitute 190 for N.

$190 = \frac{1}{2}(n^2 - n)$

Carry out.

$190 = \frac{1}{2}(n^2 - n)$

$380 = n^2 - n$ Multiplying by 2

$0 = n^2 - n - 380$

$0 = (n - 20)(n + 19)$

$n - 20 = 0$ or $n + 19 = 0$

$n = 20$ or $n = -19$

Check. The solutions of the equation are 20 and -19. Since the number of people cannot be negative, -19 cannot be a solution. But 20 checks since $\frac{1}{2}(20^2 - 20) = \frac{1}{2}(400 - 20) = \frac{1}{2} \cdot 380 = 190$.

State. 20 people took part in the toast.

28. 25

29. Familiarize. We make a drawing. Let x = the length of the unknown leg. Then x + 2 = the length of the hypotenuse.

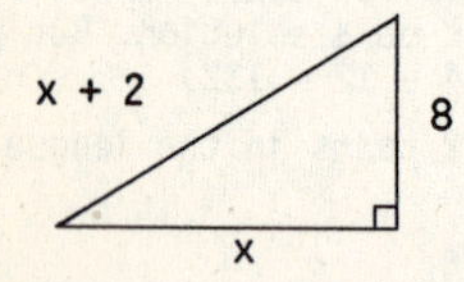

Translate. We use the Pythagorean theorem.

$a^2 + b^2 = c^2$

$8^2 + x^2 = (x + 2)^2$

Carry out. We solve the equation.

$8^2 + x^2 = (x + 2)^2$

$64 + x^2 = x^2 + 4x + 4$

$60 = 4x$ Subtracting x^2 and 4

$15 = x$

Check. When x = 15, then x + 2 = 17 and $8^2 + 15^2 = 17^2$. Thus, 15 and 17 check.

State. The lengths of the hypotenuse and the other leg are 17 ft and 15 ft, respectively.

30. Hypotenuse: 26 ft, leg: 10 ft

31. Familiarize. We label the drawing. Let x = the length of a side of the dining room.

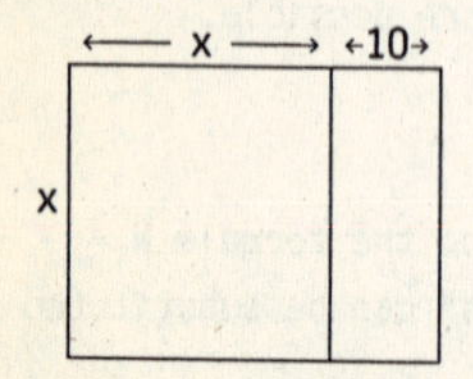

The dimensions of the dining room are x by x, and the dimensions of the kitchen are x by 10. The dimensions of the entire rectangle are x + 10 by x. Recall that the area of a rectangle is length × with.

Translate. We reword the problem.

Total area is 264 ft².

$(x + 10)x = 264$

Carry out. We solve the equation.

$(x + 10)x = 264$

$x^2 + 10x = 264$

$x^2 + 10x - 264 = 0$

$(x + 22)(x - 12) = 0$

$x + 22 = 0$ or $x - 12 = 0$

$x = -22$ or $x = 12$

Check. Since measures cannot be negative, we only consider 12. If x = 12, the areas of the entire rectangle is (10 + 12)(12), or 22·12, or 264 ft². Thus 12 checks. Another approach would be to express the area of the entire rectangle as the sum of the areas of the dining room and the kitchen:

Dining room: $x \cdot x = 12 \cdot 12$, or 144 ft²

Kitchen: $x \cdot 10 = 12 \cdot 10$, or 120 ft²

Total area: 144 ft² + 120 ft² = 264 ft²

This provides a second check.

State. The dining room is 12 ft by 12 ft, and the kitchen is 12 ft by 10 ft.

32. 4 m

33. Graph: $y = -\frac{2}{3}x + 1$

Since the equation is in the form y = mx + b, we know that the y-intercept is (0,1). We find two other solutions, using multiples of 3 for x to avoid fractions.

When x = -3, $y = -\frac{2}{3}(-3) + 1 = 2 + 1 = 3$.

When x = 3, $y = -\frac{2}{3} \cdot 3 + 1 = -2 + 1 = -1$.

x	y
0	1
-3	3
3	-1

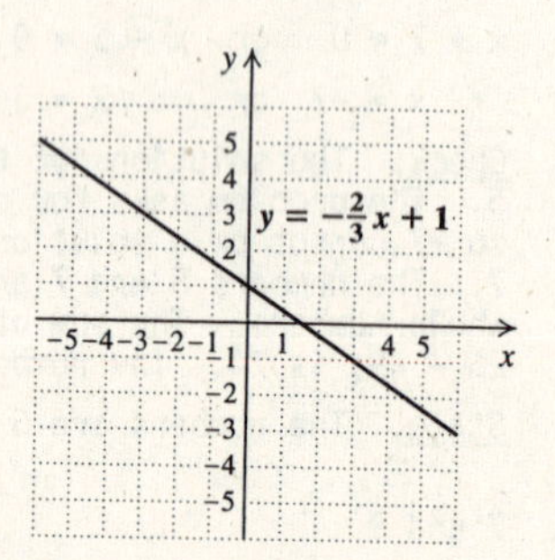

34.

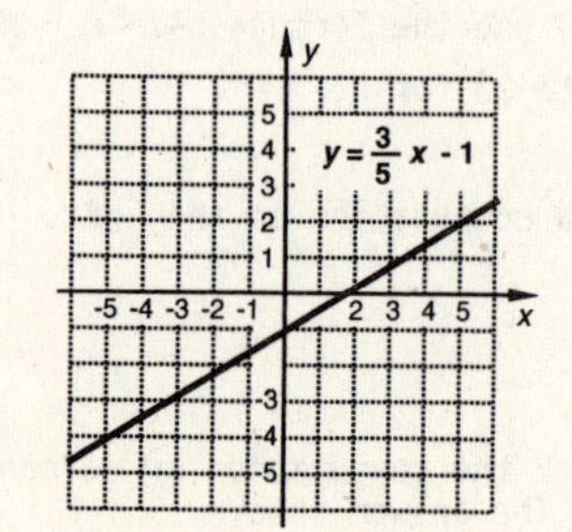

35. $7x^0 = 7 \cdot 4^0 = 7 \cdot 1 = 7$

(Any nonzero number raised to the zero power is 1.)

36. m^6n^6

37.

38. 

39. Familiarize. Using the labels shown on the drawing in the text, we let x = the width of the walk. Then the length and width of the rectangle formed by the pool and walk together are 40 + 2x and 20 + 2x, respectively.

Translate.

Area	is	length	times	width.
1500	=	(40 + 2x)	·	(20 + 2x)

Carry out.

$1500 = (40 + 2x)(20 + 2x)$

$1500 = 2(20 + x)\cdot 2(10 + x)$ Factoring 2 out of each factor on the right

$1500 = 4\cdot(20 + x)(10 + x)$

$375 = (20 + x)(10 + x)$ Dividing by 4

$375 = 200 + 30x + x^2$

$0 = x^2 + 30x - 175$

$0 = (x + 35)(x - 5)$

$x + 35 = 0$ or $x - 5 = 0$

$x = -35$ or $x = 5$

Check. The solutions of the equation are -35 and 5. Since the width of the walk cannot be negative, -35 is not a solution. When x = 5, 40 + 2x = 40 + 2·5, or 50 and 20 + 2x = 20 + 2·5, or 30. The total area of the pool and walk is 50·30, or 1500 ft^2. This checks.

State. The width of the walk is 5 ft.

40. a) 4 sec

b) $3\frac{1}{4}$ sec

41. Familiarize. Let y = the ten's digit. Then y + 4 = the one's digit and 10y + y + 4, or 11y + 4, represents the number.

Translate.

The number	plus	the product of the digits	is	58.
11y + 4	+	y(y + 4)	=	58

Carry out. We solve the equation.

$11y + 4 + y(y + 4) = 58$

$11y + 4 + y^2 + 4y = 58$

$y^2 + 15y + 4 = 58$

$y^2 + 15y - 54 = 0$

$(y + 18)(y - 3) = 0$

$y + 18 = 0$ or $y - 3 = 0$

$y = -18$ or $y = 3$

Check. Since -18 cannot be a digit of the number, we only need to check 3. When y = 3, then y + 4 = 7 and the number is 37. We see that 37 + 3·7 = 37 + 21, or 58. The result checks.

State. The number is 37.

42. 7 m

43. Familiarize. We make a drawing. Let w = the width of the piece of cardboard. Then 2w = the length.

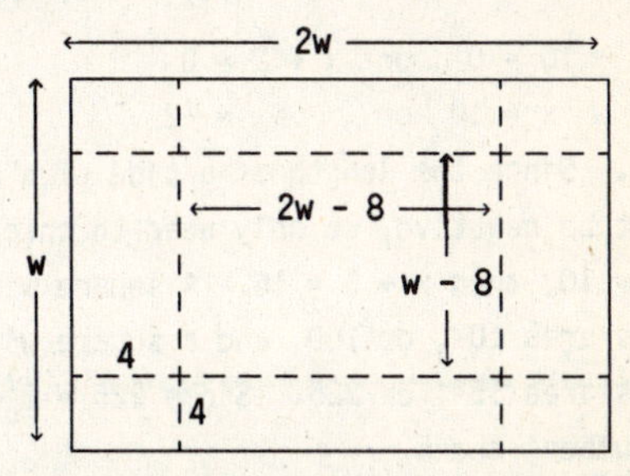

The box will have length 2w - 8, width w - 8, and height 4. Recall that the formula for volume is V = length × width × height.

Translate.

The volume	is	616 cm^3.
(2w - 8)(w - 8)(4)	=	616

Carry out. We solve the equation.

$(2w - 8)(w - 8)(4) = 616$

$(2w^2 - 24w + 64)(4) = 616$

$8w^2 - 96w + 256 = 616$

$8w^2 - 96w - 360 = 0$

$8(w^2 - 12w - 45) = 0$

$w^2 - 12w - 45 = 0$ Dividing by 8

$(w - 15)(w + 3) = 0$

$w - 15 = 0$ or $w + 3 = 0$

$w = 15$ or $w = -3$

Check. The width cannot be negative, so we only need to check 15. When w = 15, then 2w = 30 and the dimensions of the box are 30 - 8 by 15 - 8 by 4, or 22 by 7 by 4. The volume is 22·7·4, or 616.

State. The cardboard is 30 cm by 15 cm.

44. 5 in.

45. Familiarize. Let x = the length of a side of the original square. Then x + 5 = the length of a side of the new square.

Translate.

The area of the new square	is	$2\frac{1}{4}$	times	the area of the original square.
$(x + 5)^2$	=	$2\frac{1}{4}$	·	x^2

Carry out. We solve the equation.

$$(x + 5)^2 = 2\frac{1}{4} \cdot x^2$$

$$x^2 + 10x + 25 = \frac{9}{4}x^2 \quad \text{Writing } 2\frac{1}{4} \text{ as } \frac{9}{4}$$

$$4x^2 + 40x + 100 = 9x^2 \quad \text{Multiplying by 4}$$

$$0 = 5x^2 - 40x - 100$$

$$0 = 5(x^2 - 8x - 20)$$

$$0 = x^2 - 8x - 20 \quad \text{Dividing by 5}$$

$$0 = (x - 10)(x + 2)$$

$$x - 10 = 0 \quad \text{or} \quad x + 2 = 0$$

$$x = 10 \quad \text{or} \quad x = -2$$

Check. Since the length of a side of a square cannot be negative, we only need to check 10. If $x = 10$, then $x + 5 = 15$. A square with side 10 has area 10^2, or 100, and a square with side 15 has area 15^2, or 225. Since $225 = 2\frac{1}{4}(100)$, the numbers check.

State. The area of the original square is 100 cm^2, and the area of the new square is 225 cm^2.

Exercise Set 6.1

1. $\frac{-5}{2x}$

We find the real number(s) that make the denominator 0. To do so we set the denominator equal to 0 and solve for x:

$2x = 0$

$x = 0$

The expression is undefined for $x = 0$.

2. 0

3. $\frac{a + 7}{a - 8}$

Set the denominator equal to 0 and solve for a:

$a - 8 = 0$

$a = 8$

The expression is undefined for $a = 8$.

4. -7

5. $\frac{3}{2y + 5}$

Set the denominator equal to 0 and solve for y:

$2y + 5 = 0$

$2y = -5$

$y = -\frac{5}{2}$

The expression is undefined for $y = -\frac{5}{2}$.

6. 3

7. $\frac{x^2 + 11}{x^2 - 3x - 28}$

Set the denominator equal to 0 and solve for x:

$x^2 - 3x - 28 = 0$

$(x - 7)(x + 4) = 0$

$x - 7 = 0$ or $x + 4 = 0$

$x = 7$ or $x = -4$

The expression is undefined for $x = 7$ or $x = -4$.

8. 5, 2

9. $\frac{m^3 - 2m}{m^2 - 25}$

Set the denominator equal to 0 and solve for m:

$m^2 - 25 = 0$

$(m + 5)(m - 5) = 0$

$m + 5 = 0$ or $m - 5 = 0$

$m = -5$ or $m = 5$

The expression is undefined for $x = -5$ or $x = 5$.

10. -7, 7

11. $\frac{10a^3b}{30ab^2} = \frac{a^2 \cdot 10ab}{3b \cdot 10ab}$ Factoring the numerator and denominator. Note the common factor of 10ab.

$= \frac{a^2}{3b} \cdot \frac{10ab}{10ab}$ Rewriting as a product of two rational expressions

$= \frac{a^2}{3b} \cdot 1$ $\frac{10ab}{10ab} = 1$

$= \frac{a^2}{3b}$ Removing a factor of 1

12. $\frac{5y}{x^2}$

13. $\frac{35x^2y}{14x^3y^5} = \frac{5 \cdot 7x^2y}{2xy^4 \cdot 7x^2y}$

$= \frac{5}{2xy^4} \cdot \frac{7x^2y}{7x^2y}$

$= \frac{5}{2xy^4} \cdot 1$

$= \frac{5}{2xy^4}$

14. $\frac{2a^2b^5}{3}$

15. $\frac{9x + 15}{6x + 10} = \frac{3(3x + 5)}{2(3x + 5)}$

$= \frac{3}{2} \cdot \frac{3x + 5}{3x + 5}$

$= \frac{3}{2} \cdot 1$

$= \frac{3}{2}$

16. $\frac{7}{5}$

17. $\frac{a^2 - 25}{a^2 + 6a + 5} = \frac{(a + 5)(a - 5)}{(a + 5)(a + 1)}$

$= \frac{a + 5}{a + 5} \cdot \frac{a - 5}{a + 1}$

$= 1 \cdot \frac{a - 5}{a + 1}$

$= \frac{a - 5}{a + 1}$

18. $\frac{a + 2}{a - 3}$

19. $\frac{48x^4}{18x^6} = \frac{8 \cdot 6x^4}{3x^2 \cdot 6x^4}$

$= \frac{8 \cdot \cancel{6x^4}}{3x^2 \cdot \cancel{6x^4}}$

$= \frac{8}{3x^2}$

Check: Let $x = 1$.

$\frac{48x^4}{18x^6} = \frac{48 \cdot 1^4}{18 \cdot 1^6} = \frac{48}{18} = \frac{8}{3}$

$\frac{8}{3x^2} = \frac{8}{3 \cdot 1^2} = \frac{8}{3}$

The answer is probably correct.

20. $\frac{19a^2}{6}$

21. $\frac{4x - 12}{4x} = \frac{4(x - 3)}{4x}$

$= \frac{\cancel{4}(x - 3)}{\cancel{4}x}$

$= \frac{x - 3}{x}$

Check: Let x = 2.

$\frac{4x - 12}{4x} = \frac{4\cdot 2 - 12}{4\cdot 2} = \frac{-4}{8} = -\frac{1}{2}$

$\frac{x - 3}{x} = \frac{2 - 3}{2} = \frac{-1}{2} = -\frac{1}{2}$

The answer is probably correct.

22. $\frac{y - 3}{2y}$

23. $\frac{3m^2 + 3m}{6m^2 + 9m} = \frac{3m(m + 1)}{3m(2m + 3)}$

$= \frac{3m}{3m} \cdot \frac{m + 1}{2m + 3}$

$= 1 \cdot \frac{m + 1}{2m + 3}$

$= \frac{m + 1}{2m + 3}$

Check: Let m = 1.

$\frac{3m^2 + 3m}{6m^2 + 9m} = \frac{3\cdot 1^2 + 3\cdot 1}{6\cdot 1^2 + 9\cdot 1} = \frac{6}{15} = \frac{2}{5}$

$\frac{m + 1}{2m + 3} = \frac{1 + 1}{2\cdot 1 + 3} = \frac{2}{5}$

The answer is probably correct.

24. $\frac{2(2y - 1)}{5(y - 1)}$

25. $\frac{a^2 - 9}{a^2 + 5a + 6} = \frac{(a - 3)(a + 3)}{(a + 2)(a + 3)}$

$= \frac{a - 3}{a + 2} \cdot \frac{a + 3}{a + 3}$

$= \frac{a - 3}{a + 2} \cdot 1$

$= \frac{a - 3}{a + 2}$

Check: Let a = 2.

$\frac{a^2 - 9}{a^2 + 5a + 6} = \frac{2^2 - 9}{2^2 + 5\cdot 2 + 6} = \frac{-5}{20} = -\frac{1}{4}$

$\frac{a - 3}{a + 2} = \frac{2 - 3}{2 + 2} = \frac{-1}{4} = -\frac{1}{4}$

The answer is probably correct.

26. $\frac{t - 5}{t - 4}$

27. $\frac{2t^2 + 6t + 4}{4t^2 - 12t - 16} = \frac{2(t^2 + 3t + 2)}{4(t^2 - 3t - 4)}$

$= \frac{2(t + 2)(t + 1)}{2\cdot 2(t - 4)(t + 1)}$

$= \frac{2(t + 1)}{2(t + 1)} \cdot \frac{t + 2}{2(t - 4)}$

$= 1 \cdot \frac{t + 2}{2(t - 4)}$

$= \frac{t + 2}{2(t - 4)}$

Check: Let t = 1.

$\frac{2t^2 + 6t + 4}{4t^2 - 12t - 16} = \frac{2\cdot 1^2 + 6\cdot 1 + 4}{4\cdot 1^2 - 12\cdot 1 - 16} = \frac{12}{-24} = -\frac{1}{2}$

$\frac{t + 2}{2(t - 4)} = \frac{1 + 2}{2(1 - 4)} = \frac{3}{-6} = -\frac{1}{2}$

The answer is probably correct.

28. $\frac{a - 4}{2(a + 4)}$

29. $\frac{x^2 - 25}{x^2 - 10x + 25} = \frac{(x - 5)(x + 5)}{(x - 5)(x - 5)}$

$= \frac{x - 5}{x - 5} \cdot \frac{x + 5}{x - 5}$

$= 1 \cdot \frac{x + 5}{x - 5}$

$= \frac{x + 5}{x - 5}$

Check: Let x = 2.

$\frac{x^2 - 25}{x^2 - 10x + 25} = \frac{2^2 - 25}{2^2 - 10\cdot 2 + 25} = \frac{-21}{9} = -\frac{7}{3}$

$\frac{x + 5}{x - 5} = \frac{2 + 5}{2 - 5} = \frac{7}{-3} = -\frac{7}{3}$

The answer is probably correct.

30. $\frac{x + 4}{x - 4}$

31. $\frac{a^2 - 1}{a - 1} = \frac{(a - 1)(a + 1)}{a - 1}$

$= \frac{a - 1}{a - 1} \cdot \frac{a + 1}{1}$

$= 1 \cdot \frac{a + 1}{1}$

$= a + 1$

Check: Let a = 2.

$\frac{a^2 - 1}{a - 1} = \frac{2^2 - 1}{2 - 1} = \frac{3}{1} = 3$

$a + 1 = 2 + 1 = 3$

The answer is probably correct.

32. $t - 1$

33. $\frac{x^2 + 1}{x + 1}$ cannot be simplified.

Neither the numerator nor the denominator can be factored.

34. $\frac{y^2 + 4}{y + 2}$

35. $\frac{6x^2 - 54}{4x^2 - 36} = \frac{2\cdot3(x^2 - 9)}{2\cdot2(x^2 - 9)}$

$= \frac{2(x^2 - 9)}{2(x^2 - 9)} \cdot \frac{3}{2}$

$= 1 \cdot \frac{3}{2}$

$= \frac{3}{2}$

Check: Let x = 1.

$\frac{6x^2 - 54}{4x^2 - 36} = \frac{6\cdot1^2 - 54}{4\cdot1^2 - 36} = \frac{-48}{-32} = \frac{3}{2}$

$\frac{3}{2} = \frac{3}{2}$

The answer is probably correct.

36. 2

37. $\frac{6t + 12}{t^2 - t - 6} = \frac{6(t + 2)}{(t - 3)(t + 2)}$

$= \frac{6}{t - 3} \cdot \frac{t + 2}{t + 2}$

$= \frac{6}{t - 3} \cdot 1$

$= \frac{6}{t - 3}$

Check: Let t = 1.

$\frac{6t + 12}{t^2 - t - 6} = \frac{6\cdot1 + 12}{1^2 - 1 - 6} = \frac{18}{-6} = -3$

$\frac{6}{t - 3} = \frac{6}{1 - 3} = \frac{6}{-2} = -3$

The answer is probably correct.

38. $\frac{5}{y + 6}$

39. $\frac{a^2 - 10a + 21}{a^2 - 11a + 28} = \frac{(a - 7)(a - 3)}{(a - 7)(a - 4)}$

$= \frac{a - 7}{a - 7} \cdot \frac{a - 3}{a - 4}$

$= 1 \cdot \frac{a - 3}{a - 4}$

$= \frac{a - 3}{a - 4}$

Check: Let a = 2.

$\frac{a^2 - 10a + 21}{a^2 - 11a + 28} = \frac{2^2 - 10\cdot2 + 21}{2^2 - 11\cdot2 + 28} = \frac{5}{10} = \frac{1}{2}$

$\frac{a - 3}{a - 4} = \frac{2 - 3}{2 - 4} = \frac{-1}{-2} = \frac{1}{2}$

The answer is probably correct.

40. $\frac{y - 6}{y - 5}$

41. $\frac{t^2 - 4}{(t + 2)^2} = \frac{(t - 2)(t + 2)}{(t + 2)(t + 2)}$

$= \frac{t - 2}{t + 2} \cdot \frac{t + 2}{t + 2}$

$= \frac{t - 2}{t + 2} \cdot 1$

$= \frac{t - 2}{t + 2}$

Check: Let t = 1.

$\frac{t^2 - 4}{(t + 2)^2} = \frac{1^2 - 4}{(1 + 2)^2} = \frac{-3}{9} = -\frac{1}{3}$

$\frac{t - 2}{t + 2} = \frac{1 - 2}{1 + 2} = \frac{-1}{3} = -\frac{1}{3}$

The answer is probably correct.

42. $\frac{a - 3}{a + 3}$

43. $\frac{6 - x}{x - 6} = \frac{-1(-6 + x)}{x - 6}$

$= \frac{-1(x - 6)}{x - 6}$

$= -1 \cdot \frac{x - 6}{x - 6}$

$= -1\cdot1$

$= -1$

Check: Let x = 3.

$\frac{6 - x}{x - 6} = \frac{6 - 3}{3 - 6} = \frac{3}{-3} = -1$

$-1 = -1$

The answer is probably correct.

44. -1

45. $\frac{a - b}{b - a} = \frac{-1(-a + b)}{b - a}$

$= \frac{-1(b - a)}{b - a}$

$= -1 \cdot \frac{b - a}{b - a}$

$= -1\cdot1$

$= -1$

Check: Let a = 2 and b = 1.

$\frac{a - b}{b - a} = \frac{2 - 1}{1 - 2} = \frac{1}{-1} = -1$

$-1 = -1$

The answer is probably correct.

46. 1

47. $\frac{6t - 12}{2 - t} = \frac{-6(-t + 2)}{2 - t}$

$= \frac{-6(2 - t)}{2 - t}$

$= -6 \cdot \frac{2 - t}{2 - t}$

$= -6\cdot1$

$= -6$

Check: Let t = 3.

$\frac{6t - 12}{2 - t} = \frac{6\cdot3 - 12}{2 - 3} = \frac{6}{-1} = -6$

$-6 = -6$

The answer is probably correct.

48. -5

49. $\frac{a^2 - 1}{1 - a} = \frac{(a + 1)(a - 1)}{-1(-1 + a)}$

$= \frac{(a + 1)(a - 1)}{-1(a - 1)}$

$= \frac{a + 1}{-1} \cdot \frac{a - 1}{a - 1}$

$= -(a + 1)\cdot 1$

$= -a - 1$

Check: Let $a = 2$.

$\frac{a^2 - 1}{1 - a} = \frac{2^2 - 1}{1 - 2} = \frac{3}{-1} = -3$

$-a - 1 = -2 - 1 = -3$

The answer is probably correct.

50. -1

51. $x^2 + 8x + 7$

The factorization is of the form $(x + \quad)(x + \quad)$. We look for two factors of 7 whose sum is 8. The numbers we need are 1 and 7.

$x^2 + 8x + 7 = (x + 1)(x + 7)$

52. $(x - 2)(x - 7)$

53. $5x + 2y = 20$

To find the y-intercept, solve:

$2y = 20$

$y = 10$

The y-intercept is (0,10).

To find the x-intercept, solve:

$5x = 20$

$x = 4$

The x-intercept is (4,0).

We find a third point as a check. Let $x = 2$ and solve for y.

$5\cdot 2 + 2y = 20$

$10 + 2y = 20$

$2y = 10$

$y = 5$

The point (2,5) appears to line up with the intercepts, so we draw the graph.

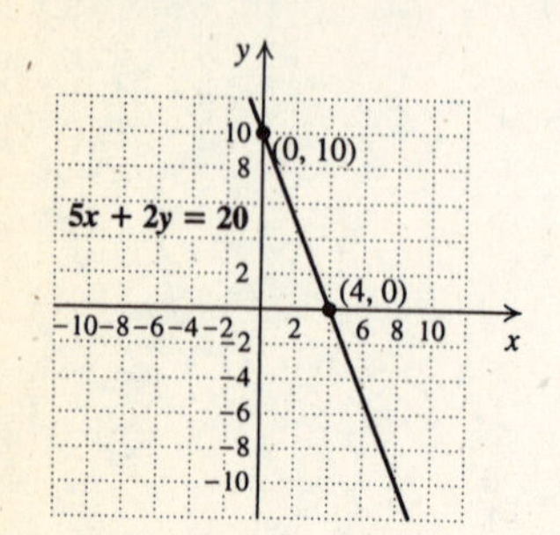

54. y-intercept: (0,-2), x-intercept: (4,0)

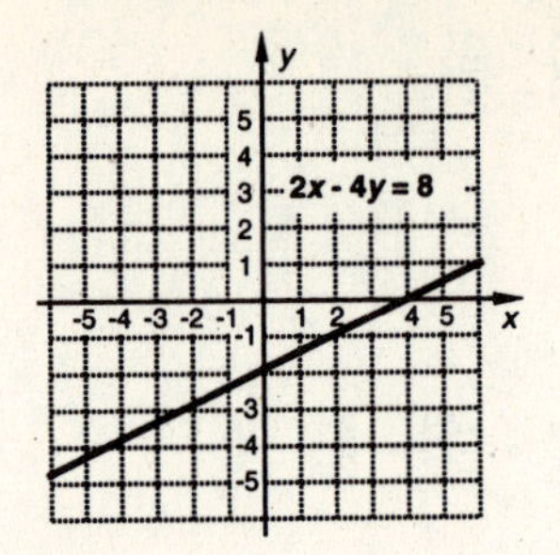

55.

56.

57. $\frac{x^4 - 16y^2}{(x^2 + 4y^2)(x - 2y)}$

$= \frac{(x^2 + 4y^2)(x + 2y)(x - 2y)}{(x^2 + 4y^2)(x - 2y)}$

$= \frac{(x^2 + 4y^2)(x - 2y)}{(x^2 + 4y^2)(x - 2y)} \cdot \frac{x + 2y}{1}$

$= x + 2y$

58. $\frac{a - b}{-a - b}$, or $\frac{b - a}{a + b}$

59. $\frac{(t^4 - 1)(t^2 - 9)(t - 9)^2}{(t^4 - 81)(t^2 + 1)(t + 1)^2}$

$= \frac{(t^2+1)(t+1)(t-1)(t+3)(t-3)(t-9)(t-9)}{(t^2+9)(t+3)(t-3)(t^2+1)(t+1)(t+1)}$

$= \frac{(t^2+1)(t+1)(t+3)(t-3)}{(t^2+1)(t+1)(t+3)(t-3)} \cdot \frac{(t-1)(t-9)(t-9)}{(t^2+9)(t+1)}$

$= \frac{(t - 1)(t - 9)(t - 9)}{(t^2 + 9)(t + 1)}$, or $\frac{(t - 1)(t - 9)^2}{(t^2 + 9)(t + 1)}$

60. 1

61. $\frac{(x^2 - y^2)(x^2 - 2xy + y^2)}{(x - y)^2(x^2 - 4xy - 5y^2)}$

$= \frac{(x + y)(x - y)(x - y)(x - y)}{(x - y)(x - y)(x - 5y)(x + y)}$

$= \frac{(x + y)(x - y)(x - y)}{(x + y)(x - y)(x - y)} \cdot \frac{x - y}{x - 5y}$

$= \frac{x - y}{x - 5y}$

62. $\frac{1}{x - 1}$

63.

Exercise Set 6.2

1. $\frac{3x}{2} \cdot \frac{x + 4}{x - 1} = \frac{3x(x + 4)}{2(x - 1)}$

2. $\dfrac{4x(x - 3)}{5(x + 2)}$

3. $\dfrac{x - 1}{x + 2} \cdot \dfrac{x + 1}{x + 2} = \dfrac{(x - 1)(x + 1)}{(x + 2)(x + 2)}$

4. $\dfrac{(x - 2)(x - 2)}{(x - 5)(x + 5)}$

5. $\dfrac{2x + 3}{4} \cdot \dfrac{x + 1}{x - 5} = \dfrac{(2x + 3)(x + 1)}{4(x - 5)}$

6. $\dfrac{(-5)(-6)}{(3x - 4)(5x + 6)}$

7. $\dfrac{a - 5}{a^2 + 1} \cdot \dfrac{a + 2}{a^2 - 1} = \dfrac{(a - 5)(a + 2)}{(a^2 + 1)(a^2 - 1)}$

8. $\dfrac{(t + 3)(t + 3)}{(t^2 - 2)(t^2 - 2)}$

9. $\dfrac{x + 1}{2 + x} \cdot \dfrac{x - 1}{x + 1} = \dfrac{(x + 1)(x - 1)}{(2 + x)(x + 1)}$

10. $\dfrac{(m^2 + 5)(m^2 - 4)}{(m + 8)(m^2 - 4)}$

11. $\dfrac{4x^3}{3x} \cdot \dfrac{14}{x} = \dfrac{4x^3 \cdot 14}{3x \cdot x}$ Multiplying the numerators and the denominators

$= \dfrac{4 \cdot x \cdot x \cdot x \cdot 14}{3 \cdot x \cdot x}$ Factoring the numerator and the denominator

$= \dfrac{x \cdot x}{x \cdot x} \cdot \dfrac{4 \cdot x \cdot 14}{3}$ Factoring the rational expression

$= 1 \cdot \dfrac{56x}{3}$

$= \dfrac{56x}{3}$ Removing a factor of 1

12. $\dfrac{12}{b^2}$

13. $\dfrac{3c}{d^2} \cdot \dfrac{4d}{6c^3} = \dfrac{3c \cdot 4d}{d^2 \cdot 6c^3}$ Multiplying the numerators and the denominators

$= \dfrac{3 \cdot c \cdot 2 \cdot 2 \cdot d}{d \cdot d \cdot 3 \cdot 2 \cdot c \cdot c^2}$ Factoring the numerator and the denominator

$= \dfrac{3 \cdot 2 \cdot c \cdot d}{3 \cdot 2 \cdot c \cdot d} \cdot \dfrac{2}{dc^2}$

$= 1 \cdot \dfrac{2}{dc^2}$

$= \dfrac{2}{dc^2}$ Removing a factor of 1

14. $\dfrac{6x}{y^2}$

15. $\dfrac{x^2 - 3x - 10}{(x - 2)^2} \cdot \dfrac{x - 2}{x - 5} = \dfrac{(x^2 - 3x - 10)(x - 2)}{(x - 2)^2(x - 5)}$

$= \dfrac{(x - 5)(x + 2)(x - 2)}{(x - 2)(x - 2)(x - 5)}$

$= \dfrac{(x - 5)(x - 2)}{(x - 5)(x - 2)} \cdot \dfrac{x + 2}{x - 2}$

$= \dfrac{x + 2}{x - 2}$

16. $\dfrac{t}{t + 2}$

17. $\dfrac{a^2 - 25}{a^2 - 4a + 3} \cdot \dfrac{2a - 5}{2a + 5} = \dfrac{(a^2 - 25)(2a - 5)}{(a^2 - 4a + 3)(2a + 5)}$

$= \dfrac{(a + 5)(a - 5)(2a - 5)}{(a - 3)(a - 1)(2a + 5)}$

(No simplification is possible.)

18. $\dfrac{(x + 3)(x + 4)(x + 1)}{(x^2 + 9)(x + 9)}$

19. $\dfrac{a^2 - 9}{a^2} \cdot \dfrac{a^2 - 3a}{a^2 + a - 12} = \dfrac{(a - 3)(a + 3)(a)(a - 3)}{a \cdot a(a + 4)(a - 3)}$

$= \dfrac{a(a - 3)}{a(a - 3)} \cdot \dfrac{(a - 3)(a + 3)}{a(a + 4)}$

$= \dfrac{(a - 3)(a + 3)}{a(a + 4)}$

20. 1

21. $\dfrac{4a^2}{3a^2 - 12a + 12} \cdot \dfrac{3a - 6}{2a} = \dfrac{4a^2(3a - 6)}{(3a^2 - 12a + 12)2a}$

$= \dfrac{2 \cdot 2 \cdot a \cdot a \cdot 3 \cdot (a - 2)}{3 \cdot (a - 2) \cdot (a - 2) \cdot 2 \cdot a}$

$= \dfrac{2 \cdot 3a(a - 2)}{2 \cdot 3a(a - 2)} \cdot \dfrac{2a}{a - 2}$

$= \dfrac{2a}{a - 2}$

22. $\dfrac{5(v - 2)}{v - 1}$

23. $\dfrac{t^2 + 2t - 3}{t^2 + 4t - 5} \cdot \dfrac{t^2 - 3t - 10}{t^2 + 5t + 6} = \dfrac{(t^2+2t-3)(t^2-3t-10)}{(t^2+4t-5)(t^2+5t+6)}$

$= \dfrac{(t+3)(t-1)(t-5)(t+2)}{(t+5)(t-1)(t+3)(t+2)}$

$= \dfrac{\cancel{(t+3)}\cancel{(t-1)}(t-5)\cancel{(t+2)}}{(t+5)\cancel{(t-1)}\cancel{(t+3)}\cancel{(t+2)}}$

$= \dfrac{t - 5}{t + 5}$

24. $\dfrac{x + 4}{x - 4}$

25. $\dfrac{5a^2 - 180}{10a^2 - 10} \cdot \dfrac{20a + 20}{2a - 12} = \dfrac{(5a^2 - 180)(20a + 20)}{(10a^2 - 10)(2a - 12)}$

$= \dfrac{5(a+6)(a-6)(2)(10)(a+1)}{10(a+1)(a-1)(2)(a-6)}$

$= \dfrac{5(a+6)\cancel{(a-6)}\cancel{(2)}\cancel{(10)}\cancel{(a+1)}}{\cancel{10(a+1)}(a-1)\cancel{(2)}\cancel{(a-6)}}$

$= \dfrac{5(a + 6)}{a - 1}$

26. $\dfrac{t + 7}{4(t - 1)}$

27. $\dfrac{x^2 - 1}{x^2 - 9} \cdot \dfrac{(x - 3)^4}{(x + 1)^2}$

$= \dfrac{(x^2 - 1)(x - 3)^4}{(x^2 - 9)(x + 1)^2}$

$= \dfrac{(x + 1)(x - 1)(x - 3)(x - 3)(x - 3)(x - 3)}{(x + 3)(x - 3)(x + 1)(x + 1)}$

$= \dfrac{\cancel{(x + 1)}(x - 1)\cancel{(x - 3)}(x - 3)(x - 3)(x - 3)}{(x + 3)\cancel{(x - 3)}\cancel{(x + 1)}(x + 1)}$

$= \dfrac{(x - 1)(x - 3)(x - 3)(x - 3)}{(x + 3)(x + 1)}$, or

$\dfrac{(x - 1)(x - 3)^3}{(x + 3)(x + 1)}$

28. $\dfrac{(x + 2)^4(x + 1)}{(x - 1)^2(x + 3)}$

29. $\dfrac{a^2 - 4}{a^2 + 2a + 1} \cdot \dfrac{a - 1}{a^4 + 1} = \dfrac{(a^2 - 4)(a - 1)}{(a^2 + 2a + 1)(a^4 + 1)}$

$= \dfrac{(a + 2)(a - 2)(a - 1)}{(a + 1)^2(a^4 + 1)}$

(No simplification is possible.)

30. $\dfrac{(a^2 + 4)(a + 3)^2}{(a - 3)^2(a^4 + 16)}$

31. $\dfrac{(t - 2)^3}{(t - 1)^3} \cdot \dfrac{t^2 - 2t + 1}{t^2 - 4t + 4}$

$= \dfrac{(t - 2)^3(t^2 - 2t + 1)}{(t - 1)^3(t^2 - 4t + 4)}$

$= \dfrac{(t - 2)(t - 2)(t - 2)(t - 1)(t - 1)}{(t - 1)(t - 1)(t - 1)(t - 2)(t - 2)}$

$= \dfrac{(t - 2)(t - 2)(t - 1)(t - 1)}{(t - 2)(t - 2)(t - 1)(t - 1)} \cdot \dfrac{t - 2}{t - 1}$

$= \dfrac{t - 2}{t - 1}$

32. $\dfrac{y + 4}{y + 2}$

33. The reciprocal of $\frac{4}{x}$ is $\frac{x}{4}$ because $\frac{4}{x} \cdot \frac{x}{4} = 1$.

34. $\dfrac{a - 1}{a + 3}$

35. The reciprocal of $x^2 - y^2$ is $\dfrac{1}{x^2 - y^2}$ because $\dfrac{x^2 - y^2}{1} \cdot \dfrac{1}{x^2 - y^2} = 1$.

36. $a + b$

37. The reciprocal of $\dfrac{x^2 + 2x - 5}{x^2 - 4x + 7}$ is $\dfrac{x^2 - 4x + 7}{x^2 + 2x - 5}$ because $\dfrac{x^2 + 2x - 5}{x^2 - 4x + 7} \cdot \dfrac{x^2 - 4x + 7}{x^2 + 2x - 5} = 1$.

38. $\dfrac{x^2 + 7xy - y^2}{x^2 - 3xy + y^2}$

39. $\dfrac{2}{5} \div \dfrac{4}{3} = \dfrac{2}{5} \cdot \dfrac{3}{4}$ Multiplying by the reciprocal

$= \dfrac{2 \cdot 3}{5 \cdot 4}$

$= \dfrac{2 \cdot 3}{5 \cdot 2 \cdot 2}$ Factoring the denominator

$= \dfrac{2}{2} \cdot \dfrac{3}{5 \cdot 2}$ Factoring the fractional expression

$= \dfrac{3}{10}$ Simplifying

40. $\dfrac{5}{4}$

41. $\dfrac{2}{x} \div \dfrac{8}{x} = \dfrac{2}{x} \cdot \dfrac{x}{8}$ Multiplying by the reciprocal

$= \dfrac{2 \cdot x}{x \cdot 8}$

$= \dfrac{2 \cdot x \cdot 1}{x \cdot 2 \cdot 4}$ Factoring the numerator and the denominator

$= \dfrac{2x}{2x} \cdot \dfrac{1}{4}$ Factoring the fractional expression

$= \dfrac{1}{4}$ Simplifying

42. $\dfrac{x^2}{6}$

43. $\dfrac{x^2}{y} \div \dfrac{x^3}{y^3} = \dfrac{x^2}{y} \cdot \dfrac{y^3}{x^3}$

$= \dfrac{x^2 \cdot y^3}{y \cdot x^3}$

$= \dfrac{x^2 \cdot y \cdot y^2}{y \cdot x^2 \cdot x}$

$= \dfrac{x^2y}{x^2y} \cdot \dfrac{y^2}{x}$

$= \dfrac{y^2}{x}$

44. $\dfrac{b}{a}$

45. $\dfrac{a + 2}{a - 3} \div \dfrac{a - 1}{a + 3} = \dfrac{a + 2}{a - 3} \cdot \dfrac{a + 3}{a - 1}$

$= \dfrac{(a + 2)(a + 3)}{(a - 3)(a - 1)}$

46. $\dfrac{y + 2}{2y}$

47. $\dfrac{x^2 - 1}{x} \div \dfrac{x + 1}{x - 1} = \dfrac{x^2 - 1}{x} \cdot \dfrac{x - 1}{x + 1}$

$= \dfrac{(x^2 - 1)(x - 1)}{x(x + 1)}$

$= \dfrac{(x - 1)(x + 1)(x - 1)}{x(x + 1)}$

$= \dfrac{x + 1}{x + 1} \cdot \dfrac{(x - 1)(x - 1)}{x}$

$= \dfrac{(x - 1)^2}{x}$

48. $4(y - 2)$

49. $\frac{x+1}{6} \div \frac{x+1}{3} = \frac{x+1}{6} \cdot \frac{3}{x+1}$

$= \frac{(x+1)\cdot 3}{6(x+1)}$

$= \frac{3(x+1)}{2\cdot 3(x+1)}$

$= \frac{3(x+1)}{3(x+1)} \cdot \frac{1}{2}$

$= \frac{1}{2}$

50. $\frac{a}{b}$

51. $(y^2 - 9) \div \frac{y^2 - 2y - 3}{y^2 + 1} = \frac{(y^2 - 9)}{1} \cdot \frac{y^2 + 1}{y^2 - 2y - 3}$

$= \frac{(y^2 - 9)(y^2 + 1)}{y^2 - 2y - 3}$

$= \frac{(y+3)(y-3)(y^2+1)}{(y-3)(y+1)}$

$= \frac{(y+3)\cancel{(y-3)}(y^2+1)}{\cancel{(y-3)}(y+1)}$

$= \frac{(y+3)(y^2+1)}{y+1}$

52. $\frac{(x-6)(x+6)}{x-1}$

53. $\frac{5x-5}{16} \div \frac{x-1}{6} = \frac{5x-5}{16} \cdot \frac{6}{x-1}$

$= \frac{(5x-5)\cdot 6}{16(x-1)}$

$= \frac{5(x-1)\cdot 2\cdot 3}{2\cdot 8(x-1)}$

$= \frac{2(x-1)}{2(x-1)} \cdot \frac{5\cdot 3}{8}$

$= \frac{15}{8}$

54. $\frac{1}{2}$

55. $\frac{-6+3x}{5} \div \frac{4x-8}{25} = \frac{-6+3x}{5} \cdot \frac{25}{4x-8}$

$= \frac{(-6+3x)\cdot 25}{5(4x-8)}$

$= \frac{3(x-2)\cdot 5\cdot 5}{5\cdot 4(x-2)}$

$= \frac{5(x-2)}{5(x-2)} \cdot \frac{3\cdot 5}{4}$

$= \frac{15}{4}$

56. 3

57. $\frac{a+2}{a-1} \div \frac{3a+6}{a-5} = \frac{a+2}{a-1} \cdot \frac{a-5}{3a+6}$

$= \frac{(a+2)(a-5)}{(a-1)(3a+6)}$

$= \frac{(a+2)(a-5)}{(a-1)\cdot 3\cdot (a+2)}$

$= \frac{a+2}{a+2} \cdot \frac{a-5}{3(a-1)}$

$= \frac{a-5}{3(a-1)}$

58. $\frac{t+1}{4(t+2)}$

59. $(x-5) \div \frac{2x^2 - 11x + 5}{4x^2 - 1} = \frac{x-5}{1} \cdot \frac{4x^2 - 1}{2x^2 - 11x + 5}$

$= \frac{(x-5)(4x^2-1)}{1\cdot(2x^2 - 11x + 5)}$

$= \frac{(x-5)(2x+1)(2x-1)}{1\cdot(2x-1)(x-5)}$

$= \frac{\cancel{(x-5)}(2x+1)\cancel{(2x-1)}}{1\cdot\cancel{(2x-1)}\cancel{(x-5)}}$

$= 2x + 1$

60. $\frac{(a+7)(a+1)}{3a-7}$

61. $\frac{x^2-4}{x} \div \frac{x-2}{x+2} = \frac{x^2-4}{x} \cdot \frac{x+2}{x-2}$

$= \frac{(x^2-4)(x+2)}{x(x-2)}$

$= \frac{(x-2)(x+2)(x+2)}{x(x-2)}$

$= \frac{x-2}{x-2} \cdot \frac{(x+2)(x+2)}{x}$

$= \frac{(x+2)^2}{x}$

62. $\frac{(x+y)^2}{x^2+y}$

63. $\frac{x^2-9}{4x+12} \div \frac{x-3}{6} = \frac{x^2-9}{4x+12} \cdot \frac{6}{x-3}$

$= \frac{(x^2-9)\cdot 6}{(4x+12)(x-3)}$

$= \frac{(x-3)(x+3)\cdot 3\cdot 2}{2\cdot 2(x+3)(x-3)}$

$= \frac{2(x-3)(x+3)}{2(x-3)(x+3)} \cdot \frac{3}{2}$

$= \frac{3}{2}$

64. $\frac{5x}{2(x+b)}$

65. $\dfrac{c^2 + 3c}{c^2 + 2c - 3} \div \dfrac{c}{c + 1} = \dfrac{c^2 + 3c}{c^2 + 2c - 3} \cdot \dfrac{c + 1}{c}$

$= \dfrac{(c^2 + 3c)(c + 1)}{(c^2 + 2c - 3)c}$

$= \dfrac{c(c + 3)(c + 1)}{(c + 3)(c - 1)c}$

$= \dfrac{c(c + 3)}{c(c + 3)} \cdot \dfrac{c + 1}{c - 1}$

$= \dfrac{c + 1}{c - 1}$

66. $\dfrac{2x}{x + 5}$

67. $\dfrac{2y^2 - 7y + 3}{2y^2 + 3y - 2} \div \dfrac{6y^2 - 5y + 1}{3y^2 + 5y - 2}$

$= \dfrac{2y^2 - 7y + 3}{2y^2 + 3y - 2} \cdot \dfrac{3y^2 + 5y - 2}{6y^2 - 5y + 1}$

$= \dfrac{(2y^2 - 7y + 3)(3y^2 + 5y - 2)}{(2y^2 + 3y - 2)(6y^2 - 5y + 1)}$

$= \dfrac{(2y - 1)(y - 3)(3y - 1)(y + 2)}{(2y - 1)(y + 2)(3y - 1)(2y - 1)}$

$= \dfrac{(2y - 1)(3y - 1)(y + 2)}{(2y - 1)(3y - 1)(y + 2)} \cdot \dfrac{y - 3}{2y - 1}$

$= \dfrac{y - 3}{2y - 1}$

68. $\dfrac{x + 3}{x - 5}$

69. $\dfrac{c^2 + 10c + 21}{c^2 - 2c - 15} \div (c^2 + 2c - 35)$

$= \dfrac{c^2 + 10c + 21}{c^2 - 2c - 15} \cdot \dfrac{1}{c^2 + 2c - 35}$

$= \dfrac{(c^2 + 10c + 21) \cdot 1}{(c^2 - 2c - 15)(c^2 + 2c - 35)}$

$= \dfrac{(c + 7)(c + 3)}{(c - 5)(c + 3)(c + 7)(c - 5)}$

$= \dfrac{(c + 7)(c + 3)}{(c + 7)(c + 3)} \cdot \dfrac{1}{(c - 5)(c - 5)}$

$= \dfrac{1}{(c - 5)^2}$

70. $\dfrac{1}{1 + 2z - z^2}$

71. $\dfrac{(t + 5)^3}{(t - 5)^3} \div \dfrac{(t + 5)^2}{(t - 5)^2}$

$= \dfrac{(t + 5)^3}{(t - 5)^3} \cdot \dfrac{(t - 5)^2}{(t + 5)^2}$

$= \dfrac{(t + 5)^3(t - 5)^2}{(t - 5)^3(t + 5)^2}$

$= \dfrac{(t + 5)^2(t - 5)^2}{(t + 5)^2(t - 5)^2} \cdot \dfrac{t + 5}{t - 5}$

$= \dfrac{t + 5}{t - 5}$

72. $\dfrac{y - 3}{y + 3}$

73. Familiarize. Let x = the number.

Translate.

Sixteen	more than	the square of a number	is	eight times the number.
16	+	x^2	=	$8x$

Carry out.

$16 + x^2 = 8x$

$x^2 - 8x + 16 = 0$

$(x - 4)(x - 4) = 0$

$x - 4 = 0$ or $x - 4 = 0$

$x = 4$ or $x = 4$

Check. The square of 4, which is 16, plus 16 is 32, and eight times 4 is 32. The number checks.

State. The number is 4.

74. $2x^2 + 16$

75. $(8x^3 - 3x^2 + 7) - (8x^2 + 3x - 5) =$

$8x^3 - 3x^2 + 7 - 8x^2 - 3x + 5 =$

$8x^3 - 11x^2 - 3x + 12$

76. $0.06y^3 - 0.09y^2 + 0.01y - 1$

77. ◆

78. ◆

79. $\dfrac{2a^2 - 5ab}{c - 3d} \div (4a^2 - 25b^2)$

$= \dfrac{2a^2 - 5ab}{c - 3d} \cdot \dfrac{1}{4a^2 - 25b^2}$

$= \dfrac{a(2a - 5b)}{(c - 3d)(2a + 5b)(2a - 5b)}$

$= \dfrac{2a - 5b}{2a - 5b} \cdot \dfrac{a}{(c - 3d)(2a + 5b)}$

$= \dfrac{a}{(c - 3d)(2a + 5b)}$

80. 1

81. $\dfrac{3a^2 - 5ab - 12b^2}{3ab + 4b^2} \div (3b^2 - ab)$

$= \dfrac{3a^2 - 5ab - 12b^2}{3ab + 4b^2} \cdot \dfrac{1}{3b^2 - ab}$

$= \dfrac{(3a + 4b)(a - 3b)}{b(3a + 4b) \cdot b(3b - a)}$

$= \dfrac{(3a + 4b)(-1)(3b - a)}{b(3a + 4b) \cdot b(3b - a)}$

$= \dfrac{(3a + 4b)(3b - a)}{(3a + 4b)(3b - a)} \cdot \dfrac{-1}{b \cdot b}$

$= -\dfrac{1}{b^2}$

82. $\dfrac{1}{(x + y)^2}$

83. $xy \cdot \frac{y^2 - 4xy}{y - x} \div \frac{16x^2y^2 - y^4}{4x^2 - 3xy - y^2}$

$= \frac{xy}{1} \cdot \frac{y^2 - 4xy}{y - x} \cdot \frac{4x^2 - 3xy - y^2}{16x^2y^2 - y^4}$

$= \frac{x\cdot y\cdot y\cdot (y - 4x)(4x + y)(x - y)}{(y - x)\cdot y\cdot y\cdot (4x - y)(4x + y)}$

$= \frac{x\cdot y\cdot y\cdot (-1)(4x - y)(4x + y)(-1)(y - x)}{(y - x)\cdot y\cdot y\cdot (4x - y)(4x + y)}$

$= \frac{y\cdot y(4x - y)(4x + y)(y - x)}{y\cdot y(4x - y)(4x + y)(y - x)} \cdot \frac{x(-1)(-1)}{1}$

$= x$

84. $\frac{(z + 4)^3}{(z - 4)^3}$

85. $\frac{x^2 - x + xy - y}{x^2 + 6x - 7} \div \frac{x^2 + 2xy + y^2}{4x + 4y}$

$= \frac{x^2 - x + xy - y}{x^2 + 6x - 7} \cdot \frac{4x + 4y}{x^2 + 2xy + y^2}$

$= \frac{x(x - 1) + y(x - 1)}{x^2 + 6x - 7} \cdot \frac{4x + 4y}{x^2 + 2xy + y^2}$

$= \frac{(x - 1)(x + y)\cdot 4(x + y)}{(x + 7)(x - 1)(x + y)(x + y)}$

$= \frac{(x - 1)(x + y)(x + y)}{(x - 1)(x + y)(x + y)} \cdot \frac{4}{x + 7}$

$= \frac{4}{x + 7}$

86. $\frac{x(x^2 + 1)}{3(x + y - 1)}$

87. $\frac{t^4 - 1}{t^4 - 81} \cdot \frac{t^2 - 9}{t^2 + 1} \div \frac{(t + 1)^2}{(t - 9)^2}$

$= \frac{t^4 - 1}{t^4 - 81} \cdot \frac{t^2 - 9}{t^2 + 1} \cdot \frac{(t - 9)^2}{(t + 1)^2}$

$= \frac{(t^4 - 1)(t^2 - 9)(t - 9)^2}{(t^4 - 81)(t^2 + 1)(t + 1)^2}$

$= \frac{(t^2 + 1)(t + 1)(t - 1)(t + 3)(t - 3)(t - 9)(t - 9)}{(t^2 + 9)(t + 3)(t - 3)(t^2 + 1)(t + 1)(t + 1)}$

$= \frac{\cancel{(t^2 + 1)}\cancel{(t + 1)}(t - 1)\cancel{(t + 3)}\cancel{(t - 3)}(t - 9)(t - 9)}{(t^2 + 9)\cancel{(t + 3)}\cancel{(t - 3)}\cancel{(t^2 + 1)}\cancel{(t + 1)}(t + 1)}$

$= \frac{(t - 1)(t - 9)(t - 9)}{(t^2 + 9)(t + 1)}$, or

$\frac{(t - 1)(t - 9)^2}{(t^2 + 9)(t + 1)}$

88. 1

89. $\left[\frac{y^2 + 5y + 6}{y^2} \cdot \frac{3y^3 + 6y^2}{y^2 - y - 12}\right] \div \frac{y^2 - y}{y^2 - 2y - 8}$

$= \frac{y^2 + 5y + 6}{y^2} \cdot \frac{3y^3 + 6y^2}{y^2 - y - 12} \cdot \frac{y^2 - 2y - 8}{y^2 - y}$

$= \frac{(y + 3)(y + 2)(3y^2)(y + 2)(y - 4)(y + 2)}{y^2(y - 4)(y + 3)(y)(y - 1)}$

$= \frac{y^2(y - 4)(y + 3)}{y^2(y - 4)(y + 3)} \cdot \frac{3(y + 2)(y + 2)(y + 2)}{y(y - 1)}$

$= \frac{3(y + 2)^3}{y(y - 1)}$

90. $\frac{a - 3b}{c}$

Exercise Set 6.3

1. $\frac{3}{x} + \frac{5}{x} = \frac{8}{x}$ Adding numerators

2. $\frac{13}{a^2}$

3. $\frac{x}{15} + \frac{2x + 1}{15} = \frac{3x + 1}{15}$ Adding numerators

4. $\frac{4a - 4}{7}$

5. $\frac{2}{a + 3} + \frac{4}{a + 3} = \frac{6}{a + 3}$

6. $\frac{13}{x + 2}$

7. $\frac{9}{a + 6} - \frac{5}{a + 6} = \frac{4}{a + 6}$ Subtracting numerators

8. $\frac{6}{x + 7}$

9. $\frac{3y + 9}{2y} - \frac{y + 1}{2y} = \frac{3y + 9 - (y + 1)}{2y}$

$= \frac{3y + 9 - y - 1}{2y}$ Removing parentheses

$= \frac{2y + 8}{2y}$

$= \frac{2(y + 4)}{2y}$

$= \frac{\cancel{2}(y + 4)}{\cancel{2}\cdot y}$

$= \frac{y + 4}{y}$

10. $\frac{t + 4}{4t}$

11. $\frac{9x + 5}{x + 1} + \frac{2x + 3}{x + 1} = \frac{11x + 8}{x + 1}$ Adding numerators

12. $\frac{5a + 9}{a + 4}$

13. $\frac{9x + 5}{x + 1} - \frac{2x + 3}{x + 1} = \frac{9x + 5 - (2x + 3)}{x + 1}$

$= \frac{9x + 5 - 2x - 3}{x + 1}$

$= \frac{7x + 2}{x + 1}$

14. $\frac{a - 5}{a + 4}$

15. $\frac{a^2}{a - 4} + \frac{a - 20}{a - 4} = \frac{a^2 + a - 20}{a - 4}$

$= \frac{(a + 5)(a - 4)}{a - 4}$

$= \frac{(a + 5)\cancel{(a - 4)}}{\cancel{a - 4}}$

$= a + 5$

16. $x + 2$

17. $\frac{x^2}{x-2} - \frac{6x-8}{x-2} = \frac{x^2-(6x-8)}{x-2}$

$= \frac{x^2-6x+8}{x-2}$

$= \frac{(x-4)(x-2)}{x-2}$

$= \frac{(x-4)\cancel{(x-2)}}{\cancel{x-2}}$

$= x-4$

18. $a - 5$

19. $\frac{t^2+4t}{t-1} + \frac{2t-7}{t-1} = \frac{t^2+6t-7}{t-1}$

$= \frac{(t+7)(t-1)}{t-1}$

$= \frac{(t+7)\cancel{(t-1)}}{\cancel{t-1}}$

$= t+7$

20. $y + 6$

21. $\frac{x+1}{x^2+5x+6} + \frac{2}{x^2+5x+6} = \frac{x+3}{x^2+5x+6}$

$= \frac{x+3}{(x+3)(x+2)}$

$= \frac{\cancel{x+3}}{\cancel{(x+3)}(x+2)}$

$= \frac{1}{x+2}$

22. $\frac{1}{x-1}$

23. $\frac{a^2+3}{a^2+5a-6} - \frac{4}{a^2+5a-6} = \frac{a^2-1}{a^2+5a-6}$

$= \frac{(a+1)(a-1)}{(a+6)(a-1)}$

$= \frac{(a+1)\cancel{(a-1)}}{(a+6)\cancel{(a-1)}}$

$= \frac{a+1}{a+6}$

24. $\frac{a+3}{a-4}$

25. $\frac{t^2-3t}{t^2+6t+9} + \frac{2t-12}{t^2+6t+9} = \frac{t^2-t-12}{t^2+6t+9}$

$= \frac{(t-4)(t+3)}{(t+3)^2}$

$= \frac{(t-4)\cancel{(t+3)}}{(t+3)\cancel{(t+3)}}$

$= \frac{t-4}{t+3}$

26. $\frac{y-5}{y+4}$

27. $\frac{2x^2+3}{x^2-6x+5} - \frac{(x^2-5x+9)}{x^2-6x+5} = \frac{2x^2+3-(x^2-5x+9)}{x^2-6x+5}$

$= \frac{2x^2+3-x^2+5x-9}{x^2-6x+5}$

$= \frac{x^2+5x-6}{x^2-6x+5}$

$= \frac{(x+6)(x-1)}{(x-5)(x-1)}$

$= \frac{(x+6)\cancel{(x-1)}}{(x-5)\cancel{(x-1)}}$

$= \frac{x+6}{x-5}$

28. $\frac{x+5}{x-6}$

29. $\frac{3x}{8} + \frac{x}{-8} = \frac{3x}{8} + \frac{-x}{8}$ Since $\frac{a}{-b} = \frac{-a}{b}$

$= \frac{3x+(-x)}{8}$

$= \frac{2x}{8}$

$= \frac{2x}{2 \cdot 4}$

$= \frac{\cancel{2}x}{\cancel{2}\cdot 4}$

$= \frac{x}{4}$

30. $\frac{2a}{3}$

31. $\frac{3}{t} + \frac{4}{-t} = \frac{3}{t} + \frac{-4}{t}$

$= \frac{3+(-4)}{t}$

$= \frac{-1}{t}$

$= -\frac{1}{t}$

32. $\frac{3}{a}$

33. $\frac{2x+7}{x-6} + \frac{3x}{6-x} = \frac{2x+7}{x-6} + \frac{3x}{-(x-6)}$

$= \frac{2x+7}{x-6} + \frac{-3x}{x-6}$

$= \frac{(2x+7)+(-3x)}{x-6}$

$= \frac{-x+7}{x-6}$

34. $\frac{x+3}{4x-3}$

35. $\frac{a}{6} - \frac{7a}{-6} = \frac{a}{6} - \left(-\frac{7a}{6}\right)$ Since $\frac{a}{-b} = -\frac{a}{b}$

$= \frac{a}{6} + \frac{7a}{6}$

$= \frac{8a}{6}$

$= \frac{2\cdot 4a}{2\cdot 3}$

$= \frac{2\cdot 4a}{2\cdot 3}$

$= \frac{4a}{3}$

36. $\frac{3x}{4}$

37. $\frac{5}{a} - \frac{8}{-a} = \frac{5}{a} - \left(-\frac{8}{a}\right)$ Since $\frac{a}{-b} = -\frac{a}{b}$

$= \frac{5}{a} + \frac{8}{a}$

$= \frac{5 + 8}{a}$

$= \frac{13}{a}$

38. $\frac{7}{t}$

39. $\frac{x}{4} - \frac{3x - 5}{-4} = \frac{x}{4} - \left(-\frac{3x - 5}{4}\right)$

$= \frac{x}{4} + \frac{3x - 5}{4}$

$= \frac{x + 3x - 5}{4}$

$= \frac{4x - 5}{4}$

40. $\frac{4}{x - 1}$

41. $\frac{y^2}{y - 3} + \frac{9}{3 - y} = \frac{y^2}{y - 3} + \frac{9}{-(y - 3)}$

$= \frac{y^2}{y - 3} + \frac{-9}{y - 3}$

$= \frac{y^2 + (-9)}{y - 3}$

$= \frac{y^2 - 9}{y - 3}$

$= \frac{(y + 3)(y - 3)}{y - 3}$

$= \frac{y + 3}{1} \cdot \frac{y - 3}{y - 3}$

$= y + 3$

42. $t + 2$

43. $\frac{b - 7}{b^2 - 16} + \frac{7 - b}{16 - b^2} = \frac{b - 7}{b^2 - 16} + \frac{7 - b}{-(b^2 - 16)}$

$= \frac{b - 7}{b^2 - 16} + \frac{-(7 - b)}{b^2 - 16}$

$= \frac{b - 7}{b^2 - 16} + \frac{b - 7}{b^2 - 16}$

$= \frac{(b - 7) + (b - 7)}{b^2 - 16}$

$= \frac{2b - 14}{b^2 - 16}$

44. 0

45. $\frac{3 - 2t}{t^2 - 5t + 4} + \frac{2 - 3t}{t^2 - 5t + 4} = \frac{5 - 5t}{t^2 - 5t + 4}$

$= \frac{-5(-1 + t)}{(t - 4)(t - 1)}$

$= \frac{-5\cancel{(t - 1)}}{(t - 4)\cancel{(t - 1)}}$

$= \frac{-5}{t - 4}$

46. $\frac{-5}{x - 4}$

47. $\frac{3 - x}{x - 7} - \frac{2x - 5}{7 - x} = \frac{3 - x}{x - 7} - \frac{2x - 5}{-(x - 7)}$

$= \frac{3 - x}{x - 7} - \left(-\frac{2x - 5}{x - 7}\right)$

$= \frac{3 - x}{x - 7} + \frac{2x - 5}{x - 7}$

$= \frac{3 - x + 2x - 5}{x - 7}$

$= \frac{x - 2}{x - 7}$

48. $\frac{t^2 + 4}{t - 2}$

49. $\frac{x - 8}{x^2 - 16} - \frac{x - 8}{16 - x^2} = \frac{x - 8}{x^2 - 16} - \frac{x - 8}{-(x^2 - 16)}$

$= \frac{x - 8}{x^2 - 16} - \left(-\frac{x - 8}{x^2 - 16}\right)$

$= \frac{x - 8}{x^2 - 16} + \frac{x - 8}{x^2 - 16}$

$= \frac{x - 8 + x - 8}{x^2 - 16}$

$= \frac{2x - 16}{x^2 - 16}$

50. $\frac{4}{x^2 - 25}$

51. $\frac{4 - x}{x - 9} - \frac{3x - 8}{9 - x} = \frac{4 - x}{x - 9} - \frac{3x - 8}{-(x - 9)}$

$= \frac{4 - x}{x - 9} - \left(-\frac{3x - 8}{x - 9}\right)$

$= \frac{4 - x}{x - 9} + \frac{3x - 8}{x - 9}$

$= \frac{4 - x + 3x - 8}{x - 9}$

$= \frac{2x - 4}{x - 9}$

52. $\dfrac{x - 2}{x - 7}$

53. $\dfrac{5 - 3x}{x^2 - 2x + 1} - \dfrac{x + 1}{x^2 - 2x + 1} = \dfrac{5 - 3x - (x + 1)}{x^2 - 2x + 1}$

$= \dfrac{5 - 3x - x - 1}{x^2 - 2x + 1}$

$= \dfrac{4 - 4x}{x^2 - 2x + 1}$

$= \dfrac{-4(-1 + x)}{(x - 1)^2}$

$= \dfrac{-4\cancel{(x - 1)}}{\cancel{(x - 1)}(x - 1)}$

$= \dfrac{-4}{x - 1}$

54. $\dfrac{-1}{x - 1}$, or $\dfrac{1}{1 - x}$

55. Graph: $y = -1$

Any ordered pair $(x,-1)$ is a solution, so the graph is a line parallel to the x-axis with y-intercept $(0,-1)$.

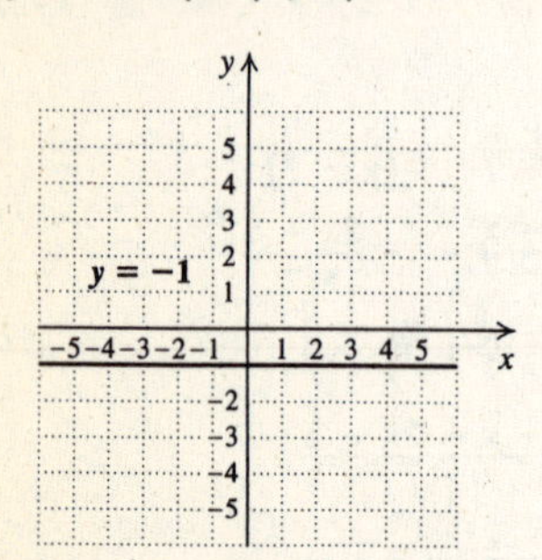

56.

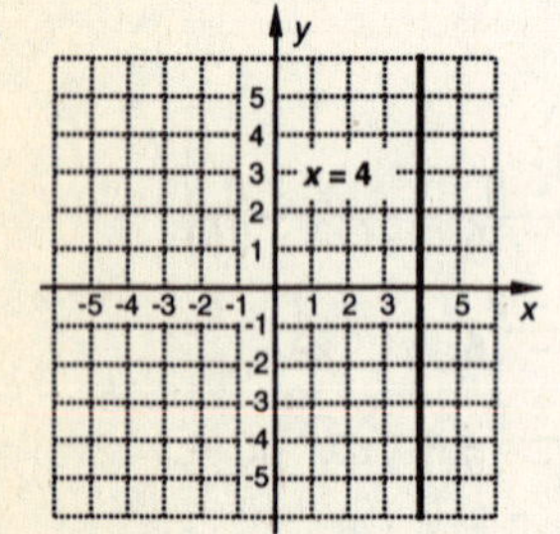

57. Graph: $y = x - 1$

The equation is in the form $y = mx + b$, so we know the y-intercept is $(0,-1)$. We find two other solutions.

When $x = -3$, $y = -3 - 1 = -4$.

When $x = 4$, $y = 4 - 1 = 3$.

x	y
0	-1
-3	-4
4	3

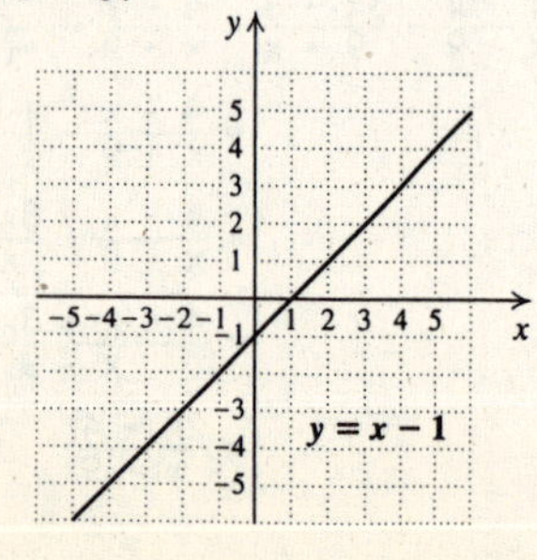

58.

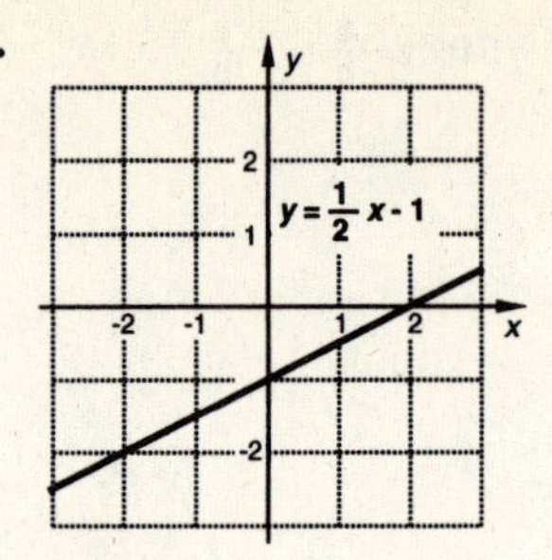

59.

60. 

61. $\dfrac{3(2x + 5)}{x - 1} - \dfrac{3(2x - 3)}{1 - x} + \dfrac{6x - 1}{x - 1}$

$= \dfrac{3(2x + 5)}{x - 1} - \dfrac{3(2x - 3)}{-(x - 1)} + \dfrac{6x - 1}{x - 1}$

$= \dfrac{3(2x + 5)}{x - 1} - \left[-\dfrac{3(2x - 3)}{x - 1}\right] + \dfrac{6x - 1}{x - 1}$

$= \dfrac{3(2x + 5)}{x - 1} + \dfrac{3(2x - 3)}{x - 1} + \dfrac{6x - 1}{x - 1}$

$= \dfrac{6x + 15 + 6x - 9 + 6x - 1}{x - 1}$

$= \dfrac{18x + 5}{x - 1}$

62. $\dfrac{-2x + 4y}{x - y}$

63. $\dfrac{x - y}{x^2 - y^2} + \dfrac{x + y}{x^2 - y^2} - \dfrac{2x}{x^2 - y^2}$

$= \dfrac{x - y + x + y - 2x}{x^2 - y^2}$

$= \dfrac{0}{x^2 - y^2}$

$= 0$

64. $\dfrac{-x + 3y}{x - y}$

65. $\dfrac{10}{2y - 1} - \dfrac{6}{1 - 2y} + \dfrac{y}{2y - 1} + \dfrac{y - 4}{1 - 2y}$

$= \dfrac{10}{2y - 1} - \dfrac{6}{-(2y - 1)} + \dfrac{y}{2y - 1} + \dfrac{y - 4}{-(2y - 1)}$

$= \dfrac{10}{2y - 1} - \left[-\dfrac{6}{2y - 1}\right] + \dfrac{y}{2y - 1} + \dfrac{-(y - 4)}{2y - 1}$

$= \dfrac{10}{2y - 1} + \dfrac{6}{2y - 1} + \dfrac{y}{2y - 1} + \dfrac{4 - y}{2y - 1}$

$= \dfrac{10 + 6 + y + 4 - y}{2y - 1}$

$= \dfrac{20}{2y - 1}$

66. $\dfrac{x^2 + 3}{(3 - 2x)(x - 3)}$

67. $\dfrac{x}{(x - y)(y - z)} - \dfrac{y}{(y - x)(z - y)}$

$= \dfrac{x}{(x - y)(y - z)} - \dfrac{x}{(-1)(x - y)(-1)(y - z)}$

$= \dfrac{x}{(x - y)(y - z)} - \dfrac{x}{(x - y)(y - z)}$

$= \dfrac{x - x}{(x - y)(y - z)}$

$= \dfrac{0}{(x - y)(y - z)}$

$= 0$

68. $\dfrac{x + y}{x - y}$

69. $\dfrac{x^2}{3x^2 - 5x - 2} - \dfrac{2x}{3x + 1} \cdot \dfrac{1}{x - 2}$

$= \dfrac{x^2}{(3x + 1)(x - 2)} - \dfrac{2x}{(3x + 1)(x - 2)}$

$= \dfrac{x^2 - 2x}{(3x + 1)(x - 2)}$

$= \dfrac{x(x - 2)}{(3x + 1)(x - 2)}$

$= \dfrac{x}{3x + 1} \cdot \dfrac{x - 2}{x - 2}$

$= \dfrac{x}{3x + 1}$

70. $\dfrac{30}{(x + 4)(x - 3)}$

71. 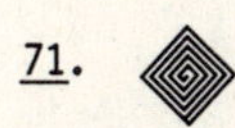

Exercise Set 6.4

1. $12 = 2 \cdot 2 \cdot 3$
   $27 = 3 \cdot 3 \cdot 3$
   LCM $= 2 \cdot 2 \cdot 3 \cdot 3 \cdot 3$, or 108

2. 30

3. $8 = 2 \cdot 2 \cdot 2$
   $9 = 3 \cdot 3$
   LCM $= 2 \cdot 2 \cdot 2 \cdot 3 \cdot 3$, or 72

4. 60

5. $6 = 2 \cdot 3$
   $9 = 3 \cdot 3$
   $21 = 3 \cdot 7$
   LCM $= 2 \cdot 3 \cdot 3 \cdot 7$, or 126

6. 360

7. $24 = 2 \cdot 2 \cdot 2 \cdot 3$
   $36 = 2 \cdot 2 \cdot 3 \cdot 3$
   $40 = 2 \cdot 2 \cdot 2 \cdot 5$
   LCM $= 2 \cdot 2 \cdot 2 \cdot 3 \cdot 3 \cdot 5$, or 360

8. 60

9. $28 = 2 \cdot 2 \cdot 7$
   $42 = 2 \cdot 3 \cdot 7$
   $60 = 2 \cdot 2 \cdot 3 \cdot 5$
   LCM $= 2 \cdot 2 \cdot 3 \cdot 5 \cdot 7$, or 420

10. 500

11. $24 = 2 \cdot 2 \cdot 2 \cdot 3$
    $18 = 2 \cdot 3 \cdot 3$
    LCM $= 2 \cdot 2 \cdot 2 \cdot 3 \cdot 3$, or 72

$\dfrac{7}{24} + \dfrac{11}{18} = \dfrac{7}{2 \cdot 2 \cdot 2 \cdot 3} \cdot \dfrac{3}{3} + \dfrac{11}{2 \cdot 3 \cdot 3} \cdot \dfrac{2 \cdot 2}{2 \cdot 2}$

$= \dfrac{21}{2 \cdot 2 \cdot 2 \cdot 3 \cdot 3} + \dfrac{44}{2 \cdot 2 \cdot 2 \cdot 3 \cdot 3}$

$= \dfrac{65}{72}$

12. $\dfrac{59}{300}$

13. $\dfrac{1}{6} + \dfrac{3}{40} + \dfrac{2}{75}$

$= \dfrac{1}{2 \cdot 3} + \dfrac{3}{2 \cdot 2 \cdot 2 \cdot 5} + \dfrac{2}{3 \cdot 5 \cdot 5}$

LCM $= 2 \cdot 2 \cdot 2 \cdot 3 \cdot 5 \cdot 5$, or 600

$= \dfrac{1}{2 \cdot 3} \cdot \dfrac{2 \cdot 2 \cdot 5 \cdot 5}{2 \cdot 2 \cdot 5 \cdot 5} + \dfrac{3}{2 \cdot 2 \cdot 2 \cdot 5} \cdot \dfrac{3 \cdot 5}{3 \cdot 5} + \dfrac{2}{3 \cdot 5 \cdot 5} \cdot \dfrac{2 \cdot 2 \cdot 2}{2 \cdot 2 \cdot 2}$

$= \dfrac{100 + 45 + 16}{2 \cdot 2 \cdot 2 \cdot 3 \cdot 5 \cdot 5}$

$= \dfrac{161}{600}$

14. $\dfrac{71}{120}$

15. $\dfrac{2}{15} + \dfrac{5}{9} + \dfrac{3}{20}$

$= \dfrac{2}{3 \cdot 5} + \dfrac{5}{3 \cdot 3} + \dfrac{3}{2 \cdot 2 \cdot 5}$

LCM is $2 \cdot 2 \cdot 3 \cdot 3 \cdot 5$, or 180

$= \dfrac{2}{3 \cdot 5} \cdot \dfrac{2 \cdot 2 \cdot 3}{2 \cdot 2 \cdot 3} + \dfrac{5}{3 \cdot 3} \cdot \dfrac{2 \cdot 2 \cdot 5}{2 \cdot 2 \cdot 5} + \dfrac{3}{2 \cdot 2 \cdot 5} \cdot \dfrac{3 \cdot 3}{3 \cdot 3}$

$= \dfrac{24 + 100 + 27}{2 \cdot 2 \cdot 3 \cdot 3 \cdot 5}$

$= \dfrac{151}{180}$

16. $\dfrac{23}{180}$

17. $6x^2 = 2 \cdot 3 \cdot x \cdot x$
    $12x^3 = 2 \cdot 2 \cdot 3 \cdot x \cdot x \cdot x$
    LCM $= 2 \cdot 2 \cdot 3 \cdot x \cdot x \cdot x$, or $12x^3$

18. $8a^2b^2$

19. $2x^2 = 2 \cdot x \cdot x$

$6xy = 2 \cdot 3 \cdot x \cdot y$

$18y^2 = 2 \cdot 3 \cdot 3 \cdot y \cdot y$

LCM = $2 \cdot 3 \cdot 3 \cdot x \cdot x \cdot y \cdot y$, or $18x^2y^2$

20. c^3d^2

21. $2(y - 3) = 2 \cdot (y - 3)$

$6(y - 3) = 2 \cdot 3 \cdot (y - 3)$

LCM = $2 \cdot 3 \cdot (y - 3)$, or $6(y - 3)$

22. $8(x - 1)$

23. $t,\ t + 2,\ t - 2$

The expressions are not factorable, so the LCM is their product:

LCM = $t(t + 2)(t - 2)$

24. $x(x + 3)(x - 3)$

25. $x^2 - 4 = (x + 2)(x - 2)$

$x^2 + 5x + 6 = (x + 3)(x + 2)$

LCM = $(x + 2)(x - 2)(x + 3)$

26. $(x + 2)(x + 1)(x - 2)$

27. $t^3 + 4t^2 + 4t = t(t^2 + 4t + 4) = t(t + 2)(t + 2)$

$t^2 - 4t = t(t - 4)$

LCM = $t(t + 2)(t + 2)(t - 4) = t(t + 2)^2(t - 4)$

28. $y^2(y + 1)(y - 1)$

29. $9a^5b^2 = 3 \cdot 3 \cdot a \cdot a \cdot a \cdot a \cdot a \cdot b \cdot b$

$6ab^6 = 2 \cdot 3 \cdot a \cdot b \cdot b \cdot b \cdot b \cdot b \cdot b$

LCM = $2 \cdot 3 \cdot 3 \cdot a \cdot a \cdot a \cdot a \cdot a \cdot b \cdot b \cdot b \cdot b \cdot b \cdot b = 18a^5b^6$

30. $30a^4b^8$

31. $10x^2y = 2 \cdot 5 \cdot x \cdot x \cdot y$

$6y^2z = 2 \cdot 3 \cdot y \cdot y \cdot z$

$5xz^3 = 5 \cdot x \cdot z \cdot z \cdot z$

LCM = $2 \cdot 3 \cdot 5 \cdot x \cdot x \cdot y \cdot y \cdot z \cdot z \cdot z = 30x^2y^2z^3$

32. $24x^3y^5z^2$

33. $a + 1 = a + 1$

$(a - 1)^2 = (a - 1)(a - 1)$

$a^2 - 1 = (a + 1)(a - 1)$

LCM = $(a + 1)(a - 1)(a - 1) = (a + 1)(a - 1)^2$

34. $2(x + y)^2(x - y)$

35. $m^2 - 5m + 6 = (m - 3)(m - 2)$

$m^2 - 4m + 4 = (m - 2)(m - 2)$

LCM = $(m - 3)(m - 2)(m - 2) = (m - 3)(m - 2)^2$

36. $(2x + 1)(x + 2)(x - 1)$

37. $2 + 3x = 2 + 3x$

$4 - 9x^2 = (2 + 3x)(2 - 3x)$

$2 - 3x = 2 - 3x$

LCM = $(2 + 3x)(2 - 3x)$

38. $(3 + 2x)(3 - 2x)$

39. $10v^2 + 30v = 10v(v + 3) = 2 \cdot 5 \cdot v(v + 3)$

$5v^2 + 35v + 60 = 5(v^2 + 7v + 12)$

$= 5(v + 4)(v + 3)$

LCM = $2 \cdot 5 \cdot v(v + 3)(v + 4) = 10v(v + 3)(v + 4)$

40. $12a(a + 2)(a + 3)$

41. $9x^3 - 9x^2 - 18x = 9x(x^2 - x - 2)$

$= 3 \cdot 3 \cdot x(x - 2)(x + 1)$

$6x^5 - 24x^4 + 24x^3 = 6x^3(x^2 - 4x + 4)$

$= 2 \cdot 3 \cdot x \cdot x \cdot x(x - 2)(x - 2)$

LCM = $2 \cdot 3 \cdot 3 \cdot x \cdot x \cdot x(x - 2)(x - 2)(x + 1) =$ $18x^3(x - 2)^2(x + 1)$

42. $x^3(x + 2)^2(x - 2)$

43. $x^5 + 4x^4 + 4x^3 = x^3(x^2 + 4x + 4)$

$= x \cdot x \cdot x(x + 2)(x + 2)$

$3x^2 - 12 = 3(x^2 - 4) = 3(x + 2)(x - 2)$

$2x + 4 = 2(x + 2)$

LCM = $2 \cdot 3 \cdot x \cdot x \cdot x(x + 2)(x + 2)(x - 2)$

$= 6x^3(x + 2)^2(x - 2)$

44. $10x^3(x + 1)^2(x - 1)$

45. $6x^5 = 2 \cdot 3 \cdot x \cdot x \cdot x \cdot x \cdot x$

$12x^3 = 2 \cdot 2 \cdot 3 \cdot x \cdot x \cdot x$

The LCD is $2 \cdot 2 \cdot 3 \cdot x \cdot x \cdot x \cdot x \cdot x$, or $12x^5$.

The factor of the LCD that is missing from the first denominator is 2. We multiply by 1 using 2/2:

$$\frac{7}{6x^5} \cdot \frac{2}{2} = \frac{14}{12x^5}$$

The second denominator is missing two factors of x, or x^2. We multiply by 1 using x^2/x^2:

$$\frac{y}{12x^3} \cdot \frac{x^2}{x^2} = \frac{x^2y}{12x^5}$$

46. $\frac{3a^3}{10a^6}, \frac{2b}{10a^6}$

47. $2a^2b = 2 \cdot a \cdot a \cdot b$

$8ab^2 = 2 \cdot 2 \cdot 2 \cdot a \cdot b \cdot b$

The LCD is $2 \cdot 2 \cdot 2 \cdot a \cdot a \cdot b \cdot b$, or $8a^2b^2$.

We multiply the first expression by $\frac{4b}{4b}$ to obtain the LCD:

$$\frac{3}{2a^2b} \cdot \frac{4b}{4b} = \frac{12b}{8a^2b^2}$$

We multiply the second expression by a/a to obtain the LCD:

$$\frac{5}{8ab^2} \cdot \frac{a}{a} = \frac{5a}{8a^2b^2}$$

48. $\frac{21y}{9x^4y^3}, \frac{4x^3}{9x^4y^3}$

49. The LCD is $(x + 2)(x - 2)(x + 3)$. (See Exercise 25.)

$$\frac{x + 1}{x^2 - 4} = \frac{x + 1}{(x + 2)(x - 2)} \cdot \frac{x + 3}{x + 3} = \frac{(x + 1)(x + 3)}{(x + 2)(x - 2)(x + 3)}$$

$$\frac{x - 2}{x^2 + 5x + 6} = \frac{x - 2}{(x + 3)(x + 2)} \cdot \frac{x - 2}{x - 2} = \frac{(x - 2)^2}{(x + 3)(x + 2)(x - 2)}$$

50. $\frac{(x - 4)(x + 8)}{(x + 3)(x - 3)(x + 8)}, \frac{(x + 2)(x - 3)}{(x + 3)(x - 3)(x + 8)}$

51. The LCD is $t(t + 2)(t - 2)$. (See Exercise 23.)

$$\frac{3}{t} = \frac{3}{t} \cdot \frac{(t + 2)(t - 2)}{(t + 2)(t - 2)} = \frac{3(t + 2)(t - 2)}{t(t + 2)(t - 2)}$$

$$\frac{4}{t + 2} = \frac{4}{t + 2} \cdot \frac{t(t - 2)}{t(t - 2)} = \frac{4t(t - 2)}{t(t + 2)(t - 2)}$$

$$\frac{t}{t - 2} = \frac{t}{t - 2} \cdot \frac{t(t + 2)}{t(t + 2)} = \frac{t^2(t + 2)}{t(t - 2)(t + 2)}$$

52. $\frac{(x + 3)(x - 3)}{x(x + 3)(x - 3)}, \frac{-2x(x - 3)}{x(x + 3)(x - 3)}, \frac{x^3(x + 3)}{x(x - 3)(x + 3)}$

53. $2x - 3 = 2x - 3$

$4x^2 - 9 = (2x + 3)(2x - 3)$

$2x + 3 = 2x + 3$

LCD $= (2x + 3)(2x - 3)$

$$\frac{x + 1}{2x - 3} = \frac{x + 1}{2x - 3} \cdot \frac{2x + 3}{2x + 3} = \frac{(x + 1)(2x + 3)}{(2x - 3)(2x + 3)}$$

$\frac{x - 2}{4x^2 - 9} = \frac{x - 2}{(2x + 3)(2x - 3)}$ already has the LCD.

$$\frac{x + 1}{2x + 3} = \frac{x + 1}{2x + 3} \cdot \frac{2x - 3}{2x - 3} = \frac{(x + 1)(2x - 3)}{(2x + 3)(2x - 3)}$$

54. $\frac{3x(x - 2)}{3x^3(x + 2)^2(x - 2)}, \frac{3x^3(x + 2)}{3x^3(x + 2)^2(x - 2)}$

55. $x^2 - 19x + 60$

Since the last term is positive and the middle term is negative, we look for a pair of negative factors of 60 whose sum is -19. The numbers we need are -4 and -15.

$x^2 - 19x + 60 = (x - 4)(x - 15)$

56. $(x + 12)(x - 3)$

57. The shaded area has dimensions $x - 6$ by $x - 3$. Then the area is $(x - 6)(x - 3)$, or $x^2 - 9x + 18$.

58. $s^2 - \pi r^2$

59.

60.

61. $72 = 2\cdot2\cdot2\cdot3\cdot3$

$90 = 2\cdot3\cdot3\cdot5$

$96 = 2\cdot2\cdot2\cdot2\cdot2\cdot3$

LCM $= 2\cdot2\cdot2\cdot2\cdot2\cdot3\cdot3\cdot5$, or 1440

62. $120(x + 1)(x - 1)^2$, or $120(x + 1)(x - 1)(1 - x)$

63. The time it takes the joggers to meet again at the starting place is the LCM of the times it takes them to complete one round of the course.

$6 = 2\cdot3$

$8 = 2\cdot2\cdot2$

LCM $= 2\cdot2\cdot2\cdot3$, or 24

It takes 24 min.

64.

Exercise Set 6.5

1. $\frac{2}{x} + \frac{5}{x^2} = \frac{2}{x} + \frac{5}{x\cdot x}$ LCD $= x\cdot x$, or x^2

$$= \frac{2}{x} \cdot \frac{x}{x} + \frac{5}{x\cdot x} = \frac{2x + 5}{x^2}$$

2. $\frac{4x + 8}{x^2}$

3. $6r = 2\cdot3\cdot r$, $8r = 2\cdot2\cdot2\cdot r$ } LCD $= 2\cdot2\cdot2\cdot3\cdot r$, or $24r$

$$\frac{5}{6r} - \frac{7}{8r} = \frac{5}{6r} \cdot \frac{4}{4} - \frac{7}{8r} \cdot \frac{3}{3} = \frac{20 - 21}{24r} = \frac{-1}{24r}$$

4. $\frac{-29}{18t}$

5. $xy^2 = x\cdot y\cdot y$, $x^2y = x\cdot x\cdot y$ } LCD $= x\cdot x\cdot y\cdot y$, or x^2y^2

$$\frac{4}{xy^2} + \frac{6}{x^2y} = \frac{4}{xy^2} \cdot \frac{x}{x} + \frac{6}{x^2y} \cdot \frac{y}{y} = \frac{4x + 6y}{x^2y^2}$$

6. $\frac{2d^2 + 7c}{c^2d^3}$

7. $9t^3 = 3\cdot3\cdot t\cdot t\cdot t$, $6t^2 = 2\cdot3\cdot t\cdot t$ } LCD $= 2\cdot3\cdot3\cdot t\cdot t\cdot t$, or $18t^3$

$$\frac{2}{9t^3} - \frac{1}{6t^2} = \frac{2}{9t^3}\cdot\frac{2}{2} - \frac{1}{6t^2}\cdot\frac{3t}{3t}$$
$$= \frac{4 - 3t}{18t^3}$$

8. $\frac{-2xy - 18}{3x^2y^3}$

9. LCD = 24 (See Example 1.)

$$\frac{x + 5}{8} + \frac{x - 3}{12} = \frac{x + 5}{8}\cdot\frac{3}{3} + \frac{x - 3}{12}\cdot\frac{2}{2}$$
$$= \frac{3(x + 5)}{24} + \frac{2(x - 3)}{24}$$
$$= \frac{3x + 15}{24} + \frac{2x - 6}{24}$$
$$= \frac{5x + 9}{24}\quad \text{Adding numerators}$$

10. $\frac{5x + 7}{18}$

11. $6 = 2\cdot3$, $3 = 3$ } LCD = 6

$$\frac{x - 2}{6} - \frac{x + 1}{3} = \frac{x - 2}{6} - \frac{x + 1}{3}\cdot\frac{2}{2}$$
$$= \frac{x - 2}{6} - \frac{2x + 2}{6}$$
$$= \frac{x - 2 - (2x + 2)}{6}$$
$$= \frac{x - 2 - 2x - 2}{6}$$
$$= \frac{-x - 4}{6}, \text{ or } \frac{-(x + 4)}{6}$$

12. $\frac{a + 8}{4}$

13. $16a = 2\cdot2\cdot2\cdot2\cdot a$, $4a^2 = 2\cdot2\cdot a\cdot a$ } LCD $= 2\cdot2\cdot2\cdot2\cdot a\cdot a$, or $16a^2$

$$\frac{a + 4}{16a} + \frac{3a + 4}{4a^2} = \frac{a + 4}{16a}\cdot\frac{a}{a} + \frac{3a + 4}{4a^2}\cdot\frac{4}{4}$$
$$= \frac{a^2 + 4a}{16a^2} + \frac{12a + 16}{16a^2}$$
$$= \frac{a^2 + 16a + 16}{16a^2}$$

14. $\frac{5a^2 + 7a - 3}{9a^2}$

15. $3z = 3\cdot z$, $4z = 2\cdot2\cdot z$ } LCD $= 2\cdot2\cdot3\cdot z$, or $12z$

$$\frac{4z - 9}{3z} - \frac{3z - 8}{4z} = \frac{4z - 9}{3z}\cdot\frac{4}{4} - \frac{3z - 8}{4z}\cdot\frac{3}{3}$$
$$= \frac{16z - 36}{12z} - \frac{9z - 24}{12z}$$
$$= \frac{16z - 36 - (9z - 24)}{12z}$$
$$= \frac{16z - 36 - 9z + 24}{12z}$$
$$= \frac{7z - 12}{12z}$$

16. $\frac{-7x - 13}{4x}$

17. LCD $= x^2y^2$ (See Exercise 5.)

$$\frac{x + y}{xy^2} + \frac{3x + y}{x^2y} = \frac{x + y}{xy^2}\cdot\frac{x}{x} + \frac{3x + y}{x^2y}\cdot\frac{y}{y}$$
$$= \frac{x(x + y) + y(3x + y)}{x^2y^2}$$
$$= \frac{x^2 + xy + 3xy + y^2}{x^2y^2}$$
$$= \frac{x^2 + 4xy + y^2}{x^2y^2}$$

18. $\frac{c^2 + 3cd - d^2}{c^2d^2}$

19. $3xt^2 = 3\cdot x\cdot t\cdot t$, $x^2t = x\cdot x\cdot t$ } LCD $= 3\cdot x\cdot x\cdot t\cdot t$, or $3x^2t^2$

$$\frac{4x + 2t}{3xt^2} - \frac{5x - 3t}{x^2t}$$
$$= \frac{4x + 2t}{3xt^2}\cdot\frac{x}{x} - \frac{5x - 3t}{x^2t}\cdot\frac{3t}{3t}$$
$$= \frac{4x^2 + 2tx}{3x^2t^2} - \frac{15xt - 9t^2}{3x^2t^2}$$
$$= \frac{4x^2 + 2tx - (15xt - 9t^2)}{3x^2t^2}$$
$$= \frac{4x^2 + 2tx - 15xt + 9t^2}{3x^2t^2}$$
$$= \frac{4x^2 - 13xt + 9t^2}{3x^2t^2}$$

(Although $4x^2 - 13x + 9t^2$ can be factored as $(4x - 9t)(x - t)$, doing so will not enable us to simplify the result.)

20. $\frac{3y^2 - 3xy - 6x^2}{2x^2y^2}$

(Although $3y^2 - 3xy - 6x^2$ can be factored as $3(y - 2x)(y + x)$, doing so will not enable us to simplify the result.)

21. The denominators do not factor, so the LCD is their product, $(x - 2)(x + 2)$.

$$\frac{3}{x-2} + \frac{3}{x+2} = \frac{3}{x-2} \cdot \frac{x+2}{x+2} + \frac{3}{x+2} \cdot \frac{x-2}{x-2}$$
$$= \frac{3(x+2) + 3(x-2)}{(x-2)(x+2)}$$
$$= \frac{3x + 6 + 3x - 6}{(x-2)(x+2)}$$
$$= \frac{6x}{(x-2)(x+2)}$$

22. $\frac{4x}{(x-1)(x+1)}$

23. $\frac{5}{x+5} - \frac{3}{x-5}$ LCD = $(x + 5)(x - 5)$

$$= \frac{5}{x+5} \cdot \frac{x-5}{x-5} - \frac{3}{x-5} \cdot \frac{x+5}{x+5}$$
$$= \frac{5x - 25}{(x+5)(x-5)} - \frac{3x + 15}{(x+5)(x-5)}$$
$$= \frac{5x - 25 - (3x + 15)}{(x+5)(x-5)}$$
$$= \frac{5x - 25 - 3x - 15}{(x+5)(x-5)}$$
$$= \frac{2x - 40}{(x+5)(x-5)}$$

(Although $2x - 40$ can be factored as $2(x - 20)$, doing so will not enable us to simplify the result.)

24. $\frac{-z^2 + 5z}{(z-1)(z+1)}$

(Although $-z^2 + 5z$ can be factored as $-z(z - 5)$, doing so will not enable us to simplify the result.)

25. $\left.\begin{array}{l} 3x = 3 \cdot x \\ x + 1 = x + 1 \end{array}\right\}$ LCD = $3x(x + 1)$

$$\frac{3}{x+1} + \frac{2}{3x} = \frac{3}{x+1} \cdot \frac{3x}{3x} + \frac{2}{3x} \cdot \frac{x+1}{x+1}$$
$$= \frac{9x + 2(x+1)}{3x(x+1)}$$
$$= \frac{9x + 2x + 2}{3x(x+1)}$$
$$= \frac{11x + 2}{3x(x+1)}$$

26. $\frac{11x + 15}{4x(x+5)}$

27. $\frac{3}{2t^2 - 2t} - \frac{5}{2t - 2}$

$$= \frac{3}{2t(t-1)} - \frac{5}{2(t-1)} \quad \text{LCD} = 2t(t-1)$$
$$= \frac{3}{2t(t-1)} - \frac{5}{2(t-1)} \cdot \frac{t}{t}$$
$$= \frac{3}{2t(t-1)} - \frac{5t}{2t(t-1)}$$
$$= \frac{3 - 5t}{2t(t-1)}$$

28. $\frac{14 - 3x}{(x+2)(x-2)}$

29. $\left.\begin{array}{l} x^2 - 16 = (x + 4)(x - 4) \\ x - 4 = x - 4 \end{array}\right\}$ LCD = $(x + 4)(x - 4)$

$$\frac{2x}{x^2 - 16} + \frac{x}{x-4} = \frac{2x}{(x+4)(x-4)} + \frac{x}{x-4} \cdot \frac{x+4}{x+4}$$
$$= \frac{2x + x(x+4)}{(x+4)(x-4)}$$
$$= \frac{2x + x^2 + 4x}{(x+4)(x-4)}$$
$$= \frac{x^2 + 6x}{(x+4)(x-4)}$$

(Although $x^2 + 6x$ can be factored as $x(x + 6)$, doing so will not enable us to simplify the result.)

30. $\frac{x^2 - x}{(x+5)(x-5)}$

(Although $x^2 - x$ can be factored as $x(x - 1)$, doing so will not enable us to simplify the result.)

31. $\frac{6}{z+4} - \frac{2}{3z+12} = \frac{6}{z+4} - \frac{2}{3(z+4)}$

LCD = $3(z + 4)$

$$= \frac{6}{z+4} \cdot \frac{3}{3} - \frac{2}{3(z+4)}$$
$$= \frac{18}{3(z+4)} - \frac{2}{3(z+4)}$$
$$= \frac{16}{3(z+4)}$$

32. $\frac{4t - 5}{4(t-3)}$

33. $\frac{3}{x-1} + \frac{2}{(x-1)^2} = \frac{3}{x-1} \cdot \frac{x-1}{x-1} + \frac{2}{(x-1)^2}$

LCD = $(x - 1)^2$

$$= \frac{3(x-1) + 2}{(x-1)^2}$$
$$= \frac{3x - 3 + 2}{(x-1)^2}$$
$$= \frac{3x - 1}{(x-1)^2}$$

34. $\frac{2x + 10}{(x+3)^2}$

(Although $2x + 10$ can be factored as $2(x + 5)$, doing so will not enable us to simplify the result.)

35. $\frac{2t}{t^2 - 9} - \frac{3}{t - 3} = \frac{2t}{(t+3)(t-3)} - \frac{3}{t-3}$

LCD = $(t + 3)(t - 3)$

$$= \frac{2t}{(t+3)(t-3)} - \frac{3}{t-3} \cdot \frac{t+3}{t+3}$$
$$= \frac{2t - 3(t+3)}{(t+3)(t-3)}$$
$$= \frac{2t - 3t - 9}{(t+3)(t-3)}$$
$$= \frac{-t - 9}{(t+3)(t-3)}$$

36. $\dfrac{6 - 20x}{15x(x + 1)}$

(Although $6 - 20x$ can be factored as $2(3 - 10x)$ or as $-2(10x - 3)$, doing so will not enable us to simplify the result.)

37. $\dfrac{4a}{5a - 10} + \dfrac{3a}{10a - 20} = \dfrac{4a}{5(a - 2)} + \dfrac{3a}{2\cdot 5(a - 2)}$

LCD $= 2\cdot 5(a - 2)$

$= \dfrac{4a}{5(a - 2)}\cdot\dfrac{2}{2} + \dfrac{3a}{2\cdot 5(a - 2)}$

$= \dfrac{8a + 3a}{10(a - 2)}$

$= \dfrac{11a}{10(a - 2)}$

38. $\dfrac{9a}{4(a - 5)}$

39. $\dfrac{a}{x + a} - \dfrac{a}{x - a}$ LCD $= (x + a)(x - a)$

$= \dfrac{a}{x + a}\cdot\dfrac{x - a}{x - a} - \dfrac{a}{x - a}\cdot\dfrac{x + a}{x + a}$

$= \dfrac{ax - a^2}{(x + a)(x - a)} - \dfrac{ax + a^2}{(x + a)(x - a)}$

$= \dfrac{ax - a^2 - (ax + a^2)}{(x + a)(x - a)}$

$= \dfrac{ax - a^2 - ax - a^2}{(x + a)(x - a)}$

$= \dfrac{-2a^2}{(x + a)(x - a)}$

40. $\dfrac{t^2 + 2ty - y^2}{(y - t)(y + t)}$

41. $\dfrac{x + 4}{x} + \dfrac{x}{x + 4} = \dfrac{x + 4}{x}\cdot\dfrac{x + 4}{x + 4} + \dfrac{x}{x + 4}\cdot\dfrac{x}{x}$

LCD $= x(x + 4)$

$= \dfrac{(x + 4)^2 + x^2}{x(x + 4)}$

$= \dfrac{x^2 + 8x + 16 + x^2}{x(x + 4)}$

$= \dfrac{2x^2 + 8x + 16}{x(x + 4)}$

(Although $2x^2 + 8x + 16$ can be factored as $2(x^2 + 4x + 8)$, doing so will not enable us to simplify the result.)

42. $\dfrac{2x^2 - 10x + 25}{x(x - 5)}$

43. $\dfrac{x}{x^2 + 5x + 6} - \dfrac{2}{x^2 + 3x + 2}$

$= \dfrac{x}{(x + 3)(x + 2)} - \dfrac{2}{(x + 2)(x + 1)}$

LCD $= (x + 3)(x + 2)(x + 1)$

$= \dfrac{x}{(x + 3)(x + 2)}\cdot\dfrac{x + 1}{x + 1} - \dfrac{2}{(x + 2)(x + 1)}\cdot\dfrac{x + 3}{x + 3}$

$= \dfrac{x^2 + x}{(x + 3)(x + 2)(x + 1)} - \dfrac{2x + 6}{(x + 3)(x + 2)(x + 1)}$

$= \dfrac{x^2 + x - (2x + 6)}{(x + 3)(x + 2)(x + 1)}$

$= \dfrac{x^2 + x - 2x - 6}{(x + 3)(x + 2)(x + 1)}$

$= \dfrac{x^2 - x - 6}{(x + 3)(x + 2)(x + 1)}$

$= \dfrac{(x - 3)\cancel{(x + 2)}}{(x + 3)\cancel{(x + 2)}(x + 1)}$

$= \dfrac{x - 3}{(x + 3)(x + 1)}$

44. $\dfrac{x - 6}{(x + 6)(x + 4)}$

45. $\dfrac{x}{x^2 + 2x + 1} + \dfrac{1}{x^2 + 5x + 4}$

$= \dfrac{x}{(x + 1)(x + 1)} + \dfrac{1}{(x + 1)(x + 4)}$

LCD $= (x + 1)^2(x + 4)$

$= \dfrac{x}{(x + 1)(x + 1)}\cdot\dfrac{x + 4}{x + 4} + \dfrac{1}{(x + 1)(x + 4)}\cdot\dfrac{x + 1}{x + 1}$

$= \dfrac{x(x + 4) + 1\cdot(x + 1)}{(x + 1)^2(x + 4)} = \dfrac{x^2 + 4x + x + 1}{(x + 1)^2(x + 4)}$

$= \dfrac{x^2 + 5x + 1}{(x + 1)^2(x + 4)}$

46. $\dfrac{12a - 11}{(a + 2)(a - 1)(a - 3)}$

47. $\dfrac{x}{x^2 + 15x + 56} - \dfrac{6}{x^2 + 13x + 42}$

$= \dfrac{x}{(x + 7)(x + 8)} - \dfrac{6}{(x + 6)(x + 7)}$

LCD $= (x + 7)(x + 8)(x + 6)$

$= \dfrac{x}{(x + 7)(x + 8)}\cdot\dfrac{x + 6}{x + 6} - \dfrac{6}{(x + 6)(x + 7)}\cdot\dfrac{x + 8}{x + 8}$

$= \dfrac{x^2 + 6x}{(x + 7)(x + 8)(x + 6)} - \dfrac{6x + 48}{(x + 7)(x + 8)(x + 6)}$

$= \dfrac{x^2 + 6x - (6x + 48)}{(x + 7)(x + 8)(x + 6)}$

$= \dfrac{x^2 + 6x - 6x - 48}{(x + 7)(x + 8)(x + 6)}$

$= \dfrac{x^2 - 48}{(x + 7)(x + 8)(x + 6)}$

48. $\dfrac{-8x - 88}{(x + 1)(x + 16)(x + 8)}$

(Although $-8x - 88$ can be factored as $-8(x + 11)$, doing so will not enable us to simplify the result.)

49. $\dfrac{10}{x^2 + x - 6} + \dfrac{3x}{x^2 - 4x + 4}$

$= \dfrac{10}{(x + 3)(x - 2)} + \dfrac{3x}{(x - 2)(x - 2)}$

LCD $= (x + 3)(x - 2)^2$

$= \dfrac{10}{(x + 3)(x - 2)} \cdot \dfrac{x - 2}{x - 2} + \dfrac{3x}{(x - 2)(x - 2)} \cdot \dfrac{x + 3}{x + 3}$

$= \dfrac{10(x - 2) + 3x(x + 3)}{(x + 3)(x - 2)^2} = \dfrac{10x - 20 + 3x^2 + 9x}{(x + 3)(x - 2)^2}$

$= \dfrac{3x^2 + 19x - 20}{(x + 3)(x - 2)^2}$

50. $\dfrac{5z + 12}{(z - 3)(z + 2)(z + 3)}$

51. $\dfrac{y + 2}{y - 7} + \dfrac{3 - y}{49 - y^2} = \dfrac{y + 2}{y - 7} + \dfrac{3 - y}{(7 + y)(7 - y)}$

$= \dfrac{y + 2}{y - 7} + \dfrac{3 - y}{-(y + 7)(y - 7)}$

[$7 + y = y + 7$;
$-1(7 - y) = y - 7$]

$= \dfrac{y + 2}{y - 7} + \dfrac{-(3 - y)}{(y + 7)(y - 7)}$

LCD $= (y + 7)(y - 7)$

$= \dfrac{y + 2}{y - 7} \cdot \dfrac{y + 7}{y + 7} + \dfrac{-(3 - y)}{(y + 7)(y - 7)}$

$= \dfrac{(y + 2)(y + 7) - (3 - y)}{(y + 7)(y - 7)}$

$= \dfrac{y^2 + 9y + 14 - 3 + y}{(y + 7)(y - 7)}$

$= \dfrac{y^2 + 10y + 11}{(y + 7)(y - 7)}$

52. $\dfrac{p^2 + 7p + 1}{(p + 5)(p - 5)}$

53. $\dfrac{8x}{16 - x^2} - \dfrac{5}{x - 4}$

$= \dfrac{8x}{(4 + x)(4 - x)} - \dfrac{5}{x - 4}$ $\quad$ $4 - x$ and $x - 4$ are opposites

$= \dfrac{8x}{(4 + x)(4 - x)} - \dfrac{5}{-(4 - x)}$

$= \dfrac{8x}{(4 + x)(4 - x)} - \dfrac{-5}{4 - x}$ $\quad$ LCD $= (4 + x)(4 - x)$

$= \dfrac{8x}{(4 + x)(4 - x)} - \dfrac{-5}{4 - x} \cdot \dfrac{4 + x}{4 + x}$

$= \dfrac{8x}{(4 + x)(4 - x)} - \dfrac{-20 - 5x}{(4 - x)(4 + x)}$

$= \dfrac{8x - (-20 - 5x)}{(4 + x)(4 - x)}$

$= \dfrac{8x + 20 + 5x}{(4 + x)(4 - x)}$

$= \dfrac{13x + 20}{(4 + x)(4 - x)}$

54. $\dfrac{9x + 12}{(x + 3)(x - 3)}$

(Although $9x + 12$ can be factored as $3(3x + 4)$, doing so will not enable us to simplify the result.)

55. $\dfrac{a}{a^2 - 1} + \dfrac{2a}{a - a^2}$

$= \dfrac{a}{(a + 1)(a - 1)} + \dfrac{2a}{a(1 - a)}$

$= \dfrac{a}{(a + 1)(a - 1)} + \dfrac{2a}{-a(a - 1)}$

$= \dfrac{a}{(a + 1)(a - 1)} + \dfrac{-2a}{a(a - 1)}$ $\quad$ Since $\dfrac{a}{-b} = \dfrac{-a}{b}$;

LCD $= a(a + 1)(a - 1)$

$= \dfrac{a}{(a + 1)(a - 1)} \cdot \dfrac{a}{a} + \dfrac{-2a}{a(a - 1)} \cdot \dfrac{a + 1}{a + 1}$

$= \dfrac{a^2 - 2a(a + 1)}{a(a + 1)(a - 1)}$

$= \dfrac{a^2 - 2a^2 - 2a}{a(a + 1)(a - 1)}$

$= \dfrac{-a^2 - 2a}{a(a + 1)(a - 1)}$

$= \dfrac{a(-a - 2)}{a(a + 1)(a - 1)}$

$= \dfrac{-a - 2}{(a + 1)(a - 1)}$

56. $\dfrac{-3x^2 + 7x + 4}{3(x + 2)(2 - x)}$, or $\dfrac{3x^2 - 7x - 4}{3(x + 2)(x - 2)}$

57. $\dfrac{4x}{x^2 - y^2} - \dfrac{6}{y - x}$

$= \dfrac{4x}{(x + y)(x - y)} - \dfrac{6}{-(x - y)}$

$= \dfrac{4x}{(x + y)(x - y)} - \dfrac{-6}{x - y}$ $\quad$ Since $\dfrac{a}{-b} = \dfrac{-a}{b}$;

LCD $= (x + y)(x - y)$

$= \dfrac{4x}{(x + y)(x - y)} - \dfrac{-6}{x - y} \cdot \dfrac{x + y}{x + y}$

$= \dfrac{4x}{(x + y)(x - y)} - \dfrac{-6x - 6y}{(x + y)(x - y)}$

$= \dfrac{4x - (-6x - 6y)}{(x + y)(x - y)}$

$= \dfrac{4x + 6x + 6y}{(x + y)(x - y)}$

$= \dfrac{10x + 6y}{(x + y)(x - y)}$

(Although $10x + 6y$ can be factored as $2(5x + 3y)$, doing so does not enable us to simplify the result.)

58. $\dfrac{a - 2}{(a + 3)(a - 3)}$, or $\dfrac{-a + 2}{(3 + a)(3 - a)}$

59. $\dfrac{4y}{y^2 - 1} - \dfrac{2}{y} - \dfrac{2}{y + 1}$

$= \dfrac{4y}{(y + 1)(y - 1)} - \dfrac{2}{y} - \dfrac{2}{y + 1}$

LCD = $y(y + 1)(y - 1)$

$= \dfrac{4y}{(y+1)(y-1)} \cdot \dfrac{y}{y} - \dfrac{2}{y} \cdot \dfrac{(y+1)(y-1)}{(y+1)(y-1)} - \dfrac{2}{y+1} \cdot \dfrac{y(y-1)}{y(y-1)}$

$= \dfrac{4y^2 - (2y^2 - 2) - (2y^2 - 2y)}{y(y + 1)(y - 1)}$

$= \dfrac{4y^2 - 2y^2 + 2 - 2y^2 + 2y}{y(y + 1)(y - 1)}$

$= \dfrac{2y + 2}{y(y + 1)(y - 1)}$

$= \dfrac{2\cancel{(y + 1)}}{y\cancel{(y + 1)}(y - 1)}$

$= \dfrac{2}{y(y - 1)}$

60. $\dfrac{2x - 3}{2 - x}$

61. $\dfrac{2z}{1 - 2z} + \dfrac{3z}{2z + 1} - \dfrac{3}{4z^2 - 1}$

$= \dfrac{2z}{-(2z - 1)} + \dfrac{3z}{2z + 1} - \dfrac{3}{(2z + 1)(2z - 1)}$

$= \dfrac{-2z}{2z - 1} + \dfrac{3z}{2z + 1} - \dfrac{3}{(2z - 1)(2z + 1)}$

LCD = $(2z - 1)(2z + 1)$

$= \dfrac{-2z}{2z-1} \cdot \dfrac{2z+1}{2z+1} + \dfrac{3z}{2z+1} \cdot \dfrac{2z-1}{2z-1} - \dfrac{3}{(2z-1)(2z + 1)}$

$= \dfrac{(-4z^2 - 2z) + (6z^2 - 3z) - 3}{(2z - 1)(2z + 1)}$

$= \dfrac{2z^2 - 5z - 3}{(2z - 1)(2z + 1)}$

$= \dfrac{(z - 3)\cancel{(2z + 1)}}{(2z - 1)\cancel{(2z + 1)}}$

$= \dfrac{z - 3}{2z - 1}$

62. 0

63. $\dfrac{5}{3 - 2x} + \dfrac{3}{2x - 3} - \dfrac{x - 3}{2x^2 - x - 3}$

$= \dfrac{5}{-(2x - 3)} + \dfrac{3}{2x - 3} - \dfrac{x - 3}{(2x - 3)(x + 1)}$

$= \dfrac{-5}{2x - 3} + \dfrac{3}{2x - 3} - \dfrac{x - 3}{(2x - 3)(x + 1)}$

LCD = $(2x - 3)(x + 1)$

$= \dfrac{-5}{2x - 3} \cdot \dfrac{x + 1}{x + 1} + \dfrac{3}{2x - 3} \cdot \dfrac{x + 1}{x + 1} - \dfrac{x - 3}{(2x - 3)(x + 1)}$

$= \dfrac{(-5x - 5) + (3x + 3) - (x - 3)}{(2x - 3)(x + 1)}$

$= \dfrac{-5x - 5 + 3x + 3 - x + 3}{(2x - 3)(x + 1)}$

$= \dfrac{-3x + 1}{(2x - 3)(x + 1)}$

64. $\dfrac{2}{r + s}$

65. $\dfrac{3}{2c - 1} - \dfrac{1}{c + 2} - \dfrac{5}{2c^2 + 3c - 2}$

$= \dfrac{3}{2c - 1} - \dfrac{1}{c + 2} - \dfrac{5}{(2c - 1)(c + 2)}$

LCD = $(2c - 1)(c + 2)$

$= \dfrac{3}{2c - 1} \cdot \dfrac{c + 2}{c + 2} - \dfrac{1}{c + 2} \cdot \dfrac{2c - 1}{2c - 1} - \dfrac{5}{(2c - 1)(c + 2)}$

$= \dfrac{(3c + 6) - (2c - 1) - 5}{(2c - 1)(c + 2)}$

$= \dfrac{3c + 6 - 2c + 1 - 5}{(2c - 1)(c + 2)}$

$= \dfrac{\cancel{c + 2}}{(2c - 1)\cancel{(c + 2)}}$

$= \dfrac{1}{2c - 1}$

66. $\dfrac{2y^2 + 2y - 7}{(2y + 3)(y - 1)}$

67. $\dfrac{1}{x + y} - \dfrac{1}{x - y} + \dfrac{2x}{x^2 - y^2}$

$= \dfrac{1}{x + y} - \dfrac{1}{x - y} + \dfrac{2x}{(x + y)(x - y)}$

LCD = $(x + y)(x - y)$

$= \dfrac{1}{x + y} \cdot \dfrac{x - y}{x - y} - \dfrac{1}{x - y} \cdot \dfrac{x + y}{x + y} + \dfrac{2x}{(x + y)(x - y)}$

$= \dfrac{x - y - (x + y) + 2x}{(x + y)(x - y)}$

$= \dfrac{x - y - x - y + 2x}{(x + y)(x - y)}$

$= \dfrac{2x - 2y}{(x + y)(x - y)}$

$= \dfrac{2\cancel{(x - y)}}{(x + y)\cancel{(x - y)}}$

$= \dfrac{2}{x + y}$

68. $\dfrac{4b}{(a + b)(a - b)}$

69. Graph: $y = \frac{1}{2}x - 5$

Since the equation is in the form $y = mx + b$, we know the y-intercept is $(0,-5)$. We find two other solutions, substituting multiples of 2 for x to avoid fractions.

When $x = 2$, $y = \frac{1}{2} \cdot 2 - 5 = 1 - 5 = -4$.

When $x = 4$, $y = \frac{1}{2} \cdot 4 - 5 = 2 - 5 = -3$.

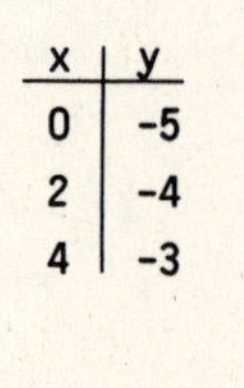

x	y
0	-5
2	-4
4	-3

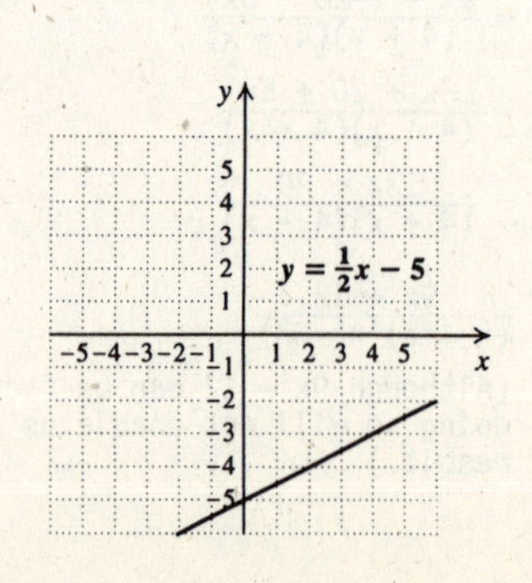

70.

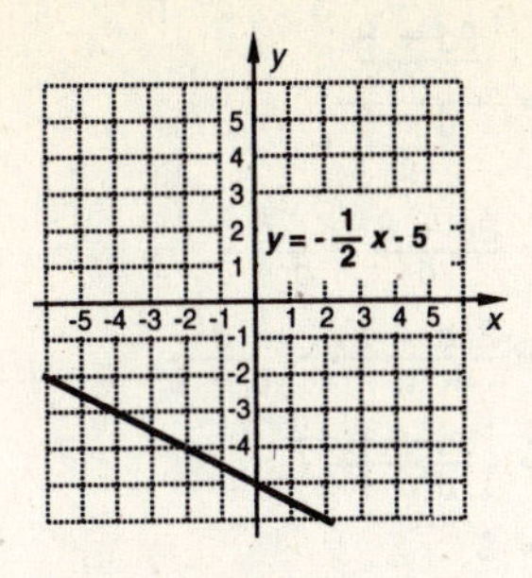

71. Graph: $y = 3$

All solutions are of the form $(x,3)$. The graph is a line parallel to the x-axis with y-intercept $(0,3)$.

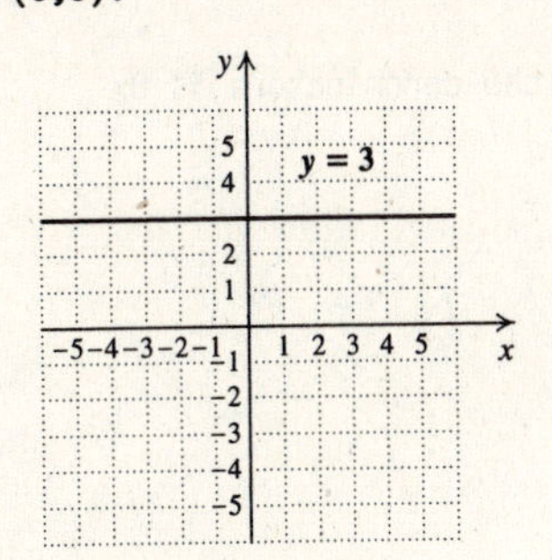

72.

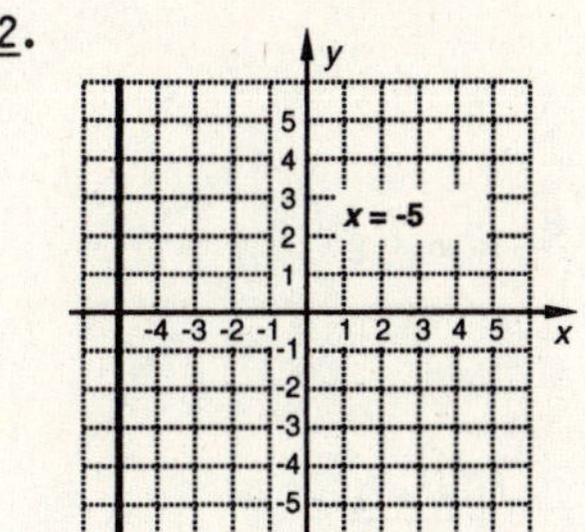

73.

74.

75. To find the perimeter we add the lengths of the sides:

$$\frac{y+4}{3} + \frac{y+4}{3} + \frac{y-2}{5} + \frac{y-2}{5} \qquad \text{LCD} = 3\cdot5$$

$$= \frac{y+4}{3}\cdot\frac{5}{5} + \frac{y+4}{3}\cdot\frac{5}{5} + \frac{y-2}{5}\cdot\frac{3}{3} + \frac{y-2}{5}\cdot\frac{3}{3}$$

$$= \frac{5y + 20 + 5y + 20 + 3y - 6 + 3y - 6}{3\cdot5}$$

$$= \frac{16y + 28}{15}$$

To find the area we multiply the length and the width:

$$\left(\frac{y+4}{3}\right)\left(\frac{y-2}{5}\right) = \frac{(y+4)(y-2)}{3\cdot5} = \frac{y^2 + 2y - 8}{15}$$

76. $P = \frac{10x - 14}{(x+4)(x-5)}$, or $\frac{10x - 14}{x^2 - x - 20}$;

$A = \frac{6}{x^2 - x - 20}$

77. $\frac{5}{z+2} + \frac{4z}{z^2 - 4} + 2$

$$= \frac{5}{z+2} + \frac{4z}{(z+2)(z-2)} + \frac{2}{1} \qquad \text{LCD} = (z+2)(z-2)$$

$$= \frac{5}{z+2}\cdot\frac{z-2}{z-2} + \frac{4z}{(z+2)(z-2)} + \frac{2}{1}\cdot\frac{(z+2)(z-2)}{(z+2)(z-2)}$$

$$= \frac{5z - 10 + 4z + 2(z^2 - 4)}{(z+2)(z-2)}$$

$$= \frac{5z - 10 + 4z + 2z^2 - 8}{(z+2)(z-2)} = \frac{2z^2 + 9z - 18}{(z+2)(z-2)}$$

$$= \frac{(2z-3)(z+6)}{(z+2)(z-2)}$$

78. $\frac{11z^4 - 22z^2 + 6}{(z^2+2)(z^2-2)(2z^2-3)}$

79. $\frac{1}{2xy - 6x + ay - 3a} - \frac{ay + xy}{(a^2 - 4x^2)(y^2 - 6y + 9)}$

$$= \frac{1}{(2x+a)(y-3)} - \frac{ay + xy}{(a+2x)(a-2x)(y-3)(y-3)}$$

LCD $= (a+2x)(a-2x)(y-3)^2$

$$= \frac{1}{(2x+a)(y-3)}\cdot\frac{(a-2x)(y-3)}{(a-2x)(y-3)} - \frac{ay + xy}{(a+2x)(a-2x)(y-3)(y-3)}$$

$$= \frac{ay - 3a - 2xy + 6x - (ay + xy)}{(a+2x)(a-2x)(y-3)^2}$$

$$= \frac{-3a - 3xy + 6x}{(a+2x)(a-2x)(y-3)^2}$$

80. $\frac{x^2 + xy - x^3 + x^2y - xy^2 + y^3}{(x^2+y^2)(x+y)^2(x-y)}$

81. Answers may vary. $\frac{a}{a-b} + \frac{3b}{b-a}$

Exercise Set 6.6

1. $\frac{1 + \frac{9}{16}}{1 - \frac{3}{4}}$ LCD of the denominators is 16

$$= \frac{1 + \frac{9}{16}}{1 - \frac{3}{4}}\cdot\frac{16}{16}$$

$$= \frac{\left(1 + \frac{9}{16}\right)16}{\left(1 - \frac{3}{4}\right)16}$$

$$= \frac{1(16) + \frac{9}{16}(16)}{1(16) - \frac{3}{4}(16)} = \frac{16 + 9}{16 - 12} = \frac{25}{4}$$

2. $\frac{5}{2}$

3. $\dfrac{1 - \frac{3}{5}}{1 + \frac{1}{5}} = \dfrac{1 \cdot \frac{5}{5} - \frac{3}{5}}{1 \cdot \frac{5}{5} + \frac{1}{5}} = \dfrac{\frac{5 - 3}{5}}{\frac{5 + 1}{5}}$

$= \dfrac{\frac{2}{5}}{\frac{6}{5}}$

$= \frac{2}{5} \cdot \frac{5}{6} = \frac{2}{6} = \frac{1}{3}$

4. $-\frac{65}{18}$

5. $\dfrac{\frac{1}{x} + 3}{\frac{1}{x} - 5}$ LCD of the denominators is x

$= \dfrac{\frac{1}{x} + 3}{\frac{1}{x} - 5} \cdot \dfrac{x}{x}$

$= \dfrac{\left(\frac{1}{x} + 3\right)x}{\left(\frac{1}{x} - 5\right)x}$

$= \dfrac{\frac{1}{x} \cdot x + 3 \cdot x}{\frac{1}{x} \cdot x - 5 \cdot x} = \dfrac{1 + 3x}{1 - 5x}$

6. $\dfrac{9 + 3s^2}{4s^2}$

7. $\dfrac{\frac{1}{2} + \frac{3}{4}}{\frac{5}{8} - \frac{5}{6}} = \dfrac{\frac{1}{2} \cdot \frac{2}{2} + \frac{3}{4}}{\frac{5}{8} \cdot \frac{3}{3} - \frac{5}{6} \cdot \frac{4}{4}} = \dfrac{\frac{2 + 3}{4}}{\frac{15 - 20}{24}}$

$= \dfrac{\frac{5}{4}}{\frac{-5}{24}}$

$= \frac{5}{4} \cdot \frac{24}{-5}$

$= -6$

8. $-\frac{4}{39}$

9. $\dfrac{\frac{2}{y} + \frac{1}{2y}}{y + \frac{y}{2}}$ LCD of the denominators is 2y

$= \dfrac{\frac{2}{y} + \frac{1}{2y}}{y + \frac{y}{2}} \cdot \dfrac{2y}{2y}$

$= \dfrac{\left(\frac{2}{y} + \frac{1}{2y}\right)2y}{\left(y + \frac{y}{2}\right)2y}$

$= \dfrac{\frac{2}{y}(2y) + \frac{1}{2y}(2y)}{y(2y) + \frac{y}{2}(2y)} = \dfrac{4 + 1}{2y^2 + y^2} = \dfrac{5}{3y^2}$

10. $\dfrac{2x + 1}{x}$

11. $\dfrac{8 + \frac{8}{d}}{1 + \frac{1}{d}} = \dfrac{8 \cdot \frac{d}{d} + \frac{8}{d}}{1 \cdot \frac{d}{d} + \frac{1}{d}} = \dfrac{\frac{8d + 8}{d}}{\frac{d + 1}{d}}$

$= \dfrac{8d + 8}{d} \cdot \dfrac{d}{d + 1}$

$= \dfrac{8(d + 1)}{d} \cdot \dfrac{d}{d + 1}$

$= \dfrac{8\cancel{d(d + 1)}}{1 \cdot \cancel{d(d + 1)}}$

$= 8$

12. $\dfrac{3(2b - 3)}{b(6 - b)}$

13. $\dfrac{\frac{x}{8} - \frac{8}{x}}{\frac{1}{8} + \frac{1}{x}}$ LCD of the denominators is 8x

$= \dfrac{\frac{x}{8} - \frac{8}{x}}{\frac{1}{8} + \frac{1}{x}} \cdot \dfrac{8x}{8x}$

$= \dfrac{\left(\frac{x}{8} - \frac{8}{x}\right)8x}{\left(\frac{1}{8} + \frac{1}{x}\right)8x}$

$= \dfrac{\frac{x}{8}(8x) - \frac{8}{x}(8x)}{\frac{1}{8}(8x) + \frac{1}{x}(8x)}$

$= \dfrac{x^2 - 64}{x + 8} = \dfrac{\cancel{(x + 8)}(x - 8)}{\cancel{x + 8}} = x - 8$

14. $\dfrac{4 + m^2}{m^2 - 4}$

15. $\dfrac{1 + \frac{1}{y}}{1 - \frac{1}{y^2}} = \dfrac{1 \cdot \frac{y}{y} + \frac{1}{y}}{1 \cdot \frac{y^2}{y^2} - \frac{1}{y^2}} = \dfrac{\frac{y + 1}{y}}{\frac{y^2 - 1}{y^2}}$

$= \dfrac{y + 1}{y} \cdot \dfrac{y^2}{y^2 - 1}$

$= \dfrac{\cancel{(y + 1)}\cancel{y} \cdot y}{\cancel{y}\cancel{(y + 1)}(y - 1)} = \dfrac{y}{y - 1}$

16. $\dfrac{1 - q}{q}$

17. $\dfrac{\frac{1}{5} - \frac{1}{a}}{\frac{5 - a}{5}}$ LCD of the denominators is 5a

$= \dfrac{\frac{1}{5} - \frac{1}{a}}{\frac{5 - a}{5}} \cdot \dfrac{5a}{5a}$

$= \dfrac{\left(\frac{1}{5} - \frac{1}{a}\right)5a}{\left(\frac{5 - a}{5}\right)5a}$

$= \dfrac{\frac{1}{5}(5a) - \frac{1}{a}(5a)}{a(5 - a)} = \dfrac{a - 5}{5a - a^2}$

$= \dfrac{a - 5}{-a(-5 + a)} = \dfrac{\cancel{a - 5}}{-a\cancel{(a - 5)}} = -\dfrac{1}{a}$

18. $\dfrac{2x - 1}{2}$

19. $\dfrac{\frac{x}{x - y}}{\frac{x^2}{x^2 - y^2}} = \dfrac{x}{x - y} \cdot \dfrac{x^2 - y^2}{x^2}$ Multiplying by the reciprocal of the divisor

$= \dfrac{x}{x - y} \cdot \dfrac{(x - y)(x + y)}{x \cdot x}$

$= \dfrac{\cancel{x}\cancel{(x - y)}(x + y)}{\cancel{x} \cdot x\cancel{(x - y)}}$

$= \dfrac{x + y}{x}$

20. $x - y$

21. $\dfrac{\frac{3}{m} + \frac{2}{m^3}}{\frac{4}{m^2} - \frac{3}{m}}$ LCD of the denominator is m^3

$= \dfrac{\frac{3}{m} + \frac{2}{m^3}}{\frac{4}{m^2} - \frac{3}{m}} \cdot \dfrac{m^3}{m^3}$

$= \dfrac{\frac{3}{m} \cdot m^3 + \frac{2}{m^3} \cdot m^3}{\frac{4}{m^2} \cdot m^3 - \frac{3}{m} \cdot m^3}$

$= \dfrac{3m^2 + 2}{4m - 3m^2}$

22. $\dfrac{1}{a(a - b)}$

23. $\dfrac{\frac{5}{4x^3} - \frac{3}{8x}}{\frac{3}{2x} + \frac{3}{4x^3}} = \dfrac{\frac{5}{4x^3} \cdot \frac{2}{2} - \frac{3}{8x} \cdot \frac{x^2}{x^2}}{\frac{3}{2x} \cdot \frac{2x^2}{2x^2} + \frac{3}{4x^3}}$

$= \dfrac{\frac{10 - 3x^2}{8x^3}}{\frac{6x^2 + 3}{4x^3}}$

$= \dfrac{10 - 3x^2}{8x^3} \cdot \dfrac{4x^3}{6x^2 + 3}$

$= \dfrac{4x^3(10 - 3x^2)}{2 \cdot 4x^3 \cdot 3(2x^2 + 1)}$

$= \dfrac{10 - 3x^2}{6(2x^2 + 1)}$, or

$= \dfrac{10 - 3x^2}{12x^2 + 6}$

24. $\dfrac{15(4 - a^3)}{14a^2(9 + 2a)}$, or $\dfrac{60 - 15a^3}{126a^2 + 28a^3}$

25. $\dfrac{\frac{a}{6b^3} + \frac{4}{9b^2}}{\frac{5}{6b} - \frac{1}{9b^3}}$ LCD of the denominators is $18b^3$

$= \dfrac{\frac{a}{6b^3} + \frac{4}{9b^2}}{\frac{5}{6b} - \frac{1}{9b^3}} \cdot \dfrac{18b^3}{18b^3}$

$= \dfrac{\frac{a}{6b^3} \cdot 18b^3 + \frac{4}{9b^2} \cdot 18b^3}{\frac{5}{6b} \cdot 18b^3 - \frac{1}{9b^3} \cdot 18b^3}$

$= \dfrac{3a + 8b}{15b^2 - 2}$

26. $\dfrac{2xy - 3y^3}{xy^3 + 30}$

27. $\dfrac{\frac{2}{x^2y} + \frac{3}{xy^2}}{\frac{2}{xy^3} + \frac{1}{x^2y}} = \dfrac{\frac{2}{x^2y} \cdot \frac{y}{y} + \frac{3}{xy^2} \cdot \frac{x}{x}}{\frac{2}{xy^3} \cdot \frac{x}{x} + \frac{1}{x^2y} \cdot \frac{y^2}{y^2}}$

$= \dfrac{\frac{2y + 3x}{x^2y^2}}{\frac{2x + y^2}{x^2y^3}}$

$= \dfrac{2y + 3x}{x^2y^2} \cdot \dfrac{x^2y^3}{2x + y^2}$

$= \dfrac{\cancel{x^2y^2} \cdot y(2y + 3x)}{\cancel{x^2y^2}(2x + y^2)}$

$= \dfrac{y(2y + 3x)}{2x + y^2}$, or $\dfrac{2y^2 + 3xy}{2x + y^2}$

28. $\dfrac{5a^2 + 2b^3}{b^3(5 - 3a^2)}$, or $\dfrac{5a^2 + 2b^3}{5b^3 - 3a^2b^3}$

29. $\dfrac{3 - \frac{2}{a^4}}{2 + \frac{3}{a^3}} = \dfrac{3 - \frac{2}{a^4}}{2 + \frac{3}{a^3}} \cdot \dfrac{a^4}{a^4}$ LCD of the denominators is a^4

$= \dfrac{3 \cdot a^4 - \frac{2}{a^4} \cdot a^4}{2 \cdot a^4 + \frac{3}{a^3} \cdot a^4}$

$= \dfrac{3a^4 - 2}{2a^4 + 3a}$

30. $\dfrac{2x^4 - 3x^2}{2x^4 + 3}$

31. $\dfrac{x + \frac{3}{x}}{x - \frac{2}{x}} = \dfrac{x \cdot \frac{x}{x} + \frac{3}{x}}{x \cdot \frac{x}{x} - \frac{2}{x}}$

$= \dfrac{\frac{x^2 + 3}{x}}{\frac{x^2 - 2}{x}}$

$= \dfrac{x^2 + 3}{x} \cdot \dfrac{x}{x^2 - 2}$

$= \dfrac{x(x^2 + 3)}{x(x^2 - 2)}$

$= \dfrac{x^2 + 3}{x^2 - 2}$

32. $\dfrac{t^2 - 2}{t^2 + 5}$

33. $\dfrac{5 + \frac{3}{x^2y}}{\frac{3 + x}{x^3y}} = \dfrac{5 + \frac{3}{x^2y}}{\frac{3 + x}{x^3y}} \cdot \dfrac{x^3y}{x^3y}$ LCD of the denominators is x^3y

$= \dfrac{5 \cdot x^3y + \frac{3}{x^2y} \cdot x^3y}{\frac{3 + x}{x^3y} \cdot x^3y}$

$= \dfrac{5x^3y + 3x}{3 + x}$

34. $\dfrac{7a^2b^3 - 5a}{b^2(4 + a)}$, or $\dfrac{7a^2b^3 - 5a}{4b^2 + ab^2}$

35. $\dfrac{\frac{x + 5}{x^2}}{\frac{2}{x} - \frac{3}{x^2}} = \dfrac{\frac{x + 5}{x^2}}{\frac{2}{x} \cdot \frac{x}{x} - \frac{3}{x^2}}$

$= \dfrac{\frac{x + 5}{x^2}}{\frac{2x - 3}{x^2}}$

$= \dfrac{x + 5}{x^2} \cdot \dfrac{x^2}{2x - 3}$

$= \dfrac{x^2(x + 5)}{x^2(2x - 3)}$

$= \dfrac{x + 5}{2x - 3}$

36. $\dfrac{a - 7}{a(3 + 2a)}$, or $\dfrac{a - 7}{3a + 2a^2}$

37. $\dfrac{x - 3 + \frac{2}{x}}{x - 4 + \frac{3}{x}} = \dfrac{x \cdot \frac{x}{x} - 3 \cdot \frac{x}{x} + \frac{2}{x}}{x \cdot \frac{x}{x} - 4 \cdot \frac{x}{x} + \frac{3}{x}}$

$= \dfrac{\frac{x^2 - 3x + 2}{x}}{\frac{x^2 - 4x + 3}{x}}$

$= \dfrac{x^2 - 3x + 2}{x} \cdot \dfrac{x}{x^2 - 4x + 3}$

$= \dfrac{(x - 2)(x - 1)}{x} \cdot \dfrac{x}{(x - 3)(x - 1)}$

$= \dfrac{x(x - 1)}{x(x - 1)} \cdot \dfrac{x - 2}{x - 3}$

$= \dfrac{x - 2}{x - 3}$

38. $\dfrac{a + b}{a - b}$

39. $\dfrac{a + 5 - \frac{3}{a}}{a - 3 + \frac{5}{a}} = \dfrac{a + 5 - \frac{3}{a}}{a - 3 + \frac{5}{a}} \cdot \dfrac{a}{a}$ LCD of the denominators is a

$= \dfrac{a \cdot a + 5 \cdot a - \frac{3}{a} \cdot a}{a \cdot a - 3 \cdot a + \frac{5}{a} \cdot a}$

$= \dfrac{a^2 + 5a - 3}{a^2 - 3a + 5}$

40. $\dfrac{20x^2 - 30x}{15x^2 + 105x - 12}$

41. $(5x^4 - 6x^3 + 23x^2 - 79x + 24) - (-18x^4 - 56x^3 + 84x - 17) = 5x^4 - 6x^3 + 23x^2 - 79x + 24 + 18x^4 + 56x^3 - 84x + 17 = 23x^4 + 50x^3 + 23x^2 - 163x + 41$

42. 14 yd

43.

44.

45. $\dfrac{1}{\frac{2}{x-1} - \frac{1}{3x-2}} = \dfrac{1}{\frac{2}{x-1} - \frac{1}{3x-2}} \cdot \dfrac{(x-1)(3x-2)}{(x-1)(3x-2)}$

$= \dfrac{(x - 1)(3x - 2)}{\left[\frac{2}{x - 1} - \frac{1}{3x - 2}\right](x - 1)(3x - 2)}$

$= \dfrac{(x - 1)(3x - 2)}{\frac{2}{x-1}(x-1)(3x-2) - \frac{1}{3x-2}(x-1)(3x-2)}$

$= \dfrac{(x - 1)(3x - 2)}{2(3x - 2) - (x - 1)}$

$= \dfrac{(x - 1)(3x - 2)}{6x - 4 - x + 1}$

$= \dfrac{(x - 1)(3x - 2)}{5x - 3}$

46. $\frac{ac}{bd}$

47. $\dfrac{\frac{a}{b}-\frac{c}{d}}{\frac{b}{a}-\frac{d}{c}} = \dfrac{\frac{a}{b}\cdot\frac{d}{d}-\frac{c}{d}\cdot\frac{b}{b}}{\frac{b}{a}\cdot\frac{c}{c}-\frac{d}{c}\cdot\frac{a}{a}} = \dfrac{\frac{ad-bc}{bd}}{\frac{bc-ad}{ac}}$

$= \frac{ad-bc}{bd}\cdot\frac{ac}{bc-ad}$

$= \frac{-(bc-ad)}{bd}\cdot\frac{ac}{bc-ad}$

$= \frac{-ac\cancel{(bc-ad)}}{bd\cancel{(bc-ad)}}$

$= -\frac{ac}{bd}$

48. x^5

49. $1+\dfrac{1}{1+\dfrac{1}{1+\frac{1}{x}}} = 1+\dfrac{1}{1+\dfrac{1}{\frac{x+1}{x}}}$

$= 1+\dfrac{1}{1+\frac{x}{x+1}}$

$= 1+\dfrac{1}{\frac{x+1+x}{x+1}}$

$= 1+\dfrac{1}{\frac{2x+1}{x+1}}$

$= 1+\frac{x+1}{2x+1}$

$= \frac{2x+1+x+1}{2x+1}$

$= \frac{3x+2}{2x+1}$

50. $\frac{-2z(5z-2)}{(2+z)(-13z+6)}$

Exercise Set 6.7

1. $\frac{3}{8}+\frac{4}{5}=\frac{x}{20}$, LCD = 40

$40\left(\frac{3}{8}+\frac{4}{5}\right) = 40\cdot\frac{x}{20}$

$40\cdot\frac{3}{8}+40\cdot\frac{4}{5} = 40\cdot\frac{x}{20}$

$15+32 = 2x$

$47 = 2x$

$\frac{47}{2} = x$

Check:

$\frac{3}{8}+\frac{4}{5}=\frac{x}{20}$	
$\frac{3}{8}+\frac{4}{5}$?	$\dfrac{\frac{47}{2}}{20}$
$\frac{15}{40}+\frac{32}{40}$	$\frac{47}{2}\cdot\frac{1}{20}$
$\frac{47}{40}$	$\frac{47}{40}$ TRUE

This checks, so the solution is $\frac{47}{2}$.

2. $\frac{57}{2}$

3. $\frac{2}{3}-\frac{5}{6}=\frac{1}{x}$, LCD = 6x

$6x\left(\frac{2}{3}-\frac{5}{6}\right) = 6x\cdot\frac{1}{x}$

$6x\cdot\frac{2}{3}-6x\cdot\frac{5}{6} = 6x\cdot\frac{1}{x}$

$4x-5x = 6$

$-x = 6$

$x = -6$

Check:

$\frac{2}{3}-\frac{5}{6}=\frac{1}{x}$	
$\frac{2}{3}-\frac{5}{6}$?	$\frac{1}{-6}$
$\frac{4}{6}-\frac{5}{6}$	$-\frac{1}{6}$
$-\frac{1}{6}$	$-\frac{1}{6}$ TRUE

This checks, so the solution is -6.

4. $-\frac{40}{19}$

5. $\frac{1}{6}+\frac{1}{8}=\frac{1}{t}$, LCD = 24t

$24t\left(\frac{1}{6}+\frac{1}{8}\right) = 24t\cdot\frac{1}{t}$

$24t\cdot\frac{1}{6}+24t\cdot\frac{1}{8} = 24t\cdot\frac{1}{t}$

$4t+3t = 24$

$7t = 24$

$t = \frac{24}{7}$

Check:

$\frac{1}{6}+\frac{1}{8}=\frac{1}{t}$	
$\frac{1}{6}+\frac{1}{8}$?	$\dfrac{1}{\frac{24}{7}}$
$\frac{4}{24}+\frac{3}{24}$	$1\cdot\frac{7}{24}$
$\frac{7}{24}$	$\frac{7}{24}$ TRUE

This checks, so the solution is $\frac{24}{7}$.

6. $\frac{40}{9}$

7. $x + \frac{4}{x} = -5$, LCD = x

$$x\left(x + \frac{4}{x}\right) = x(-5)$$
$$x \cdot x + x \cdot \frac{4}{x} = x(-5)$$
$$x^2 + 4 = -5x$$
$$x^2 + 5x + 4 = 0$$
$$(x + 4)(x + 1) = 0$$
$$x + 4 = 0 \quad \text{or} \quad x + 1 = 0$$
$$x = -4 \quad \text{or} \quad x = -1$$

Check: $x + \frac{4}{x} = -5$: $-4 + \frac{4}{-4} \;?\; -5$; $-4 - 1$; $-5 \mid -5$ TRUE

$x + \frac{4}{x} = -5$: $-1 + \frac{4}{-1} \mid -5$; $-1 - 4$; $-5 \mid -5$ TRUE

Both of these check, so the two solutions are -4 and -1.

8. -3, -1

9. $\frac{x}{4} - \frac{4}{x} = 0$, LCD = 4x

$$4x\left(\frac{x}{4} - \frac{4}{x}\right) = 4x \cdot 0$$
$$4x \cdot \frac{x}{4} - 4x \cdot \frac{4}{x} = 4x \cdot 0$$
$$x^2 - 16 = 0$$
$$(x + 4)(x - 4) = 0$$
$$x + 4 = 0 \quad \text{or} \quad x - 4 = 0$$
$$x = -4 \quad \text{or} \quad x = 4$$

Check: $\frac{x}{4} - \frac{4}{x} = 0$: $\frac{-4}{4} - \frac{4}{-4} \;?\; 0$; $-1 + 1$; $0 \mid 0$ TRUE

$\frac{x}{4} - \frac{4}{x} = 0$: $\frac{4}{4} - \frac{4}{4} \;?\; 0$; $1 - 1$; $0 \mid 0$ TRUE

Both of these check, so the two solutions are -4 and 4.

10. -5, 5

11. $\frac{5}{x} = \frac{6}{x} - \frac{1}{3}$, LCD = 3x

$$3x \cdot \frac{5}{x} = 3x\left(\frac{6}{x} - \frac{1}{3}\right)$$
$$3x \cdot \frac{5}{x} = 3x \cdot \frac{6}{x} - 3x \cdot \frac{1}{3}$$
$$15 = 18 - x$$
$$-3 = -x$$
$$3 = x$$

Check: $\frac{5}{x} = \frac{6}{x} - \frac{1}{3}$: $\frac{5}{3} \;?\; \frac{6}{3} - \frac{1}{3}$; $\frac{5}{3} \mid \frac{5}{3}$ TRUE

This checks, so the solution is 3.

12. 2

13. $\frac{5}{3x} + \frac{3}{x} = 1$, LCD = 3x

$$3x\left(\frac{5}{3x} + \frac{3}{x}\right) = 3x \cdot 1$$
$$3x \cdot \frac{5}{3x} + 3x \cdot \frac{3}{x} = 3x \cdot 1$$
$$5 + 9 = 3x$$
$$14 = 3x$$
$$\frac{14}{3} = x$$

Check: $\frac{5}{3x} + \frac{3}{x} = 1$: $\frac{5}{3 \cdot \frac{14}{3}} + \frac{3}{\frac{14}{3}} \;?\; 1$; $\frac{5}{14} + \frac{9}{14}$; $\frac{14}{14}$; $1 \mid 1$ TRUE

This checks, so the solution is $\frac{14}{3}$.

14. $\frac{23}{4}$

15. $\frac{x - 7}{x + 2} = \frac{1}{4}$, LCD = 4(x + 2)

$$4(x + 2) \cdot \frac{x - 7}{x + 2} = 4(x + 2) \cdot \frac{1}{4}$$
$$4(x - 7) = x + 2$$
$$4x - 28 = x + 2$$
$$3x = 30$$
$$x = 10$$

Check: $\frac{x - 7}{x + 2} = \frac{1}{4}$: $\frac{10 - 7}{10 + 2} \;?\; \frac{1}{4}$; $\frac{3}{12}$; $\frac{1}{4} \mid \frac{1}{4}$ TRUE

16. 5

17. $\frac{2}{x + 1} = \frac{1}{x - 2}$, LCD = (x+1)(x-2)

$$(x + 1)(x - 2) \cdot \frac{2}{x + 1} = (x + 1)(x - 2) \cdot \frac{1}{x - 2}$$
$$2(x - 2) = x + 1$$
$$2x - 4 = x + 1$$
$$x = 5$$

This checks, so the solution is 5.

18. $-\frac{13}{2}$

19. $$\frac{x}{6} - \frac{x}{10} = \frac{1}{6}, \quad \text{LCD} = 30$$
$$30\left[\frac{x}{6} - \frac{x}{10}\right] = 30 \cdot \frac{1}{6}$$
$$30 \cdot \frac{x}{6} - 30 \cdot \frac{x}{10} = 30 \cdot \frac{1}{6}$$
$$5x - 3x = 5$$
$$2x = 5$$
$$x = \frac{5}{2}$$
This checks, so the solution is $\frac{5}{2}$.

20. 3

21. $$\frac{x+3}{3} - 1 = \frac{x-1}{2}, \quad \text{LCD} = 6$$
$$\frac{x+1}{3} - \frac{x-1}{2} = 1$$
$$6\left[\frac{x+1}{3} - \frac{x-1}{2}\right] = 6 \cdot 1$$
$$6 \cdot \frac{x+1}{3} - 6 \cdot \frac{x-1}{2} = 6 \cdot 1$$
$$2(x+1) - 3(x-1) = 6$$
$$2x + 2 - 3x + 3 = 6$$
$$-x + 5 = 6$$
$$-x = 1$$
$$x = -1$$
This checks, so the solution is -1.

22. -2

23. $$\frac{a-3}{3a+2} = \frac{1}{5}, \quad \text{LCD} = 5(3a+2)$$
$$5(3a+2) \cdot \frac{a-3}{3a+2} = 5(3a+2) \cdot \frac{1}{5}$$
$$5(a-3) = 3a + 2$$
$$5a - 15 = 3a + 2$$
$$2a = 17$$
$$a = \frac{17}{2}$$
This checks, so the solution is $\frac{17}{2}$.

24. $\frac{9}{2}$

25. $$\frac{x-1}{x-5} = \frac{4}{x-5}, \quad \text{LCD} = x - 5$$
$$(x-5) \cdot \frac{x-1}{x-5} = (x-5) \cdot \frac{4}{x-5}$$
$$x - 1 = 4$$
$$x = 5$$
Check: $\frac{x-1}{x-5} = \frac{4}{x-5}$

$\frac{5-1}{5-5}$	$\frac{4}{5-5}$	
$\frac{4}{0}$	$\frac{4}{0}$	UNDEFINED

The number 5 is not a solution because it makes a denominator zero. Thus, there is no solution.

26. No solution

27. $$\frac{2}{x+3} = \frac{5}{x}, \quad \text{LCD} = x(x+3)$$
$$x(x+3) \cdot \frac{2}{x+3} = x(x+3) \cdot \frac{5}{x}$$
$$2x = 5(x+3)$$
$$2x = 5x + 15$$
$$-15 = 3x$$
$$-5 = x$$
This checks, so the solution is -5.

28. -16

29. $$\frac{x-2}{x-3} = \frac{x-1}{x+1}, \quad \text{LCD} = (x-3)(x+1)$$
$$(x-3)(x+1) \cdot \frac{x-2}{x-3} = (x-3)(x+1) \cdot \frac{x-1}{x+1}$$
$$(x+1)(x-2) = (x-3)(x-1)$$
$$x^2 - x - 2 = x^2 - 4x + 3$$
$$-x - 2 = -4x + 3$$
$$3x = 5$$
$$x = \frac{5}{3}$$
This checks, so the solution is $\frac{5}{3}$.

30. $\frac{1}{5}$

31. $$\frac{1}{x+3} + \frac{1}{x-3} = \frac{1}{x^2-9}, \quad \text{LCD} = (x+3)(x-3)$$
$$(x+3)(x-3)\left[\frac{1}{x+3} + \frac{1}{x-3}\right] = (x+3)(x-3) \cdot \frac{1}{(x+3)(x-3)}$$
$$(x-3) + (x+3) = 1$$
$$2x = 1$$
$$x = \frac{1}{2}$$
This checks, so the solution is $\frac{1}{2}$.

32. -3

33. $$\frac{x}{x+4} - \frac{4}{x-4} = \frac{x^2+16}{x^2-16}, \quad \text{LCD} = (x+4)(x-4)$$
$$(x+4)(x-4)\left[\frac{x}{x+4} - \frac{4}{x-4}\right] = (x+4)(x-4) \cdot \frac{x^2+16}{(x+4)(x-4)}$$
$$x(x-4) - 4(x+4) = x^2 + 16$$
$$x^2 - 4x - 4x - 16 = x^2 + 16$$
$$-8x - 16 = 16$$
$$-8x = 32$$
$$x = -4$$
The number -4 is not a solution because it makes a denominator zero. Thus, there is no solution.

34. 2

35. $\frac{-3}{y-7} = \frac{-10-y}{7-y}$ $y - 7$ and $7 - y$ are opposites

$\frac{-3}{y-7} = \frac{-10-y}{-(y-7)}$

$\frac{-3}{y-7} = \frac{-(-10-y)}{y-7}$ Since $\frac{a}{-b} = \frac{-a}{b}$

$\frac{-3}{y-7} = \frac{10+y}{y-7}$, LCD = $y - 7$

$(y-7)\left[\frac{-3}{y-7}\right] = (y-7)\left[\frac{10+y}{y-7}\right]$

$-3 = 10 + y$

$-13 = y$

This checks, so the solution is -13.

36. No solution

37. $(a^2b^5)^{-3} = a^{2(-3)}b^{5(-3)} = a^{-6}b^{-15}$, or $\frac{1}{a^6b^{15}}$

38. x^8y^{12}

39. $\left[\frac{2x}{t^2}\right]^4 = \frac{(2x)^4}{(t^2)^4} = \frac{2^4x^4}{t^8} = \frac{16x^4}{t^8}$

40. $\frac{w^4}{y^6}$

41. ◈

42. ◈

43. $\frac{4}{y-2} - \frac{2y-3}{y^2-4} = \frac{5}{y+2}$, LCD = $(y+2)(y-2)$

$(y+2)(y-2)\left[\frac{4}{y-2} - \frac{2y-3}{y^2-4}\right] = (y+2)(y-2) \cdot \frac{5}{y+2}$

$4(y+2) - (2y-3) = 5(y-2)$

$4y + 8 - 2y + 3 = 5y - 10$

$2y + 11 = 5y - 10$

$21 = 3y$

$7 = y$

This checks, so the solution is 7.

44. $-\frac{1}{6}$

45. $\frac{12-6x}{x^2-4} = \frac{3x}{x+2} - \frac{2x-3}{x-2}$, LCD = $(x+2)(x-2)$

$(x+2)(x-2) \cdot \frac{12-6x}{(x+2)(x-2)} = (x+2)(x-2)\left[\frac{3x}{x+2} - \frac{2x-3}{x-2}\right]$

$12 - 6x = 3x(x-2) - (x+2)(2x-3)$

$12 - 6x = 3x^2 - 6x - 2x^2 - x + 6$

$0 = x^2 - x - 6$

$0 = (x-3)(x+2)$

$x - 3 = 0$ or $x + 2 = 0$

$x = 3$ or $x = -2$

The number 3 is a solution, but -2 is not because it makes denominators 0.

46. 0, -1

47. $4a - 3 = \frac{a+13}{a+1}$, LCD = $a + 1$

$(a+1)(4a-3) = (a+1) \cdot \frac{a+13}{a+1}$

$4a^2 + a - 3 = a + 13$

$4a^2 - 16 = 0$

$4(a+2)(a-2) = 0$

$a + 2 = 0$ or $a - 2 = 0$

$a = -2$ or $a = 2$

Both of these check, so the two solutions are -2 and 2.

48. 2

49. $\frac{y^2-4}{y+3} = 2 - \frac{y-2}{y+3}$, LCD = $y + 3$

$(y+3) \cdot \frac{y^2-4}{y+3} = (y+3)\left[2 - \frac{y-2}{y+3}\right]$

$y^2 - 4 = 2(y+3) - (y-2)$

$y^2 - 4 = 2y + 6 - y + 2$

$y^2 - 4 = y + 8$

$y^2 - y - 12 = 0$

$(y-4)(y+3) = 0$

$y - 4 = 0$ or $y + 3 = 0$

$y = 4$ or $y = -3$

The number 4 is a solution, but -3 is not because it makes a denominator zero.

50. -6

51.

52.

Exercise Set 6.8

1. Familiarize. Let x = the number.

Translate.

A number,	minus	twice its reciprocal	is	1.
x	$-$	$2 \cdot \frac{1}{x}$	$=$	1

Carry out. We solve the equation.

$x - \frac{2}{x} = 1$

$x\left[x - \frac{2}{x}\right] = x \cdot 1$ Multiplying by the LCD

$x^2 - 2 = x$

$x^2 - x - 2 = 0$

$(x-2)(x+1) = 0$

$x - 2 = 0$ or $x + 1 = 0$

$x = 2$ or $x = -1$

Check. Twice the reciprocal of 2 is $2 \cdot \frac{1}{2}$, or 1. Since $2 - 1 = 1$, the number 2 is a solution. Twice the reciprocal of -1 is $2(-1)$, or -2. Since $-1 - (-2) = -1 + 2$, or 1, the number -1 is a solution.

State. The solutions are 2 and -1.

2. 4, -1

3. Familiarize. Let x = the number.

Translate. We reword the problem.

A number plus its reciprocal is 2.

$$x + \frac{1}{x} = 2$$

Carry out. We solve the equation.

$$x + \frac{1}{x} = 2$$

$$x\left(x + \frac{1}{x}\right) = x \cdot 2 \quad \text{Multiplying by the LCD}$$

$$x^2 + 1 = 2x$$

$$x^2 - 2x + 1 = 0$$

$$(x - 1)(x - 1) = 0$$

$x - 1 = 0$ or $x - 1 = 0$

$x = 1$ or $x = 1$

Check. The reciprocal of 1 is 1. Since $1 + 1 = 2$, the number 1 is a solution.

State. The solution is 1.

4. 1, 5

5. Familiarize. The job takes David 4 hours working alone and Sierra 5 hours working alone. Then in 1 hour David does $\frac{1}{4}$ of the job and Sierra does $\frac{1}{5}$ of the job. Working together, they can do $\frac{1}{4} + \frac{1}{5}$, or $\frac{9}{20}$ of the job in 1 hour. In two hours, David does $2\left(\frac{1}{4}\right)$ of the job and Sierra does $2\left(\frac{1}{5}\right)$ of the job. Working together they can do $2\left(\frac{1}{4}\right) + 2\left(\frac{1}{5}\right)$, or $\frac{9}{10}$ of the job in 2 hours. In 3 hours they can do $3\left(\frac{1}{4}\right) + 3\left(\frac{1}{5}\right)$, or $\frac{27}{20}$ or $1\frac{7}{20}$ of the job which is more of the job than needs to be done. The answer is somewhere between 2 hr and 3 hr.

Translate. If they work together t hours, then David does $t\left(\frac{1}{4}\right)$ of the job and Sierra does $t\left(\frac{1}{5}\right)$ of the job. We want some number t such that

$$t\left(\frac{1}{4}\right) + t\left(\frac{1}{5}\right) = 1.$$

Carry out. We solve the equation.

$$\frac{t}{4} + \frac{t}{5} = 1, \quad \text{LCD} = 20$$

$$20\left(\frac{t}{4} + \frac{t}{5}\right) = 20 \cdot 1$$

$$5t + 4t = 20$$

$$9t = 20$$

$$t = \frac{20}{9}, \text{ or } 2\frac{2}{9}$$

Check. The check can be done by repeating the computations. We also have a partial check in that we expected from our familiarization step that the answer would be between 2 hr and 3 hr.

State. Working together, it takes them $2\frac{2}{9}$ hr to complete the job.

6. $6\frac{6}{7}$ hr

7. Familiarize. The job takes Vern 45 min working alone and Nina 60 min working alone. Then in 1 minute Vern does $\frac{1}{45}$ of the job and Nina does $\frac{1}{60}$ of the job. Working together, they can do $\frac{1}{45} + \frac{1}{60}$, or $\frac{7}{180}$ of the job in 1 minute. In 20 minutes, Vern does $\frac{20}{45}$ of the job and Nina does $\frac{20}{60}$ of the job. Working together, they can do $\frac{20}{45} + \frac{20}{60}$, or $\frac{7}{9}$ of the job. In 30 minutes, they can do $\frac{30}{45} + \frac{30}{60}$, or $\frac{7}{6}$ of the job which is more of the job than needs to be done. The answer is somewhere between 20 minutes and 30 minutes.

Translate. If they work together t minutes, then Vern does $t\left(\frac{1}{45}\right)$ of the job and Nina does $t\left(\frac{1}{60}\right)$ of the job. We want some number t such that

$$t\left(\frac{1}{45}\right) + t\left(\frac{1}{60}\right) = 1.$$

Carry out. We solve the equation.

$$\frac{t}{45} + \frac{t}{60} = 1, \quad \text{LCD} = 180$$

$$180\left(\frac{t}{45} + \frac{t}{60}\right) = 180 \cdot 1$$

$$4t + 3t = 180$$

$$7t = 180$$

$$t = \frac{180}{7}, \text{ or } 25\frac{5}{7}$$

Check. The check can be done by repeating the computations. We also have a partial check in that we expected from our familiarization step that the answer would be between 20 minutes and 30 minutes.

State. It would take them $25\frac{5}{7}$ minutes to complete the job working together.

8. $1\frac{5}{7}$ hr

9. Familiarize. The job takes Rory 12 hours working alone and Mira 9 hours working alone. Then in 1 hour Rory does $\frac{1}{12}$ of the job and Mira does $\frac{1}{9}$ of the job. Working together they can do $\frac{1}{12} + \frac{1}{9}$, or $\frac{7}{36}$ of the job in 1 hour. In two hours, Rory does $2\left(\frac{1}{12}\right)$ of the job and Mira does $2\left(\frac{1}{9}\right)$ of the job. Working together they can do $2\left(\frac{1}{12}\right) + 2\left(\frac{1}{9}\right)$, or $\frac{14}{36}$ of the job in two hours. In 3 hours they can do $3\left(\frac{1}{12}\right) + 3\left(\frac{1}{9}\right)$, or $\frac{21}{36}$ of the job. In 5 hours, they can do $\frac{35}{36}$ of the job. In 6 hours, they can do $\frac{42}{36}$, or $1\frac{1}{6}$ of the job which is more of the job than needs to be done. The answer is somewhere between 5 hr and 6 hr.

Translate. If they work together t hours, then Rory does $t\left(\frac{1}{12}\right)$ of the job and Mira does $t\left(\frac{1}{9}\right)$ of the job. We want some number t such that

$$t\left(\frac{1}{12}\right) + t\left(\frac{1}{9}\right) = 1.$$

Carry out. We solve the equation

$$\frac{t}{12} + \frac{t}{9} = 1, \quad \text{LCM} = 36$$

$$36\left(\frac{t}{12} + \frac{t}{9}\right) = 36 \cdot 1$$

$$3t + 4t = 36$$

$$7t = 36$$

$$t = \frac{36}{7}, \text{ or } 5\frac{1}{7}$$

Check. The check can be done by repeating the computations. We also have a partial check in that we expected from our familiarization step that the answer would be between 5 hr and 6 hr.

State. Working together, it takes them $5\frac{1}{7}$ hr to complete the job.

10. $10\frac{2}{7}$ hr

11. Familiarize. Let t = the time it takes Red Bryck and Lotta Mudd to do the job working together.

Translate. We use the work principle.

$$\frac{t}{6} + \frac{t}{8} = 1$$

Carry out. We solve the equation.

$$\frac{t}{6} + \frac{t}{8} = 1, \quad \text{LCD} = 24$$

$$24\left(\frac{t}{6} + \frac{t}{8}\right) = 24 \cdot 1$$

$$4t + 3t = 24$$

$$7t = 24$$

$$t = \frac{24}{7}, \text{ or } 3\frac{3}{7}$$

Check. We repeat the computations.

State. It would take them $3\frac{3}{7}$ hr to do the job working together.

12. $11\frac{1}{9}$ min

13. Familiarize. We complete the table shown in the text.

$$d = r \cdot t$$

	Distance	Speed	Time
Slow car	150	r	t
Fast car	350	r + 40	t

Translate. We can replace the t's in the table above using the formula t = d/r.

	Distance	Speed	Time
Slow car	150	r	$\frac{150}{r}$
Fast car	350	r + 40	$\frac{350}{r + 40}$

Since the times are the same for both cars, we have the equation

$$\frac{150}{r} = \frac{350}{r + 40}.$$

Carry out. We multiply by the LCD, r(r + 40).

$$r(r + 40) \cdot \frac{150}{r} = r(r + 40) \cdot \frac{350}{r + 40}$$

$$150(r + 40) = 350r$$

$$150r + 6000 = 350r$$

$$6000 = 200r$$

$$30 = r$$

Check. If r is 30 km/h, then r + 40 is 70 km/h. The time for the slow car is 150/30, or 5 hr. The time for the fast car is 350/70, or 5 hr. The times are the same. The values check.

State. The speed of the slow car is 30 km/h, and the speed of the fast car is 70 km/h.

14. Slow car: 50 km/h, fast car: 80 km/h

15. Familiarize. We complete the table shown in the text.

$$d = r \cdot t$$

	Distance	Speed	Time
Freight	330	r - 14	t
Passenger	400	r	t

Translate. We can replace the t's in the table above using the formula t = d/r.

	Distance	Speed	Time
Freight	330	r - 14	$\frac{330}{r - 14}$
Passenger	400	r	$\frac{400}{r}$

Since the times are the same for both trains, we have the equation

$$\frac{330}{r - 14} = \frac{400}{r}.$$

Carry out. We multiply by the LCD, $r(r - 14)$.

$$r(r - 14) \cdot \frac{330}{r - 14} = r(r - 14) \cdot \frac{400}{r}$$
$$330r = 400(r - 14)$$
$$330r = 400r - 5600$$
$$-70r = -5600$$
$$r = 80$$

Then substitute 80 for r in either equation to find t:

$$t = \frac{400}{r}$$
$$t = \frac{400}{80} \quad \text{Substituting 80 for r}$$
$$t = 5$$

Check. If $r = 80$, then $r - 14 = 66$. In 5 hr the freight train travels $66 \cdot 5$, or 330 km, and the passenger train travels $80 \cdot 5$, or 400 km. The values check.

State. The speed of the passenger train is 80 km/h. The speed of the freight train is 66 km/h.

16. Passenger train: 80 km/h, freight train: 65 km/h

17. Familiarize. We let t = the number of hours Dexter rode and organize the given information in a table.

	Distance	Speed	Time
Dexter	16	r	t
Gail	50	r	t + 3

Translate. We can replace the r's in the table above using the formula $r = d/t$.

	Distance	Speed	Time
Dexter	16	$\frac{16}{t}$	t
Gail	50	$\frac{50}{t + 3}$	t + 3

Since the speeds are the same for both riders, we have the equation

$$\frac{16}{t} = \frac{50}{t + 3}.$$

Carry out. We multiply by the LCD, $t(t + 3)$.

$$t(t + 3) \cdot \frac{16}{t} = t(t + 3) \cdot \frac{50}{t + 3}$$
$$16(t + 3) = 50t$$
$$16t + 48 = 50t$$
$$48 = 34t$$
$$\frac{24}{17} = t \quad \text{Dividing by 34 and simplifying}$$

Check. If $t = \frac{24}{17}$, then $t + 3 = \frac{24}{17} + 3$, or $\frac{75}{17}$, and $\frac{16}{t} = \frac{16}{\frac{24}{17}}$, or $\frac{34}{3}$. In $\frac{24}{17}$ hr, Dexter rides $\frac{34}{3} \cdot \frac{24}{17}$, or 16 mi, and in $\frac{75}{17}$ hr Gail rides $\frac{34}{3} \cdot \frac{75}{17}$, or 50 mi. The values check.

State. Dexter rides $\frac{24}{17}$, or $1\frac{7}{17}$ hr.

18. 2 hr

19. Familiarize. Let t = the time it takes Caledonia to drive to town and organize the given information in a table.

	Distance	Speed	Time
Caledonia	15	r	t
Manley	20	r	t + 1

Translate. We can replace the r's in the table above using the formula $r = d/t$.

	Distance	Speed	Time
Caledonia	15	$\frac{15}{t}$	t
Manley	20	$\frac{20}{t + 1}$	t + 1

Since the speeds are the same, we have the equation

$$\frac{15}{t} = \frac{20}{t + 1}.$$

Carry out. We multiply by the LCD, $t(t + 1)$.

$$t(t + 1) \cdot \frac{15}{t} = t(t + 1) \cdot \frac{20}{t + 1}$$
$$15(t + 1) = 20t$$
$$15t + 15 = 20t$$
$$15 = 5t$$
$$3 = t$$

Check. If $t = 3$, then $t + 1 = 3 + 1$, or 4, and $\frac{15}{t} = \frac{15}{3}$, or 5. In 3 hr, Caledonia drives $5 \cdot 3$, or 15 mi, and in 4 hr, Manley drives $5 \cdot 4$, or 20 mi. The values check.

State. It takes Caledonia 3 hr to drive to town.

20. $1\frac{1}{3}$ hr

21. $\frac{54 \text{ days}}{6 \text{ days}} = 9$

22. $16 \frac{\text{mi}}{\text{gal}}$

23. $\frac{4.6 \text{ km}}{2 \text{ hr}} = 2.3$ km/h

24. 186,000 mi/sec

25. Familiarize. The coffee beans from 14 trees are required to produce 7.7 kilograms of coffee, and we wish to find how many trees are required to produce 320 kilograms of coffee. We can set up ratios:

$$\frac{T}{320} \quad \frac{14}{7.7}$$

Translate. Assuming the two ratios are the same, we can translate to a proportion.

$$\begin{matrix}\text{Trees} \longrightarrow \\ \text{Kilograms} \longrightarrow\end{matrix} \frac{T}{320} = \frac{14}{7.7} \begin{matrix}\longleftarrow \text{Trees} \\ \longleftarrow \text{Kilograms}\end{matrix}$$

Carry out. We solve the proportion.

We multiply by 320 to get T alone.

$$320 \cdot \frac{T}{320} = 320 \cdot \frac{14}{7.7}$$

$$T = \frac{4480}{7.7}, \text{ or } \frac{4480}{\frac{77}{10}}$$

$$T = 581\frac{9}{11}$$

Check.

$$\frac{581\frac{9}{11}}{320} = \frac{\frac{6400}{11}}{320} = \frac{6400}{11} \cdot \frac{1}{320} = \frac{320 \cdot 20}{11 \cdot 320} = \frac{20}{11} \text{ and}$$

$$\frac{14}{7.7} = \frac{14}{\frac{77}{10}} = \frac{14}{1} \cdot \frac{10}{77} = \frac{7 \cdot 2 \cdot 10}{7 \cdot 11} = \frac{20}{11}.$$

The ratios are the same.

State. 582 trees are required to produce 320 kg of coffee. (We round to the nearest whole number.)

26. 200

27. Familiarize. Wanda walked 234 kilometers in 14 days, and we wish to find how far she would walk in 42 days. We can set up ratios:

$$\frac{K}{42} \quad \frac{234}{14}$$

Translate. Assuming the rates are the same, we can translate to a proportion.

$$\begin{matrix}\text{Kilometers} \longrightarrow \\ \text{Days} \longrightarrow\end{matrix} \frac{K}{42} = \frac{234}{14} \begin{matrix}\longleftarrow \text{Kilometers} \\ \longleftarrow \text{Days}\end{matrix}$$

Carry out. We solve the proportion.

We multiply by 42 to get K alone.

$$42 \cdot \frac{K}{42} = 42 \cdot \frac{234}{14}$$

$$K = \frac{9828}{14}$$

$$K = 702$$

Check.

$$\frac{702}{42} \approx 16.7 \qquad \frac{234}{14} \approx 16.7$$

The ratios are the same.

State. Wanda would walk 702 kilometers in 42 days.

28. $21\frac{2}{3}$ cups

29. Familiarize. 10 cm^3 of human blood contains 1.2 grams of hemoglobin, and we wish to find how many grams of hemoglobin are contained in 16 cm^3 of the same blood. We can set up ratios:

$$\frac{H}{16} \quad \frac{1.2}{10}$$

Translate. Assuming the ratios are the same, we can translate to a proportion.

$$\begin{matrix}\text{Grams} \longrightarrow \\ cm^3 \longrightarrow\end{matrix} \frac{H}{16} = \frac{1.2}{10} \begin{matrix}\longleftarrow \text{Grams} \\ \longleftarrow cm^3\end{matrix}$$

Carry out. We solve the proportion.

We multiply by 16 to get H alone.

$$16 \cdot \frac{H}{16} = 16 \cdot \frac{1.2}{10}$$

$$H = \frac{19.2}{10}$$

$$H = 1.92$$

Check.

$$\frac{1.92}{16} = 0.12 \qquad \frac{1.2}{10} = 0.12$$

The ratios are the same.

State. Thus 16 cm^3 of the same blood would contain 1.92 grams of hemoglobin.

30. 216

31. We write a proportion and then solve it.

$$\frac{b}{6} = \frac{7}{4}$$

$$b = \frac{7}{4} \cdot 6$$

$$b = \frac{42}{4}, \text{ or } 10.5$$

[Note that the proportions $\frac{6}{b} = \frac{4}{7}$, $\frac{b}{7} = \frac{6}{4}$, or $\frac{7}{b} = \frac{4}{6}$ could also be used.]

32. 6.75

33. We write a proportion and then solve it.

$\frac{4}{f} = \frac{6}{4}$

$4 \cdot 4 = 6f$ Cross multiplying

$\frac{4 \cdot 4}{6} = f$ Dividing by 6

$\frac{8}{3} = f$ Simplifying

[One of the following proportions could also be used: $\frac{f}{4} = \frac{4}{6}$, $\frac{4}{f} = \frac{9}{6}$, $\frac{f}{4} = \frac{6}{9}$, $\frac{4}{9} = \frac{f}{6}$, $\frac{9}{4} = \frac{6}{f}$]

34. 7.5

35. We write a proportion and then solve it.

$\frac{8}{5} = \frac{10}{n}$

$8n = 50$ Cross multiplying

$n = \frac{50}{8}$, or 6.25

[One of the following proportions could also be used: $\frac{5}{8} = \frac{n}{10}$, $\frac{8}{10} = \frac{5}{n}$, $\frac{10}{8} = \frac{n}{5}$]

36. $\frac{35}{3}$

37. Familiarize. The ratio of trout tagged to the total number of trout in the lake, T, is $\frac{112}{T}$. Of the 82 trout caught later, there were 32 trout tagged. The ratio of trout tagged to trout caught is $\frac{32}{82}$.

Translate. Assuming the two ratios are the same, we can translate to a proportion.

Trout tagged originally ⟶ $\frac{112}{T} = \frac{32}{82}$ ⟵ Tagged trout caught later
Trout in lake ⟶ ⟵ Trout caught later

Carry out. We solve the proportion.

We multiply by the LCD, 82T.

$82T \cdot \frac{112}{T} = 82T \cdot \frac{32}{82}$

$82 \cdot 112 = T \cdot 32$

$9184 = 32T$

$\frac{9184}{32} = T$

$287 = T$

Check.

$\frac{112}{287} \approx 0.39$ $\frac{32}{82} \approx 0.39$

The ratios are the same.

State. We estimate that there are 287 trout in the lake.

38. 954

39. Familiarize. The ratio of deer tagged to the total number of deer in the forest, D, is $\frac{612}{D}$. Of the 244 deer caught later, 72 are tagged. The ratio of tagged deer to deer caught is $\frac{72}{244}$.

Translate. We translate to a proportion.

Deer originally tagged ⟶ $\frac{612}{D} = \frac{72}{244}$ ⟵ Tagged deer caught later
Deer in forest ⟶ ⟵ Deer caught later

Carry out. We solve the proportion. We multiply by the LCD, 244D.

$244D \cdot \frac{612}{D} = 244D \cdot \frac{72}{244}$

$244 \cdot 612 = D \cdot 72$

$149{,}328 = 72D$

$\frac{149{,}328}{72} = D$

$2074 = D$

Check.

$\frac{612}{2074} \approx 0.295$ $\frac{72}{244} \approx 0.295$

The ratios are the same.

State. We estimate that there are 2074 deer in the forest.

40. 42

41. Familiarize. Let D = the number of "duds" in a sample of 320 firecrackers. We set up two ratios:

$\frac{9}{144}$ $\frac{D}{320}$

Translate. We write a proportion.

"Duds" ⟶ $\frac{9}{144} = \frac{D}{320}$ ⟵ "Duds"
Firecrackers ⟶ ⟵ Firecrackers

Carry out. We solve the proportion. We multiply by 320 to get D alone.

$320 \cdot \frac{9}{144} = 320 \cdot \frac{D}{320}$

$\frac{2880}{144} = D$

$20 = D$

Check.

$\frac{9}{144} = 0.0625$ $\frac{20}{320} = 0.0625$

The ratios are the same.

State. You would expect 20 "duds" in a sample of 320 firecrackers.

42. 225

43. Familiarize. The ratio of the weight of an object on the moon to the weight of an object on the earth is 0.16 to 1.

a) We wish to find out how much a 12-ton rocket would weigh on the moon.

b) We wish to find out how much a 180-lb astronaut would weigh on the moon.

We can set up ratios.

$\frac{0.16}{1}$ $\frac{T}{12}$ $\frac{P}{180}$

Translate. Assuming the ratios are the same, we can translate to proportions.

a) Wgt. on moon $\longrightarrow$ $\frac{0.16}{1} = \frac{T}{12}$ $\longleftarrow$ Wgt. on moon
Wgt. on earth $\longrightarrow$ $\longleftarrow$ Wgt. on earth

b) Wgt. on moon $\longrightarrow$ $\frac{0.16}{1} = \frac{P}{180}$ $\longleftarrow$ Wgt. on moon
Wgt. on earth $\longrightarrow$ $\longleftarrow$ Wgt. on earth

Carry out. We solve each proportion.

a) $\frac{0.16}{1} = \frac{T}{12}$ b) $\frac{0.16}{1} = \frac{P}{180}$

$12(0.16) = T$ $180(0.16) = P$

$1.92 = T$ $28.8 = P$

Check.

$\frac{0.16}{1} = 0.16$ $\frac{1.92}{12} = 0.16$ $\frac{28.8}{180} = 0.16$

The ratios are the same.

State.

a) A 12-ton rocket would weigh 1.92 tons on the moon.

b) A 180-lb astronaut would weigh 28.8 lb on the moon.

44. a) 4.8 tons

b) 48 lb

45. Familiarize. Let x represent the numerator. Then 104 - x represents the denominator. The ratio is $\frac{x}{104 - x}$.

Translate. The ratios are equal.

$$\frac{x}{104 - x} = \frac{9}{17}$$

Carry out. We solve the proportion.

We multiply by the LCD, $17(104 - x)$.

$$17(104 - x) \cdot \frac{x}{104 - x} = 17(104 - x) \cdot \frac{9}{17}$$
$$17x = 9(104 - x)$$
$$17x = 936 - 9x$$
$$26x = 936$$
$$x = \frac{936}{26}$$
$$x = 36$$

Check. If $x = 36$, then $104 - x = 68$. The ratio is $\frac{36}{68}$. If we multiply $\frac{9}{17}$ by $\frac{4}{4}$, a form of 1, we get $\frac{36}{68}$. The ratios are equal.

State. The equal ratio is $\frac{36}{68}$.

46. 11

47. $(x + 2) - (x + 1) = x + 2 - x - 1 = 1$

48. $x^2 - 1$

49. $(4y^3 - 5y^2 + 7y - 24) - (-9y^3 + 9y^2 - 5y + 49)$
$= 4y^3 - 5y^2 + 7y - 24 + 9y^3 - 9y^2 + 5y - 49$
$= 13y^3 - 14y^2 + 12y - 73$

50. 25,704 ft^2

51.

52.

53. Familiarize. Let x represent the numerator and x + 1 represent the denominator of the original fraction. The fraction is $\frac{x}{x + 1}$. If 2 is subtracted from the numerator and the denominator, the resulting fraction is $\frac{x - 2}{x + 1 - 2}$, or $\frac{x - 2}{x - 1}$.

Translate.

The resulting fraction is $\frac{1}{2}$.

$$\frac{x - 2}{x - 1} = \frac{1}{2}$$

Carry out. We solve the equation.

$$\frac{x - 2}{x - 1} = \frac{1}{2}, \quad \text{LCD} = 2(x - 1)$$
$$2(x - 1) \cdot \frac{x - 2}{x - 1} = 2(x - 1) \cdot \frac{1}{2}$$
$$2(x - 2) = x - 1$$
$$2x - 4 = x - 1$$
$$x = 3$$

Check. If $x = 3$, then $x + 1 = 4$ and the original fraction is $\frac{3}{4}$. If 2 is subtracted from both numerator and denominator, the resulting fraction is $\frac{3 - 2}{4 - 2}$, or $\frac{1}{2}$. The value checks.

State. The original fraction was $\frac{3}{4}$.

54. Ann: 6 hr, Betty: 12 hr

55. Familiarize. We organize the information in a table. Let r = the speed of the current and t = the time it takes to travel upstream.

$d = r \cdot t$

	Distance	Speed	Time
Upstream	24	$10 - r$	t
Downstream	24	$10 + r$	$5 - t$

Translate. From the rows of the table we get two equations:

$$24 = (10 - r)t$$
$$24 = (10 + r)(5 - t)$$

We solve each equation for t and set the results equal:

Solving $24 = (10 - r)t$ for t: $t = \frac{24}{10 - r}$

Solving $24 = (10 + r)(5 - t)$ for t:

$$\frac{24}{10 + r} = 5 - t$$
$$t = 5 - \frac{24}{10 + r}$$

Then $\frac{24}{10 - r} = 5 - \frac{24}{10 + r}$.

Carry out. We first multiply on both sides of the equation by the LCD, $(10 - r)(10 + r)$:

$$(10-r)(10+r) \cdot \frac{24}{10 - r} = (10-r)(10+r)\left(5 - \frac{24}{10 + r}\right)$$
$$24(10 + r) = 5(10-r)(10+r) - 24(10-r)$$
$$240 + 24r = 500 - 5r^2 - 240 + 24r$$
$$240 + 24r = 260 - 5r^2 + 24r$$
$$5r^2 - 20 = 0$$
$$5(r^2 - 4) = 0$$
$$5(r + 2)(r - 2) = 0$$

$r + 2 = 0$ or $r - 2 = 0$

$r = -2$ or $r = 2$

Check. We only check 2 since the speed of the current cannot be negative. If $r = 2$, then the speed upstream is $10 - 2$, or 8 mph and the time is $\frac{24}{8}$, or 3 hours. If $r = 2$, then the speed downstream is $10 + 2$, or 12 mph and the time is $\frac{24}{12}$, or 2 hours. The sum of 3 hr and 2 hr is 5 hr. This checks.

State. The speed of the current is 2 mph.

56. $75 + 25$

57. Find a second proportion:

$\frac{A}{B} = \frac{C}{D}$ Given

$\frac{D}{A} \cdot \frac{A}{B} = \frac{D}{A} \cdot \frac{C}{D}$ Multiplying by $\frac{D}{A}$

$\frac{D}{B} = \frac{C}{A}$

Find a third proportion:

$\frac{A}{B} = \frac{C}{D}$ Given

$\frac{B}{C} \cdot \frac{A}{B} = \frac{B}{C} \cdot \frac{C}{D}$ Multiplying by $\frac{B}{C}$

$\frac{A}{C} = \frac{B}{D}$

Find a fourth proportion:

$\frac{A}{B} = \frac{C}{D}$ Given

$\frac{DB}{AC} \cdot \frac{A}{B} = \frac{DB}{AC} \cdot \frac{C}{D}$ Multiplying by $\frac{DB}{AC}$

$\frac{D}{C} = \frac{B}{A}$

58. $27\frac{3}{11}$ minutes after 5:00

59. Familiarize. The job takes Rosina 8 days working alone and Ng 10 days working alone. Let x represent the number of days it would take Oscar working alone. Then in 1 day Rosina does $\frac{1}{8}$ of the job, Ng does $\frac{1}{10}$ of the job, and Oscar does $\frac{1}{x}$ of the job. In 1 day they would complete $\frac{1}{8} + \frac{1}{10} + \frac{1}{x}$ of the job, and in 3 days they would complete $3\left(\frac{1}{8} + \frac{1}{10} + \frac{1}{x}\right)$, or $\frac{3}{8} + \frac{3}{10} + \frac{3}{x}$.

Translate. The amount done in 3 days is one entire job, so we have

$$\frac{3}{8} + \frac{3}{10} + \frac{3}{x} = 1.$$

Carry out. We solve the equation.

$$\frac{3}{8} + \frac{3}{10} + \frac{3}{x} = 1, \quad \text{LCD} = 40x$$
$$40x\left(\frac{3}{8} + \frac{3}{10} + \frac{3}{x}\right) = 40x \cdot 1$$
$$40x \cdot \frac{3}{8} + 40x \cdot \frac{3}{10} + 40x \cdot \frac{3}{x} = 40x$$
$$15x + 12x + 120 = 40x$$
$$120 = 13x$$
$$\frac{120}{13} = x$$

Check. If it takes Oscar $\frac{120}{13}$, or $9\frac{3}{13}$ days, to complete the job, then in one day Oscar does $\frac{1}{\frac{120}{13}}$, or $\frac{13}{120}$, of the job, and in 3 days he does $3\left(\frac{13}{120}\right)$, or $\frac{13}{40}$, of the job. The portion of the job done by Rosina, Ng, and Oscar in 3 days is $\frac{3}{8} + \frac{3}{10} + \frac{13}{40} = \frac{15}{40} + \frac{12}{40} + \frac{13}{40} = \frac{40}{40} = 1$ entire job. The answer checks.

State. It will take Oscar $9\frac{3}{13}$ days to write the program working alone.

60. Michelle: 6 hr, Sal: 3 hr, Kristen: 4 hr

61. Familiarize. We organize the information in a table. Let r = the speed on the first part of the trip and t = the time driven at that speed.

$$d = r \cdot t$$

	Distance	Speed	Time
First part	30	r	t
Second part	20	$r + 15$	$1 - t$

Translate. From the rows of the table we obtain two equations:

$$30 = rt$$

$$20 = (r + 15)(1 - t)$$

We solve each equation for t and set the results equal:

Solving $30 = rt$ for t: $t = \frac{30}{r}$

Solving $20 = (r + 15)(1 - t)$ for t:

$$\frac{20}{r + 15} = 1 - t$$

$$t = 1 - \frac{20}{r + 15}$$

Then $\frac{30}{r} = 1 - \frac{20}{r + 15}$.

Carry out. We first multiply the equation by the LCD, $r(r + 15)$:

$$r(r + 15) \cdot \frac{30}{r} = r(r + 15)\left[1 - \frac{20}{r + 15}\right]$$

$$30(r + 15) = r(r + 15) - 20r$$

$$30r + 450 = r^2 + 15r - 20r$$

$$0 = r^2 - 35r - 450$$

$$0 = (r - 45)(r + 10)$$

$$r - 45 = 0 \quad or \quad r + 10 = 0$$

$$r = 45 \quad or \quad r = -10$$

Check. Since the speed cannot be negative, we only check 45. If $r = 45$, then the time for the first part is $\frac{30}{45}$, or $\frac{2}{3}$ hr. If $r = 45$, then $r + 15 = 60$ and the time for the second part is $\frac{20}{60}$, or $\frac{1}{3}$ hr. The total time is $\frac{2}{3} + \frac{1}{3}$, or 1 hour. The value checks.

State. The speed for the first 30 miles was 45 mph.

62. 

63. Let p = the width of the pond. We have similar triangles:

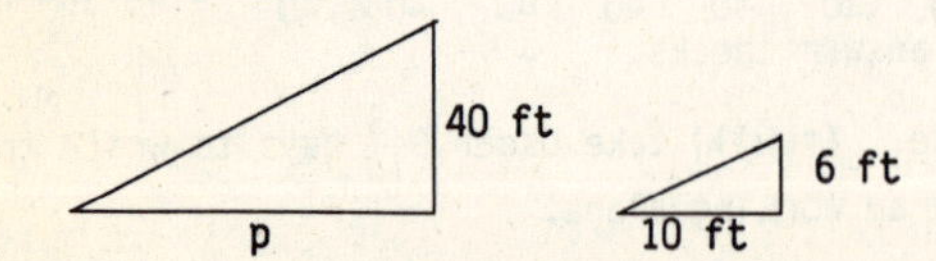

We write a proportion and solve it.

$$\frac{p}{10} = \frac{40}{6}$$

$$p = \frac{40 \cdot 10}{6} \quad \text{Multiplying by 10}$$

$$p = \frac{400}{6}$$

$$p = \frac{200}{3}, \text{ or } 66\frac{2}{3}$$

The pond is $66\frac{2}{3}$ ft wide.

Exercise Set 6.9

1. $S = 2\pi rh$

$\frac{S}{2\pi h} = r$ Multiplying by $\frac{1}{2\pi h}$

2. $t = \frac{A - P}{Pr}$

3. $A = \frac{1}{2}bh$

$2A = bh$ Multiplying by 2

$\frac{2A}{h} = b$ Multiplying by $\frac{1}{h}$

4. $g = \frac{2s}{t^2}$

5. $\frac{1}{180} = \frac{n - 2}{s}$

$\frac{s}{180} = n - 2$ Multiplying by s

$\frac{s}{180} + 2 = n$, Adding 2

or $\frac{s + 360}{180} = n$

6. $a = \frac{2S - n\ell}{n}$

7. $V = \frac{1}{3}k(B + b + 4M)$

$3V = k(B + b + 4M)$ Multiplying by 3

$3V = kB + kb + 4kM$ Removing parentheses

$3V - kB - 4kM = kb$ Adding $-kb - 4kM$

$\frac{3V - kB - 4kM}{k} = b$ Multiplying by $\frac{1}{k}$

8. $P = \frac{A}{1 + rt}$

9. $r\ell - rS = L$

$r(\ell - S) = L$ Factoring out r

$r = \frac{L}{\ell - S}$ Dividing by $\ell - S$

10. $m = \frac{T}{g - f}$

11. $A = \frac{1}{2}h(b_1 + b_2)$

$2A = h(b_1 + b_2)$ Multiplying by 2

$\frac{2A}{b_1 + b_2} = h$ Multiplying by $\frac{1}{b_1 + b_2}$

12. $h = \frac{S}{2\pi r} - r$, or $\frac{S - 2\pi r^2}{2\pi r}$

13. $ab = ac + d$

$ab - ac = d$ Subtracting ac to get all terms involving a on one side

$a(b - c) = d$ Factoring out a

$a = \frac{d}{b - c}$ Dividing by b - c

14. $n = \frac{p}{p - m}$

15. $\frac{r}{p} = q$

$r = pq$ Multiplying by p

$\frac{r}{q} = p$ Dividing by q

16. $v = \frac{m}{n}$

17. $a + b = \frac{c}{d}$

$d(a + b) = c$ Multiplying by d

$d = \frac{c}{a + b}$ Dividing by a + b

18. $n = \frac{m}{p - q}$

19. $\frac{x - y}{z} = p + q$

$x - y = z(p + q)$ Multiplying by z

$\frac{x - y}{p + q} = z$ Dividing by p + q

20. $t = \frac{M - g}{r + s}$

21. $\frac{1}{p} + \frac{1}{q} = \frac{1}{f}$

$pqf\left[\frac{1}{p} + \frac{1}{q}\right] = pqf \cdot \frac{1}{f}$ Multiplying by the LCD, pqf, to clear fractions

$pqf \cdot \frac{1}{p} + pqf \cdot \frac{1}{q} = pq$

$qf + pf = pq$

$f(q + p) = pq$

$f = \frac{pq}{q + p}$

22. $R = \frac{r_1 r_2}{r_2 + r_1}$

23. $r = \frac{v^2pL}{a}$

$ar = v^2pL$ Multiplying by a

$\frac{ar}{v^2L} = p$ Multiplying by $\frac{1}{v^2L}$

24. $\ell = \frac{P - 2w}{2}$

25. $\frac{a}{c} = n + bn$

$\frac{a}{c} = n(1 + b)$ Factoring out n

$\frac{a}{c(1 + b)} = n$ Multiplying by $\frac{1}{1 + b}$

26. $a = \frac{Q}{M(b - c)}$

27. $S = \frac{a + 2b}{3b}$

$3bS = a + 2b$ Multiplying by 3b

$3bS - 2b = a$ Adding -2b

$b(3S - 2) = a$ Factoring out b

$b = \frac{a}{3S - 2}$ Multiplying by $\frac{1}{3S - 2}$

28. $a = \frac{b}{K - C}$

29. $C = \frac{5}{9}(F - 32)$

$9C = 5(F - 32)$ Multiplying by 9

$9C = 5F - 160$ Removing parentheses

$9C + 160 = 5F$ Adding 160

$\frac{9C + 160}{5} = F$ Multiplying by $\frac{1}{5}$

30. $r^3 = \frac{3V}{4\pi}$

31. $f = \frac{gm - t}{m}$

$mf = gm - t$ Multiplying by m

$mf + t = gm$ Adding t

$\frac{mf + t}{m} = g$ Multiplying by $\frac{1}{m}$

32. $a = r\ell - Sr + S\ell$

33. $f = \frac{gm - t}{m}$

$fm = gm - t$ Multiplying by m

$fm - gm = -t$ Subtracting gm

$m(f - g) = -t$ Factoring out m

$m = \frac{-t}{f - g}$ Dividing by f - g

or $m = \frac{t}{g - f}$ Since $\frac{-a}{b} = \frac{a}{-b}$

34. $r = \frac{S\ell - a}{S - \ell}$

35.
$$C = \frac{a + Kb}{a}$$
$aC = a + Kb$ Multiplying by a
$aC - a = Kb$
$a(C - 1) = Kb$ Factoring out a
$$a = \frac{Kb}{C - 1}$$

36. $t = \frac{Ph}{Q - 1}$

37.
$$-\frac{3}{5}x = \frac{9}{20}$$
$$-\frac{5}{3}\left(-\frac{3}{5}x\right) = -\frac{5}{3}\cdot\frac{9}{20}$$
$$x = -\frac{45}{60},\text{ or } -\frac{3}{4}$$

38. y-intercept: (0,6), x-intercept: (8,0)

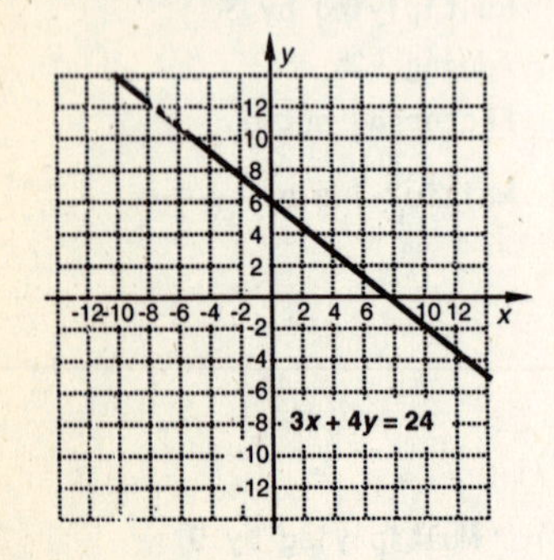

39. $x^2 - 13x - 30$

We look for two factors of -30 whose sum is -13. The numbers we need are -15 and 2.

$x^2 - 13x - 30 = (x - 15)(x + 2)$

40. $-3x^3 - 5x^2 + 5$

41.

42.

43.
$$u = -F\left(E - \frac{P}{T}\right)$$
$u = -EF + \frac{FP}{T}$ Removing parentheses
$T\cdot u = T\left(-EF + \frac{FP}{T}\right)$ Multiplying by T
$Tu = -EFT + FP$
$Tu + EFT = FP$ Adding EFT
$T(u + EF) = FP$ Factoring out T
$T = \frac{FP}{u + EF}$ Dividing by u + EF

44. $n_2 = \frac{n_1p_2R + p_1p_2n_1}{p_1p_2 - p_1R}$

45. When C = F, we have
$$C = \frac{5}{9}(C - 32)$$
$$9C = 5(C - 32)$$
$$9C = 5C - 160$$
$$4C = -160$$
$$C = -40°$$

At -40° the Fahrenheit and Celsius readings are the same.

46. N decreases when c increases; N increases when c decreases.

Exercise Set 7.1

1. We can use any two points on the line, such as (0,1) and (3,3).

$m = \frac{\text{change in } y}{\text{change in } x}$

$= \frac{3-1}{3-0} = \frac{2}{3}$

2. $\frac{3}{4}$

3. We can use any two points on the line, such as (0,-2) and (2,0).

$m = \frac{\text{change in } y}{\text{change in } x}$

$= \frac{0-(-2)}{2-0} = \frac{2}{2} = 1$

4. 2

5. We can use any two points on the line such as (0,2) and (3,3).

$m = \frac{\text{change in } y}{\text{change in } x}$

$= \frac{3-2}{3-0} = \frac{1}{3}$

6. $\frac{3}{2}$

7. We can use any two points on the line, such as (-3,0) and (-2,3).

$m = \frac{\text{change in } y}{\text{change in } x}$

$= \frac{3-0}{-2-(-3)} = \frac{3}{1} = 3$

8. $\frac{1}{3}$

9. We can use any two points on the line, such as (0,3) and (4,1).

$m = \frac{\text{change in } y}{\text{change in } x}$

$= \frac{1-3}{4-0} = \frac{-2}{4} = -\frac{1}{2}$

10. -1

11. We can use any two points on the line, such as (-1,3) and (3,-3).

$m = \frac{\text{change in } y}{\text{change in } x}$

$= \frac{-3-3}{3-(-1)} = \frac{-6}{4} = -\frac{3}{2}$

12. $-\frac{1}{3}$

13. We can use any two points on the line, such as (2,4) and (4,0).

$m = \frac{\text{change in } y}{\text{change in } x}$

$= \frac{0-4}{4-2} = \frac{-4}{2} = -2$

14. 0

15. This is a vertical line, so the slope is undefined. If we did not recognize this, we could use any two points on the line and attempt to compute the slope. We use (2,0) and (2,2).

$m = \frac{\text{change in } y}{\text{change in } x}$

$= \frac{2-0}{2-2}$

$= \frac{2}{0}$ (undefined)

16. $-\frac{1}{4}$

17. (3,2) and (-1,5)

$m = \frac{5-2}{-1-3} = \frac{3}{-4} = -\frac{3}{4}$

18. $\frac{2}{3}$

19. (-2,4) and (3,0)

$m = \frac{4-0}{-2-3} = \frac{4}{-5} = -\frac{4}{5}$

20. $-\frac{5}{6}$

21. (4,0) and (5,7)

$m = \frac{0-7}{4-5} = \frac{-7}{-1} = 7$

22. $\frac{2}{3}$

23. (0,8) and (-3,10)

$m = \frac{8-10}{0-(-3)} = \frac{8-10}{0+3} = \frac{-2}{3} = -\frac{2}{3}$

24. $-\frac{1}{2}$

25. (3,-2) and (5,-6)

$m = \frac{-2-(-6)}{3-5} = \frac{-2+6}{3-5} = \frac{4}{-2} = -2$

26. $-\frac{11}{8}$

27. $\left(-2,\frac{1}{2}\right)$ and $\left(-5,\frac{1}{2}\right)$

$m = \frac{\frac{1}{2}-\frac{1}{2}}{-2-(-5)} = \frac{\frac{1}{2}-\frac{1}{2}}{-2+5} = \frac{0}{3} = 0$

28. 0

29. (9,-4) and (9,-7)

$m = \frac{-4 - (-7)}{9 - 9} = \frac{-4 + 7}{9 - 9} = \frac{3}{0}$ (undefined)

The slope is undefined.

30. Undefined

31. (-1,5) and (4,5)

$m = \frac{5 - 5}{4 - (-1)} = \frac{0}{5} = 0$

32. Undefined

33. The line $x = -8$ is a vertical line. The slope is undefined.

34. Undefined

35. The line $y = 2$ is a horizontal line. A horizontal line has slope 0.

36. 0

37. The line $x = 9$ is a vertical line. The slope is undefined.

38. Undefined

39. The line $y = -9$ is a horizontal line. A horizontal line has slope 0.

40. 0

41. Grade = slope = $\frac{\text{vertical change}}{\text{horizontal change}} = \frac{920.58}{13,740} =$ $0.067 = 6.7\%$

42. $\frac{12}{41}$, or about 29%

43. Grade = slope = $\frac{\text{vertical change}}{\text{horizontal change}} = \frac{0.4}{5} =$ $0.08 = 8\%$

44. $\frac{28}{129}$, or about 22%

45. Grade = $\frac{\text{vertical change}}{\text{horizontal change}} = \frac{158.4}{5280} = 0.03 = 3\%$

46. 0.045

47. Grade = $\frac{\text{vertical change}}{\text{horizontal change}}$

$0.12 = \frac{v}{5}$

$0.6 = v$ Multiplying by 5

The end of the treadmill is set at 0.6 ft vertically.

48. 30 ft

49. $5x(9x - 3) = 5x \cdot 9x - 5x \cdot 3 = 45x^2 - 15x$

50. $x^3 + x^2 - 11x + 10$

51. $(x - 7)(x + 7) = x^2 - 7^2 = x^2 - 49$

52. $25x^2 + 20x + 4$

53. ◆

54. ◆

55. If the line never enters the second quadrant and is nonvertical, then two points on the line are (3,4) and (a,0), $0 \le a < 3$. The slope is of the form $m = \frac{4 - 0}{3 - a}$, or $\frac{4}{3 - a}$, $0 \le a < 3$, so $m \ge \frac{4}{3}$. Then the numbers the line could have for its slope are $\{m | m \ge \frac{4}{3}\}$.

56. $-\frac{6}{5} \le m \le 0$

57. Note that 40 min = 40 min $\cdot \frac{1 \text{ hr}}{60 \text{ min}} = \frac{40}{60}$ hr = $\frac{2}{3}$ hr.

Rate = $\frac{\text{change in number of candles produced}}{\text{corresponding change in time}}$

$= \frac{64 - 46 \text{ candles}}{\frac{2}{3} \text{ hr}}$

$= \frac{18}{\frac{2}{3}} \frac{\text{candles}}{\text{hr}}$

$= \frac{18}{1} \cdot \frac{3}{2}$ candles per hour

= 27 candles per hour

58. 3.6 bushels per hour.

59. ◆

60. ◆

Exercise Set 7.2

1. Slope $\frac{2}{5}$; y-intercept (0,1)

We plot (0,1) and from there move up 2 units and right 5 units. This locates the point (5,3). We plot (5,3) and draw a line passing through (0,1) and (5,3).

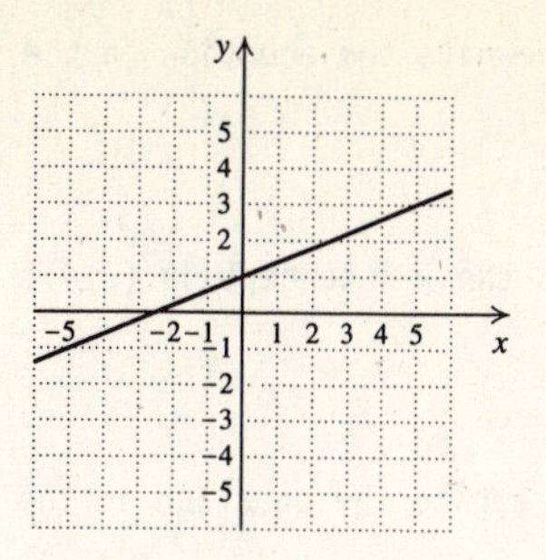

2.

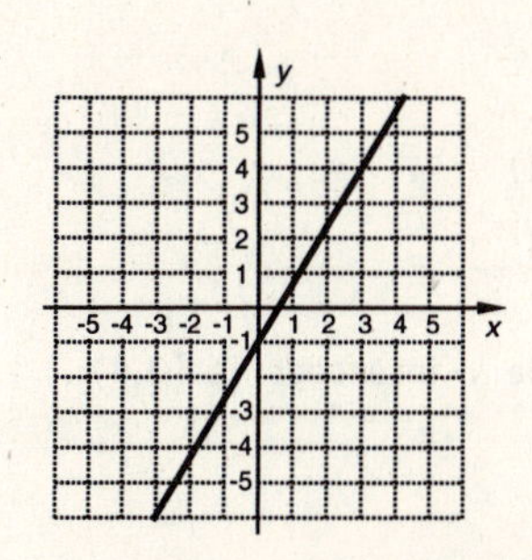

3. Slope $\frac{5}{2}$; y-intercept (0,-3)

We plot (0,-3) and from there move up 5 units and right 2 units. This locates the point (2,2). We plot (2,2) and draw a line passing through (0,-3) and (2,2).

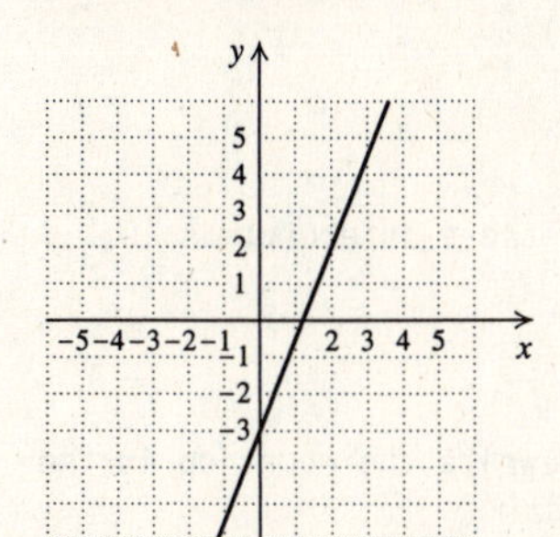

4.

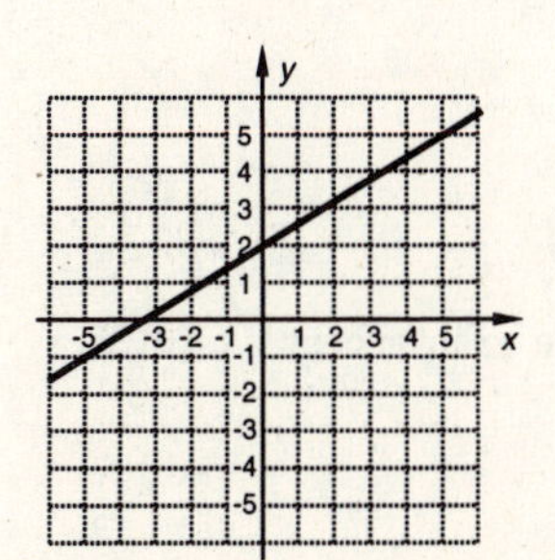

5. Slope $-\frac{3}{4}$; y-intercept (0,5)

We plot (0,5). We can think of the slope as $\frac{-3}{4}$, so from (0,5) we move down 3 units and right 4 units. This locates the point (4,2). We plot (4,2) and draw a line passing through (0,5) and (4,2).

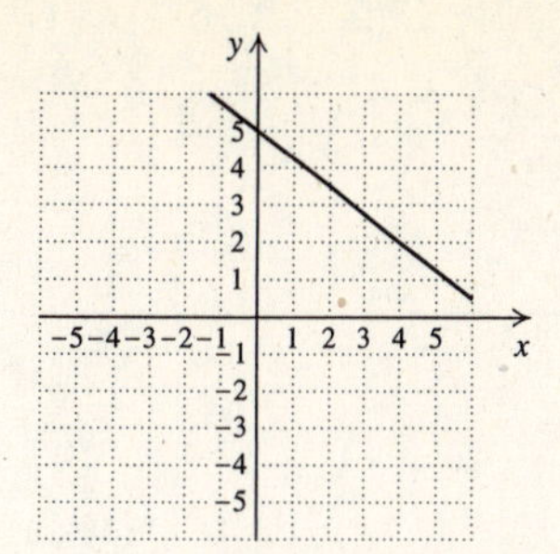

6.

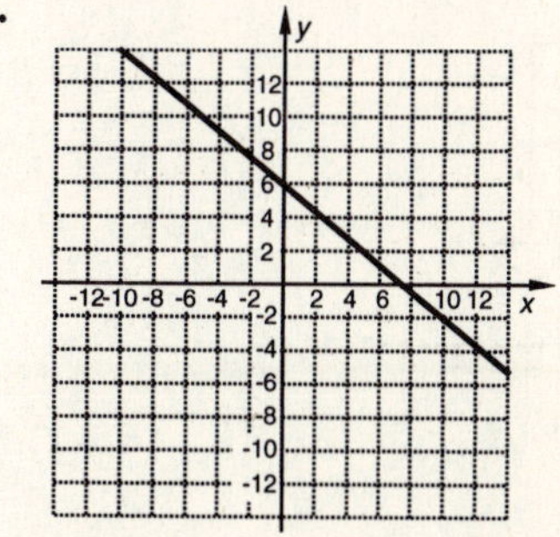

7. Slope 2; y-intercept (0,-4)

We plot (0,-4). We can think of the slope as $\frac{2}{1}$, so from (0,-4) we move up 2 units and right 1 unit. This locates the point (1,-2). We plot (1,-2) and draw a line passing through (0,-4) and (1,-2).

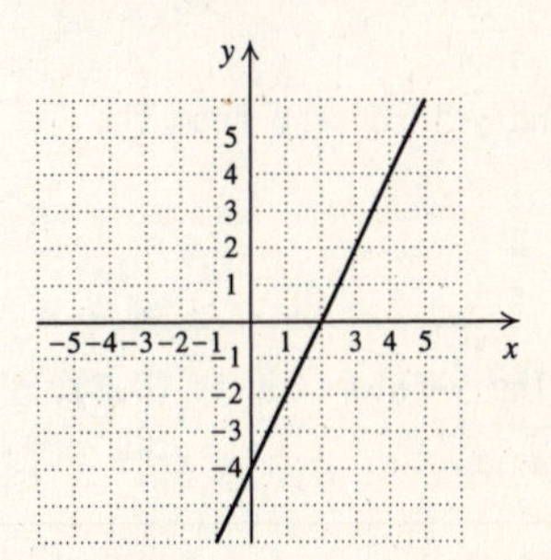

8.

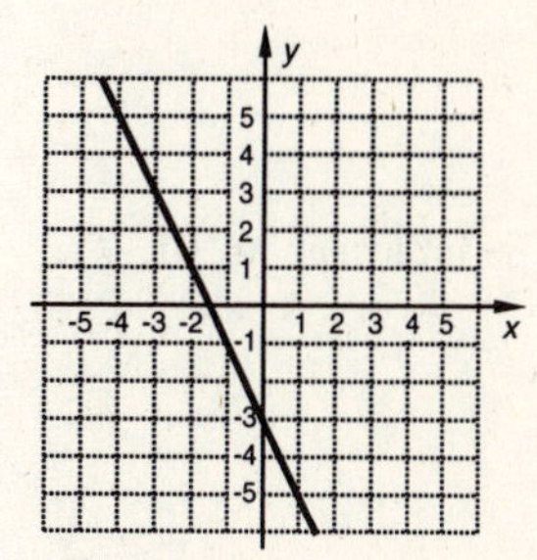

9. Slope -3; y-intercept (0,2)

We plot (0,2). We can think of the slope as $\frac{-3}{1}$, so from (0,2) we move down 3 units and right 1 unit. This locates the point (1,-1). We plot (1,-1) and draw a line passing through (0,2) and (1,-1).

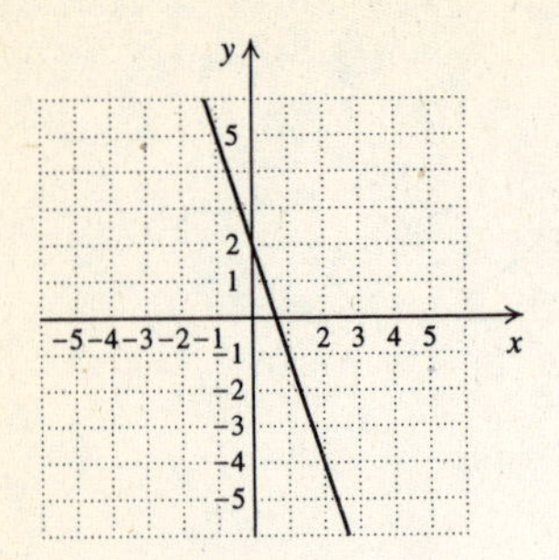

10.

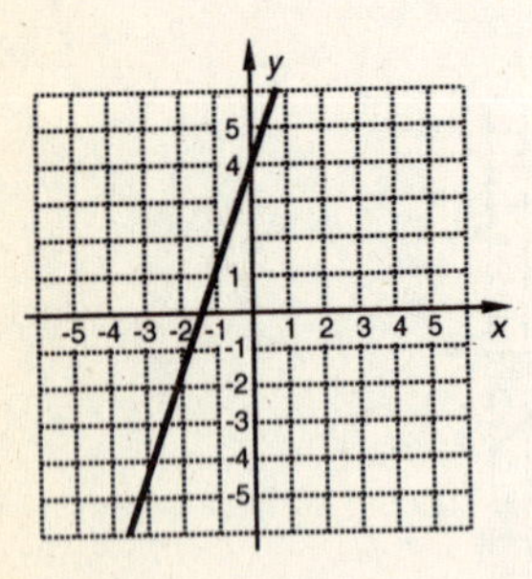

11. We read the slope and y-intercept from the equation.

$$y = \frac{3}{7}x + 6$$

The slope is $\frac{3}{7}$. The y-intercept is (0,6).

12. $-\frac{3}{8}$, (0,7)

13. We read the slope and y-intercept from the equation.

$$y = -\frac{5}{6}x + 2$$

The slope is $-\frac{5}{6}$. The y-intercept is (0,2).

14. $\frac{7}{2}$, (0,4)

15. $y = \frac{9}{4}x - 7$

$y = \frac{9}{4}x + (-7)$

The slope is $\frac{9}{4}$, and the y-intercept is (0,-7).

16. $\frac{2}{9}$, (0,-1)

17. $y = -\frac{2}{5}x$

$y = -\frac{2}{5}x + 0$

The slope is $-\frac{2}{5}$, and the y-intercept is (0,0).

18. $\frac{4}{3}$, (0,0)

19. We solve for y to rewrite the equation in the form y = mx + b.

$-2x + y = 4$

$y = 2x + 4$

The slope is 2, and the y-intercept is (0,4).

20. 5, (0,5)

21. We solve for y to rewrite the equation in the form y = mx + b.

$4x - 3y = -12$

$-3y = -4x - 12$

$y = -\frac{1}{3}(-4x - 12)$

$y = \frac{4}{3}x + 4$

The slope is $\frac{4}{3}$, and the y-intercept is (0,4).

22. $\frac{1}{2}$, $\left(0,-\frac{9}{2}\right)$

23. We solve for y to rewrite the equation in the form y = mx + b.

$x - 3y = -2$

$-3y = -x - 2$

$y = -\frac{1}{3}(-x - 2)$

$y = \frac{1}{3}x + \frac{2}{3}$

The slope is $\frac{1}{3}$, and the y-intercept is $\left(0,\frac{2}{3}\right)$.

24. -1, (0,7)

25. We solve for y to rewrite the equation in the form y = mx + b.

$-2x + 4y = 8$

$4y = 2x + 8$

$y = \frac{1}{4}(2x + 8)$

$y = \frac{1}{2}x + 2$

The slope is $\frac{1}{2}$, and the y-intercept is (0,2).

26. $\frac{5}{7}$, $\left(0,\frac{2}{7}\right)$

27. y = 5

We can rewrite this as $y = 0\cdot x + 5$. The slope is 0, and the y-intercept is (0,5).

28. 0, (0,4)

29. We use the slope-intercept equation, substituting 5 for m and 6 for b:

$y = mx + b$

$y = 5x + 6$

30. $y = -4x - 2$

31. We use the slope-intercept equation, substituting $\frac{7}{8}$ for m and -1 for b:

$$y = mx + b$$
$$y = \frac{7}{8}x - 1$$

32. $y = \frac{5}{7}x + 4$

33. We use the slope-intercept equation, substituting $-\frac{5}{3}$ for m and -8 for b:

$$y = mx + b$$
$$y = -\frac{5}{3}x - 8$$

34. $y = \frac{3}{4}x + 23$

35. We use the slope-intercept equation, substituting -2 for m and 3 for b:

$$y = mx + b$$
$$y = -2x + 3$$

36. $y = 7x - 6$

37. $y = \frac{2}{3}x + 7$: The slope is $\frac{2}{3}$, and the y-intercept is (0,7).

$y = \frac{2}{3}x - 5$: The slope is $\frac{2}{3}$, and the y-intercept is (0,-5).

Since both lines have slope $\frac{2}{3}$ but different y-intercepts, the equations represent parallel lines.

38. No

39. The equation $y = 2x - 5$ represents a line with slope 2 and y-intercept (0,-5). We rewrite the second equation in slope-intercept form.

$$4x + 2y = 9$$
$$2y = -4x + 9$$
$$y = \frac{1}{2}(-4x + 9)$$
$$y = -2x + \frac{9}{2}$$

The slope is -2, and the y-intercept is $\frac{9}{2}$.

Since the lines have different slopes, the equations do not represent parallel lines.

40. Yes

41. Rewrite each equation in slope-intercept form.

$$3x + 4y = 8$$
$$4y = -3x + 8$$
$$y = \frac{1}{4}(-3x + 8)$$
$$y = -\frac{3}{4}x + 2$$

The slope is $-\frac{3}{4}$, and the y-intercept is (0,2).

$$7 - 12y = 9x$$
$$-12y = 9x - 7$$
$$y = -\frac{1}{12}(9x - 7)$$
$$y = -\frac{3}{4}x + \frac{7}{12}$$

The slope is $-\frac{3}{4}$, and the y-intercept is $\left(0,\frac{7}{12}\right)$.

Since both lines have slope $-\frac{3}{4}$ but different y-intercepts, the equations represent parallel lines.

42. No

43. $y = \frac{3}{5}x + 2$

First we plot the y-intercept (0,2). We can start at the y-intercept and use the slope, $\frac{3}{5}$, to find another point. We move up 3 units and right 5 units to get a new point (5,5). Thinking of the slope as $\frac{-3}{-5}$ we can start at (0,2) and move down 3 units and left 5 units to get another point (-5,-1).

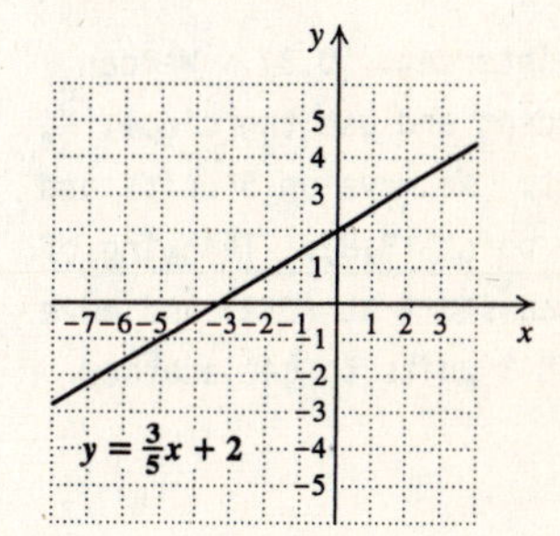

44.

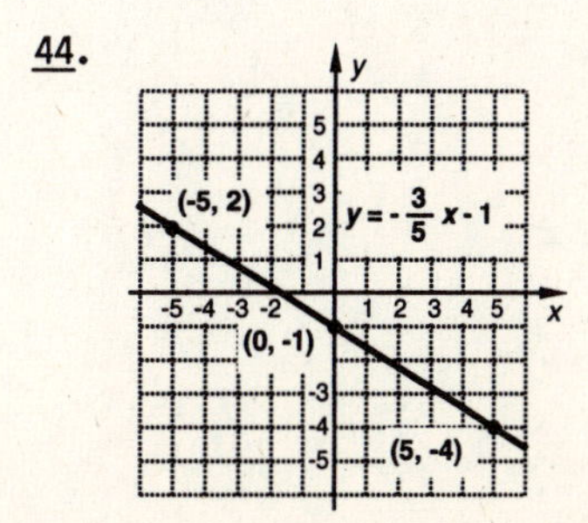

45. $y = -\frac{3}{5}x + 4$

First we plot the y-intercept (0,4). We can start at the y-intercept and, thinking of the slope as $\frac{-3}{5}$, find another point by moving down 3 units and right 5 units to the point (5,1). Thinking of the slope as $\frac{3}{-5}$ we can start at (0,4) and move up 3 units and left 5 units to get another point (-5,7).

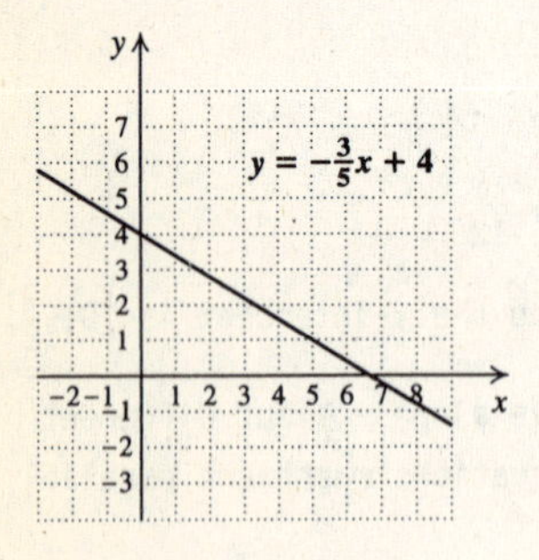

46.

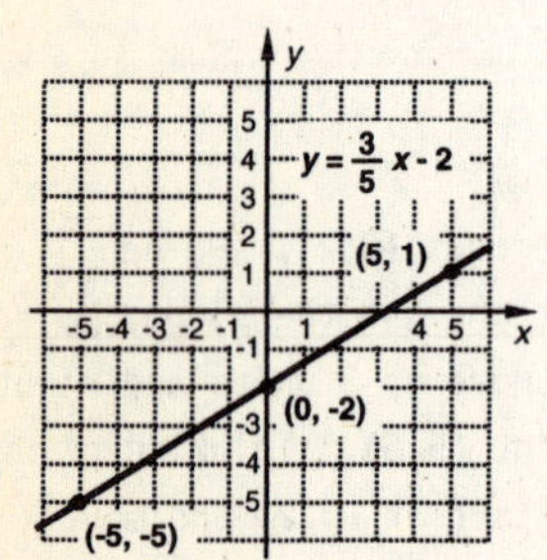

47. $y = \frac{5}{3}x + 3$

First we plot the y-intercept (0,3). We can start at the y-intercept and use the slope, $\frac{5}{3}$, to find another point. We move up 5 units and right 3 units to the point (3,8). Thinking of the slope as $\frac{-5}{-3}$ we can start at (0,3) and move down 5 units and left 3 units to get another point (-3,-2).

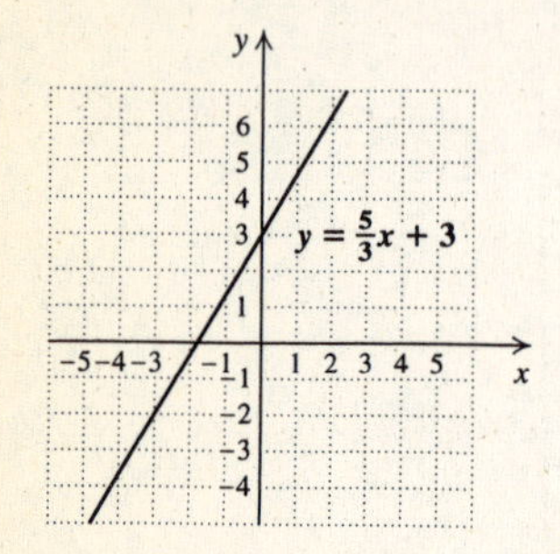

48.

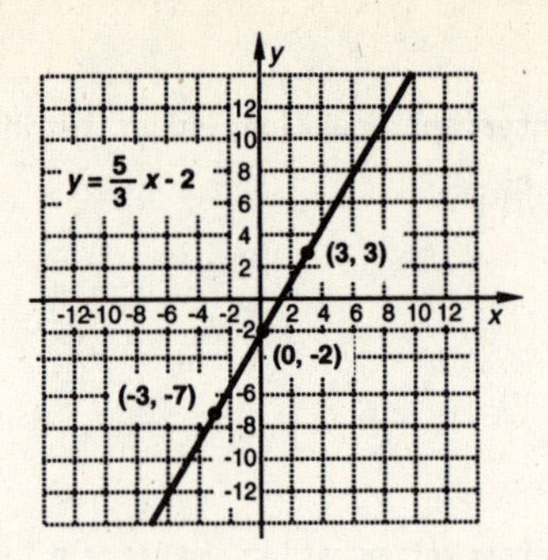

49. $y = -\frac{3}{2}x - 2$

First we plot the y-intercept (0,-2). We can start at the y-intercept and, thinking of the slope as $\frac{-3}{2}$, find another point by moving down 3 units and right 2 units to the point (2,-5). Thinking of the slope as $\frac{3}{-2}$ we can start at (0,-2) and move up 3 units and left 2 units to get another point (-2,1).

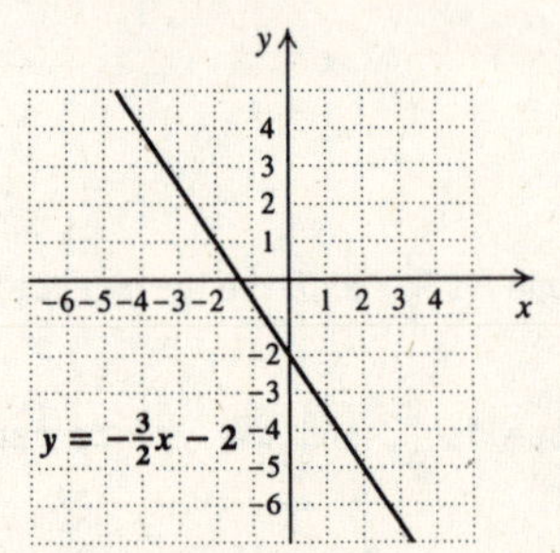

50.

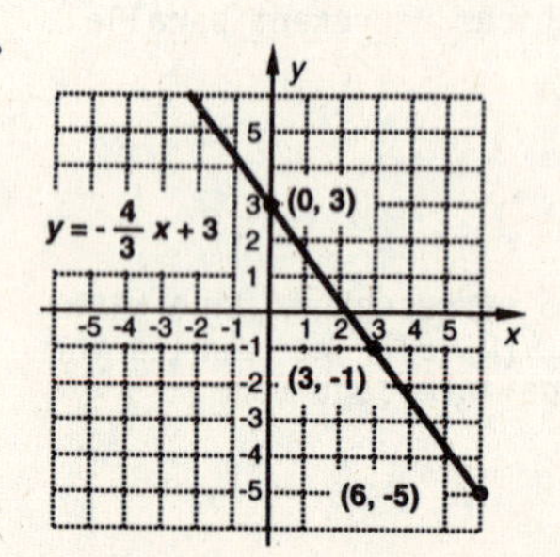

51. We first rewrite the equation in slope-intercept form.

$$2x + y = 1$$
$$y = -2x + 1$$

Now we plot the y-intercept (0,1). We can start at the y-intercept and, thinking of the slope as $\frac{-2}{1}$, find another point by moving down 2 units and right 1 unit to the point (1,-1). In a similar manner, we can move from the point (1,-1) to find a third point (2,-3).

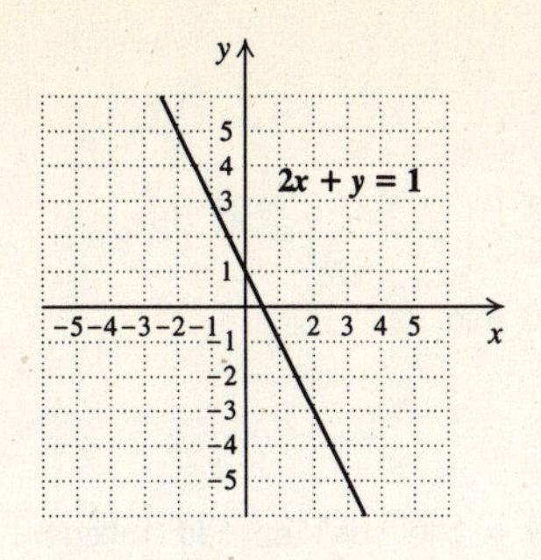

52.

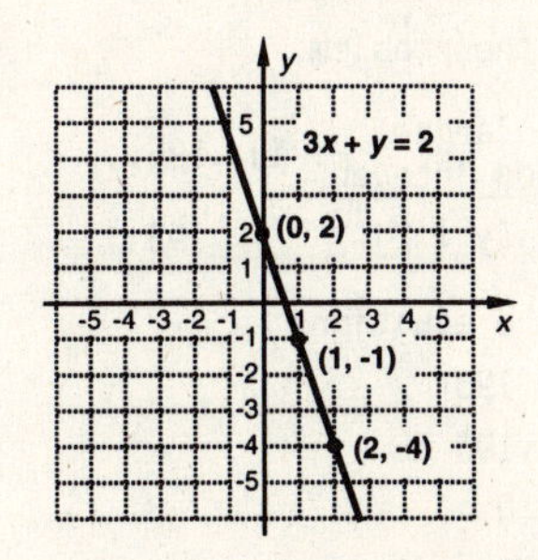

53. We first rewrite the equation in slope-intercept form.

$$3x - y = 4$$
$$-y = -3x + 4$$
$$y = 3x - 4 \quad \text{Multiplying by } -1$$

Now we plot the y-intercept (0,-4). We can start at the y-intercept and, thinking of the slope as $\frac{3}{1}$, find another point by moving up 3 units and right 1 unit to the point (1,-1). In a similar manner, we can move from the point (1,-1) to find a third point (2,2).

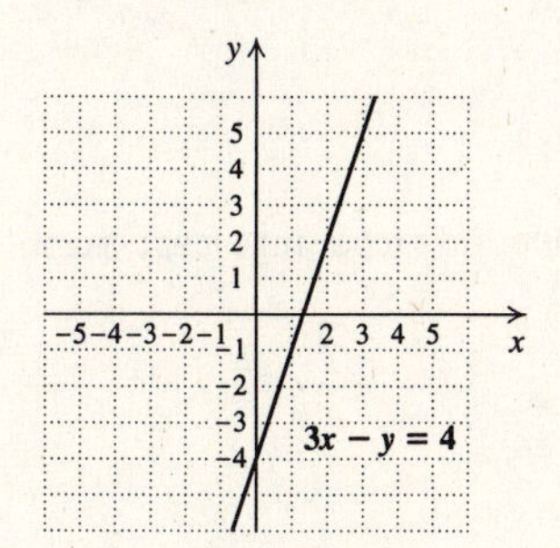

54.

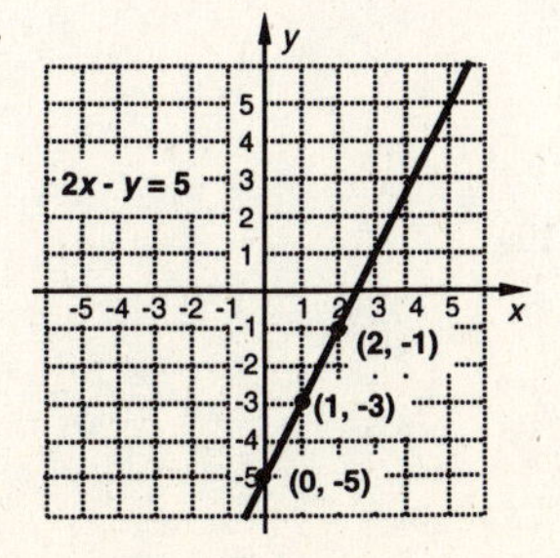

55. We first rewrite the equation in slope-intercept form.

$$2x + 3y = 9$$
$$3y = -2x + 9$$
$$y = \frac{1}{3}(-2x + 9)$$
$$y = -\frac{2}{3}x + 3$$

Now we plot the y-intercept (0,3). We can start at the y-intercept and, thinking of the slope as $\frac{-2}{3}$, find another point by moving down 2 units and right 3 units to the point (3,1). Thinking of the slope as $\frac{2}{-3}$ we can start at (0,3) and move up 2 units and left 3 units to get another point (-3,5).

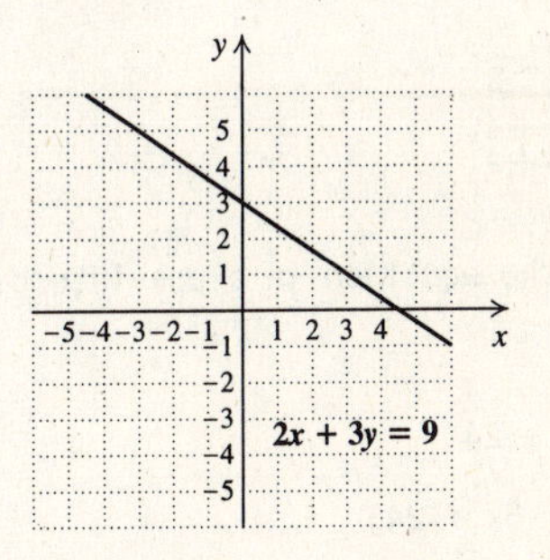

56.

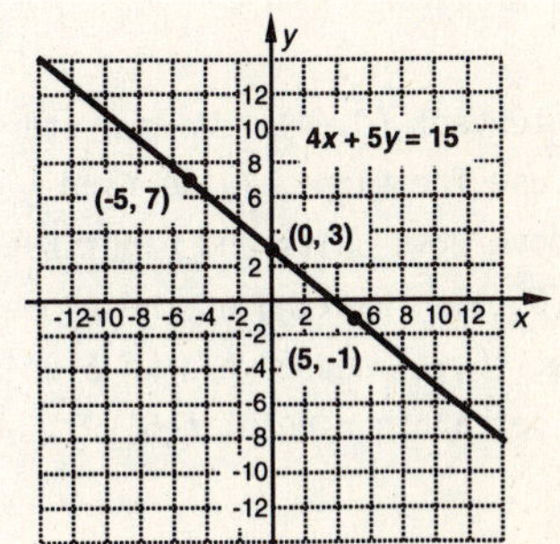

57. We first rewrite the equation in slope-intercept form.

$$x - 4y = 12$$
$$-4y = -x + 12$$
$$y = -\frac{1}{4}(-x + 12)$$
$$y = \frac{1}{4}x - 3$$

Now we plot the y-intercept (0,-3). We can start at the y-intercept and use the slope, $\frac{1}{4}$, to find another point. We move up 1 unit and right 4 units to the point (4,-2). Thinking of the slope as $\frac{-1}{-4}$ we can start at (0,-3) and move down 1 unit and left 4 units to get another point (-4,-4).

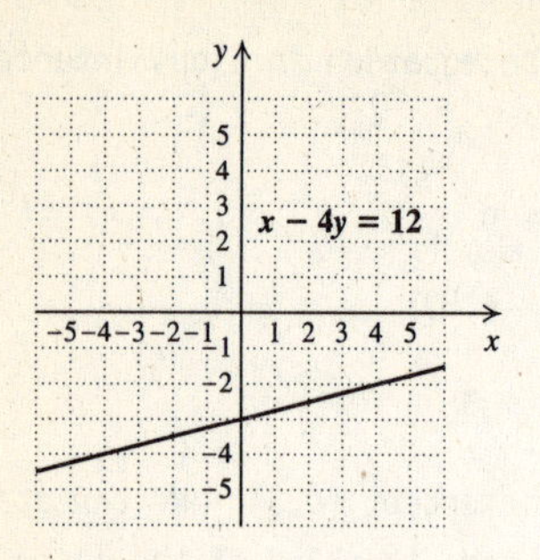

58.

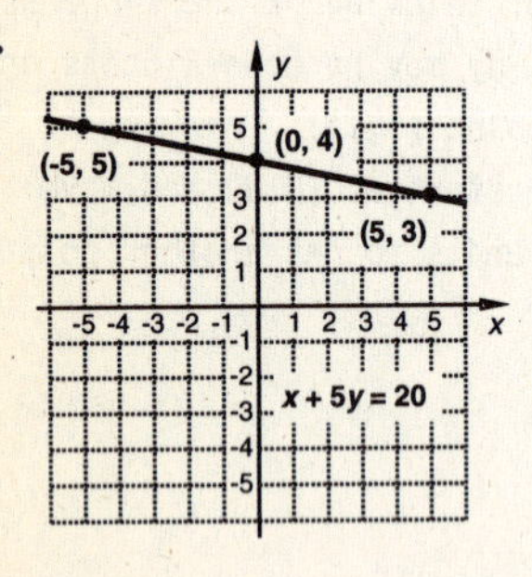

59. We first rewrite the equation in slope-intercept form.

$$5x - 6y = 24$$
$$-6y = -5x + 24$$
$$y = -\frac{1}{6}(-5x + 24)$$
$$y = \frac{5}{6}x - 4$$

Now we plot the y-intercept (0,-4). We can start the y-intercept and use the slope, $\frac{5}{6}$, to find another point. We move up 5 units and right 6 units to the point (6,1). Thinking of the slope as $\frac{-5}{-6}$ we can start at (0,-4) and move down 5 units and left 6 units to get another point (-6,-9).

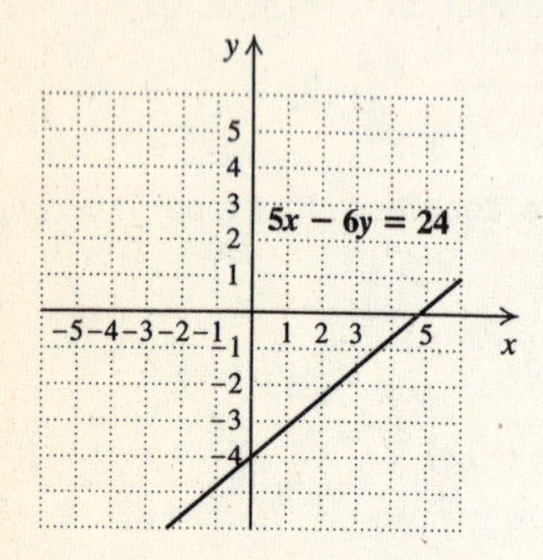

60.

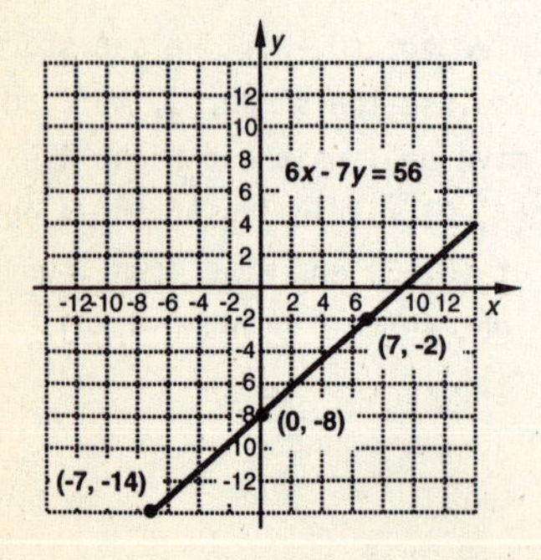

61. $2x^2 + 6x = 0$

$2x(x + 3) = 0$

$2x = 0$ or $x + 3 = 0$

$x = 0$ or $x = -3$

The solutions are 0 and -3.

62. $x(x + 7)(x - 2)$

63. Familiarize. Let y = the smaller odd integer. Then y + 2 = the larger odd integer.

Translate. We reword the problem.

Smaller odd integer	times	larger odd integer	is	195.
y	$\cdot$	$(y + 2)$	$=$	195

Carry out. We solve the equation.

$$y(y + 2) = 195$$
$$y^2 + 2y = 195$$
$$y^2 + 2y - 195 = 0$$
$$(y + 15)(y - 13) = 0$$

$y + 15 = 0$ or $y - 13 = 0$

$y = -15$ or $y = 13$

Check. If y = -15, then y + 2 = -13. These are consecutive odd integers and their product is (-15)(-13), or 195. This pair checks. If y = 13 then y + 2 = 15. These are consecutive odd integers and their product is 13·15, or 195. This pair checks also.

State. The integers are -15 and -13 or 13 and 15.

64. 11 or -1

65.

66.

67. Rewrite each equation in slope-intercept form.

$$3y = 5x - 3$$
$$y = \frac{1}{3}(5x - 3)$$
$$y = \frac{5}{3}x - 1$$

The slope is $\frac{5}{3}$.

$$3x + 5y = 10$$
$$5y = -3x + 10$$
$$y = \frac{1}{5}(-3x + 10)$$
$$y = -\frac{3}{5}x + 2$$

The slope is $-\frac{3}{5}$.

Since $\frac{5}{3}\left(-\frac{3}{5}\right) = -1$, the equations represent perpendicular lines.

68. Yes

69. Rewrite each equation in slope-intercept form.

$3x + 5y = 10$

$y = -\frac{3}{5}x + 2$ (See Exercise 67.)

The slope is $-\frac{3}{5}$.

$15x + 9y = 18$

$9y = -15x + 18$

$y = \frac{1}{9}(-15x + 18)$

$y = -\frac{5}{3}x + 2$

The slope is $-\frac{5}{3}$.

Since $-\frac{3}{5}\left(-\frac{5}{3}\right) = 1 \neq -1$, the equations do not represent perpendicular lines.

70. Yes

71. Since $x = 5$ represents a vertical line and $y = \frac{1}{2}$ represents a horizontal line, the equations represent perpendicular lines.

72. Yes

73. See the answer section in the text.

74. $N = \frac{3}{2}t + 4$

75. Rewrite each equation in slope-intercept form.

$3x - 2y = 8$

$-2y = -3x + 8$

$y = -\frac{1}{2}(-3x + 8)$

$y = \frac{3}{2}x - 4$

The slope is $\frac{3}{2}$.

$2y + 3x = -4$

$2y = -3x - 4$

$y = \frac{1}{2}(-3x - 4)$

$y = -\frac{3}{2}x - 2$

The y-intercept is (0,-2).

We write an equation of the line with slope $\frac{3}{2}$ and y-intercept (0,-2):

$y = \frac{3}{2}x - 2$

Exercise Set 7.3

1. $y - y_1 = m(x - x_1)$
We substitute 5 for m, 2 for x_1, and 5 for y_1.
$y - 5 = 5(x - 2)$

2. $y - 0 = -2(x - (-3))$

3. $y - y_1 = m(x - x_1)$
We substitute $\frac{3}{4}$ for m, 2 for x_1, and 4 for y_1.
$y - 4 = \frac{3}{4}(x - 2)$

4. $y - 2 = -1\left(x - \frac{1}{2}\right)$

5. $y - y_1 = m(x - x_1)$
We substitute 1 for m, 2 for x_1, and -6 for y_1.
$y - (-6) = 1(x - 2)$

6. $y - (-2) = 6(x - 4)$

7. $y - y_1 = m(x - x_1)$
We substitute -3 for m, -3 for x_1, and 0 for y_1.
$y - 0 = -3(x - (-3))$

8. $y - 3 = -3(x - 0)$

9. $y - y_1 = m(x - x_1)$
We substitute $\frac{2}{3}$ for m, 5 for x_1, and 6 for y_1.
$y - 6 = \frac{2}{3}(x - 5)$

10. $y - 7 = \frac{5}{6}(x - 2)$

11. First we write the equation in point-slope form.

$y - y_1 = m(x - x_1)$

$y - 7 = 2(x - 3)$ Substituting

Next we find an equivalent equation of the form $y = mx + b$.

$y - 7 = 2(x - 3)$

$y - 7 = 2x - 6$

$y = 2x + 1$

12. $y = 4x + 1$

13. We first write the equation in point-slope form and then find an equivalent equation of the form $y = mx + b$.

$y - y_1 = m(x - x_1)$

$y - 5 = -1(x - 4)$ Substituting

$y - 5 = -x + 4$

$y = -x + 9$

14. $y = x - 5$

15. We first write the equation in point-slope form and then find an equivalent equation of the form $y = mx + b$.

$$y - y_1 = m(x - x_1)$$
$$y - 3 = \frac{1}{2}(x - (-2))$$
$$y - 3 = \frac{1}{2}(x + 2)$$
$$y - 3 = \frac{1}{2}x + 1$$
$$y = \frac{1}{2}x + 4$$

16. $y = -\frac{1}{2}x - 1$

17. We first write the equation in point-slope form and then find an equivalent equation of the form $y = mx + b$.

$$y - y_1 = m(x - x_1)$$
$$y - (-5) = -\frac{1}{3}(x - (-6))$$
$$y + 5 = -\frac{1}{3}(x + 6)$$
$$y + 5 = -\frac{1}{3}x - 2$$
$$y = -\frac{1}{3}x - 7$$

18. $y = \frac{1}{5}x + 8$

19. We first write the equation in point-slope form and then find an equivalent equation of the form $y = mx + b$.

$$y - (-3) = \frac{5}{4}(x - 4)$$
$$y + 3 = \frac{5}{4}x - 5$$
$$y = \frac{5}{4}x - 8$$

20. $y = \frac{4}{3}x + 12$

21. $(-6,1)$ and $(2,3)$

First we find the slope.

$$m = \frac{1 - 3}{-6 - 2} = \frac{-2}{-8} = \frac{1}{4}$$

Then we use the point-slope equation.

$$y - y_1 = m(x - x_1)$$

We substitute $\frac{1}{4}$ for m, -6 for x_1, and 1 for y_1.

$$y - 1 = \frac{1}{4}(x - (-6))$$
$$y - 1 = \frac{1}{4}(x + 6)$$
$$y - 1 = \frac{1}{4}x + \frac{3}{2}$$
$$y = \frac{1}{4}x + \frac{5}{2}$$

We also could substitute $\frac{1}{4}$ for m, 2 for x_1, and 3 for y_1.

$$y - 3 = \frac{1}{4}(x - 2)$$
$$y - 3 = \frac{1}{4}x - \frac{1}{2}$$
$$y = \frac{1}{4}x + \frac{5}{2}$$

22. $y = x + 4$

23. $(0,4)$ and $(4,2)$

First we find the slope.

$$m = \frac{4 - 2}{0 - 4} = \frac{2}{-4} = -\frac{1}{2}$$

Then we use the point-slope equation.

$$y - y_1 = m(x - x_1)$$

We substitute $-\frac{1}{2}$ for m, 0 for x_1, and 4 for y_1.

$$y - 4 = -\frac{1}{2}(x - 0)$$
$$y - 4 = -\frac{1}{2}x$$
$$y = -\frac{1}{2}x + 4$$

24. $y = \frac{1}{2}x$

25. $(3,2)$ and $(1,5)$

First we find the slope.

$$m = \frac{2 - 5}{3 - 1} = \frac{-3}{2} = -\frac{3}{2}$$

Then we use the point-slope equation.

$$y - y_1 = m(x - x_1)$$

We substitute $-\frac{3}{2}$ for m, 3 for x_1, and 2 for y_1.

$$y - 2 = -\frac{3}{2}(x - 3)$$
$$y - 2 = -\frac{3}{2}x + \frac{9}{2}$$
$$y = -\frac{3}{2}x + \frac{13}{2}$$

26. $y = x + 5$

27. $(5,0)$ and $(0,-2)$

We first find the slope.

$$m = \frac{0 - (-2)}{5 - 0} = \frac{2}{5}$$

Then we use the point-slope equation.

$$y - y_1 = m(x - x_1)$$

We substitute $\frac{2}{5}$ for m, 5 for x_1, and 0 for y_1.

$$y - 0 = \frac{2}{5}(x - 5)$$
$$y = \frac{2}{5}x - 2$$

28. $y = \frac{5}{3}x + \frac{4}{3}$

29. (-2,-4) and (2,-1)

We first find the slope.

$m = \frac{-4 - (-1)}{-2 - 2} = \frac{-4 + 1}{-2 - 2} = \frac{-3}{-4} = \frac{3}{4}$

Then we use the point-slope equation.

$y - y_1 = m(x - x_1)$

We substitute $\frac{3}{4}$ for m, -2 for x_1, and -4 for y_1.

$y - (-4) = \frac{3}{4}(x - (-2))$

$y + 4 = \frac{3}{4}(x + 2)$

$y + 4 = \frac{3}{4}x + \frac{3}{2}$

$y = \frac{3}{4}x - \frac{5}{2}$

30. $y = -4x - 7$

31. $y - 5 = \frac{1}{2}(x - 3)$ Point-slope form

The line has slope $\frac{1}{2}$ and passes through (3,5). We plot (3,5) and then find a second point by moving up 1 unit and right 2 units to (5,6). We draw the line through these points.

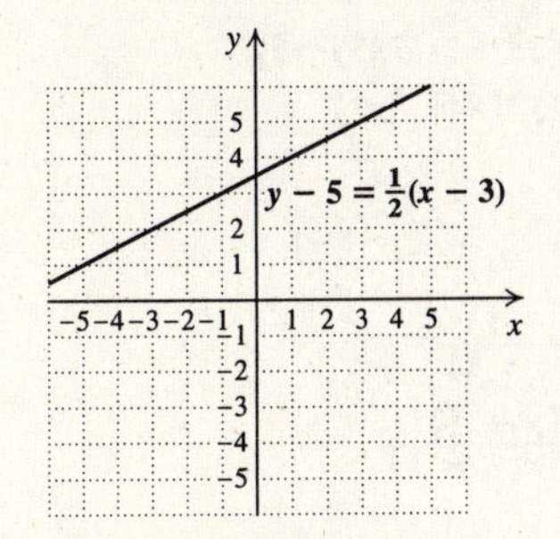

32.

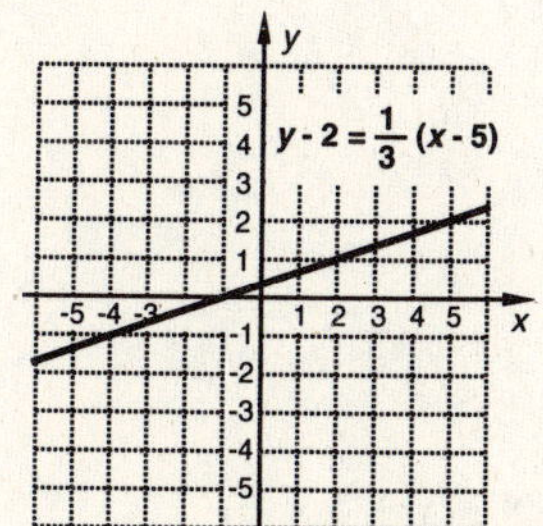

33. $y - 3 = -\frac{1}{2}(x - 5)$ Point-slope form

The line has slope $-\frac{1}{2}$, or $\frac{1}{-2}$ and passes through (5,3). We plot (5,3) and then find a second point by moving up 1 unit and left 2 units to (3,4). We draw the line through these points.

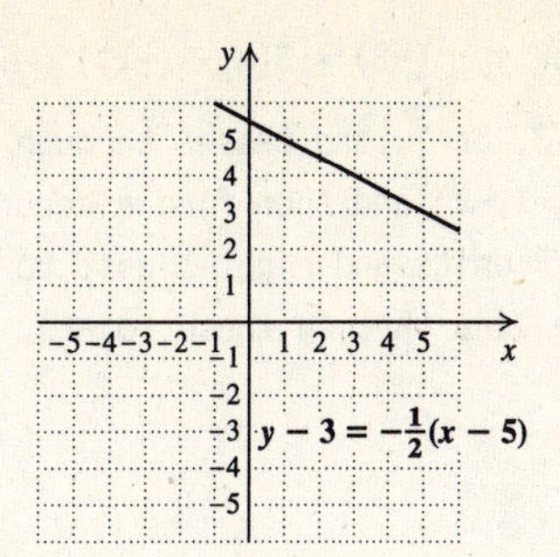

34.

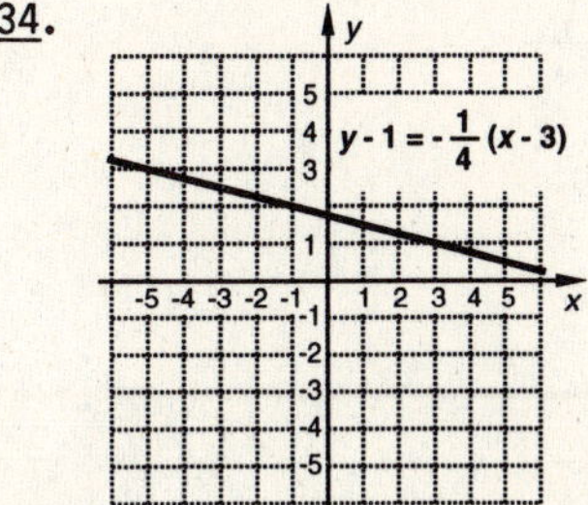

35. $y + 5 = \frac{1}{2}(x - 3)$, or $y - (-5) = \frac{1}{2}(x - 3)$

The line has slope $\frac{1}{2}$ and passes through (3,-5). We plot (3,-5) and then find a second point by moving up 1 unit and right 2 units to (5,-4). We draw the line through these points.

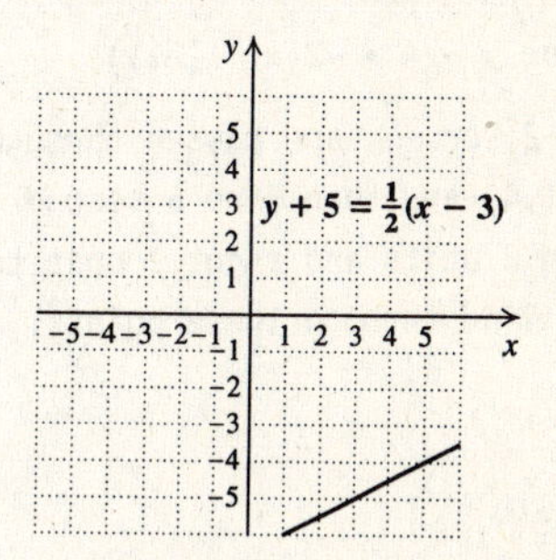

36.

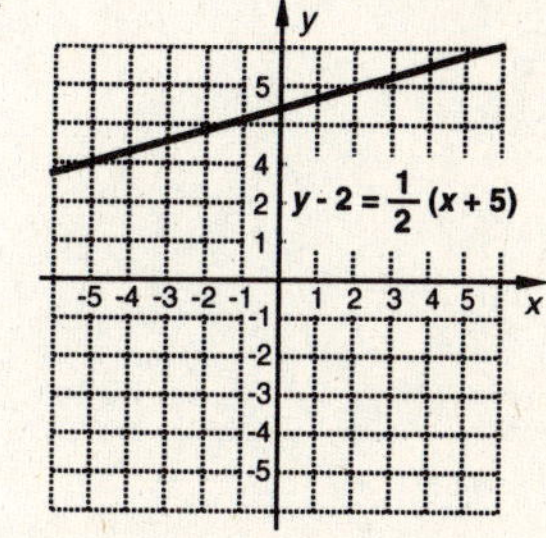

37. $y + 2 = 3(x + 1)$, or $y - (-2) = 3(x - (-1))$

The line has slope 3, or $\frac{3}{1}$, and passes through $(-1,-2)$. We plot $(-1,-2)$ and then find a second point by moving up 3 units and right 1 unit to $(0,1)$. We draw the line through these points.

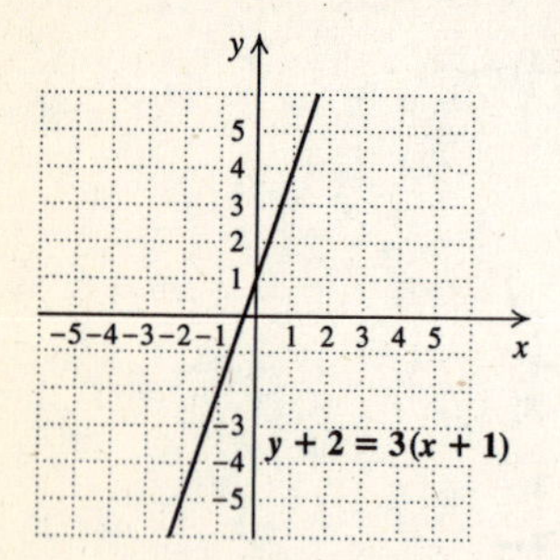

38.

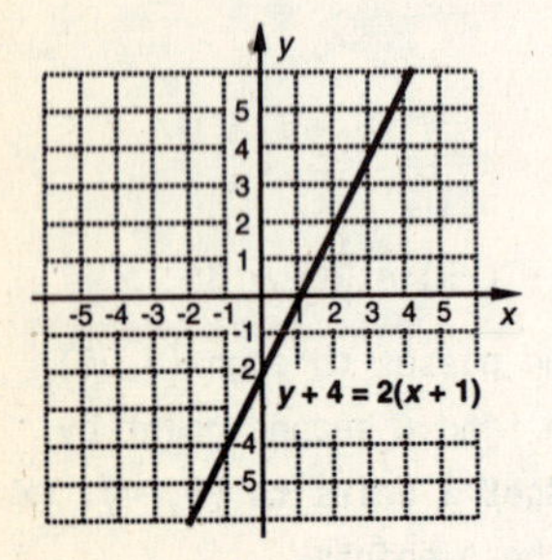

39. $y - 4 = -2(x + 1)$, or $y - 4 = -2(x - (-1))$

The line has slope -2, or $\frac{-2}{1}$, and passes through $(-1,4)$. We plot $(-1,4)$ and then find a second point by moving down 2 units and right 1 unit to $(0,2)$. We draw the line through these points.

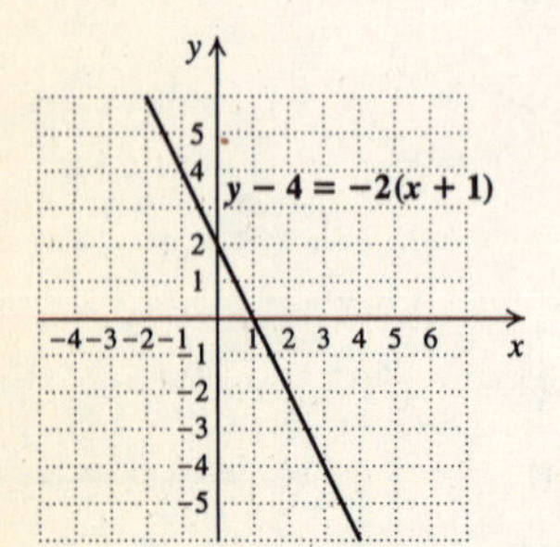

40.

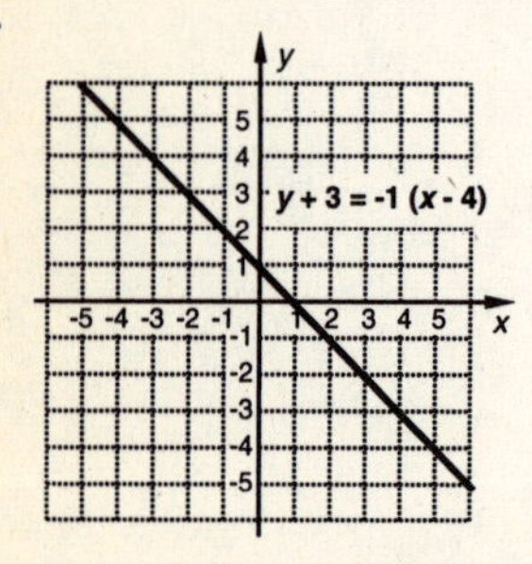

41. $y + 3 = -(x + 2)$, or $y - (-3) = -1(x - (-2))$

The line has slope -1, or $\frac{-1}{1}$, and passes through $(-2,-3)$. We plot $(-2,-3)$ and then find a second point by moving down 1 unit and right 1 unit to $(-1,-4)$. We draw the line through these points.

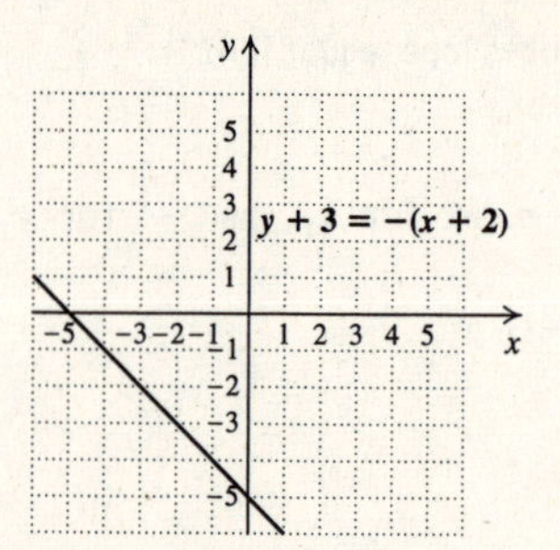

42.

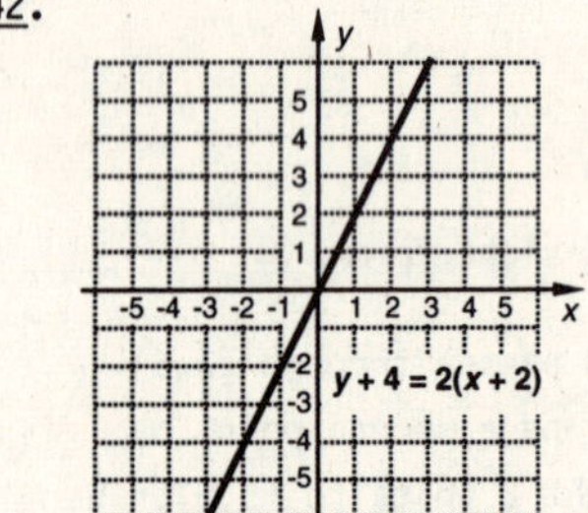

43. $7x^3y^2 + 35x^2y^6 = 7x^2y^2 \cdot x + 7x^2y^2 \cdot 5y^4$
$= 7x^2y^2(x + 5y^4)$

44. $(3x - 1)(5x^2 + 1)$

45. $\frac{5x^2 + 5x}{10x^3 - 10x^2} = \frac{5x(x + 1)}{10x^2(x - 1)}$
$= \frac{\cancel{5x}(x + 1)}{\cancel{5} \cdot 2 \cdot \cancel{x} \cdot x(x - 1)}$
$= \frac{x + 1}{2x(x - 1)}$

46. $\frac{x - 5}{x + 2}$

47.

48.

49. First find the slope of $3x - y + 4 = 0$.

$3x - y + 4 = 0$
$3x + 4 = y$

The slope is 3.

Then find an equation of the line containing $(2,-3)$ and having slope 3.

$y - y_1 = m(x - x_1)$

We substitute 3 for m, 2 for x_1, and -3 for y_1.

$$y - (-3) = 3(x - 2)$$
$$y + 3 = 3x - 6$$
$$y = 3x - 9$$

50. $y = \frac{1}{5}x - 2$

51. First we find the slope of the given line:

$$4x - 8y = 12$$
$$-8y = -4x + 12$$
$$y = \frac{1}{2}x - \frac{3}{2}$$

The slope is $\frac{1}{2}$.

Then we use the point-slope equation to find the equation of a line with slope $\frac{1}{2}$ containing the point (-2,0):

$$y - y_1 = m(x - x_1)$$
$$y - 0 = \frac{1}{2}(x - (-2))$$
$$y = \frac{1}{2}(x + 2)$$
$$y = \frac{1}{2}x + 1$$

52. 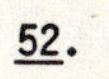

Exercise Set 7.4

1. We use alphabetical order of variables. We replace x by -3 and y by -5.

$$\begin{array}{c|l} -x - 3y < 18 & \\ \hline -(-3) - 3(-5) \;?\; 18 & \\ 3 + 15 & \\ 18 & 18 \quad \text{FALSE} \end{array}$$

Since 18 < 18 is false, (-3,-5) is not a solution.

2. Yes

3. We use alphabetical order of variables. We substitute $\frac{1}{2}$ for x and $-\frac{1}{4}$ for y.

$$\begin{array}{c|l} 7y - 9x > -3 & \\ \hline 7\left(-\frac{1}{4}\right) - 9 \cdot \frac{1}{2} \;?\; -3 & \\ -\frac{7}{4} - \frac{9}{2} & \\ -\frac{7}{4} - \frac{18}{4} & \\ -\frac{25}{4} & \\ -6\frac{1}{4} & -3 \quad \text{FALSE} \end{array}$$

Since $-6\frac{1}{4} > -3$ is false, $\left(\frac{1}{2}, -\frac{1}{4}\right)$ is not a solution.

4. Yes

5. Graph $y > -3x$.

First graph the line $y = -3x$. Two points on the line are (0,0) and (-1,3). We draw a dashed line since the inequality symbol is >. Then we test the point (2,4), which is not a point on the line.

$$\begin{array}{c|l} y > -3x & \\ \hline 4 \;?\; -3 \cdot 2 & \\ 4 & -6 \quad \text{TRUE} \end{array}$$

Since (2,4) is a solution of the inequality, we shade the region above the boundary line.

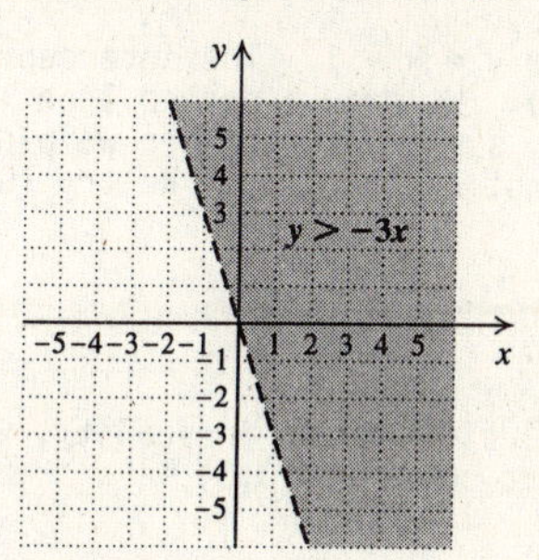

6.

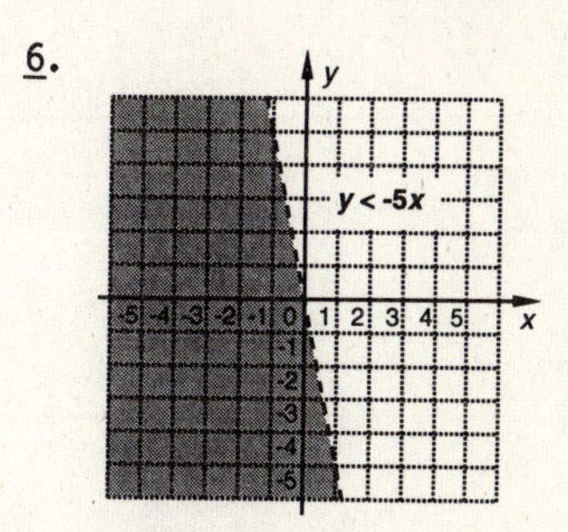

7. Graph $y \leqslant x - 3$.

First graph the line $y = x - 3$. The intercepts are (0,-3) and (3,0). We draw a solid line since the inequality symbol is ⩽. Then we pick a test point that is not on the line. We try (0,0).

$$\begin{array}{c|l} y \leqslant x - 3 & \\ \hline 0 \;?\; 0 - 3 & \\ 0 & -3 \quad \text{FALSE} \end{array}$$

We see that (0,0) is not a solution of the inequality, so we shade the region that does not contain (0,0).

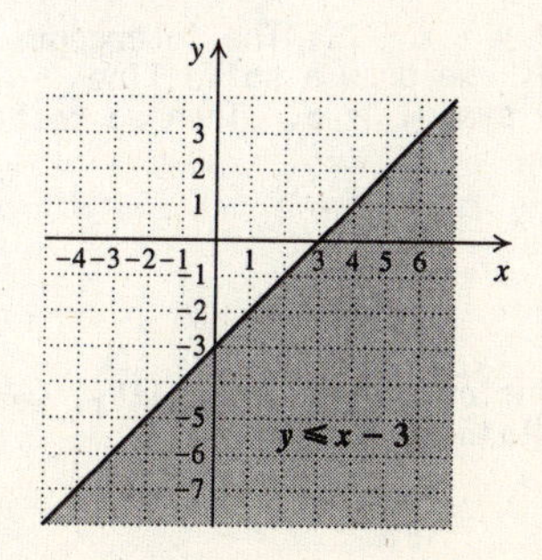

8.

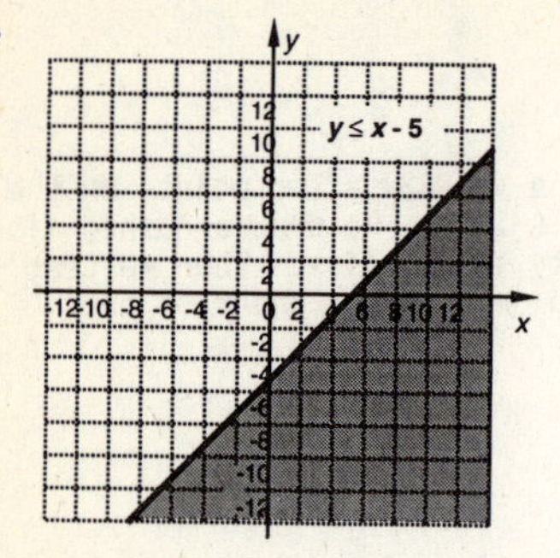

9. Graph $y < x + 1$.

First graph the line $y = x + 1$. The intercepts are (0,1) and (-1,0). We draw a dashed line since the inequality symbol is <. Then we pick a test point that is not on the line. We try (0,0).

$$\begin{array}{l|l} \multicolumn{2}{l}{y < x + 1} \\ \hline 0 \; ? & 0 + 1 \\ 0 & 1 \qquad \text{TRUE} \end{array}$$

Since (0,0) is a solution of the inequality, we shade the region that contains (0,0).

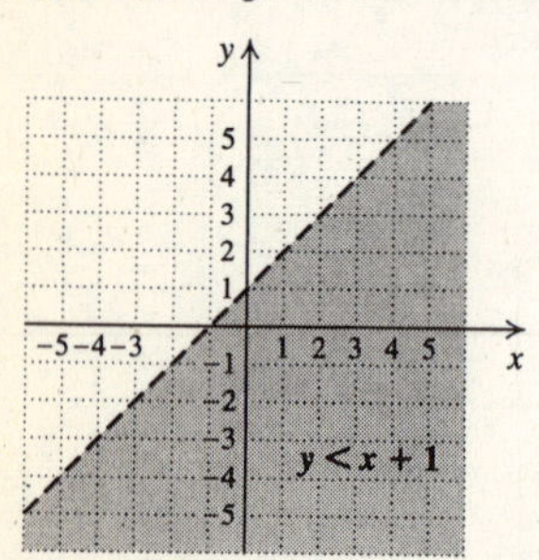

10.

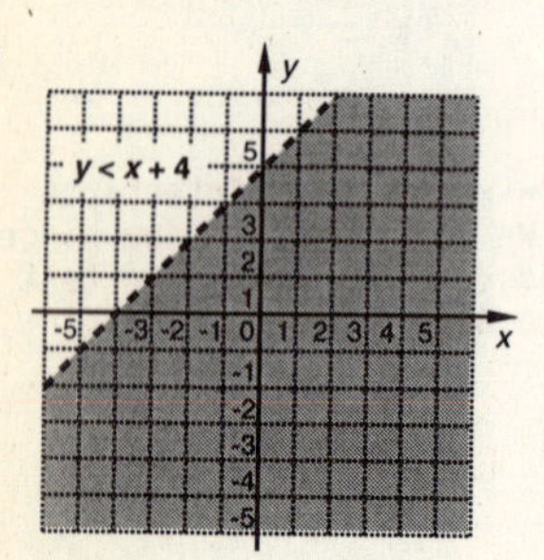

11. Graph $y \geqslant x - 2$.

First graph the line $y = x - 2$. The intercepts are (0,-2) and (2,0). We draw a solid line since the inequality symbol is ⩾. Then we test the point (0,0).

$$\begin{array}{l|l} \multicolumn{2}{l}{y \geqslant x - 2} \\ \hline 0 \; ? & 0 - 2 \\ 0 & -2 \qquad \text{TRUE} \end{array}$$

Since (0,0) is a solution of the inequality, we shade the region containing (0,0).

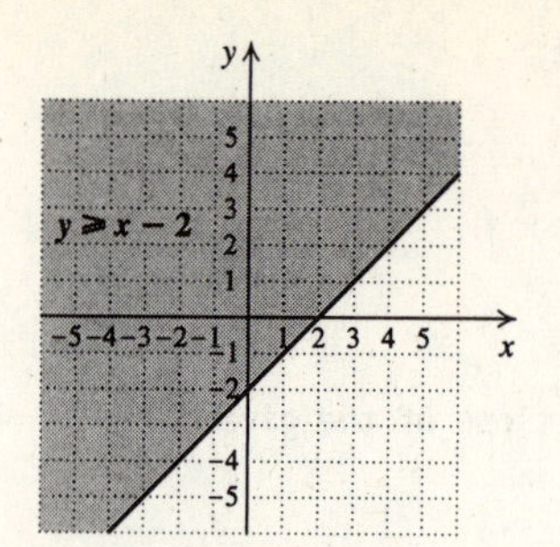

12.

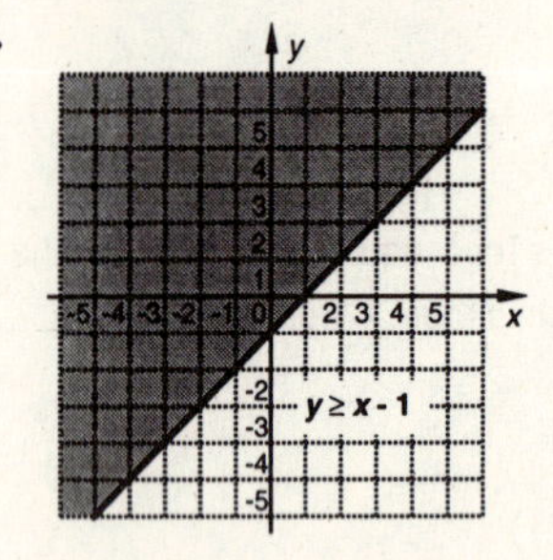

13. Graph $y \leqslant 2x - 1$.

First graph the line $y = 2x - 1$. The intercepts are (0,-1) and $\left(\frac{1}{2},0\right)$. We draw a solid line since the inequality symbol is ⩽. Then we test the point (0,0).

$$\begin{array}{l|l} \multicolumn{2}{l}{y \leqslant 2x - 1} \\ \hline 0 \; ? & 2 \cdot 0 - 1 \\ 0 & -1 \qquad \text{FALSE} \end{array}$$

Since (0,0) is not a solution of the inequality, we shade the region that does not contain (0,0).

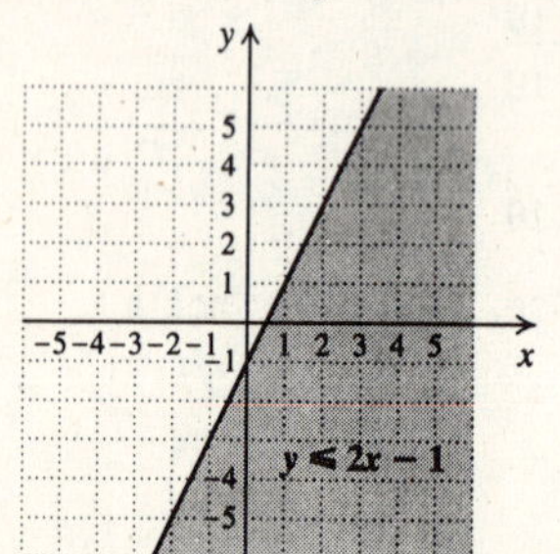

14.

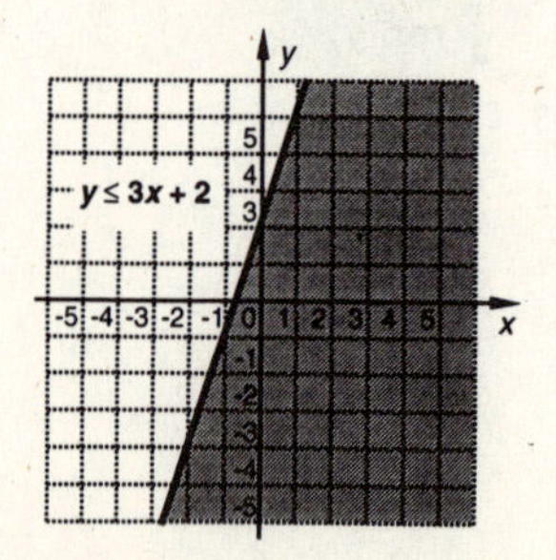

15. Graph $x + y \leqslant 3$.

First graph the line $x + y = 3$. The intercepts are (0,3) and (3,0). We draw a solid line since the inequality symbol is $\leqslant$. Then we test the point (0,0).

$$\begin{array}{c|l} x + y \leqslant 3 & \\ \hline 0 + 0 \;?\; 3 & \\ 0 & 3 \quad \text{TRUE} \end{array}$$

Since (0,0) is a solution of the inequality, we shade the region that contains (0,0).

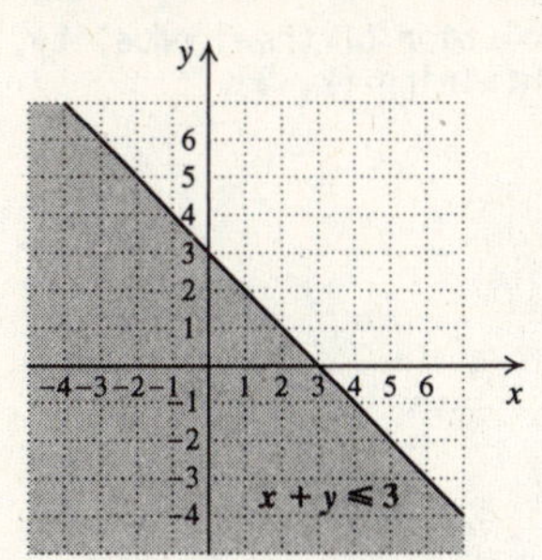

16.

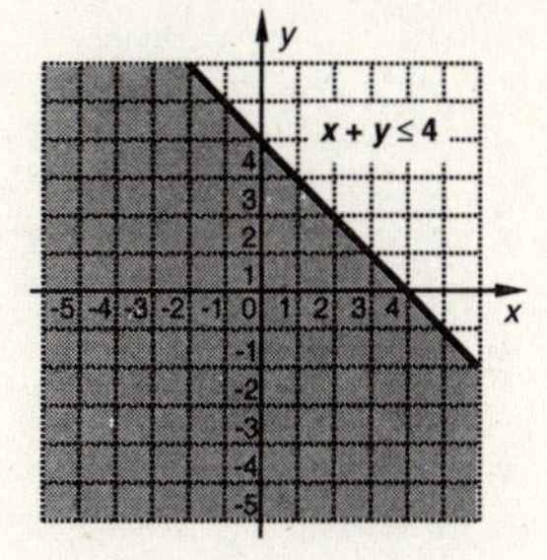

17. Graph $x - y > 7$.

First graph the line $x - y = 7$. The intercepts are (0,-7) and (7,0). We draw a dashed line since the inequality symbol is $>$. Then we test the point (0,0).

$$\begin{array}{c|l} x - y > 7 & \\ \hline 0 - 0 \;?\; 7 & \\ 0 & 7 \quad \text{FALSE} \end{array}$$

Since (0,0) is not a solution of the inequality, we shade the region that does not contain (0,0).

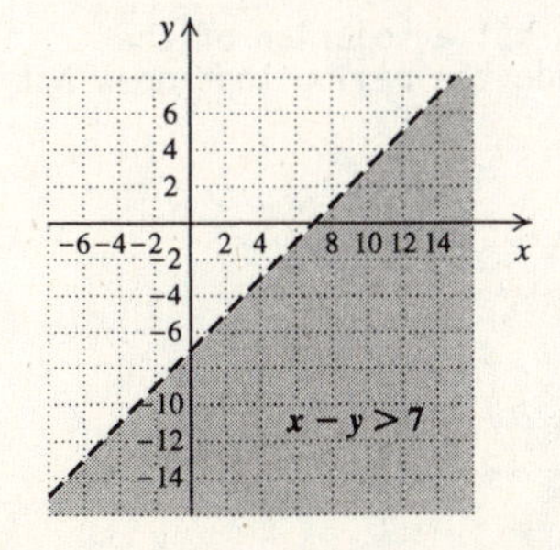

18.

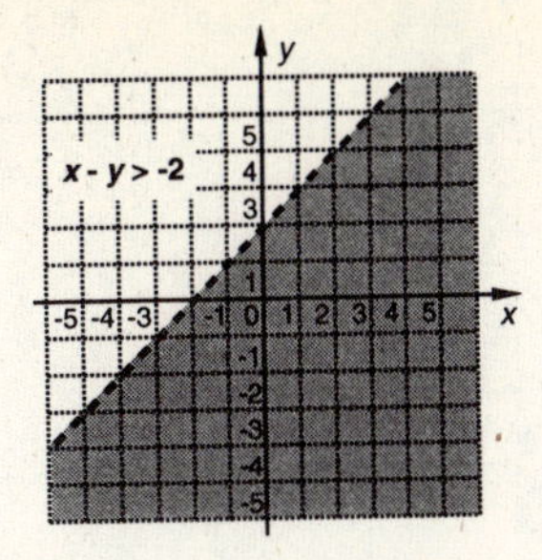

19. Graph $x - 3y < 6$.

First graph the line $x - 3y = 6$. The intercepts are (0,-2) and (6,0). We draw a dashed line since the inequality symbol is $<$. Then we test the point (0,0).

$$\begin{array}{c|l} x - 3y < 6 & \\ \hline 0 - 3\cdot 0 \;?\; 6 & \\ 0 & 6 \quad \text{TRUE} \end{array}$$

Since (0,0) is a solution of the inequality, we shade the region containing (0,0).

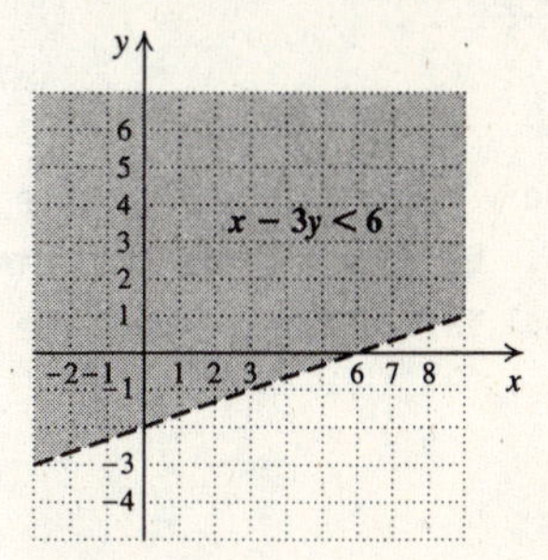

20.

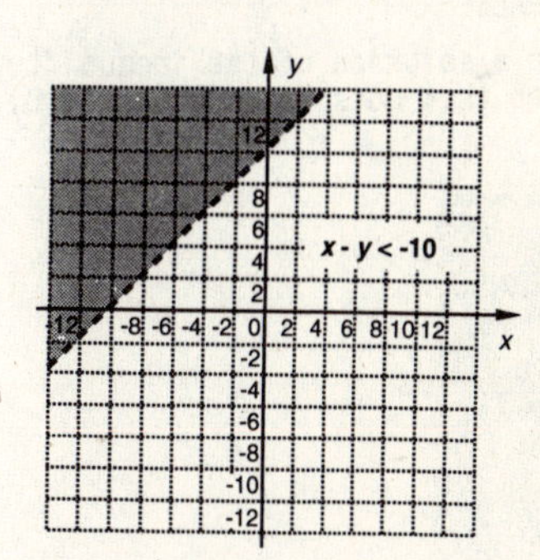

21. Graph $2x + 3y \leqslant 12$.

First graph the line $2x + 3y = 12$. The intercepts are (0,4) and (6,0). We draw a solid line since the inequality symbol is $\leqslant$. Then we test the point (0,0).

$$\begin{array}{c|l} 2x + 3y \leqslant 12 & \\ \hline 2\cdot 0 + 3\cdot 0 \;?\; 12 & \\ 0 & 12 \quad \text{TRUE} \end{array}$$

Since (0,0) is a solution of the inequality, we shade the region containing (0,0).

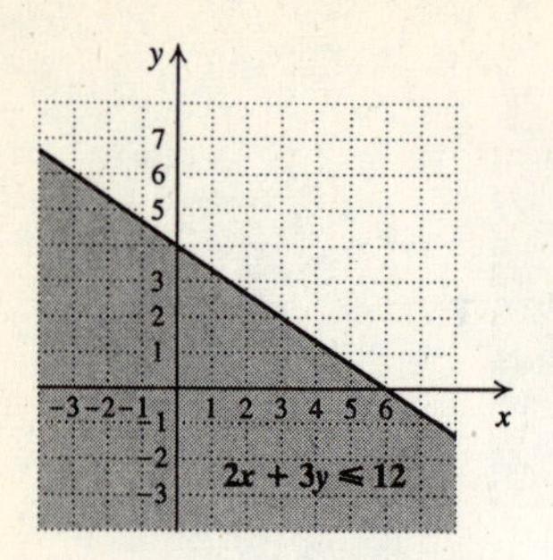

22.

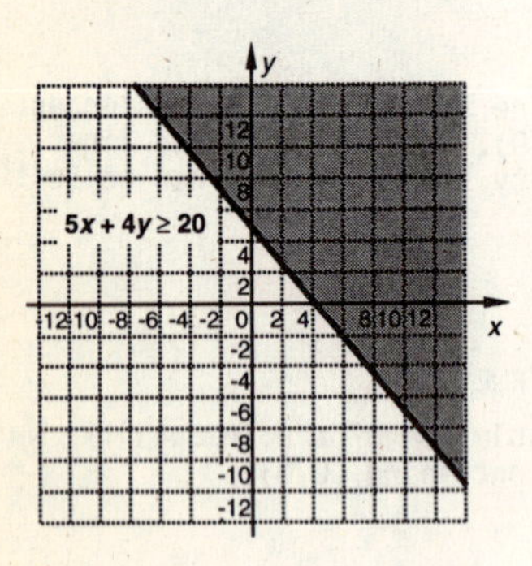

23. Graph $y \geq 1 - 2x$.

First graph the line $y = 1 - 2x$. The intercepts are (0,1) and $\left(\frac{1}{2},0\right)$. We draw a solid line since the inequality symbol is $\geq$. Then we test the point (0,0).

$$\begin{array}{c|l} y \geq 1 - 2x & \\ \hline 0 \; ? \; 1 - 2\cdot 0 & \\ 0 & 1 \quad \text{FALSE} \end{array}$$

Since (0,0) is not a solution of the inequality, we shade the region that does not contain (0,0).

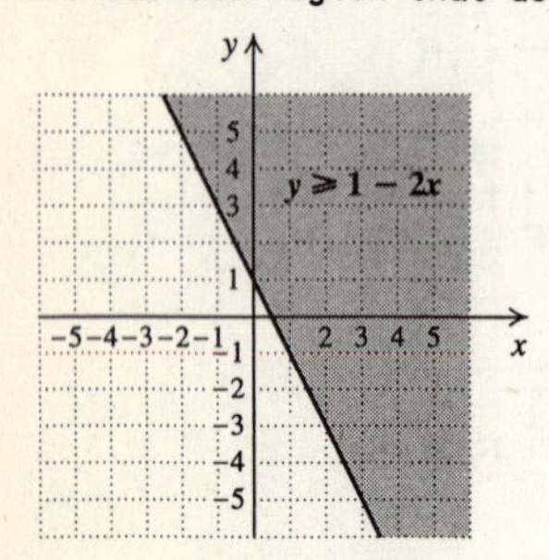

24.

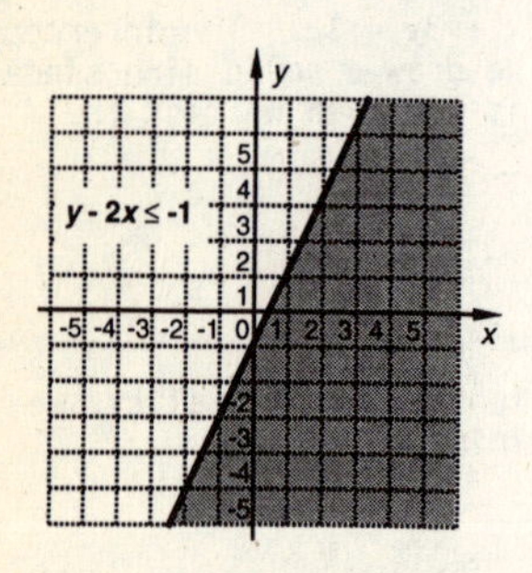

25. Graph $y + 4x > 0$.

First graph the line $y + 4x = 0$, or $y = -4x$. Two points on the line are (0,0) and (1,-4). We draw a dashed line, since the inequality symbol is >. Then we test the point (2,-3), which is not a point on the line.

$$\begin{array}{c|l} y + 4x > 0 & \\ \hline -3 + 4\cdot 2 \; ? \; 0 & \\ -3 + 8 & \\ 5 & 0 \quad \text{TRUE} \end{array}$$

Since (2,-3) is a solution of the inequality, we shade the region containing (2,-3).

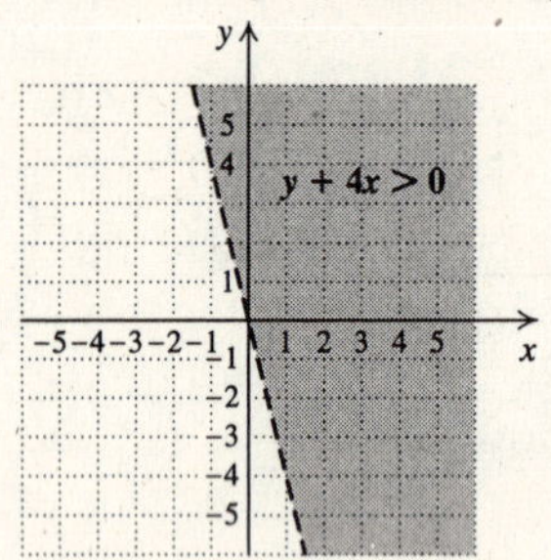

26.

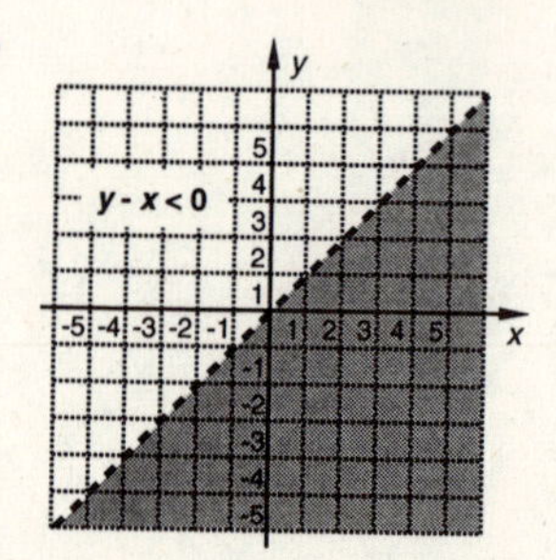

27. Graph $x > 2y$.

First graph the line $x = 2y$, or $y = \frac{1}{2}x$. Two points on the line are (0,0) and (4,2). We draw a dashed line since the inequality symbol is >. Then we pick a test point that is not on the line. We try (-2,1).

$$\begin{array}{c|l} x > 2y & \\ \hline -2 \; ? \; 2\cdot 1 & \\ -2 & 2 \quad \text{FALSE} \end{array}$$

We see that (-2,1) is not a solution of the inequality, so we shade the region that does not contain (-2,1).

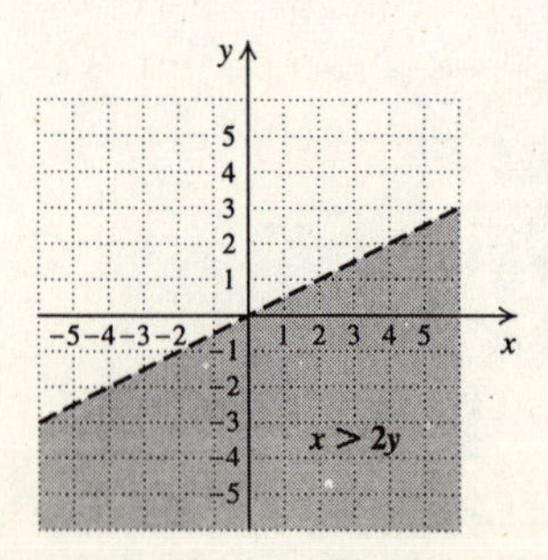

28.

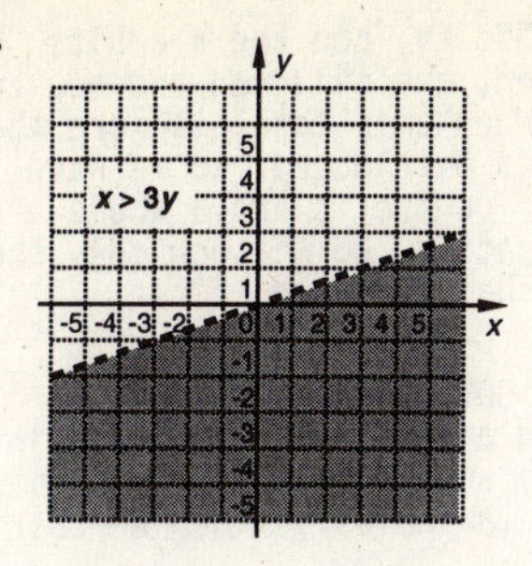

29. Graph $x < 4$.

Graph the line $x = 4$ using a dashed line since the inequality symbol is $<$. Then use $(-1,2)$ as a test point. We can write the inequality as $x + 0y < 4$.

$$\begin{array}{c|l} \multicolumn{2}{c}{x + 0y < 4} \\ \hline -1 + 0\cdot 2 & ?\ 4 \\ -1 & 4 \quad \text{TRUE} \end{array}$$

Since $(-1,2)$ is a solution of the inequality, we shade the region containing $(-1,2)$.

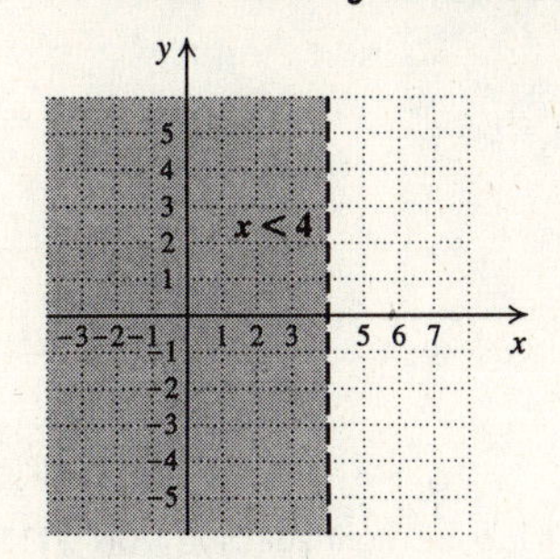

30.

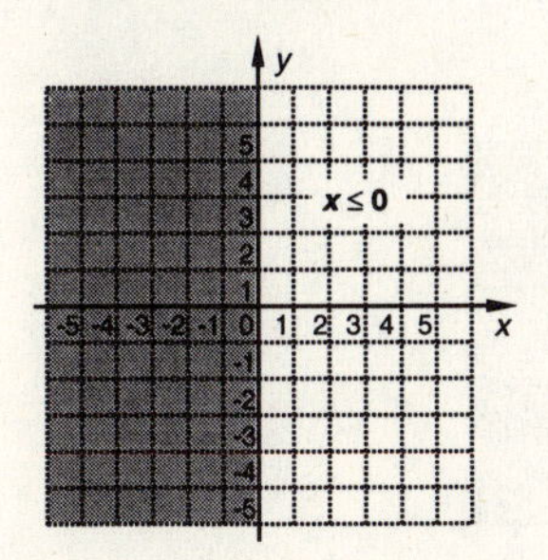

31. Graph $x \geq 3$.

Graph the line $x = 3$ using a solid line since the inequality symbol is $\geq$. Then use $(4,-3)$ as a test point. We can write the inequality as $x + 0y \geq 3$.

$$\begin{array}{c|l} \multicolumn{2}{c}{x + 0y \geq 3} \\ \hline 4 + 0(-3) & ?\ 3 \\ 4 & 3 \quad \text{TRUE} \end{array}$$

Since $(4,-3)$ is a solution of the inequality, we shade the region containing $(4,-3)$.

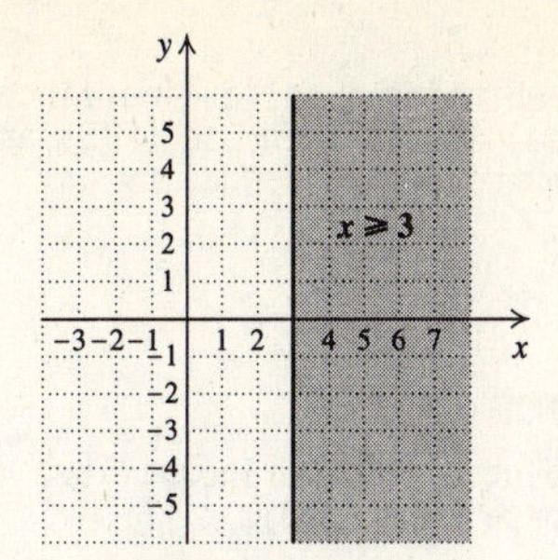

32.

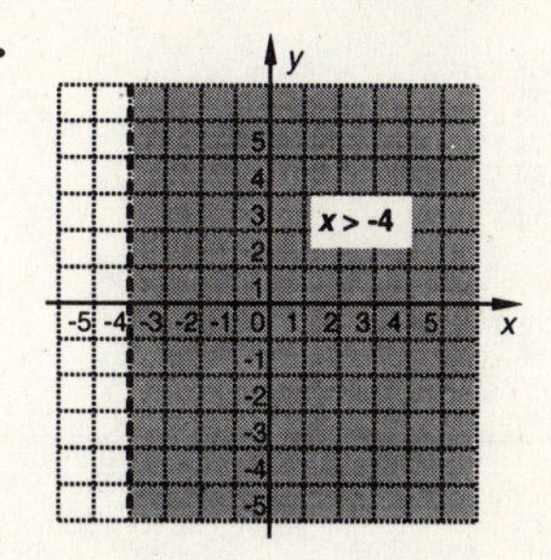

33. Graph $y \leq 3$.

Graph the line $y = 3$ using a solid line since the inequality symbol is $\leq$. Then use $(1,-2)$ as a test point. We can write the inequality as $0x + y \leq 3$.

$$\begin{array}{c|l} \multicolumn{2}{c}{0x + y \leq 3} \\ \hline 0\cdot 1 + (-2) & ?\ 3 \\ -2 & 3 \quad \text{TRUE} \end{array}$$

Since $(1,-2)$ is a solution of the inequality, we shade the region containing $(1,-2)$.

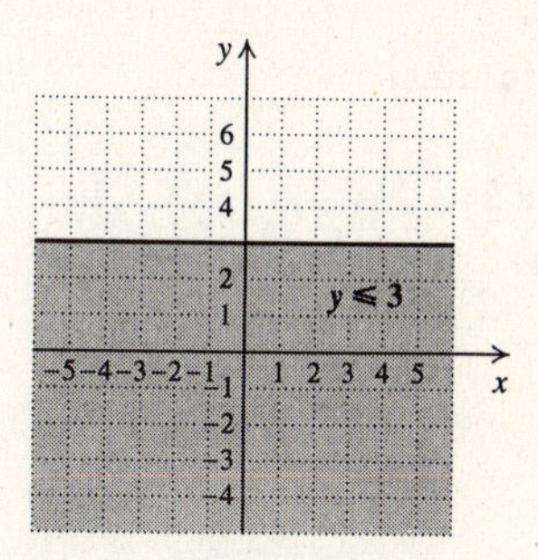

34.

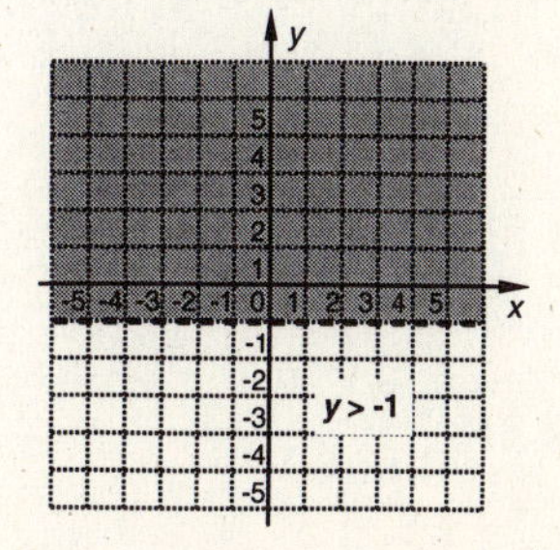

35. Graph $y \geqslant -5$.

Graph the line $y = -5$ using a solid line since the inequality symbol is $\geqslant$. Then use (2,3) as a test point. We can write the inequality as $0x + y \geqslant -5$.

$$\begin{array}{c|l} \multicolumn{2}{c}{0x + y \geqslant -5} \\ \hline 0\cdot 2 + 3 & ?\ -5 \\ 3 & -5 \quad \text{TRUE} \end{array}$$

Since (2,3) is a solution of the inequality, we shade the region containing (2,3).

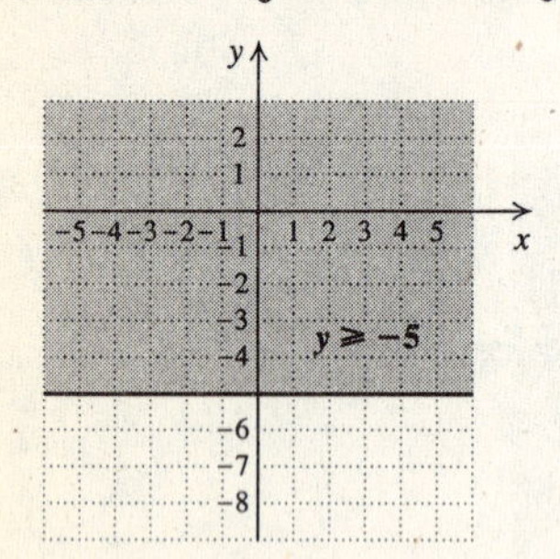

36.

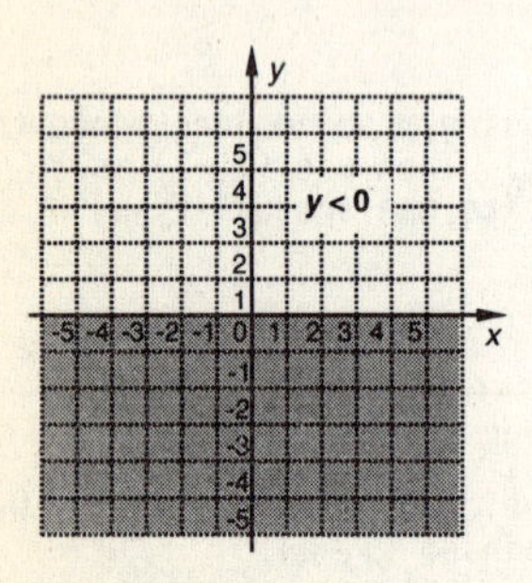

37. We multiply using columns.

$$\begin{array}{rl} 2x^2 - x - 1 & \\ x + 3 & \\ \hline 6x^2 - 3x - 3 & \text{Multiplying by 3} \\ 2x^3 - x^2 - x \quad & \text{Multiplying by x} \\ \hline 2x^3 + 5x^2 - 4x - 3 & \text{Adding} \end{array}$$

38. $9x^2 - 25$

39. $3a^3 + 18a^2 - 4a - 24$

$= 3a^2(a + 6) - 4(a + 6)$ Factoring by grouping

$= (a + 6)(3a^2 - 4)$

40. $\dfrac{x - 7}{x + 5}$

41.

42.

43. The c children weigh 35c kg, and the a adults weigh 75a kg. Together, the children and adults weigh 35c + 75a kg. When this total is more than 1000 kg the elevator is overloaded, so we have $35c + 75a > 1000$. (Of course, c and a would also have to be nonnegative, so we show only the portion of the graph that is in the first quadrant.)

To graph $35c + 75a > 1000$, we first graph $35c + 75a = 1000$ using a dashed line. Two points on the line are (4,20) and (5,11). (We are using alphabetical order of variables.) Then we test the point (0,0).

$$\begin{array}{c|l} \multicolumn{2}{c}{35c + 75a > 1000} \\ \hline 35\cdot 0 + 75\cdot 0 & ?\ 1000 \\ 0 & 1000 \qquad \text{FALSE} \end{array}$$

Since (0,0) is not a solution of the inequality, we shade the region that does not contain (0,0).

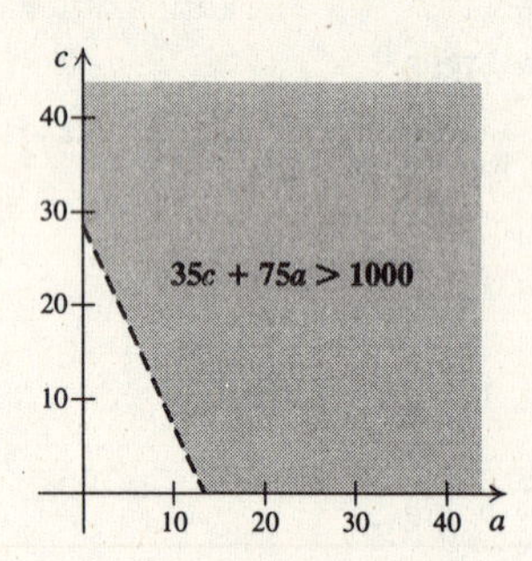

44. $2w + t \geqslant 60$

(Since w and t must also be nonnegative, we will show only the portion of the graph that is in the first quadrant.)

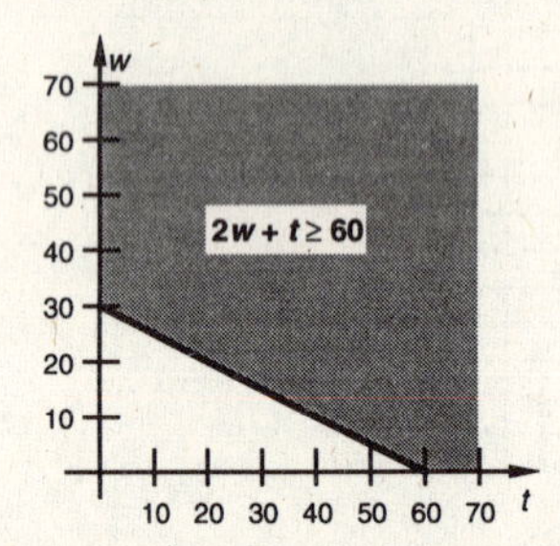

45. First find the equation of the line containing the points (2,0) and (0,-2). The slope is

$$\frac{-2 - 0}{0 - 2} = \frac{-2}{-2} = 1.$$

We know that the y-intercept is (0,-2), so we write the equation using the slope-intercept form.

$y = mx + b$

$y = 1\cdot x + (-2)$

$y = x - 2$

Since the line is dashed, the inequality symbol will be < or >. To determine which, we substitute the coordinates of a point in the shaded region. We will use (0,0).

y	x - 2
0 ?	0 - 2
0	-2

Since 0 > -2 is true, the correct symbol is >. The inequality is $y > x - 2$.

46. $x \geqslant -2$

47. Graph $xy \leqslant 0$.

From the principle of zero products, we know that $xy = 0$ when $x = 0$ or $y = 0$. Therefore, the graph contains the lines $x = 0$ and $y = 0$, or the y- and x-axes. Also, $xy < 0$ when x and y have different signs. This is the case for all points in the second quadrant (x is negative and y is positive) and in the fourth quadrants (x is positive and y is negative). Thus, we shade the second and fourth quadrants.

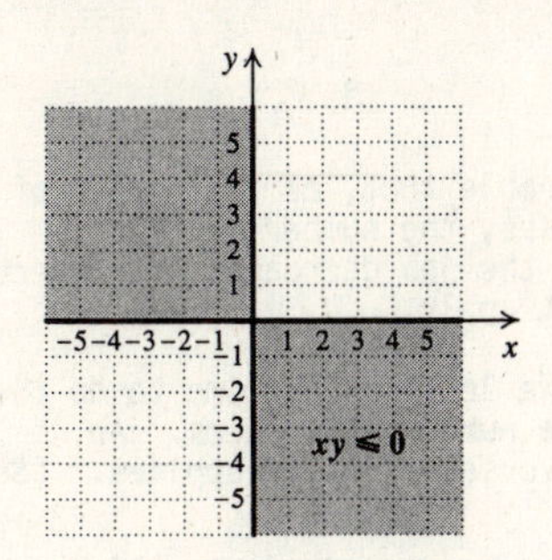

48.

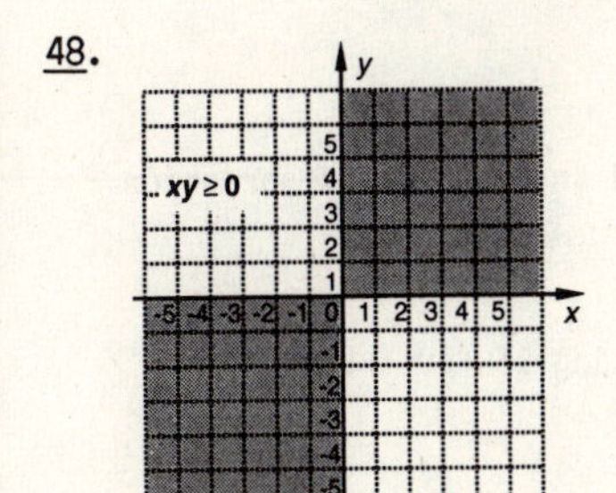

Exercise Set 7.5

1. We substitute to find k.

$y = kx$ y varies directly as x

$28 = k \cdot 7$ Substituting 28 for y and 7 for x

$\frac{28}{7} = k$

$4 = k$ k is the constant of variation

The equation of variation is $y = 4x$.

2. $y = 3.75x$

3. We substitute to find k.

$y = kx$ y varies directly as x

$0.7 = k \cdot 0.4$ Substituting 0.7 for y and 0.4 for x

$\frac{0.7}{0.4} = k$

$\frac{7}{4} = k$, or $k = 1.75$

The equation of variation is $y = 1.75x$.

4. $y = 1.6x$

5. We substitute to find k.

$y = kx$

$400 = k \cdot 125$ Substituting 400 for y and 125 for x

$\frac{400}{125} = k$

$\frac{16}{5} = k$, or $k = 3.2$

The equation of variation is $y = 3.2x$.

6. $y = 3.6x$

7. We substitute to find k.

$y = kx$

$200 = k \cdot 300$ Substituting 200 for y and 300 for x

$\frac{200}{300} = k$

$\frac{2}{3} = k$

The equation of variation is $y = \frac{2}{3}x$.

8. $y = \frac{25}{3}x$

9. We substitute to find k.

$y = \frac{k}{x}$

$25 = \frac{k}{3}$ Substituting 25 for y and 3 for x

$75 = k$ k is the constant of variation

The equation of variation is $y = \frac{75}{x}$.

10. $y = \frac{90}{x}$

11. We substitute to find k.

$y = \frac{k}{x}$

$8 = \frac{k}{10}$ Substituting 8 for y and 10 for x

$80 = k$

The equation of variation is $y = \frac{80}{x}$.

12. $y = \frac{70}{x}$

13. We substitute to find k.

$y = \frac{k}{x}$

$0.125 = \frac{k}{8}$ Substituting 0.125 for y and 8 for x

$1 = k$

The equation of variation is $y = \frac{1}{x}$.

14. $y = \frac{1}{x}$

15. We substitute to find k.

$y = \frac{k}{x}$

$42 = \frac{k}{25}$ Substituting 42 for y and 25 for x

$1050 = k$

The equation of variation is $y = \frac{1050}{x}$.

16. $y = \frac{2100}{x}$

17. We substitute to find k.

$y = \frac{k}{x}$

$0.2 = \frac{k}{0.3}$ Substituting 0.2 for y and 0.3 for x

$0.06 = k$

The equation of variation is $y = \frac{0.06}{x}$.

18. $y = \frac{0.24}{x}$

19. a) It seems reasonable that, as the number of hours of production increases, the number of compact-disc players produced will increase, so direct variation might apply.

b) Familiarize. Let H = the number of hours the production line is working, and let P = the number of compact-disc players produced. An equation of direct variation applies. (See part (a)).

Translate. Since direct variation applies we write the equation

$P = kH.$

Carry out. First we find an equation of variation.

$P = kH$

$15 = k \cdot 8$ Substituting 8 for H and 15 for P

$\frac{15}{8} = k$

The equation of variation is $P = \frac{15}{8}H$.

When H = 37,

$P = \frac{15}{8}H$

$P = \frac{15}{8} \cdot 37$ Substituting 37 for H

$P = \frac{555}{8}$

$P \approx 69$

Check. In addition to repeating the computations, we can do some reasoning. The number of hours increased from 8 to 37. Similarly, the number of compact disc players produced increased from 15 to about 69. The ratios 15/8 and 69/37 are about the same value: about 1.9.

State. About 69 compact-disc players can be produced in 37 hr.

20. a) Direct

b) $218.75

21. a) It seems reasonable that, as the number of workers increases, the number of hours required to do the job decreases, so inverse variation might apply.

b) Familiarize. We let T = the time to do the job and N = the number of workers. An equation of inverse variation applies. (See part (a)).

Translate. Since inverse variation applies we write the equation

$T = \frac{k}{N}.$

Carry out. First find an equation of variation.

$T = \frac{k}{N}$

$16 = \frac{k}{2}$ Substituting 16 for T and 2 for N

$32 = k$

The equation of variation is $T = \frac{32}{N}$.

When N = 6,

$T = \frac{32}{N}$

$T = \frac{32}{6}$ Substituting 6 for N

$T = 5\frac{1}{3}$

Check. The check might be done by repeating the computations. We might also analyze the results. The number of people increased from 2 to 6. The time decreased from 16 hours to $5\frac{1}{3}$ hours. This is what we would expect with inverse variation. We also note that the products (16)(2) and $\left(5\frac{1}{3}\right)(6)$ are both 32.

State. It would take 6 people $5\frac{1}{3}$ hours to do the job.

22. a) Inverse

b) $4\frac{1}{2}$ hr

23. Familiarize and Translate. The problem states that we have direct variation between the variables P and H. Thus, an equation $P = kH$ applies.

Carry out. First find an equation of variation.

$P = kH$

$78.75 = k\cdot 15$ Substituting 78.75 for P and 15 for H

$\frac{78.75}{15} = k$

$5.25 = k$

The equation of variation is $P = 5.25H$.

When $H = 35$,

$P = 5.25H$

$P = 5.25(35)$ Substituting 35 for H

$P = 183.75$

Check. This check might be done by repeating the computations. We might also do some reasoning about the answer. The paycheck increased from \$78.75 to \$183.75. Similarly, the hours increased from 15 to 35. The ratios 15/78.75 and 35/183.75 are the same value: about 0.19.

State. For 35 hours work, the paycheck is \$183.75.

24. 16,445

25. Familiarize and Translate. This problem states that we have direct variation between S and W. Thus, an equation $S = kW$ applies.

Carry out. First find an equation of variation.

$S = kW$

$40 = k\cdot 14$ Substituting 40 for S and 14 for W

$\frac{40}{14} = k$

$\frac{20}{7} = k$

The equation of variation is $S = \frac{20}{7}W$.

When $N = 8$,

$S = \frac{20}{7}W$

$S = \frac{20}{7}\cdot 8$ Substituting 8 for W

$S = \frac{160}{7}$, or $22\frac{6}{7}$

Check. A check can always be done by repeating the computations. We can also do some reasoning about the answer. The number of servings decreased from 40 to $22\frac{6}{7}$. Similarly, the weight decreased from 14 kg to 8 kg. The ratios 14/40 and $8/22\frac{6}{7}$ are the same value: 0.35.

State. $22\frac{6}{7}$ servings can be obtained from an 8-kg turkey.

26. $93\frac{1}{3}$

27. Familiarize and Translate. The problem states that we have inverse variation between the variables V and P. Thus, an equation $V = \frac{k}{P}$ applies.

Carry out. First find an equation of variation.

$V = \frac{k}{P}$

$200 = \frac{k}{32}$ Substituting 200 for V and 32 for P

$6400 = k$

The equation of variation is $V = \frac{6400}{P}$.

When $P = 20$,

$V = \frac{6400}{P}$

$V = \frac{6400}{20}$ Substituting 20 for P

$V = 320$

Check. Checking can be done by repeating the computations. We can also analyze the results. The pressure decreased from 32 km/cm^2 to 20 km/cm^2. The volume increased from 200 cm^3 to 320 cm^3. This is what we would expect with inverse variation. We also note that the products $200\cdot 32$ and $320\cdot 20$ are both 6400.

State. The volume is 320 cm^3 under a pressure of 20 km/cm^2.

28. $3\frac{5}{9}$ amperes

29. Familiarize and Translate. The problem states that we have direct variation between the variables M and E. Thus, an equation $M = kE$ applies.

Carry out. First find an equation of variation.

$M = kE$

$28.6 = k\cdot 171.6$ Substituting 28.6 for M and 171.6 for E

$286 = 1716k$ Clearing decimals

$\frac{286}{1716} = k$

$\frac{1}{6} = k$

The equation of variation is $M = \frac{1}{6}E$.

When $E = 110$,

$M = \frac{1}{6}E$

$M = \frac{1}{6}\cdot 110$ Substituting 110 for E

$M = \frac{110}{6}$, or $18\frac{1}{3}$

Check. In addition to repeating the computations we can do some reasoning. The weight on the earth decreased from 171.6 lb to 110 lb. Similarly, the weight on the moon decreased from 28.6 lb to $18\frac{1}{3}$ lb. The ratios 171.6/28.6 and $110/18\frac{1}{3}$ are the same value: 6.

State. A 110-lb person would weigh $18\frac{1}{3}$ lb on the moon.

30. 66.88 lb

31. Familiarize and Translate. The problem states that we have inverse variation between the variables P and W. Thus, an equation $P = \frac{k}{W}$ applies.

Carry out. First find an equation of variation.

$P = \frac{k}{W}$

$660 = \frac{k}{1.6}$ Substituting 660 for P and 1.6 for W

$1056 = k$

The equation of variation is $P = \frac{1056}{W}$.

When P = 440,

$P = \frac{1056}{W}$

$440 = \frac{1056}{W}$

$W = \frac{1056}{440}$ Multiplying by $\frac{W}{440}$

$W = 2.4$

Check. We can repeat the computations. We also note that the products 660(1.6) and 440(2.4) are both 1056.

State. A tone that has a pitch of 440 vibrations per second has a wavelength of 2.4 ft.

32. 54 min

33. Familiarize and Translate. This problem states that we have direct variation between the variables c and n. Thus, an equation c = kn, applies.

Carry out. First we find an equation of variation.

$c = kn$

$14 = k\cdot 720$ Substituting 14 for c and 720 (the number of hours in 30 days) for n

$\frac{14}{720} = k$

$\frac{7}{360} = k$

The equation of variation is $c = \frac{7}{360}n$.

Use the equation of variation to find the operating cost for 1 day and 1 hr.

For 1 day: 1 day is equivalent to 24 hr, so we substitute 24 for n.

$c = \frac{7}{360}\cdot 24$

$c = \frac{168}{360}$, or $c \approx 0.467$

For 1 hr: $c = \frac{7}{360}n$

$c = \frac{7}{360}\cdot 1$ Substituting 1 for n

$c \approx 0.019$

Check. In addition to repeating the computations we can do some reasoning. The hours decreased from 720 to 24 and from 720 to 1. Similarly, the cost decreased from \$14 to \$0.467 and from \$14 to \$0.019.

State. It cost \$0.467, or 46.7¢, to operate a television for 1 day. It cost \$0.019, or 1.9¢, to operate a television for 1 hr.

34. \$4500

35. Familiarize and Translate. The problem states that we have inverse variation between the variables m and n. Thus, an equation $m = \frac{k}{n}$ applies.

Carry out. First we find an equation of variation.

$m = \frac{k}{n}$

$2.5 = \frac{k}{16}$ Substituting 2.5 for m and 16 for n

$40 = k$

The equation of variation is $m = \frac{40}{n}$.

When m = 4,

$m = \frac{40}{n}$

$4 = \frac{40}{n}$

$n = 10$ Multiplying by $\frac{n}{4}$

Check. We can repeat the computations. We also note that the products (2.5)(16) and $4\cdot 10$ are both 40.

State. 10 questions would appear on a quiz in which students have 4 minutes per question.

36. 5

37. ◈

38. ◈

39. ◈

40. ◈

41. $P = kS$, where k = the number of sides
In this case $k = 8$, so we have $P = 8S$.

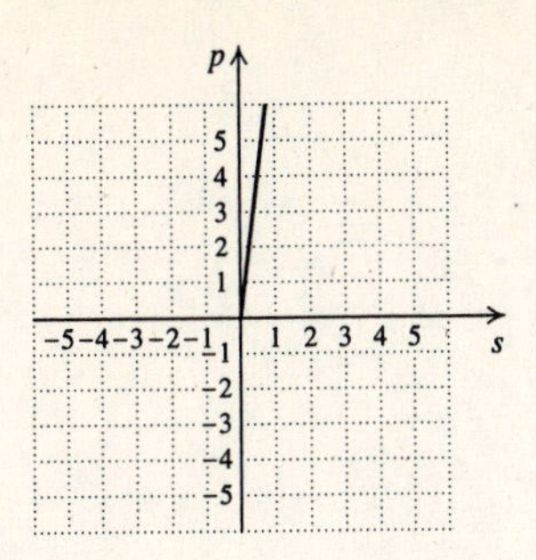

42. $C = 2\pi r$

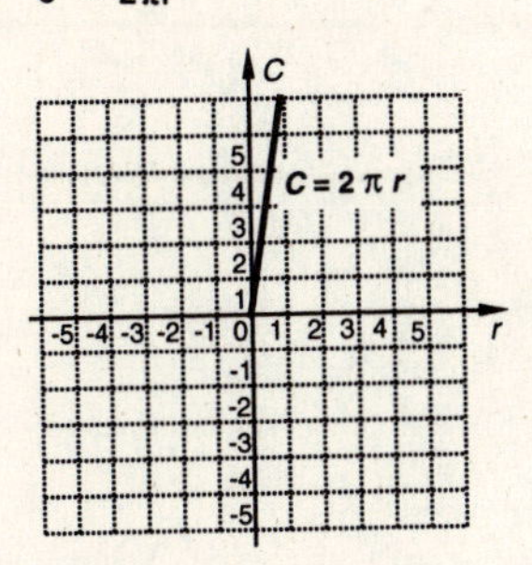

43. $B = kN$

44. $C = kA$

45. $$p = kq$$
$$\frac{1}{k} \cdot p = \frac{1}{k} \cdot kq$$
$$\frac{1}{k} \cdot p = q, \text{ or } q = \frac{1}{k}p$$
Since k is a constant, so is $\frac{1}{k}$, and q varies directly as p.

46. π

47. $S = kv^2$

48. $P^2 = kt$

49. $I = \frac{k}{d^2}$

50. $D = \frac{k}{V}$

51. $V = kr^3$

52. $P = kv^3$

53.

Exercise Set 8.1

1. We check by substituting alphabetically 3 for x and 2 for y.

$2x + 3y = 12$		
$2\cdot3 + 3\cdot2$? 12		
$6 + 6$		
12	12	TRUE

$x - 4y = -5$		
$3 - 4\cdot2$? -5		
$3 - 8$		
-5	-5	TRUE

The ordered pair (3,2) is a solution of each equation. Therefore it is a solution of the system of equations.

2. Yes

3. We check by substituting alphabetically 3 for a and 2 for b.

$3b - 2a = 0$		
$3\cdot2 - 2\cdot3$? 0		
$6 - 6$		
0	0	TRUE

$b + 2a = 15$		
$2 + 2\cdot3$? 15		
$2 + 6$		
8	15	FALSE

The ordered pair (3,2) is not a solution of $b + 2a = 15$. Therefore it is not a solution of the system of equations.

4. Yes

5. We check by substituting alphabetically 15 for x and 20 for y.

$3x - 2y = 5$		
$3\cdot15 - 2\cdot20$? 5		
$45 - 40$		
5	5	TRUE

$6x - 5y = -10$		
$6\cdot15 - 5\cdot20$? -10		
$90 - 100$		
-10	-10	TRUE

The ordered pair (15,20) is a solution of each equation. Therefore it is a solution of the system of equations.

6. Yes

7. We check by substituting alphabetically -1 for x and 1 for y.

$x = -1$		
-1 ? -1		TRUE

$x - y = -2$		
$-1 - 1$? -2		
-2	-2	TRUE

The ordered pair (-1,1) is a solution of each equation. Therefore it is a solution of the system of equations.

8. No

9. We check by substituting alphabetically 12 for x and 3 for y.

$y = \frac{1}{4}x$		
3 ? $\frac{1}{4}\cdot 12$		
3	3	TRUE

$3x - y = 33$		
$3\cdot12 - 3$? 33		
$36 - 3$		
33	33	TRUE

The ordered pair (12,3) is a solution of each equation. Therefore it is a solution of the system of equations.

10. Yes

11. We check by substituting alphabetically 10 for x and -3 for y.

$x + 2y = 4$		
$10 + 2(-3)$? 4		
$10 - 6$		
4	4	TRUE

$18 = (x - 7)y$		
18 ? $(10 - 7)(-3)$		
	$3(-3)$	
18	-9	FALSE

The ordered pair (10,-3) is not a solution of $18 = (x - 7)y$. Therefore it is not a solution of the system of equations.

12. Yes

13. We graph the equations.

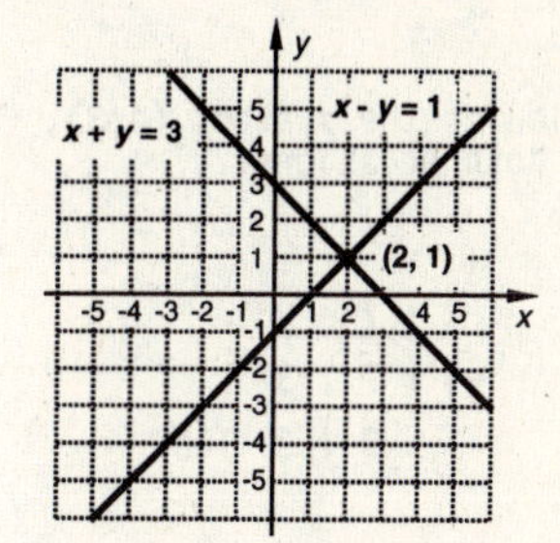

The "apparent" solution of the system, (2,1), should be checked in both equations.

Check:

$x + y = 3$		
$2 + 1$? 3		
3	3	TRUE

$x - y = 1$		
$2 - 1$? 1		
1	1	TRUE

The solution is (2,1).

14. (4,2)

15. We graph the equations.

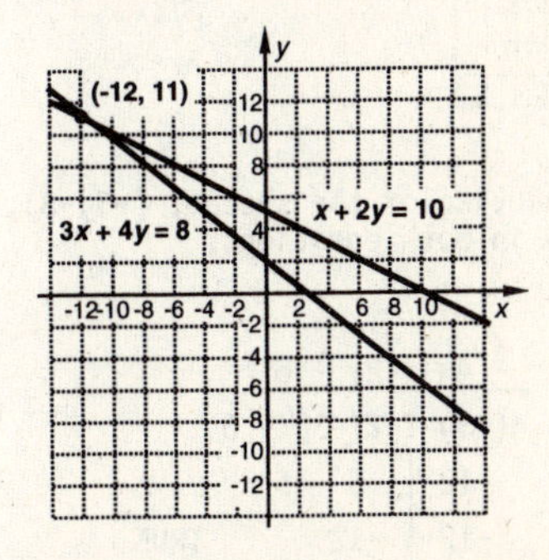

The "apparent" solution of the system, (-12,11), should be checked in both equations.

Check:

$$\begin{array}{c|c} x + 2y = 10 & \\ \hline -12 + 2\cdot 11 \ ? & 10 \\ -12 + 22 & \\ 10 & 10 \quad \text{TRUE} \end{array} \qquad \begin{array}{c|c} 3x + 4y = 8 & \\ \hline 3(-12) + 4(11) \ ? & 8 \\ -36 + 44 & \\ 8 & 8 \quad \text{TRUE} \end{array}$$

The solution is (-12,11).

16. (-8,-7)

17. We graph the equations.

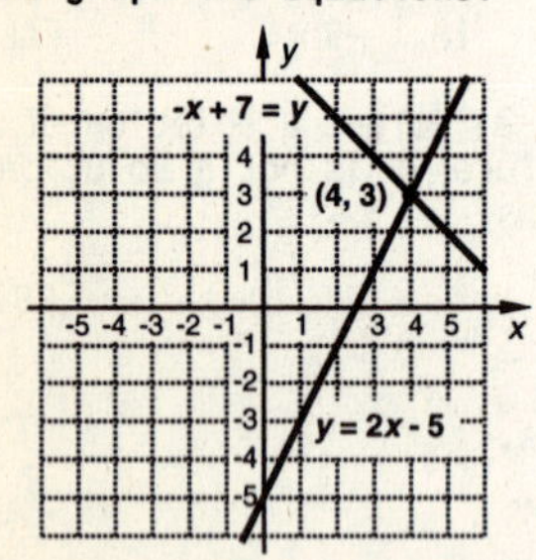

The "apparent" solution of the system, (4,3), should be checked in both equations.

Check:

$$\begin{array}{c|c} y = 2x - 5 & \\ \hline 3 \ ? & 2\cdot 4 - 5 \\ & 8 - 5 \\ 3 & 3 \quad \text{TRUE} \end{array} \qquad \begin{array}{c|c} -x + 7 = y & \\ \hline -4 + 7 \ ? & 3 \\ 3 & 3 \quad \text{TRUE} \end{array}$$

The solution is (4,3).

18. (-5,3)

19. We graph the equations.

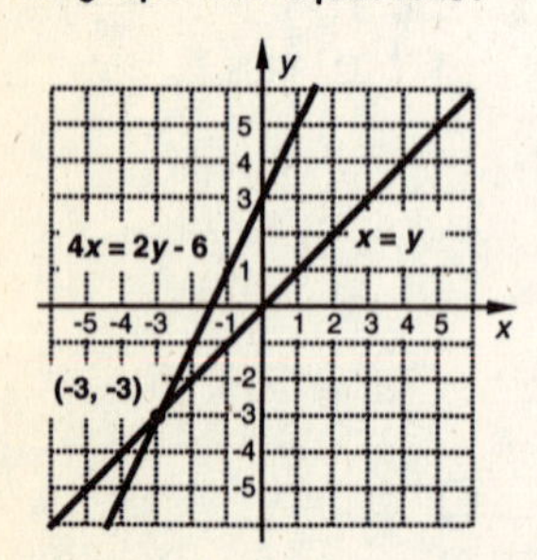

The "apparent" solution of the system, (-3,-3), should be checked in both equations.

Check:

$$\begin{array}{c|c} x = y & \\ \hline -3 \ ? & -3 \quad \text{TRUE} \end{array} \qquad \begin{array}{c|c} 4x = 2y - 6 & \\ \hline 4(-3) \ ? & 2(-3) - 6 \\ -12 & -6 - 6 \\ -12 & -12 \quad \text{TRUE} \end{array}$$

The solution is (-3,-3).

20. (-6,-2)

21. We graph the equations.

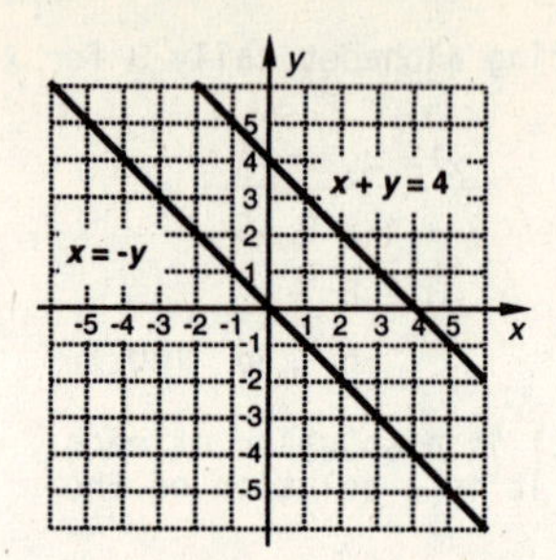

The lines are parallel. There is no solution.

22. Infinitely many solutions

23. We graph the equations.

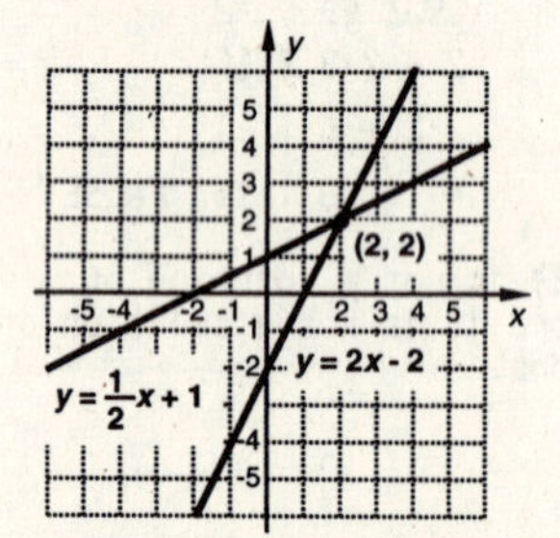

The "apparent" solution of the system, (2,2), should be checked in both equations.

Check:

$$\begin{array}{c|c} y = \frac{1}{2}x + 1 & \\ \hline 2 \ ? & \frac{1}{2}\cdot 2 + 1 \\ & 1 + 1 \\ 2 & 2 \quad \text{TRUE} \end{array} \qquad \begin{array}{c|c} y = 2x - 2 & \\ \hline 2 \ ? & 2\cdot 2 - 2 \\ & 4 - 2 \\ 2 & 2 \quad \text{TRUE} \end{array}$$

The solution is (2,2).

24. (1,-3)

25. We graph the equations.

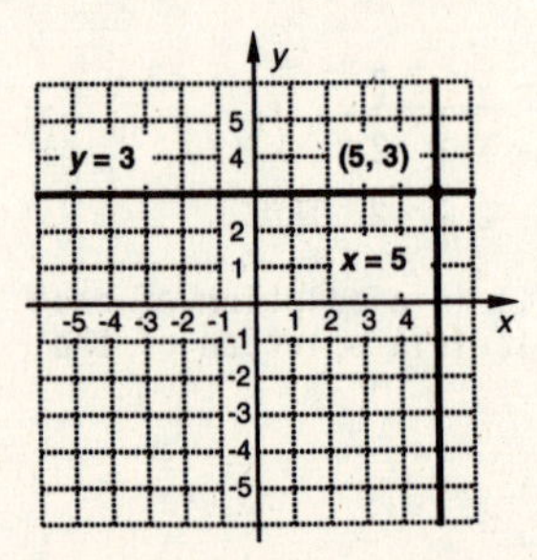

The "apparent" solution of the system, (5,3), should be checked in both equations.

Check:

$$\begin{array}{c|c} y = 3 & \\ \hline 3 \ ? & 3 \quad \text{TRUE} \end{array} \qquad \begin{array}{c|c} x = 5 & \\ \hline 5 \ ? & 5 \quad \text{TRUE} \end{array}$$

The solution is (5,3).

26. $\left[\frac{1}{3},1\right]$

27. We graph the equations.

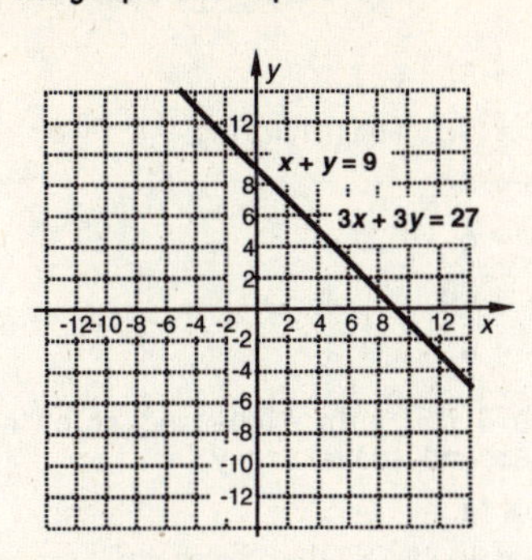

The lines coincide. The system has infinitely many solutions.

28. No solution

29. We graph the equations.

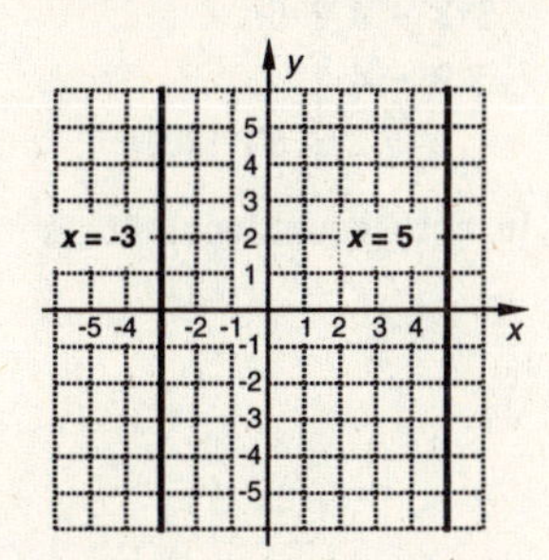

The lines are parallel. The system has no solution.

30. (5,-3)

31. $\frac{x+2}{x-4} - \frac{x+1}{x+4}$, LCD is $(x-4)(x+4)$

$= \frac{x+2}{x-4} \cdot \frac{x+4}{x+4} - \frac{x+1}{x+4} \cdot \frac{x-4}{x-4}$

$= \frac{(x+2)(x+4) - (x+1)(x-4)}{(x-4)(x+4)}$

$= \frac{x^2 + 6x + 8 - (x^2 - 3x - 4)}{(x-4)(x+4)}$

$= \frac{x^2 + 6x + 8 - x^2 + 3x + 4}{(x-4)(x+4)}$

$= \frac{9x + 12}{(x-4)(x+4)}$

(Although $9x + 12$ can be factored as $3(3x + 4)$, doing so will not enable us to further simplify the result.)

32. $\frac{2x + 5}{x + 3}$

33. The polynomial has three terms, so it is a trinomial.

34. Binomial

35. 

36.

37. Systems in which the graphs of the equations coincide contain dependent equations. This is the case in Exercises 22 and 27.

38. The systems in Exercises 13-20, 22-27, and 30 are consistent.

39. Systems in which the graphs of the equations are parallel are inconsistent. This is the case in Exercises 21, 28, and 29.

40. The systems in Exercises 13-21, 23-26, and 28-30 contain independent equations.

41. Answers may vary. Any two equations with (2,-4) as a solution will do.

$2x - y = 8,$

$x + 3y = -10$

42. Answers may vary. One possibility is $2x - y = 8$

43. (2,-3) is a solution of $Ax - 3y = 13$. Substitute 2 for x and -3 for y and solve for A.

$Ax - 3y = 13$

$A \cdot 2 - 3(-3) = 13$

$2A + 9 = 13$

$2A = 4$

$A = 2$

(2,-3) is a solution of $x - By = 8$. Substitute 2 for x and -3 for y and solve for B.

$x - By = 8$

$2 - B(-3) = 8$

$2 + 3B = 8$

$3B = 6$

$B = 2$

44.

45. (41.5, 17.1)

Exercise Set 8.2

1. $x + y = 4,$ (1)
$y = 2x + 1$ (2)

We substitute $2x + 1$ for y in equation (1) and solve for x.

$x + y = 4$ (1)
$x + (2x + 1) = 4$ Substituting
$3x + 1 = 4$
$3x = 3$
$x = 1$

Next we substitute 1 for x in either equation of the original system and solve for y.

$x + y = 4$ (1)
$1 + y = 4$ Substituting
$y = 3$

We check the ordered pair (1,3).

$x + y = 4$	
$1 + 3$?	4
4	4 TRUE

$y = 2x + 1$	
3 ?	$2\cdot1 + 1$
	$2 + 1$
3	3 TRUE

Since (1,3) checks in both equations, it is the solution.

2. (1,9)

3. $y = x + 1,$ (1)
$2x + y = 4$ (2)

We substitute $x + 1$ for y in equation (2) and solve for x.

$2x + y = 4$ (2)
$2x + (x + 1) = 4$ Substituting
$3x + 1 = 4$
$3x = 3$
$x = 1$

Next we substitute 1 for x in either equation of the original system and solve for y.

$y = x + 1$ (1)
$y = 1 + 1$ Substituting
$y = 2$

We check the ordered pair (1,2).

$y = x + 1$	
2 ?	$1 + 1$
2	2 TRUE

$2x + y = 4$	
$2\cdot1 + 2$?	4
$2 + 2$	
4	4 TRUE

Since (1,2) checks in both equations, it is the solution.

4. (2,-4)

5. $y = 2x - 5,$ (1)
$3y - x = 5$ (2)

We substitute $2x - 5$ for y in equation (2) and solve for x.

$3y - x = 5$ (2)
$3(2x - 5) - x = 5$ Substituting
$6x - 15 - x = 5$
$5x - 15 = 5$
$5x = 20$
$x = 4$

Next we substitute 4 for x in either equation of the original system and solve for y.

$y = 2x - 5$ (1)
$y = 2\cdot4 - 5$ Substituting
$y = 8 - 5$
$y = 3$

We check the ordered pair (4,3).

$y = 2x - 5$	
3 ?	$2\cdot4 - 5$
	$8 - 5$
3	3 TRUE

$3y - x = 5$	
$3\cdot3 - 4$?	5
$9 - 4$	
5	5 TRUE

Since (4,3) checks in both equations, it is the solution.

6. (-1,-1)

7. $x = -2y,$ (1)
$x + 4y = 2$ (2)

We substitute $-2y$ for x in equation (2) and solve for y.

$x + 4y = 2$ (2)
$-2y + 4y = 2$
$2y = 2$
$y = 1$

Next we substitute 1 for y in either equation of the original system and solve for x.

$x = -2y$ (1)
$x = -2\cdot1$
$x = -2$

We check the ordered pair (-2,1).

$x = -2y$	
-2 ?	$-2\cdot1$
-2	-2 TRUE

$3y - x = 5$	
$3\cdot1 - (-2)$?	5
$3 + 2$	
5	5 TRUE

Since (-2,1) checks in both equations, it is the solution.

8. (-30,10)

9. $y = x - 6,$ (1)
$3x + 2y = 8$ (2)

We substitute $x - 6$ for y in equation (2) and solve for x.

$$3x + 2y = 8 \quad (2)$$
$$3x + 2(x - 6) = 8$$
$$3x + 2x - 12 = 8$$
$$5x - 12 = 8$$
$$5x = 20$$
$$x = 4$$

Next we substitute 4 for x in either equation of the original system and solve for y.

$$y = x - 6 \quad (1)$$
$$y = 4 - 6$$
$$y = -2$$

We check the ordered pair (4,-2).

$y = x - 6$: $-2 \;?\; 4 - 6$; $-2 \mid -2$ TRUE

$3x + 2y = 8$: $3\cdot4 + 2(-2) \;?\; 8$; $12 - 4$; $8 \mid 8$ TRUE

Since (4,-2) checks in both equations, it is the solution.

10. (-3,5)

11. $x = 2y + 1,$ (1)
$3x - 6y = 2$ (2)

We substitute $2y + 1$ for x in equation (2) and solve for y.

$$3x - 6y = 2 \quad (2)$$
$$3(2y + 1) - 6y = 2$$
$$6y + 3 - 6y = 2$$
$$3 = 2$$

We obtain a false equation, so the system has no solution.

12. Infinitely many solutions

13. $s + t = -4,$ (1)
$s - t = 2$ (2)

We solve equation (2) for s.

$$s - t = 2 \quad (2)$$
$$s = t + 2 \quad (3)$$

We substitute $t + 2$ for s in equation (1) and solve for t.

$$s + t = -4 \quad (1)$$
$$(t + 2) + t = -4 \quad \text{Substituting}$$
$$2t + 2 = -4$$
$$2t = -6$$
$$t = -3$$

Now we substitute -3 for t in either of the original equations or in equation (3) and solve for s. It is easiest to use (3).

$$s = t + 2 = -3 + 2 = -1$$

We check the ordered pair (-1,-3).

$s + t = -4$: $-1 + (-3) \;?\; -4$; $-4 \mid -4$ TRUE

$s - t = 2$: $-1 - (-3) \;?\; 2$; $-1 + 3$; $2 \mid 2$ TRUE

Since (-1,-3) checks in both equations, it is the solution.

14. (2,-4)

15. $y - 2x = -6,$ (1)
$2y - x = 5$ (2)

We solve equation (1) for y.

$$y - 2x = -6 \quad (1)$$
$$y = 2x - 6 \quad (3)$$

We substitute $2x - 6$ for y in equation (2) and solve for x.

$$2y - x = 5 \quad (2)$$
$$2(2x - 6) - x = 5 \quad \text{Substituting}$$
$$4x - 12 - x = 5$$
$$3x - 12 = 5$$
$$3x = 17$$
$$x = \frac{17}{3}$$

We substitute $\frac{17}{3}$ for x in equation (3) and compute y.

$$y = 2x - 6 = 2\left(\frac{17}{3}\right) - 6 = \frac{34}{3} - \frac{18}{3} = \frac{16}{3}.$$

The ordered pair $\left(\frac{17}{3},\frac{16}{3}\right)$ checks in both equations. It is the solution.

16. $\left(\frac{17}{3},\frac{2}{3}\right)$

17. $x - 4y = 3,$ (1)
$2x - 6 = 8y$ (2)

We solve equation (1) for x.

$$x - 4y = 3$$
$$x = 4y + 3$$

We substitute $4y + 3$ for x in equation (2) and solve for y.

$$2x - 6 = 8y \quad (2)$$
$$2(4y + 3) - 6 = 8y$$
$$8y + 6 - 6 = 8y$$
$$8y = 8y$$

The last equation is true for any choice of y, so there are infinitely many solutions.

18. Infinitely many solutions

19. $y = 2x + 5$, (1)
$y = 2x - 5$ (2)

We substitute $2x + 5$ for y in equation (2) and solve for x.

$y = 2x - 5$ (2)
$2x + 5 = 2x - 5$
$-5 = 5$ Subtracting $2x$

We obtain a false equation, so the system has no solution.

20. Infinitely many solutions

21. $2x + 3y = -2$, (1)
$2x - y = 9$ (2)

We solve equation (2) for y.

$2x - y = 9$ (2)
$2x - 9 = y$ (3)

We substitute $2x - 9$ for y in equation (1) and solve for x.

$2x + 3y = -2$ (1)
$2x + 3(2x - 9) = -2$
$2x + 6x - 27 = -2$
$8x - 27 = -2$
$8x = 25$
$x = \frac{25}{8}$

Now we substitute $\frac{25}{8}$ for x in equation (3) and compute y.

$y = 2x - 9 = 2\left(\frac{25}{8}\right) - 9 = \frac{25}{4} - \frac{36}{4} = -\frac{11}{4}$

The ordered pair $\left(\frac{25}{8},-\frac{11}{4}\right)$ checks in both equations. It is the solution.

22. (-12,11)

23. $x - y = -3$, (1)
$2x + 3y = -6$ (2)

We solve equation (1) for x.

$x - y = -3$ (1)
$x = y - 3$ (3)

We substitute $y - 3$ for x in equation (2) and solve for y.

$2x + 3y = -6$ (2)
$2(y - 3) + 3y = -6$ Substituting
$2y - 6 + 3y = -6$
$5y - 6 = -6$
$5y = 0$
$y = 0$

Now we substitute 0 for y in equation (3) and compute x.

$x = y - 3 = 0 - 3 = -3$

The ordered pair (-3,0) checks in both equations. It is the solution.

24. (4,-2)

25. $r - 2s = 0$, (1)
$4r - 3s = 15$ (2)

We solve equation (1) for r.

$r - 2s = 0$ (1)
$r = 2s$ (3)

We substitute $2s$ for r in equation (2) and solve for s.

$4r - 3s = 15$ (2)
$4(2s) - 3s = 15$ Substituting
$8s - 3s = 15$
$5s = 15$
$s = 3$

Now we substitute 3 for s in equation (3) and compute r.

$r = 2s = 2\cdot3 = 6$

The ordered pair (6,3) checks in both equations. It is the solution.

26. (1,2)

27. $x - 3y = 7$, (1)
$-4x + 12y = 28$ (2)

We solve equation (1) for x.

$x - 3y = 7$ (1)
$x = 3y + 7$ (3)

We substitute $3y + 7$ for x in equation (2) and solve for y.

$-4x + 12y = 28$ (2)
$-4(3y + 7) + 12y = 28$ Substituting
$-12y - 28 + 12y = 28$
$-28 = 28$

We obtain a false equation, so the system has no solution.

28. Infinitely many solutions

29. $x - 2y = 5$, (1)
$2y - 3x = 1$ (2)

We solve equation (1) for x.

$x - 2y = 5$
$x = 2y + 5$ (3)

We substitute $2y + 5$ for x in equation (2) and solve for y.

$$2y - 3x = 1 \quad (2)$$
$$2y - 3(2y + 5) = 1$$
$$2y - 6y - 15 = 1$$
$$-4y - 15 = 1$$
$$-4y = 16$$
$$y = -4$$

Next we substitute -4 for y in equation (3) and solve for x.

$$x = 2y + 5 = 2(-4) + 5 = -8 + 5 = -3$$

The ordered pair (-3,-4) checks in both equations. It is the solution.

30. (-2,-7)

31. $2x = y - 3, \quad (1)$
$2x = y + 5 \quad (2)$

We solve equation (1) for y.

$$2x = y - 3 \quad (1)$$
$$2x + 3 = y \quad (3)$$

We substitute 2x + 3 for y in equation (2) and solve for x.

$$2x = (2x + 3) + 5$$
$$2x = 2x + 8$$
$$0 = 8$$

We obtain a false equation, so the system has no solution.

32. No solution

33. Familiarize. We let x = the larger number and y = the smaller number.

Translate.

The sum of two numbers is 27.
$x + y = 27$

One number is 3 more than the other.
$x = 3 + y$

The resulting system is

$$x + y = 27, \quad (1)$$
$$x = 3 + y. \quad (2)$$

Carry out. We solve the system of equations.

We substitute 3 + y for x in equation (1) and solve for y.

$$x + y = 27 \quad (1)$$
$$(3 + y) + y = 27 \quad \text{Substituting}$$
$$3 + 2y = 27$$
$$2y = 24$$
$$y = 12$$

Next we substitute 12 for y in either equation of the original system and solve for x.

$$x + y = 27 \quad (1)$$
$$x + 12 = 27 \quad \text{Substituting}$$
$$x = 15$$

Check. The sum of 12 and 15 is 27. The number 15 is 3 more than 12. These numbers check.

State. The numbers are 15 and 12.

34. 19 and 17

35. Familiarize. Let x = larger number and y = the smaller number.

Translate.

The sum of two numbers is 58.
$x + y = 58$

The difference of two numbers is 16.
$x - y = 16$

The resulting system is

$$x + y = 58, \quad (1)$$
$$x - y = 16. \quad (2)$$

Carry out. We solve the system.

We solve equation (2) for x.

$$x - y = 16 \quad (2)$$
$$x = y + 16 \quad (3)$$

We substitute y + 16 for x in equation (1) and solve for y.

$$x + y = 58 \quad (1)$$
$$(y + 16) + y = 58 \quad \text{Substituting}$$
$$2y + 16 = 58$$
$$2y = 42$$
$$y = 21$$

Now we substitute 21 for y in equation (3) and compute x.

$$x = y + 16 = 21 + 16 = 37$$

Check. The sum of 37 and 21 is 58. The difference between 37 and 21, 37 - 21, is 16. The numbers check.

State. The numbers are 37 and 21.

36. 37 and 29

37. Familiarize. We let x = the larger number and y = the smaller number.

Translate.

The difference between two numbers is 16.
$x - y = 16$

Three times the larger number is seven times the smaller number.
$3x = 7y$

The resulting system is

$$x - y = 16, \quad (1)$$
$$3x = 7y. \quad (2)$$

Carry out. We solve the system.

We solve equation (1) for x.

$$x - y = 16 \quad (1)$$
$$x = y + 16 \quad (3)$$

We substitute $y + 16$ for x in equation (2) and solve for y.

$$3x = 7y \quad (2)$$
$$3(y + 16) = 7y \quad \text{Substituting}$$
$$3y + 48 = 7y$$
$$48 = 4y$$
$$12 = y$$

Next we substitute 12 for y in equation (3) and compute x.

$$x = y + 16 = 12 + 16 = 28$$

Check. The difference between 28 and 12, $28 - 12$, is 16. Three times the larger, $3 \cdot 28$ or 84, is seven times the smaller, $7 \cdot 12 = 84$. The numbers check.

State. The numbers are 28 and 12.

38. 22 and 4

39. Familiarize. Let x = one angle and y = the other angle.

Translate. Since the angles are supplementary, we have one equation.

$$x + y = 180$$

The second sentence can be translated as follows:

One angle is 30° less than twice the other.

$$x = 2y - 30$$

The resulting system is

$$x + y = 180, \quad (1)$$
$$x = 2y - 30. \quad (2)$$

Carry out. We solve the system.

We substitute $2y - 30$ for x in equation (1) and solve for y.

$$x + y = 180 \quad (1)$$
$$2y - 30 + y = 180$$
$$3y - 30 = 180$$
$$3y = 210$$
$$y = 70$$

Next we substitute 70 for y in equation (2) and solve for x.

$$x = 2y - 30 = 2 \cdot 70 - 30 = 140 - 30 = 110$$

Check. The sum of the angles is 70° + 110°, or 180°, so the angles are supplementary. If 30° is subtracted from twice 70°, we have $2 \cdot 70° - 30°$, or 110°, which is the other angle. The answer checks.

State. One angle is 70°, and the other is 110°.

40. 133°, 47°

41. Familiarize. We let x = the larger angle and y = the smaller angle.

Translate. Since the angles are complementary, we have one equation.

$$x + y = 90$$

We reword and translate the second statement.

The difference of two angles is 34°.

$$x - y = 34$$

The resulting system is

$$x + y = 90, \quad (1)$$
$$x - y = 34. \quad (2)$$

Carry out. We solve the system.

We first solve equation (2) for x.

$$x - y = 34 \quad (2)$$
$$x = y + 34 \quad (3)$$

Substitute $y + 34$ for x in equation (1) and solve for y.

$$x + y = 90 \quad (1)$$
$$y + 34 + y = 90$$
$$2y + 34 = 90$$
$$2y = 56$$
$$y = 28$$

Next we substitute 28 for y in equation (3) and solve for x.

$$x = y + 34 = 28 + 34 = 62$$

Check. The sum of the angles is 62° + 28°, or 90°, so the angles are complementary. The difference of the angles is 62° - 28°, or 34°. These numbers check.

State. The angles are 62° and 28°.

42. 32°, 58°

43. Familiarize. From the drawing in the text we see that we have a rectangle with length L and width W. The perimeter is $L + L + W + W$, or $2L + 2W$.

Translate.

The perimeter is 1280 mi.

$$2L + 2W = 1280$$

The width is 90 mi less than the length.

$$W = L - 90$$

The resulting system is

$$2L + 2W = 1280, \quad (1)$$
$$W = L - 90. \quad (2)$$

Carry out. We solve the system.

We substitute $L - 90$ for W in equation (1) and solve for L.

$$2L + 2W = 1280 \quad (1)$$
$$2L + 2(L - 90) = 1280 \quad \text{Substituting}$$
$$2L + 2L - 180 = 1280$$
$$4L - 180 = 1280$$
$$4L = 1460$$
$$L = 365$$

Now we substitute 365 for L in equation (2).

$$W = L - 90 \quad (2)$$
$$W = 365 - 90 \quad \text{Substituting}$$
$$W = 275$$

Check. A possible solution is a length of 365 mi and a width of 275 mi. The perimeter would be 2(365) + 2(275), or 730 + 550, or 1280. Also, the width is 90 mi less than the length. These numbers check.

State. The length is 365 mi, and the width is 275 mi.

44. Length: 380 mi, width: 270 mi

45. Familiarize. We make a drawing. We let ℓ = the length and w = the width.

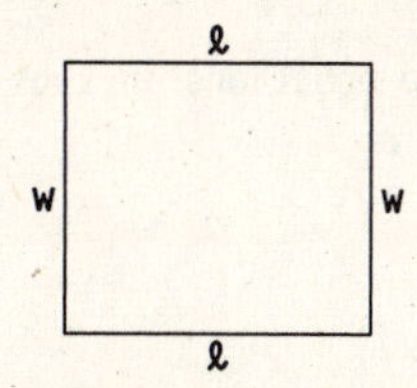

The perimeter is $\ell + \ell + w + w$, or $2\ell + 2w$.

Translate.

The perimeter is 400 m.

$$2\ell + 2w = 400$$

The length is 3m more than twice the width.

$$\ell = 3 + 2w$$

The resulting system is

$$2\ell + 2w = 400, \quad (1)$$
$$\ell = 3 + 2w. \quad (2)$$

Carry out. We solve the system.

We substitute 3 + 2w for ℓ in equation (1) and solve for w.

$$2\ell + 2w = 400 \quad (1)$$
$$2(3 + 2w) + 2w = 400 \quad \text{Substituting}$$
$$6 + 4w + 2w = 400$$
$$6w = 394$$
$$w = \frac{394}{6} = \frac{197}{3}, \text{ or } 65\frac{2}{3}$$

Now we substitute $\frac{197}{3}$ for w in equation (2).

$$\ell = 3 + 2w$$
$$\ell = 3 + 2\left(\frac{197}{3}\right)$$
$$\ell = \frac{9}{3} + \frac{394}{3}$$
$$\ell = \frac{403}{3}, \text{ or } 134\frac{1}{3}$$

Check. A possible solution is a length of $134\frac{1}{3}$ m and a width of $65\frac{2}{3}$ m. The perimeter would be $2\left(134\frac{1}{3}\right) + 2\left(65\frac{2}{3}\right)$, or $2\left(\frac{403}{3}\right) + 2\left(\frac{197}{3}\right)$, or $\frac{806}{3} + \frac{394}{3}$, or $\frac{1200}{3}$, or 400. Also, 3 more than twice the width is $3 + 2\left(65\frac{2}{3}\right)$, or $3 + 2\left(\frac{197}{3}\right)$, or $\frac{9}{3} + \frac{394}{3}$, or $\frac{403}{3}$, or $134\frac{1}{3}$, which is the length. These numbers check.

State. The length is $134\frac{1}{3}$ m, and the width is $65\frac{2}{3}$ m.

46. Length: $328\frac{1}{4}$ cm, width: $109\frac{3}{4}$ cm

47. Familiarize. Let ℓ = the length and w = the width. The perimeter is $\ell + \ell + w + w$, or $2\ell + 2w$.

Translate.

The perimeter is 340 yd.

$$2\ell + 2w = 340$$

The length is 50 yd more than the width.

$$\ell = 50 + w$$

The resulting system is

$$2\ell + 2w = 340, \quad (1)$$
$$\ell = 50 + w. \quad (2)$$

Carry out. We solve the system. We substitute 50 + w for ℓ in equation (1) and solve for w.

$$2\ell + 2w = 340 \quad (1)$$
$$2(50 + w) + 2w = 340$$
$$100 + 2w + 2w = 340$$
$$100 + 4w = 340$$
$$4w = 240$$
$$w = 60$$

Next we substitute 60 for w in equation (2) and solve for ℓ.

$$\ell = 50 + w = 50 + 60 = 110$$

Check. The perimeter is $2\cdot110 + 2\cdot60$, or 340 yd. Also 50 yd more than the width is 50 + 60, or 110 yd. The answer checks.

State. The length is 110 yd, and the width is 60 yd.

48. Length: 120 yd, width: $53\frac{1}{3}$ yd

49. $6x^2 - 13x + 6$

The possibilities are (x +)(6x +) and (2x +)(3x +). We look for a pair of factors of the last term, 6, which produce the correct middle term. Since the last term is positive and the middle term is negative, we need only consider negative pairs. The factorization is $(2x - 3)(3x - 2)$.

50. $(4p + 3)(p - 1)$

51. $4x^2 + 3x + 2$

The possibilities are (x +)(4x +) and (2x +)(2x +). We look for a pair of factors of the last term, 2, which produce the correct middle term. Since the last term and the middle term are both positive, we need only consider positive pairs. We find that there is no possibility that works. The trinomial is prime.

52. $(3a + 5)(3a - 5)$

53. 

54.

55. When the substitution method yields an equation that is true for any choice of x, the system contains dependent equations. This is the case in Exercises 12, 17, 18, 20, and 28.

56. The systems in Exercises 11, 19, 27, 31, and 32 are inconsistent.

57. $y - 2.35x = -5.97$, (1)
$2.14y - x = 4.88$ (2)

Solve equation (1) for y.

$y - 2.35x = -5.97$ (1)
$y = 2.35x - 5.97$ (3)

Substitute $2.35x - 5.97$ for y in equation (2) and solve for x.

$2.14(2.35x - 5.97) - x = 4.88$
$5.029x - 12.7758 - x = 4.88$
$4.029x = 17.6558$
$x \approx 4.382$

Substitute 4.382 for x in equation (3) and solve for y.

$y = 2.35x - 5.97 = 2.35(4.382) - 5.97 \approx 4.328$

The solution is (4.382,4.328).

58. (7,-1)

59. $\frac{x}{2} + \frac{3y}{2} = 2$, (1)
$\frac{x}{5} - \frac{y}{2} = 3$ (2)

Clear the fractions.

$x + 3y = 4$, (1a) Multiplying equation (1) by 2
$2x - 5y = 30$ (2a) Multiplying equation (2) by 10

Solve equation (1a) for x.

$x + 3y = 4$ (1a)
$x = -3y + 4$ (3)

Substitute $-3y + 4$ for x in equation (2a) and solve for y.

$2(-3y + 4) - 5y = 30$
$-6y + 8 - 5y = 30$
$-11y + 8 = 30$
$-11y = 22$
$y = -2$

Substitute -2 for y in equation (3) and solve for x.

$x = -3y + 4 = -3(-2) + 4 = 6 + 4 = 10$

The ordered pair (10,-2) checks in both equations, so it is the solution.

60. (2,-1)

61. $x + y + z = 4$, (1)
$x - 2y - z = 1$, (2)
$y = -1$ (3)

Substitute -1 for y in equations (1) and (2).

$x + y + z = 4$ (1)
$x + (-1) + z = 4$
$x + z = 5$

$x - 2y - z = 1$ (2)
$x - 2(-1) - z = 1$
$x + 2 - z = 1$
$x - z = -1$

We now have a system of two equations in two variables.

$x + z = 5$, (4)
$x - z = -1$ (5)

We solve equation (5) for x.

$x - z = -1$ (5)
$x = z - 1$ (6)

We substitute $z - 1$ for x in equation (4) and solve for z.

$x + z = 5$ (4)
$(z - 1) + z = 5$ Substituting
$2z - 1 = 5$
$2z = 6$
$z = 3$

Next we substitute 3 for z in equation (6) and compute x.

$x = z - 1 = 3 - 1 = 2$

We check the ordered triple (2,-1,3).

$x + y + z = 4$
$2 + (-1) + 3 \ ? \ 4$
4 | 4 TRUE

$x - 2y - z = 1$
$2 - 2(-1) - 3 \ ? \ 1$
$2 + 2 - 3$
1 | 1 TRUE

$y = -1$
$-1 \ ? \ -1$ TRUE

Since (2,-1,3) checks in all three equations, it is the solution.

62. (30,50,100)

63.

64. Answers may vary.
$2x + 3y = 5$,
$5x + 4y = 2$

Exercise Set 8.3

1. $x + y = 10$ (1)
$x - y = 8$ (2)
$2x = 18$ Adding
$x = 9$

Substitute 9 for x in one of the original equations and solve for y.

$x + y = 10$ (1)
$9 + y = 10$ Substituting
$y = 1$

Check:

$x + y = 10$	
$9 + 1$? 10	
10	10 TRUE

$x - y = 8$	
$9 - 1$? 8	
8	8 TRUE

Since (9,1) checks, it is the solution.

2. (5,-2)

3. $x + y = 8$ (1)
$-x + 2y = 7$ (2)
$3y = 15$ Adding
$y = 5$

Substitute 5 for y in one of the original equations and solve for x.

$x + y = 8$ (1)
$x + 5 = 8$ Substituting
$x = 3$

Check:

$x + y = 8$	
$3 + 5$? 8	
8	8 TRUE

$-x + 2y = 7$	
$-3 + 2\cdot5$? 7	
$-3 + 10$	
7	7 TRUE

Since (3,5) checks, it is the solution.

4. (5,1)

5. $3x - y = 9$ (1)
$2x + y = 6$ (2)
$5x = 15$ Adding
$x = 3$

Substitute 3 for x in one of the original equations and solve for y.

$2x + y = 6$ (2)
$2\cdot3 + y = 6$ Substituting
$6 + y = 6$
$y = 0$

Check:

$3x - y = 9$	
$3\cdot3 - 0$? 9	
$9 - 0$	
9	9 TRUE

$2x + y = 6$	
$2\cdot3 + 0$? 6	
$6 + 0$	
6	6 TRUE

Since (3,0) checks, it is the solution.

6. (2,7)

7. $4a + 3b = 7$ (1)
$-4a + b = 5$ (2)
$4b = 12$ Adding
$b = 3$

Substitute 3 for b in one of the original equations and solve for a.

$4a + 3b = 7$ (1)
$4a + 3\cdot3 = 7$ Substituting
$4a + 9 = 7$
$4a = -2$
$a = -\frac{1}{2}$

Check:

$4a + 3b = 7$	
$4\left(-\frac{1}{2}\right) + 3\cdot3$? 7	
$-2 + 9$	
7	7 TRUE

$-4a + b = 5$	
$-4\left(-\frac{1}{2}\right) + 3$? 5	
$2 + 3$	
5	5 TRUE

Since $\left(-\frac{1}{2},3\right)$ checks, it is the solution.

8. $\left(2,\frac{4}{5}\right)$

9. $8x - 5y = -9$ (1)
$3x + 5y = -2$ (2)
$11x = -11$ Adding
$x = -1$

Substitute -1 for x in one of the original equations and solve for y.

$3x + 5y = -2$ (2)
$3(-1) + 5y = -2$ Substituting
$-3 + 5y = -2$
$5y = 1$
$y = \frac{1}{5}$

Check:

$8x - 5y = -9$	
$8(-1) - 5\left(\frac{1}{5}\right)$? -9	
$-8 - 1$	
-9	-9 TRUE

$3x + 5y = -2$	
$3(-1) + 5\left(\frac{1}{5}\right)$? -2	
$-3 + 1$	
-2	-2 TRUE

Since $\left(-1,\frac{1}{5}\right)$ checks, it is the solution.

10. (-2,3)

11. $4x - 5y = 7$
$-4x + 5y = 7$
$0 = 14$ Adding

We obtain a false equation, 0 = 14, so there is no solution.

12. Infinitely many solutions

13. $-x - y = 8,$ (1)
$2x - y = -1$ (2)

We multiply by -1 on both sides of equation (1) and then add.

$x + y = -8$ Multiplying by -1
$2x - y = -1$
$3x = -9$ Adding
$x = -3$

Substitute -3 for x in one of the original equations and solve for y.

$2x - y = -1$ (2)
$2(-3) - y = -1$ Substituting
$-6 - y = -1$
$-y = 5$
$y = -5$

Check:

$-x - y = 8$: $-(-3) - (-5) ? 8$; $3 + 5$; $8 \mid 8$ TRUE

$2x - y = -1$: $2(-3) - (-5) ? -1$; $-6 + 5$; $-1 \mid -1$ TRUE

Since (-3,-5) checks, it is the solution.

14. (-1,-6)

15. $x + 3y = 19,$
$x - y = -1$

We multiply by -1 on both sides of equation (2) and then add.

$x + 3y = 19$
$-x + y = 1$ Multiplying by -1
$4y = 20$ Adding
$y = 5$

Substitute 5 for y in one of the original equations and solve for x.

$x - y = -1$ (2)
$x - 5 = -1$ Substituting
$x = 4$

Check:

$x + 3y = 19$: $4 + 3\cdot5 ? 19$; $4 + 15$; $19 \mid 19$ TRUE

$x - y = -1$: $4 - 5 ? -1$; $-1 \mid -1$ TRUE

Since (4,5) checks, it is the solution.

16. (3,1)

17. $x + y = 5,$ (1)
$5x - 3y = 17$ (2)

We multiply by 3 on both sides of equation (1) and then add.

$3x + 3y = 15$ Multiplying by 3
$5x - 3y = 17$
$8x = 32$
$x = 4$

Substitute 4 for x in one of the original equations and solve for y.

$x + y = 5$ (1)
$4 + y = 5$ Substituting
$y = 1$

Check:

$x + y = 5$: $4 + 1 ? 5$; $5 \mid 5$ TRUE

$5x - 3y = 17$: $5\cdot4 - 3\cdot1 ? 17$; $20 - 3$; $17 \mid 17$ TRUE

Since (4,1) checks, it is the solution.

18. (10,3)

19. $2w - 3z = -1,$ (1)
$3w + 4z = 24$ (2)

We use the multiplication principle with both equations and then add.

$8w - 12z = -4$ Multiplying (1) by 4
$9w + 12z = 72$ Multiplying (2) by 3
$17w = 68$ Adding
$w = 4$

Substitute 4 for w in one of the original equations and solve for z.

$3w + 4z = 24$ (2)
$3\cdot4 + 4z = 24$ Substituting
$12 + 4z = 24$
$4z = 12$
$z = 3$

Check:

$2w - 3z = -1$: $2\cdot4 - 3\cdot3 ? -1$; $8 - 9$; $-1 \mid -1$ TRUE

$3w + 4z = 24$: $3\cdot4 + 4\cdot3 ? 24$; $12 + 12$; $24 \mid 24$ TRUE

Since (4,3) checks, it is the solution.

20. (1,-1)

21. $2a + 3b = -1,$ (1)
$3a + 5b = -2$ (2)

We use the multiplication principle with both equations and then add.

$-10a - 15b = 5$ Multiplying (1) by -5
$9a + 15b = -6$ Multiplying (2) by 3
$-a = -1$ Adding
$a = 1$

Substitute 1 for a in one of the original equations and solve for b.

$2a + 3b = -1$ (1)
$2 \cdot 1 + 3b = -1$ Substituting
$3b = -3$
$b = -1$

Check:

$2a + 3b = -1$: $2 \cdot 1 + 3(-1) \; ? \; -1$; $2 - 3$; $-1 \mid -1$ TRUE

$3a + 5b = -2$: $3 \cdot 1 + 5(-1) \; ? \; -2$; $3 - 5$; $-2 \mid -2$ TRUE

Since (1,-1) checks, it is the solution.

22. (4,-1)

23. $x = 3y,$
$5x + 14 = y$

We first get each equation in the form $Ax + By = C$.

$x - 3y = 0,$ (1) Adding $-3y$
$5x - y = -14$ (2) Adding $-y - 14$

We multiply by -5 on both sides of equation (1) and add.

$-5x + 15y = 0$ Multiplying by -5
$5x - y = -14$
$14y = -14$ Adding
$y = -1$

Substitute -1 for y in equation (1) and solve for x.

$x - 3y = 0$
$x - 3(-1) = 0$ Substituting
$x + 3 = 0$
$x = -3$

Check:

$x - 3y = 0$: $-3 - 3(-1) \; ? \; 0$; $-3 + 3$; $0 \mid 0$ TRUE

$5x - y = -14$: $5(-3) - (-1) \; ? \; -14$; $-15 + 1$; $-14 \mid -14$ TRUE

Since (-3,-1) checks, it is the solution.

24. (2,5)

25. $3x - 2y = 10,$ (1)
$5x + 3y = 4$ (2)

We use the multiplication principle with both equations and add.

$9x - 6y = 30$ Multiplying (1) by 3
$10x + 6y = 8$ Multiplying (2) by 2
$19x = 38$ Adding
$x = 2$

Substitute 2 for x in one of the original equations and solve for y.

$5x + 3y = 4$ (2)
$5 \cdot 2 + 3y = 4$ Substituting
$10 + 3y = 4$
$3y = -6$
$y = -2$

Check:

$3x - 2y = 10$: $3 \cdot 2 - 2(-2) \; ? \; 10$; $6 + 4$; $10 \mid 10$ TRUE

$5x + 3y = 4$: $5 \cdot 2 + 3(-2) \; ? \; 4$; $10 - 6$; $4 \mid 4$ TRUE

Since (2,-2) checks, it is the solution.

26. (2,1)

27. $3x = 8y + 11,$
$x + 6y - 8 = 0$

We first get each equation in the form $Ax + By = C$.

$3x - 8y = 11,$ (1) Adding $-8y$
$x + 6y = 8$ (2) Adding 8

We multiply by -3 on both sides of equation (2) and add.

$3x - 8y = 11$
$-3x - 18y = -24$ Multiplying by -3
$-26y = -13$ Adding
$y = \frac{1}{2}$

Substitute $\frac{1}{2}$ for y in equation (1) and solve for x.

$3x - 8y = 11$
$3x - 8 \cdot \frac{1}{2} = 11$ Substituting
$3x - 4 = 11$
$3x = 15$
$x = 5$

Check:

$$\begin{array}{c|c} 3x - 8y = 11 & \\ \hline 3\cdot 5 - 8\cdot\frac{1}{2} & 11 \\ 15 - 4 & \\ 11 & 11 \quad \text{TRUE} \end{array} \qquad \begin{array}{c|c} x + 6y = 0 & \\ \hline 5 + 6\cdot\frac{1}{2} & 8 \\ 5 + 3 & \\ 8 & 8 \quad \text{TRUE} \end{array}$$

Since $\left(5,\frac{1}{2}\right)$ checks, it is the solution.

28. (50,18)

29. $3x - 2y = 10$, (1)

$-6x + 4y = -20$ (2)

We multiply by 2 on both sides of equation (1) and add.

$$\begin{aligned} 6x - 4y &= 20 \\ -6x + 4y &= -20 \\ \hline 0 &= 0 \end{aligned}$$

We get an equation that is always true, so there are infinitely many solutions.

30. No solution

31. $0.06x + 0.05y = 0.07$,

$0.04x - 0.03y = 0.11$

We first multiply each equation by 100 to clear the decimals.

$6x + 5y = 7$, (1)

$4x - 3y = 11$ (2)

We use the multiplication principle with both equations of the resulting system.

$18x + 15y = 21$ Multiplying (1) by 3

$20x - 15y = 55$ Multiplying (2) by 5

$38x \qquad = 76$ Adding

$x = 2$

Substitute 2 for x in equation (1) and solve for y.

$$\begin{aligned} 6x + 5y &= 7 \\ 6\cdot 2 + 5y &= 7 \\ 12 + 5y &= 7 \\ 5y &= -5 \\ y &= -1 \end{aligned}$$

Check:

$$\begin{array}{c|c} 0.06x + 0.05y = 0.07 & \\ \hline 0.06(2) + 0.05(-1) & 0.07 \\ 0.12 - 0.05 & \\ 0.07 & 0.07 \quad \text{TRUE} \end{array}$$

$$\begin{array}{c|c} 0.04x - 0.03y = 0.11 & \\ \hline 0.04(2) - 0.03(-1) & 0.11 \\ 0.08 + 0.03 & \\ 0.11 & 0.11 \quad \text{TRUE} \end{array}$$

Since (2,-1) checks, it is the solution.

32. (10,-2)

33. $x + \frac{9}{2}y = \frac{15}{4}$,

$\frac{9}{10}x - y = \frac{9}{20}$

First we clear fractions. We multiply both sides of the first equation by 4 and both sides of the second equation by 20.

$$\begin{aligned} 4\left(x + \frac{9}{2}y\right) &= 4\cdot\frac{15}{4} \\ 4x + 4\cdot\frac{9}{2}y &= 15 \\ 4x + 18y &= 15 \\ 20\left(\frac{9}{10}x - y\right) &= 20\cdot\frac{9}{20} \\ 20\cdot\frac{9}{10}x - 20y &= 9 \\ 18x - 20y &= 9 \end{aligned}$$

The resulting system is

$4x + 18y = 15$, (1)

$18x - 20y = 9$. (2)

We use the multiplication principle with both equations.

$72x + 324y = 270$ Multiplying (1) by 18

$-72x + 80y = -36$ Multiplying (2) by -4

$404y = 234$

$y = \frac{234}{404}$, or $\frac{117}{202}$

Substitute $\frac{117}{202}$ for y in (1) and solve for x.

$$\begin{aligned} 4x + 18\left(\frac{117}{202}\right) &= 15 \\ 4x + \frac{1053}{101} &= 15 \\ 4x &= \frac{462}{101} \\ x &= \frac{1}{4}\cdot\frac{462}{101} \\ x &= \frac{231}{202} \end{aligned}$$

The ordered pair $\left(\frac{231}{202},\frac{117}{202}\right)$ checks in both equations. It is the solution.

34. $\left(\frac{231}{202},\frac{117}{202}\right)$

35. Familiarize. We let m = the number of miles driven and c = the total cost of the car rental.

Translate. We reword and translate the first statement, using \$0.30 for 30¢.

\$53.95 plus 30¢ times the number of miles driven is cost.

$53.95 + 0.30 \cdot m = c$

We reword and translate the second statement using \$0.20 for 20¢.

\$54.95 plus 20¢ times $\underbrace{\text{the number of miles driven}}$ is cost.

$$54.95 + 0.20 \cdot m = c$$

We have a system of equations:

$53.95 + 0.30m = c,$

$54.95 + 0.20m = c$

Carry out. We solve the system of equations. We clear the decimals by multiplying both sides of each equation by 100.

$5395 + 30m = 100c,$ (1)

$5495 + 20m = 100c$ (2)

We multiply (1) by -1 and then add.

$$\begin{aligned} -5395 - 30m &= -100c \\ 5495 + 20m &= 100c \\ \hline 100 - 10m &= 0 \quad \text{Adding} \\ 100 &= 10m \\ 10 &= m \end{aligned}$$

Check. For 10 mi, the cost of the Avis car is $53.95 + 0.30(10)$, or $53.95 + 3$, or \$56.95. For 10 mi, the cost of the other car is $54.95 + 0.20(10)$, or $54.95 + 2$, or \$56.95, so the costs are the same when the mileage is 10.

State. When the cars are driven 10 miles, the cost will be the same.

36. 5 miles

37. Familiarize. Let x = one angle and y = the other angle.

Translate. Since the angles are supplementary, we have one equation.

$x + y = 180$

$\underbrace{\text{One angle}}$ is $\underbrace{\text{45° less than two times the other.}}$

$$x = 2y - 45$$

The resulting system is

$x + y = 180,$

$x = 2y - 45.$

Carry out. We solve the system. We will use the elimination method although we could also easily use the substitution method. First we get the second equation in the form Ax + By = C.

$x + y = 180,$ (1)

$-x + 2y = 45$ (2) Adding -x and 45

We add the equations.

$$\begin{aligned} x + y &= 180 \\ -x + 2y &= 45 \\ \hline 3y &= 225 \\ y &= 75 \end{aligned}$$

Then we substitute 75 for y in equation (1) and solve for x.

$x + y = 180$ (1)

$x + 75 = 180$

$x = 105$

Check. The sum of the angles is 75° + 105°, or 180°, so the angles are supplementary. Also, 45° less than two times the 75° angle is $2 \cdot 75° - 45°$, or 105°, the other angle. These numbers check.

State. The angles are 75° and 105°.

38. 145°, 35°

39. Familiarize. We let x = the larger angle and y = the smaller angle.

Translate. The angles are complementary so we have one equation.

$x + y = 90$

We reword and translate the second statement.

$\underbrace{\text{The difference of two angles}}$ is 26°.

$$x - y = 26$$

We have a system of equations:

$x + y = 90,$

$x - y = 26$

Carry out. We solve the system.

$$\begin{aligned} x + y &= 90, \quad (1) \\ x - y &= 26 \quad (2) \\ \hline 2x &= 116 \quad \text{Adding} \\ x &= 58 \end{aligned}$$

Now we substitute 58 for x in equation (1) and solve for y.

$x + y = 90$ (1)

$58 + y = 90$ Substituting

$y = 32$

Check. The sum of the angles is 58° + 32°, or 90°, so the angles are complementary. The difference of the angles is 58° - 32°, or 26°. These numbers check.

State. The angles are 58° and 32°.

40. 64°, 26°

41. Familiarize. We let x = the number of acres of Riesling grapes that should be planted and y = the number of acres of Chardonnay grapes that should be planted.

Translate. We reword and translate the first statement.

$\underbrace{\text{Total number of acres}}$ is 820.

$$x + y = 820$$

Now we reword and translate the second statement.

$\underbrace{\text{Acres of Chardonnay grapes}}$ is $\underbrace{\text{140 acres}}$ $\underbrace{\text{more than}}$ $\underbrace{\text{acres of Riesling grapes.}}$

$$y = 140 + x$$

We have a system of equations:

$x + y = 820,$

$y = 140 + x$

Carry out. We solve the system. We will use the elimination method, although we could also easily use the substitution method. First we get the second equation in the form $Ax + By = C$. Then we add the equations.

$$\begin{aligned} x + y &= 820, && (1) \\ -x + y &= 140 && (2) \quad \text{Adding } -x \\ \hline 2y &= 960 && \text{Adding} \\ y &= 480 \end{aligned}$$

Now we substitute 480 for y in equation (1) and solve for x.

$$\begin{aligned} x + y &= 820 && (1) \\ x + 480 &= 820 && \text{Substituting} \\ x &= 340 \end{aligned}$$

Check. The total number of acres is 340 + 480, or 820. The number of acres of Chardonnay grapes, 480, is 140 more than the number of acres of Riesling grapes, 340. These numbers check.

State. The vintner should plant 340 acres of Riesling grapes and 480 acres of Chardonnay grapes.

42. 415 acres of hay, 235 acres of oats

43. Familiarize. Let ℓ = the length of the frame and w = the width.

Translate.

The perimeter is 10 ft.

$2\ell + 2w = 10$

The length is twice the width.

$\ell = 2w$

The resulting system is

$2\ell + 2w = 10,$

$\ell = 2w.$

Carry out. We solve the system. We will use the elimination method, although we could also easily use the substitution method. First we get the second equation in the form $A\ell + Bw = C$. Then we add the equations.

$$\begin{aligned} 2\ell + 2w &= 10 && (1) \\ \ell - 2w &= 0 && (2) \\ \hline 3\ell &= 10 \\ \ell &= \frac{10}{3}, \text{ or } 3\tfrac{1}{3} \end{aligned}$$

Substitute $\frac{10}{3}$ for ℓ in equation (2) and solve for w.

$$\begin{aligned} \ell - 2w &= 0 \quad (2) \\ \frac{10}{3} - 2w &= 0 \\ \frac{10}{3} &= 2w \\ \frac{5}{3} &= w, \quad \text{Dividing by 2} \\ \text{or } 1\tfrac{2}{3} &= w \end{aligned}$$

Check. The perimeter is $2 \cdot \frac{10}{3} + 2 \cdot \frac{5}{3}$, or 10 ft. Twice the width is $2 \cdot \frac{5}{3}$, or $\frac{10}{3}$, which is the length. These numbers check.

State. The length of the frame is $\frac{10}{3}$, or $3\frac{1}{3}$ ft, and the width is $\frac{5}{3}$, or $1\frac{2}{3}$ ft.

44. 6 ft by 4 ft

45. $\frac{(a^2b^{-3})^4}{a^5b^{-6}} = \frac{a^{2(4)}b^{-3(4)}}{a^5b^{-6}} = \frac{a^8b^{-12}}{a^5b^{-6}} = a^{8-5}b^{-12-(-6)} = a^3b^{-6}$, or $\frac{a^3}{b^6}$

46. $\frac{9a^4}{4b^6}$

47. $4x^2 + 20x + 25 = (2x)^2 + 2 \cdot 2x \cdot 5 + 5^2$ Trinomial square

$= (2x + 5)^2$

48. $(3a - 4)^2$

49.

50.

51. $3(x - y) = 9,$

$x + y = 7$

First we remove parentheses in the first equation.

$$\begin{aligned} 3x - 3y &= 9, && (1) \\ x + y &= 7 && (2) \end{aligned}$$

Then we multiply equation (2) by 3 and add.

$$\begin{aligned} 3x - 3y &= 9 \\ 3x + 3y &= 21 \\ \hline 6x &= 30 \\ x &= 5 \end{aligned}$$

Now we substitute 5 for x in equation (2) and solve for y.

$$\begin{aligned} x + y &= 7 \\ 5 + y &= 7 \\ y &= 2 \end{aligned}$$

The ordered pair (5,2) checks and is the solution.

52. (1,-1)

53. $2(5a - 5b) = 10,$
$-5(6a + 2b) = 10$

Simplify both equations.

$10a - 10b = 10, \quad (1)$
$-30a - 10b = 10 \quad (2)$

Now multiply equation (2) by -1 and add.

$$\begin{aligned} 10a - 10b &= 10 \\ 30a + 10b &= -10 \\ \hline 40a \quad\quad &= 0 \\ a &= 0 \end{aligned}$$

Substitute 0 for a in equation (1) and solve for b.

$$\begin{aligned} 10\cdot 0 - 10b &= 10 \\ -10b &= 10 \\ b &= -1 \end{aligned}$$

The ordered pair (0,-1) checks and is the solution.

54. (4,0)

55. $y = ax + b, \quad (1)$
$y = x + c \quad (2)$

Substitute x + c for y in equation (1) and solve for x.

$$\begin{aligned} y &= ax + b \\ x + c &= ax + b \quad \text{Substituting} \\ x - ax &= b - c \\ (1 - a)x &= b - c \\ x &= \frac{b - c}{1 - a} \end{aligned}$$

Substitute $\frac{b - c}{1 - a}$ for x in equation (2) and simplify to find y.

$$\begin{aligned} y &= x + c \\ y &= \frac{b - c}{1 - a} + c \\ y &= \frac{b - c}{1 - a} + c \cdot \frac{1 - a}{1 - a} \\ y &= \frac{b - c + c - ac}{1 - a} \\ y &= \frac{b - ac}{1 - a} \end{aligned}$$

The ordered pair $\left(\frac{b - c}{1 - a}, \frac{b - ac}{1 - a}\right)$ checks and is the solution.

56. $\left(\frac{-b - c}{a}, 1\right)$

57. Familiarize. Let x represent the number of rabbits and y the number of pheasants in the cage. Each rabbit has one head and four feet. Thus, there are x rabbit heads and 4x rabbit feet in the cage. Each pheasant has one head and two feet. Thus, there are y pheasant heads and 2y pheasant feet in the cage.

Translate. We reword the problem.

Rabbit heads plus pheasant heads is 35.
$x + y = 35$

Rabbit feet plus pheasant feet is 94.
$4x + 2y = 94$

The resulting system is

$x + y = 35, \quad (1)$
$4x + 2y = 94. \quad (2)$

Carry out. We solve the system of equations.

We multiply equation (1) by -2 and then add.

$$\begin{aligned} -2x - 2y &= -70 \\ 4x + 2y &= 94 \\ \hline 2x \quad\quad &= 24 \quad \text{Adding} \\ x &= 12 \end{aligned}$$

Substitute 12 for x in one of the original equations and solve for y.

$$\begin{aligned} x + y &= 35 \quad (1) \\ 12 + y &= 35 \quad \text{Substituting} \\ y &= 23 \end{aligned}$$

Check. If there are 12 rabbits and 23 pheasants, the total number of heads in the cage is 12 + 23, or 35. The total number of feet in the cage is $4\cdot 12 + 2\cdot 23$, or 48 + 46, or 94. The numbers check.

State. There are 12 rabbits and 23 pheasants.

58. Patrick: 6, mother: 30

59. Familiarize. Let x = the man's age and y = his daughter's age. Five years ago their ages were x - 5 and y - 5.

Translate.

Dividing the sum of a man's age and 5 by 5 yields his daughter's age.
$\frac{x + 5}{5} = y$

The man's age 5 years ago was 8 times his daughter's 5 years ago
$x - 5 = 8(y - 5)$

We have a system of equations:

$$\frac{x + 5}{5} = y$$
$$x - 5 = 8(y - 5)$$

Carry out. Solve the system.

Multiply the first equation by 5 to clear the fraction.

$$x + 5 = 5y$$
$$x - 5y = -5$$

Simplify the second equation.

$$x - 5 = 8(y - 5)$$
$$x - 5 = 8y - 40$$
$$x - 8y = -35$$

The resulting system is

$$x - 5y = -5, \quad (1)$$
$$x - 8y = -35. \quad (2)$$

Multiply equation (2) by -1 and add.

$$x - 5y = -5$$
$$-x + 8y = 35 \quad \text{Multiplying by } -1$$
$$3y = 30 \quad \text{Adding}$$
$$y = 10$$

Substitute 10 for y in equation (1) and solve for x.

$$x - 5y = -5$$
$$x - 5 \cdot 10 = -5 \quad \text{Substituting}$$
$$x - 50 = -5$$
$$x = 45$$

Possible solution: Man is 45, daughter is 10.

Check. If 5 is added to the man's age, 5 + 45, the result is 50. If 50 is divided by 5, the result is 10, the daughter's age. Five years ago the father and daughter were 40 and 5, respectively, and $40 = 8 \cdot 5$. The numbers check.

State. The man is 45 years old; his daughter is 10 years old.

60. Base: 10 ft, height: 5 ft

Exercise Set 8.4

1. Familiarize. Let x = the number of cars and y = the number of trucks.

Translate. We reword the problem.

The number of cars	plus	the number of trucks	is	510.
x	+	y	=	510

The number of cars	is	the number of trucks	plus	190.
x	=	y	+	190

Carry out. We solve the system of equations.

$$x + y = 510, \quad (1)$$
$$x = y + 190 \quad (2)$$

We substitute y + 190 for x in equation (1) and solve for y.

$$x + y = 510$$
$$(y + 190) + y = 510 \quad \text{Substituting}$$
$$2y + 190 = 510$$
$$2y = 320$$
$$y = 160$$

Next we substitute 160 for y in one of the original equations and solve for x.

$$x = y + 190 \quad (2)$$
$$x = 160 + 190 \quad \text{Substituting}$$
$$x = 350$$

Check. If there are 350 cars and 160 trucks, then the total number of vehicles is 350 + 160, or 510. Since the number of trucks plus 190 is 160 + 190, or 350, we know that it is true that the number of cars is 190 more than the number of trucks.

State. The firm should have 350 cars and 160 trucks.

2. 11 km

3. Familiarize. Let x = the cost of one slice of pizza and y = the cost of one soda.

Translate. We reword the problem.

The cost of one slice of pizza	plus	the cost of one soda	is	\$1.99.
x	+	y	=	1.99

The cost of 3 slices of pizza	plus	the cost of 2 sodas	is	\$5.48.
3x	+	2y	=	5.48

We have translated to a system of equations:

$$x + y = 1.99,$$
$$3x + 2y = 5.48$$

Carry out. We will use elimination. We first multiply both equations by 100 to clear decimals.

$$100x + 100y = 199 \quad (1)$$
$$300x + 200y = 548 \quad (2)$$

We solve using elimination.

$$-200x - 200y = -398 \quad \text{Multiplying (1) by } -2$$
$$300x + 200y = 548 \quad (2)$$
$$100x = 150 \quad \text{Adding}$$
$$x = 1.5 \quad \text{Dividing by 100}$$

We go back to equation (1) and substitute 1.5 for x.

$$100x + 100y = 199$$
$$100(1.5) + 100y = 199$$
$$150 + 100y = 199$$
$$100y = 49$$
$$y = 0.49$$

Check. If a slice of pizza costs \$1.50 and a soda costs \$0.49, then together they cost \$1.99. Also, 3 slices of pizza and 2 sodas cost 3(\$1.50) + 2(\$0.49), or \$5.48. These numbers check.

State. One soda costs \$0.49, and one slice of pizza costs \$1.50.

4. \$0.99

5. Familiarize. Let h and s represent the number of bags of trash generated each month by the Hendersons and the Savickis, respectively.

Translate. Since the Hendersons generate two and a half times as much trash as the Savickis, we have

$$h = 2.5s.$$

To find a second equation, we reword some information.

Henderson's trash	plus	Savicki's trash	is	14 bags.
h	$+$	s	$=$	14

We have a system of equations:

$$h = 2.5s, \quad (1)$$
$$h + s = 14 \quad (2)$$

Carry out. We solve by substitution. Substitute $2.5s$ for h in equation (2) and solve for s.

$$h + s = 14$$
$$2.5s + s = 14$$
$$3.5s = 14$$
$$s = 4$$

Substitute 4 for s in equation (1) and solve for h.

$$h = 2.5s = 2.5(4) = 10$$

Check. Two and a half times 4 is 10, and 10 + 4 is 14. The numbers check.

State. Each month the Hendersons produce 10 bags of trash, and the Savickis produce 4 bags.

6. Mazza: 28 rolls, Kranepool: 8 rolls

7. Familiarize. Let t = the number of two-pointers made and f = the number of foul shots made.

Translate. To find one equation we rephrase the information about the points scored.

Points from two-pointers	plus	points from foul shots	totaled	69.
$2t$	$+$	f	$=$	69

Since a total of 41 baskets were made, we have a second equation.

$$t + f = 41$$

The resulting system is

$$2t + f = 69, \quad (1)$$
$$t + f = 41 \quad (2)$$

Carry out. We solve using elimination.

$$2t + f = 69 \quad (1)$$
$$-t - f = -41 \quad \text{Multiplying (2) by -1}$$
$$t = 28 \quad \text{Adding}$$

Substitute 28 for t in equation (2) and solve for f.

$$t + f = 41 \quad (2)$$
$$28 + f = 41$$
$$f = 13$$

Check. If the Hot Shots made 28 two-pointers and 13 foul shots, they made 28 + 13, or 41 shots for a total of $2 \cdot 28 + 13 = 56 + 13 = 69$ points. The numbers check.

State. The Hot Shots made 13 foul shots.

8. Two-point: 37, three-point: 11

9. Familiarize. Let x = the number of soft-serve and y = the number of hard-pack cones ordered.

Translate. We present the information in a table.

	Soft-serve	Hard-pack	Total
Price	\$1.25	\$1.50	
Number ordered	x	y	75
Total cost	$1.25x$	$1.50y$	104.25

From the "Number ordered" row we have

$$x + y = 75.$$

From the "Total cost" row we have

$$1.25x + 1.50y = 104.25.$$

The resulting system is

$$x + y = 75,$$
$$1.25x + 1.50y = 104.25.$$

Carry out. First we multiply the second equation by 100 to clear decimals.

$$x + y = 75, \quad (1)$$
$$125x + 150y = 10,425 \quad (2)$$

We solve using elimination.

$$-125x - 125y = -9375 \quad \text{Multiplying (1) by -125}$$
$$125x + 150y = 10,425 \quad (2)$$
$$25y = 1050 \quad \text{Adding}$$
$$y = 42$$

Substitute 42 for y in equation (1) and solve for x.

$$x + y = 75 \quad (1)$$
$$x + 42 = 75$$
$$x = 33$$

Check. If $x = 33$ and $y = 42$, a total of 75 cones were ordered. The total cost would be 33(\$1.25), or \$41.25, for soft-serve cones and 42(\$1.50), or \$63, for hard-pack cones. Then the total bill would be \$41.25 + \$63, or \$104.25. These numbers check.

State. 33 soft-serve and 42 hard-pack cones were ordered.

10. 5-cent: 240, 10-cent: 96

11. Familiarize. Let x = the number of adults and y = the number of children.

Translate. We present the information in a table.

	Adults	Children	Totals
Paid	\$4	\$3	
Number attending	x	y	429
Money taken in	$4x$	$3y$	\$1490

The last two rows of the table give us two equations.

The total number of people attending was 429, so

$$x + y = 429.$$

The total amount taken in was \$1490, so

$$4x + 3y = 1490.$$

Carry out. We use the elimination method.

$$x + y = 429, \quad (1)$$
$$4x + 3y = 1490 \quad (2)$$

We multiply on both sides of the first equation by -3 and then add.

$$-3x - 3y = -1287$$
$$4x + 3y = 1490$$
$$x = 203 \quad \text{Adding}$$

Next we substitute 203 for x in one of the original equations and solve for y.

$$x + y = 429 \quad (1)$$
$$203 + y = 429 \quad \text{Substituting}$$
$$y = 226$$

Check. The total attending was 203 adults plus 226 children, or 429. The total receipts were \$4(203) + \$3(226). This is \$812 + \$678, or \$1490. The numbers check.

State. 203 adults and 226 children attended the play.

12. Adults: 236, children: 342

13. Familiarize. Let x = the number of student tickets sold and y = the number of adult tickets sold.

Translate. We present the information in a table.

	Students	Adults	Totals
Paid	\$2	\$3	
Number attending	x	y	200
Money taken in	$2x$	$3y$	\$530

The last two rows of the table give us two equations.

The total number of tickets sold was 200, so

$$x + y = 200.$$

The total amount collected was \$530, so

$$2x + 3y = 530.$$

Carry out. We use the elimination method.

$$x + y = 200, \quad (1)$$
$$2x + 3y = 530 \quad (2)$$

We multiply on both sides of equation (1) by -2 and then add.

$$-2x - 2y = -400$$
$$2x + 3y = 530$$
$$y = 130$$

Next we substitute 130 for y in one of the original equations and solve for x.

$$x + y = 200 \quad (1)$$
$$x + 130 = 200 \quad \text{Substituting}$$
$$x = 70$$

Check. The total number of tickets sold was 70 students plus 130 adults, or 200. The total receipts were \$2(70) + \$3(130). This amount is \$140 + \$390, or \$530. The numbers check.

State. Thus, 70 student tickets and 130 adult tickets were sold.

14. Activity-card: 128, noncard: 75

15. Familiarize. Let x = the number of kilograms of cashews and y = the number of kilograms of pecans.

Translate. We present the information in a table.

	Cashews	Pecans	Mixture
Cost of nuts	\$8.00	\$9.00	\$8.40
Amount (in kg)	x	y	10
Mixture	$8x$	$9y$	\$8.40 × 10, or \$84

The last two rows of the table give us two equations.

Since the total number of kilograms is 10, we have

$$x + y = 10.$$

The value of the cashews is $8x$ (x pounds at \$8 per pound). The value of the pecans is $9y$ (y pounds at \$9 per pound). The value of the mixture is 8.40 × 10, or \$84. Thus we have

$$8x + 9y = 84.$$

Carry out. We use the elimination method.

$$x + y = 10, \quad (1)$$
$$8x + 9y = 84 \quad (2)$$

We multiply equation (1) by -8 and then add.

$$-8x - 8y = -80 \quad \text{Multiplying by -8}$$
$$8x + 9y = 84$$
$$y = 4$$

Next we substitute 4 for y in one of the original equations and solve for x.

$x + y = 10$ (1)

$x + 4 = 10$ Substituting

$x = 6$

Check. We consider $x = 6$ kg and $y = 4$ kg. The sum is 10 kg. The value of the mixture of nuts is $\$8 \cdot 6 + \$9 \cdot 4$, or $48 + $36, or $84. These values check.

State. The mixture consists of 6 kg of cashews and 4 kg of pecans.

16. Brazilian: 100 kg, Turkish: 200 kg

17. Familiarize. Let x = the amount of sunflower seed and y = the amount of rolled oats to be used.

Translate. We present the information in a table.

Type of ingredient	Sunflower seed	Rolled oats	Mixture
Cost of ingredient	$1.00	$1.35	$1.14
Amount (in pounds)	x	y	50
Mixture	$1.00x$	$1.35y$	$1.14(50), or $57

The last two rows of the table give us two equations.

Since the total amount of grass seed is 50 lb, we have

$x + y = 50.$

The value of sunflower seed is $1.00x$ (x lb at $1.00 per pound), and the value of rolled oats is $1.35y$ (y lb at $1.35 per pound). The value of the mixture is $1.14(50), or $57, so we have

$1.00x + 1.35y = 57$, or

$100x + 135y = 5700$ Clearing decimals

Carry out. We use the elimination method.

$x + y = 50,$ (1)

$100x + 135y = 5700$ (2)

We multiply equation (1) by -100 and then add.

$$\begin{aligned} -100x - 100y &= -5000 \\ 100x + 135y &= 5700 \\ \hline 35y &= 700 \\ y &= 20 \end{aligned}$$

Next we substitute 20 for y in one of the original equations and solve for x.

$x + y = 50$ (1)

$x + 20 = 50$

$x = 30$

Check. We consider $x = 30$ lb and $y = 20$ lb. The sum is 50 lb. The value of the mixture is $1.00(30) + $1.35(20), or $30 + $27, or $57. These values check.

State. 30 lb of sunflower seed and 20 lb of rolled oats should be used.

18. $2.52 per pound: 135 lb, $3.80 per pound: 345 lb

19. Familiarize. We complete the table in the text. Note that x represents the number of liters of solution A to be used and y represents the number of liters of solution B.

Type of solution	A	B	Mix
Amount of solution	x	y	100 liters
Percent of acid	50%	80%	68%
Amount of acid in solution	$0.5x$	$0.8y$	0.68×100, or 68 liters

Translate. Since the total amount of solution is 100 liters, we have

$x + y = 100.$

The amount of acid in the mixture is to be 68% of 100, or 68 liters. The amounts of acid from the two solutions are $50\%x$ and $80\%y$. Thus

$50\%x + 80\%y = 68,$

or $0.5x + 0.8y = 68,$

or $5x + 8y = 680$ Clearing decimals

Carry out. We use the elimination method.

$x + y = 100,$ (1)

$5x + 8y = 680$ (2)

We multiply equation (1) by -5 and then add.

$$\begin{aligned} -5x - 5y &= -500 \quad \text{Multiplying by } -5 \\ 5x + 8y &= 680 \\ \hline 3y &= 180 \\ y &= 60 \end{aligned}$$

Next we substitute 60 for y in one of the original equations and solve for x.

$x + y = 100$ (1)

$x + 60 = 100$ Substituting

$x = 40$

Check. We consider $x = 40$ and $y = 60$. The sum is 100. Now 50% of 40 is 20 and 80% of 60 is 48. These add up to 68. The numbers check.

State. 40 liters of solution A and 60 liters of solution B should be used.

20. $55\frac{5}{9}$ L of A, $44\frac{4}{9}$ L of B

21. Familiarize. We can arrange the information in a table. We let x represent the amount of 30% solution and y represent the amount of 50% solution.

Type of insecticide	30% solution	50% solution	Mix
Amount of solution	x	y	200 L
Percent of insecticide	30%	50%	42%
Amount of insecticide in solution	0.3x	0.5y	0.42 × 200, or 84 L

Translate. Since the total amount of solution is 200 liters, we have

$x + y = 200.$

The amount of insecticide in the mixture is to be 42% of 200, or 84 liters. The amounts of insecticide from the two solutions are 30%x and 50%y. Thus

$30\%x + 50\%y = 84,$

or $0.3x + 0.5y = 84,$

or $3x + 5y = 840$ Clearing decimals

Carry out. We use the elimination method.

$x + y = 200,$ (1)

$3x + 5y = 840$ (2)

We multiply equation (1) by -3 and then add.

$-3x - 3y = -600$ Multiplying by -3

$3x + 5y = 840$

$2y = 240$

$y = 120$

Next we substitute 120 for y in one of the original equations and solve for x.

$x + y = 200$ (1)

$x + 120 = 200$ Substituting

$x = 80$

Check. We consider x = 80 and y = 120. The sum is 200. Now 30% of 80 is 24 and 50% of 120 is 60. These add up to 84. The numbers check.

State. 80 L of the 30% solution and 120 L of the 50% solution should be used.

22. 100 L of 28% solution, 200 L of 40% solution

23. Familiarize. We present the information in a table. Let x = the amount of 80% solution and y = the amount of 30% solution.

Type of solution	80% base	30% base	Mix
Amount of solution	x	y	200 liters
Percent of base	80%	30%	62%
Amount of base in solution	0.8x	0.3y	0.62 × 200, or 124 liters

Translate. Since the total amount of solution is 200 liters, we have

$x + y = 200.$

The last row of the table gives us a second equation:

$0.8x + 0.3y = 124,$ or

$8x + 3y = 1240$ Clearing decimals

Carry out. We use the elimination method.

$x + y = 200,$ (1)

$8x + 3y = 1240$ (2)

We multiply equation (1) by -3 and then add.

$-3x - 3y = -600$

$8x + 3y = 1240$

$5x = 640$

$x = 128$

Next we substitute 128 for x in one of the original equations and solve for y.

$x + y = 200$ (1)

$128 + y = 200$ Substituting

$y = 72$

Check. Clearly, 128 L of 80% solution and 72 L of 30% solution combine to make a 200 L mixture. Also, 128(0.8) + 72(0.3) = 102.4 + 21.6 = 124. Since 124 is 62% of 200 L, the mixture is correct. The numbers check.

State. The chemist should use 128 L of 80% solution and 72 L of 30% solution.

24. 22.5 L of 50% solution, 7.5 L of 70% solution

25. Familiarize. We organize the information in a table. Let a = the number of type A questions and b = the number of type B questions.

Type of question	A	B	Mixture (Test)
Number	a	b	16
Point value	10	15	
Total points	10a	15b	180

Translate. The table gives us two equations. Since the total number of questions is 16, we have

$a + b = 16.$

The total number of points is 180, so we have

$10a + 15b = 180.$

The resulting system is

$a + b = 16,$ (1)

$10a + 15b = 180.$ (2)

Carry out. We use the elimination method.

We multiply equation (1) by -10 and then add.

$$\begin{aligned} -10a - 10b &= -160 \\ 10a + 15b &= 180 \\ \hline 5b &= 20 \\ b &= 4 \end{aligned}$$

Now we substitute 4 for b in equation (1) and solve for a.

$$\begin{aligned} a + b &= 16 \\ a + 4 &= 16 \\ a &= 12 \end{aligned}$$

Check. We consider a = 12 questions and b = 4 questions. The total number of questions is 16. The total points are $10 \cdot 12 + 15 \cdot 4$, or $120 + 60$, or 180. These values check.

State. 12 questions of type A and 4 questions of type B were answered correctly.

26. Three-fourths gold: 45 oz,
five-twelfths gold: 15 oz

27. Familiarize. Let d represent the number of dimes and q the number of quarters. Then, 10d represents the value of the dimes in cents, and 25q represents the value of the quarters in cents. The total value is \$15.25, or 1525¢. The total number of coins is 103.

Translate.

Number of dimes	plus	number of quarters	is	103.
d	+	q	=	103

Value of dimes	plus	value of quarters	is	\$15.25
10d	+	25q	=	1525

The resulting system is

$d + q = 103,$ (1)

$10d + 25q = 1525.$ (2)

Carry out. We use the elimination method.

We multiply equation (1) by -10 and then add.

$$\begin{aligned} -10d - 10q &= -1030 && \text{Multiplying by -10} \\ 10d + 25q &= 1525 \\ \hline 15q &= 495 && \text{Adding} \\ q &= 33 \end{aligned}$$

Next we substitute 33 for q in one of the original equations and solve for d.

$$\begin{aligned} d + q &= 103 && (1) \\ d + 33 &= 103 && \text{Substituting} \\ d &= 70 \end{aligned}$$

Check. The number of dimes plus the number of quarters is 70 + 33, or 103. The total value in cents is $10 \cdot 70 + 25 \cdot 33$, or $700 + 825$, or 1525. This is equal to \$15.25. This checks.

State. There are 70 dimes and 33 quarters.

28. 10 nickels, 3 quarters

29. Familiarize. Let n represent the number of nickels and d the number of dimes. Then, 5n represents the value of the nickels in cents, and 10d represents the value of the dimes in cents. The total value is \$25, or 2500 cents.

Translate.

Value of nickels	plus	value of dimes	is	\$25.
5n	+	10d	=	2500

Number of nickels	is	three	times	the number of dimes.
n	=	3	·	d

The resulting system is

$5n + 10d = 2500,$ (1)

$n = 3d.$ (2)

Carry out. We use the substitution method.

We substitute 3d for n in equation (1) and solve for d.

$$\begin{aligned} 5n + 10d &= 2500 \\ 5 \cdot 3d + 10d &= 2500 && \text{Substituting} \\ 15d + 10d &= 2500 \\ 25d &= 2500 \\ d &= 100 \end{aligned}$$

We now substitute 100 for d in one of the original equations and solve for n.

$$\begin{aligned} n &= 3d && (2) \\ n &= 3 \cdot 100 && \text{Substituting} \\ n &= 300 \end{aligned}$$

Check. The number of nickels is three times the number of dimes ($300 = 3 \cdot 100$). The total value in cents is $5 \cdot 300 + 10 \cdot 100$, or $1500 + 1000$, or 2500. This is equal to \$25. This checks.

State. There are 300 nickels and 100 dimes.

30. 32 nickels, 13 dimes

31. Familiarize. We organize the information in a table. Let x = the price per gallon of the inexpensive paint and y = the price per gallon of the expensive paint.

	Inexpensive paint	Expensive paint	Mixture
Price per gallon	x	y	
First mixture amount	9 gal	7 gal	16 gal
First mixture cost	9x	7y	\$19.70(16) or \$315.20
Second mixture amount	3 gal	5 gal	8 gal
Second mixture cost	3x	5y	\$19.825(8), or \$158.60

Translate. The third and fifth rows of the table give us two equations. The total cost of the first mixture is:

$9x + 7y = 315.20$, or

$90x + 70y = 3152$ Clearing the decimal

The total cost of the second mixture is:

$3x + 5y = 158.60$, or

$30x + 50y = 1586$ Clearing the decimal

Carry out. We use the elimination method.

$90x + 70y = 3152,$ (1)

$30x + 50y = 1586,$ (2)

We multiply equation (2) by -3 and add.

$$\begin{aligned} 90x + 70y &= 3152 \\ -90x - 150y &= -4758 \\ \hline -80y &= -1606 \\ y &= 20.075 \end{aligned}$$

Now we substitute 20.075 for y in equation (2) and solve for x.

$$\begin{aligned} 30x + 50y &= 1586 \\ 30x + 50(20.075) &= 1586 \\ 30x + 1003.75 &= 1586 \\ 30x &= 582.25 \\ x &\approx 19.408 \end{aligned}$$

Check. We check $x \approx \$19.408$ and $y = \$20.075$. The cost of the first mixture is \$19.408(9) + \$20.075(7), or about \$315.20. The cost of the second mixture is \$19.408(3) + \$20.075(5), or about \$158.60. These values check.

State. The inexpensive paint costs \$19.408 per gallon, and the expensive paint costs \$20.075 per gallon.

32. Large type: 6.68, small type: 5.32

33. $25x^2 - 81 = (5x)^2 - 9^2$

$= (5x + 9)(5x - 9)$

34. $(6 + a)(6 - a)$

35. $\frac{x^2}{x + 4} = \frac{16}{x + 4}$, LCD = x + 4

$$\begin{aligned} (x + 4) \cdot \frac{x^2}{x + 4} &= (x + 4) \cdot \frac{16}{x + 4} \\ x^2 &= 16 \\ x^2 - 16 &= 0 \\ (x + 4)(x - 4) &= 0 \end{aligned}$$

$x + 4 = 0$ or $x - 4 = 0$

$x = -4$ or $x = 4$

Check: For x = -4:

$$\frac{x^2}{x+4} = \frac{16}{x+4}$$
$$\frac{(-4)^2}{-4+4} \;?\; \frac{16}{-4+4}$$
$$\frac{16}{0} \;\Big|\; \frac{16}{0} \quad \text{UNDEFINED}$$

For x = 4:

$$\frac{x^2}{x+4} = \frac{16}{x+4}$$
$$\frac{4^2}{4+4} \;?\; \frac{16}{4+4}$$
$$\frac{16}{8} \;\Big|\; \frac{16}{8} \quad \text{TRUE}$$

Since 4 checks and -4 does not, the solution is 4.

36. 5

37. ◈

38. ◈

39. Familiarize. Let x represent the part invested at 12% and y represent the part invested at 13%. The interest earned from the 12% investment is 12%·x. The interest earned from the 13% investment is 13%·y. The total investment is \$27,000, and the total interest earned is \$3385.

Translate.

$x + y = 27,000,$

$12\%x + 13\%y = \$3385$ or $12x + 13y = 338,500$

Carry out. Multiply the first equation by -12 and add.

$$\begin{aligned} -12x - 12y &= -324,000 \\ 12x + 13y &= 338,500 \\ \hline y &= 14,500 \end{aligned}$$

Substitute 14,500 for y in the first equation and solve for x.

$$x + y = 27{,}000$$
$$x + 14{,}500 = 27{,}000$$
$$x = 12{,}500$$

Check. We consider \$12,500 invested at 12% and \$14,500 invested at 13%. The sum of the investments is \$27,000. The interest earned is 12%·12,500 + 13%·14,500, or 1500 + 1885, or \$3385. These numbers check.

State. \$12,500 was invested at 12%, and \$14,500 was invested at 13%.

40. 9%, 10.5%

41. Familiarize. Let x represent the ten's digit and y the one's digit. Then the number is $10x + y$.

Translate.

The number is 6 times the sum of its digits.

$$10x + y = 6 \cdot (x + y)$$

The ten's digit is 1 more than the one's digit.

$$x = 1 + y$$

We simplify the first equation.

$$10x + y = 6(x + y)$$
$$10x + y = 6x + 6y$$
$$4x - 5y = 0$$

The system of equations is

$$4x - 5y = 0, \quad (1)$$
$$x = 1 + y. \quad (2)$$

Carry out. We use the substitution method. We substitute $1 + y$ for x in equation (1) and solve for y.

$$4(1 + y) - 5y = 0$$
$$4 + 4y - 5y = 0$$
$$4 - y = 0$$
$$4 = y$$

Then we substitute 4 for y in equation (2) and compute x.

$$x = 1 + y = 1 + 4 = 5$$

Check. We consider the number 54. The number is 6 times the sum of the digits, 9. The ten's digit is 1 more than the one's digit. This number checks.

State. The number is 54.

42. 75

43. Familiarize. We organize the information in a table. Let x = the number of liters of skim milk and y = the number of liters of 3.2% milk.

Type of milk	4.6%	Skim	3.2% (Mixture)
Amount of milk	100 L	x	y
Percent of butterfat	4.6%	0%	3.2%
Amount of butterfat in milk	4.6% × 100, or 4.6 L	0%·x, or 0 L	3.2%y

Translate. The first and third rows of the table give us two equations.

Liters of 4.6% milk + Liters of skim milk = Liters of 3.2% milk.

$$100 + x = y$$

Amt. of butterfat in 4.6% milk + Amt. of butterfat in skim milk = Amt. of butterfat in 3.2% milk

$$4.6 + 0 = 3.2\%y$$

The resulting system is

$$100 + x = y,$$
$$4.6 = 3.2\%y, \text{ or}$$

$$100 + x = y,$$
$$4.6 = 0.032y.$$

Carry out. We solve the second equation for y.

$$4.6 = 3.2\%y$$
$$4.6 = 0.032y$$
$$\frac{4.6}{0.032} = y$$
$$143.75 = y$$

We substitute 143.75 for y in the first equation and solve for x.

$$100 + x = y$$
$$100 + x = 143.75$$
$$x = 43.75$$

Check. We consider $x = 43.75$ L and $y = 143.75$ L. The difference between 143.75 L and 43.75 L is 100 L. There is no butterfat in the skim milk. There are 4.6 liters of butterfat in the 100 liters of the 4.6% milk. Thus there are 4.6 liters of butterfat in the mixture. This checks because 3.2% of 143.75 is 4.6.

State. 43.75 L of skim milk should be added.

44. $2666\frac{2}{3}$ L

45. Familiarize. In a table we organize the information regarding the solution after some of the 30% solution is drained and replaced with pure antifreeze. We let x represent the amount of the original (30%) solution remaining, and we let y represent the amount of the 30% mixture that is drained and replaced with pure antifreeze.

Type of solution	Original (30%)	Pure antifreeze	Mixture
Amount of solution	x	y	16
Percent of antifreeze	30%	100%	50%
Amount of antifreeze in solution	0.3x	1·y, or y	0.5(16), or 8

Translate. The table gives us two equations.

Amount of solution: $x + y = 16$

Amount of antifreeze in solution: $0.3x + y = 8$, or $3x + 10y = 80$

The resulting system is

$$x + y = 16, \quad (1)$$
$$3x + 10y = 80. \quad (2)$$

Carry out. We multiply the first equation by -3 and solve for y.

$$\begin{aligned} -3x - 3y &= -48 \\ 3x + 10y &= 80 \\ \hline 7y &= 32 \\ y &= \frac{32}{7}, \text{ or } 4\tfrac{4}{7} \end{aligned}$$

Then we substitute $4\frac{4}{7}$ for y in equation (1) and solve for x.

$$x + y = 16$$
$$x + 4\tfrac{4}{7} = 16$$
$$x = 11\tfrac{3}{7}$$

Check. When $x = 11\frac{3}{7}$ L and $y = 4\frac{4}{7}$ L, the total is 16 L. The amount of antifreeze in the mixture is $0.3\left(11\frac{3}{7}\right) + 4\frac{4}{7}$, or $\frac{3}{10}\cdot\frac{80}{7} + \frac{32}{7}$, or $\frac{24}{7} + \frac{32}{7} = \frac{56}{7}$, or 8 L. This is 50% of 16 L, so the numbers check.

State. $4\frac{4}{7}$ L of the original mixture should be drained and replaced with pure antifreeze.

46. 10 \$20 workers, 5 \$25 workers

47. Familiarize. Let x represent the ten's digit and y represent the one's digit. Then the number is $10x + y$.

Translate.

One's digit	plus	the number	is	43	more than	5 times	ten's digit
y	+	10x + y	=	43	+	5	· x

Sum of the digits is 11.

$$x + y = 11$$

We simplify the first equation.

$$y + 10x + y = 43 + 5x$$
$$10x + 2y = 43 + 5x$$
$$5x + 2y = 43$$

The resulting system is

$$5x + 2y = 43, \quad (1)$$
$$x + y = 11 \quad (2)$$

Carry out. We multiply equation (2) by -2 and add.

$$\begin{aligned} 5x + 2y &= 43 \\ -2x - 2y &= -22 \\ \hline 3x &= 21 \\ x &= 7 \end{aligned}$$

Then we substitute 7 for x in equation (2) and solve for y.

$$x + y = 11$$
$$7 + y = 11$$
$$y = 4$$

Check. We check the number 74. The sum of the one's digit and the number is 4 + 74, or 78. This is 43 more than 5 times the ten's digit, or 35. The sum of the digits is 11. The number checks.

State. The number is 74.

48. 126

49. Familiarize. Let x = the cost of the bat, y = the cost of the ball, and z = the cost of the glove.

Translate.

Cost of bat	+	Cost of ball	+	Cost of glove	=	\$99.
x	+	y	+	z	=	99

Cost of bat	=	\$9.95	+	Cost of ball
x	=	9.95	+	y

Cost of glove	=	\$65.45	+	Cost of bat
z	=	65.45	+	x

Carry out. We solve the system.

$$x + y + z = 99, \quad (1)$$
$$x = 9.95 + y, \quad (2)$$
$$z = 65.45 + x. \quad (3)$$

Solve the second equation for y.

$$x = 9.95 + y$$
$$x - 9.95 = y \quad (4)$$

Substitute $x - 9.95$ for y and $65.45 + x$ for z in the first equation and solve for x.

$$x + y + z = 99$$
$$x + (x - 9.95) + (65.45 + x) = 99$$
$$3x + 55.5 = 99$$
$$3x = 43.5$$
$$x = 14.5$$

Then substitute 14.5 for x in equations (3) and (4) and compute z and y.

$$z = 65.45 + x = 65.45 + 14.5 = 79.95$$
$$y = x - 9.95 = 14.5 - 9.95 = 4.55$$

Check. The total cost is \$14.50 + \$4.55 + \$79.95, or \$99. The cost of the bat, \$14.50, is \$9.95 more than \$4.55, the cost of the ball. Also, the cost of the glove, \$79.95, is \$65.45 more than \$14.50, the cost of the bat. These numbers check.

State. The bat costs \$14.50, the ball costs \$4.55, and the glove costs \$79.95.

50. Tweedledum: 120 pounds, Tweedledee: 121 pounds

Exercise Set 8.5

1. $x + y \leqslant 1$,
   $x - y \leqslant 5$

We graph the lines $x + y = 1$ and $x - y = 5$ using solid lines. We indicate the region for each inequality by the arrows at the ends of the lines. We shade the area where the regions overlap.

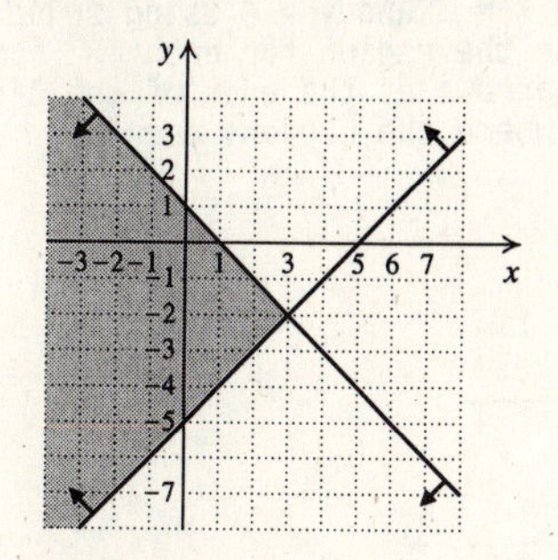

2.

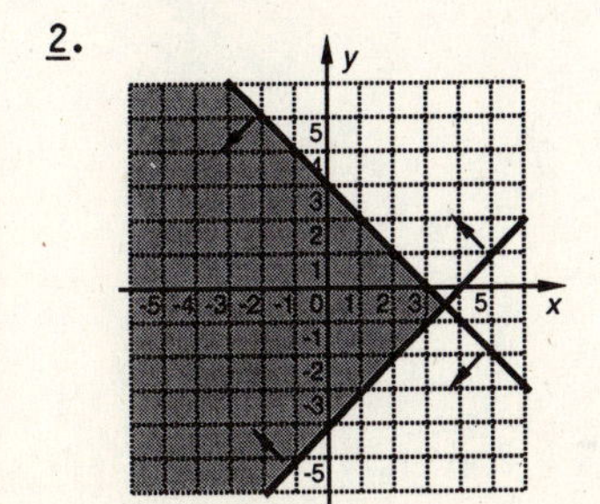

3. $y - 2x > 1$,
   $y - 2x < 3$

We graph the lines $y - 2x = 1$ and $y - 2x = 3$ using dashed lines. We indicate the region for each inequality by the arrows at the ends of the lines. We shade the area where the regions overlap.

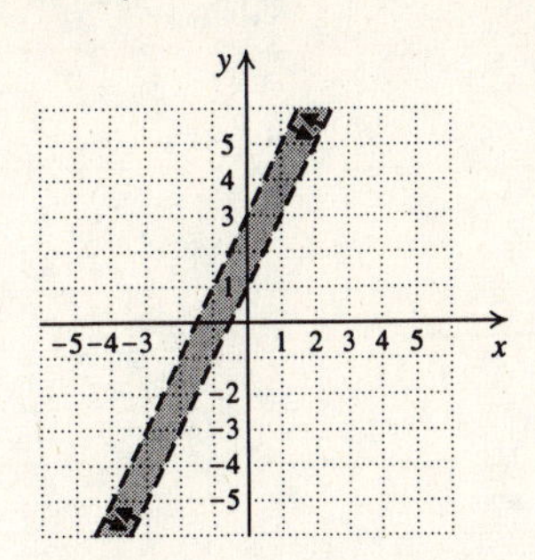

4.

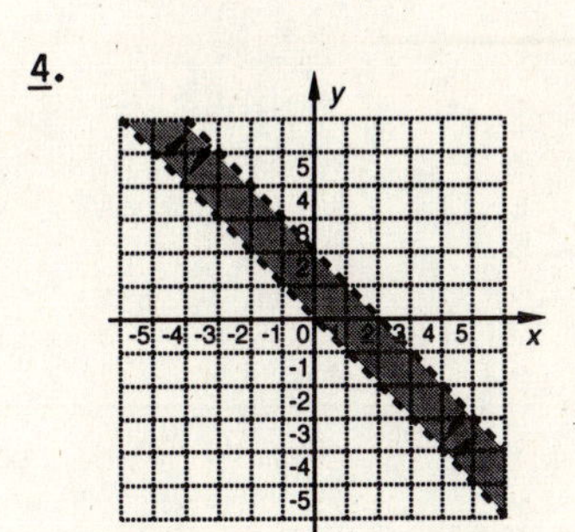

5. $y \geqslant 1$,
   $x > 2 + y$

We graph the line $y = 1$ using a solid line and the line $x = 2 + y$ using a dashed line. We indicate the region for each inequality by the arrows at the ends of the lines. We shade the area where the regions overlap.

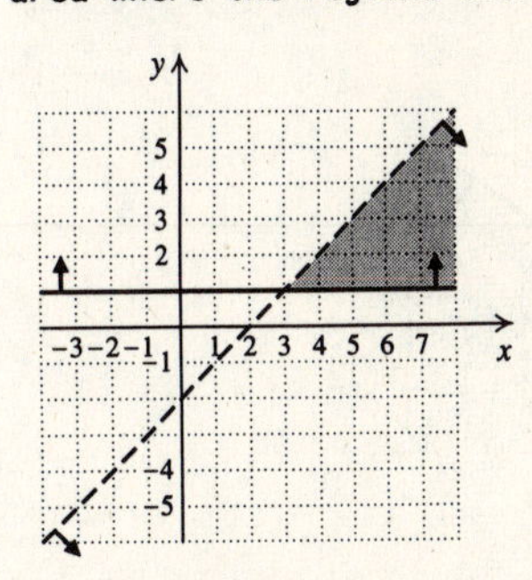

6.

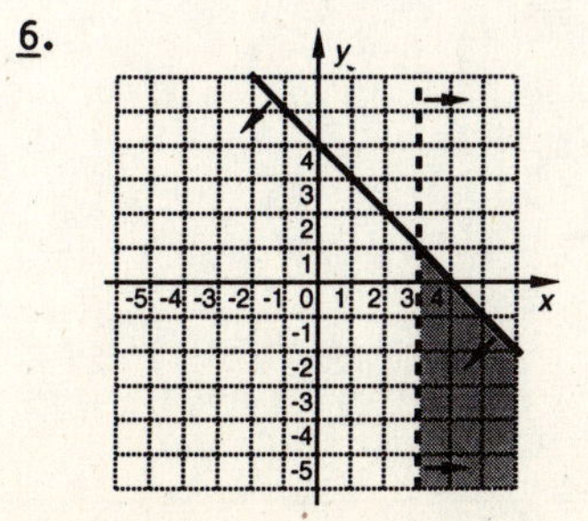

7. $y > 4x - 1$,

$y < -2x + 3$

We graph the lines $y = 4x - 1$ and $y = -2x + 3$ using dashed lines. We indicate the region for each inequality by the arrows at the ends of the lines. We shade the area where the regions overlap.

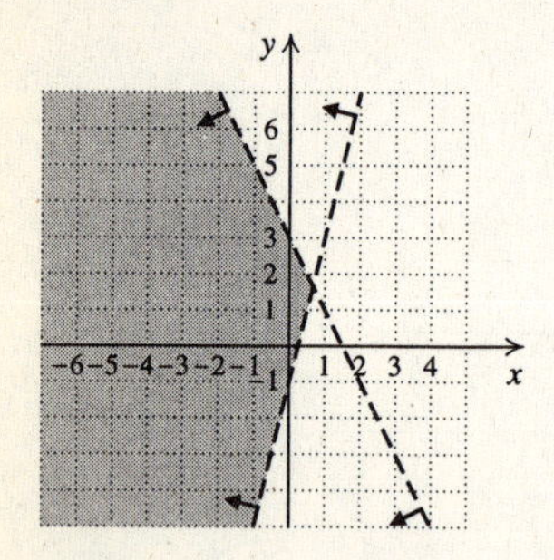

8.

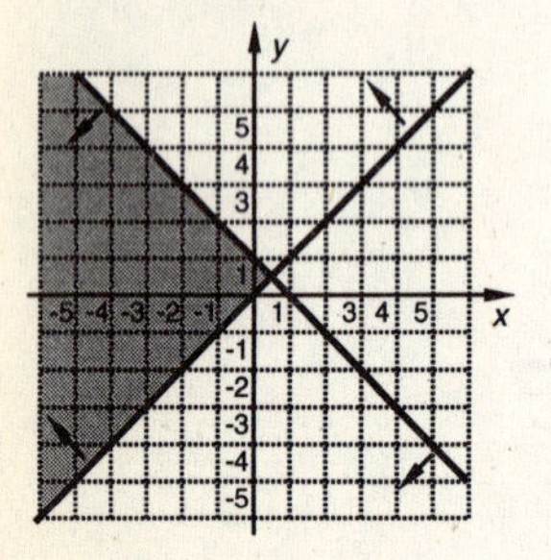

9. $2x - 3y \geq 9$,

$2y + x > 6$

We graph the line $2x - 3y = 9$ using a solid line and the line $2y + x = 6$ using a dashed line. We indicate the region for each inequality by the arrows at the ends of the lines. We shade the area where the regions overlap.

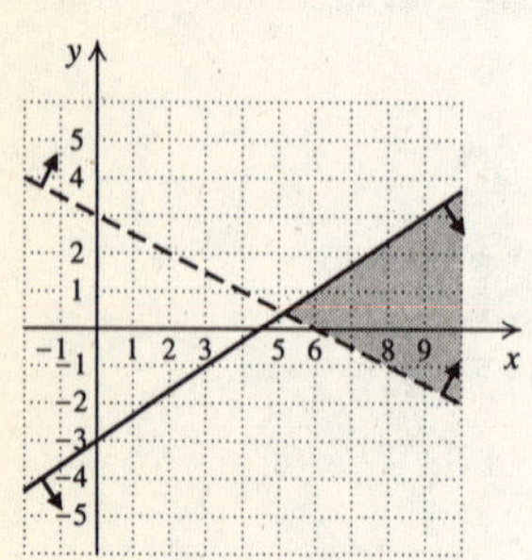

10.

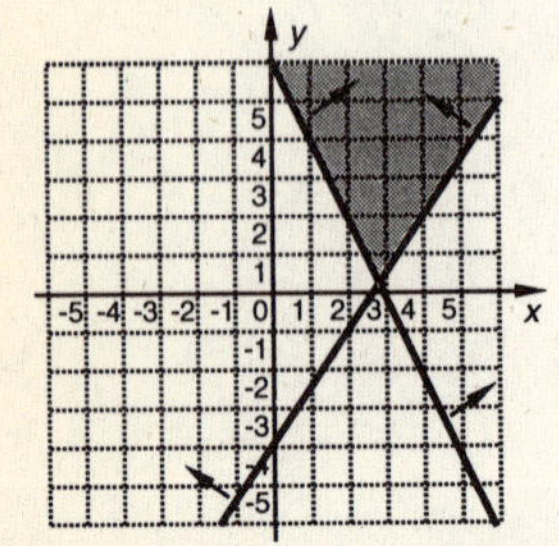

11. $y > 5x + 2$,

$y \leq 1 - x$

We graph the line $y = 5x + 2$ using a dashed line and the line $y = 1 - x$ using a solid line. We indicate the region for each inequality by the arrows at the ends of the lines. We shade the area where the regions overlap.

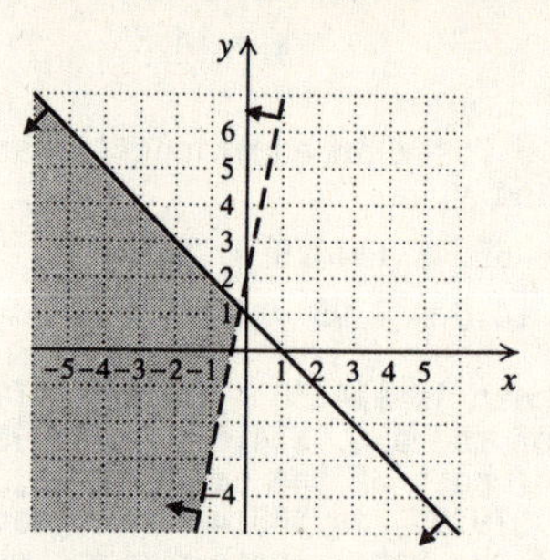

12.

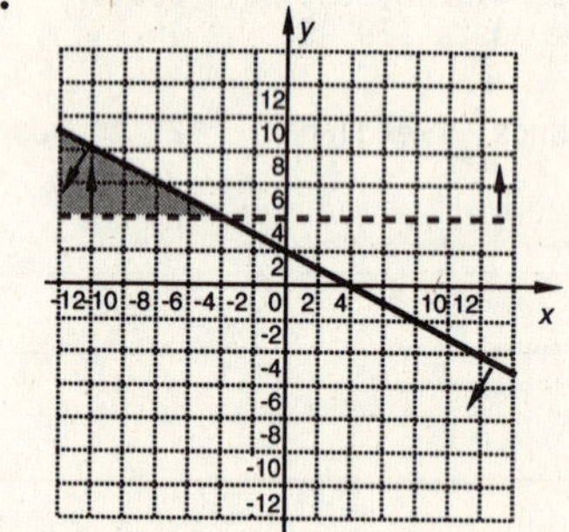

13. $x \leq 3$,

$y \leq 4$

We graph the lines $x = 3$ and $y = 4$ using solid lines. We indicate the region for each inequality by the arrows at the ends of the lines. We shade the area where the regions overlap.

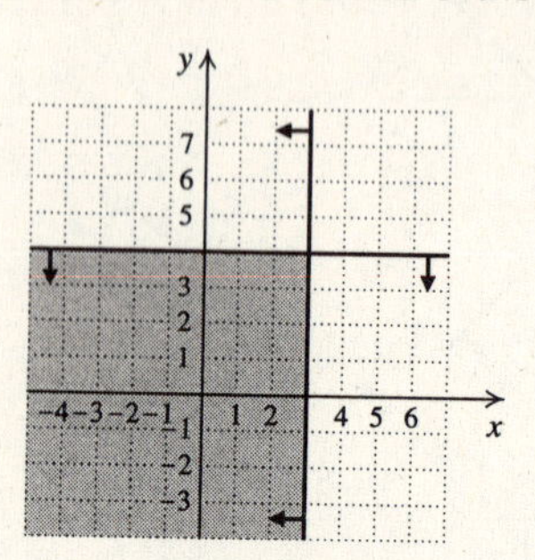

14.

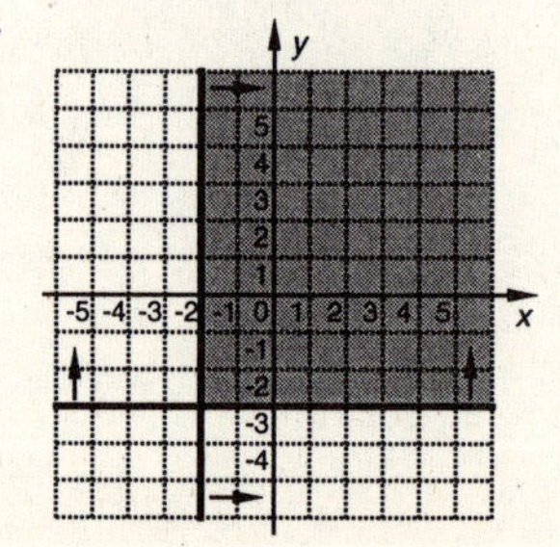

15. $x \leqslant 0,$

$y \leqslant 0$

We graph the lines $x = 0$ and $y = 0$ using solid lines. We indicate the region for each inequality by the arrows at the ends of the lines. We shade the area where the regions overlap.

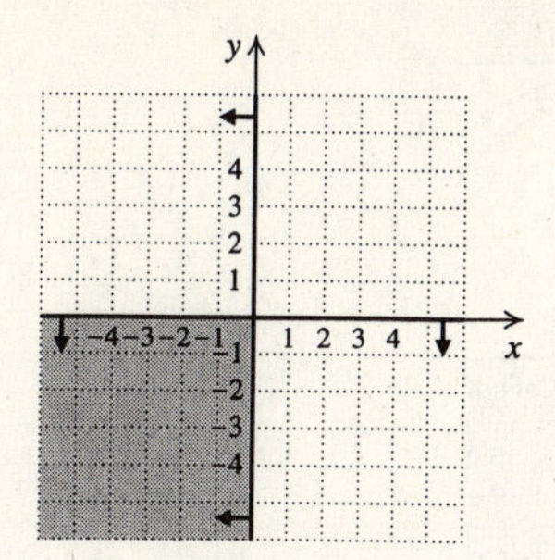

16.

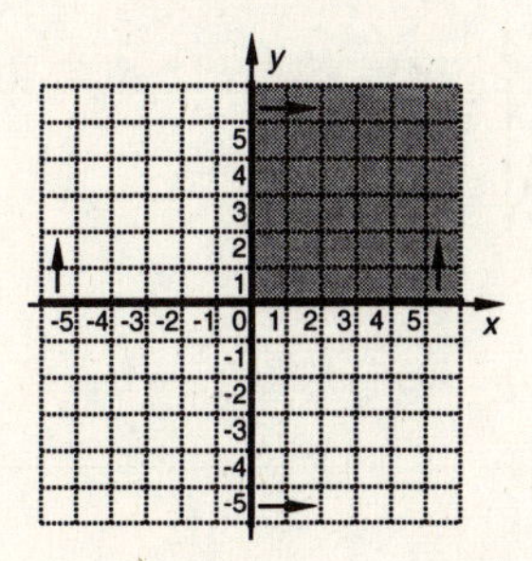

17. $x + y \leqslant 6,$

$x \geqslant 0,$

$y \geqslant 0,$

$y \leqslant 5$

We graph the lines $x + y = 6$, $x = 0$, $y = 0$, and $y = 5$ using solid lines. We indicate the region for each inequality by the arrows at the ends of the lines. We shade the area where the regions overlap.

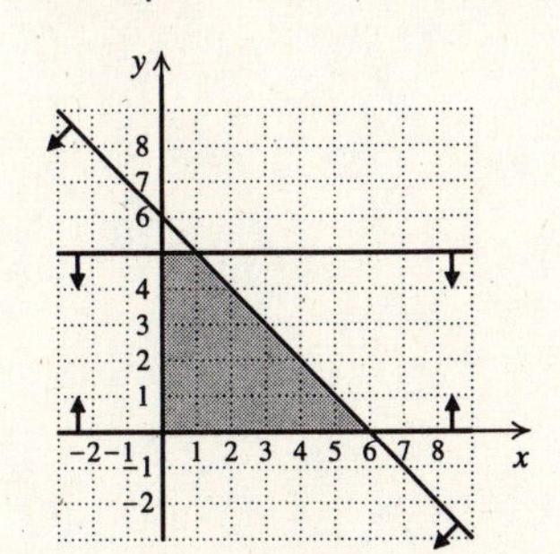

18.

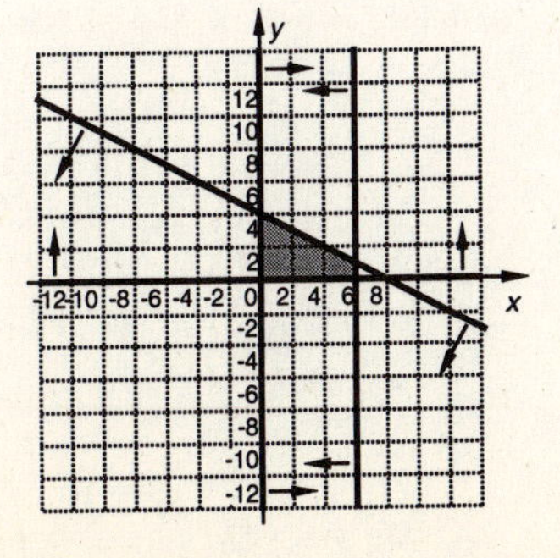

19. $y - x \geqslant 1,$

$y - x \leqslant 3,$

$x \leqslant 5,$

$x \geqslant 2$

We graph the lines $y - x = 1$, $y - x = 3$, $x = 5$, and $x = 2$ using solid lines. We indicate the region for each inequality by the arrows at the ends of the lines. We shade the area where the regions overlap.

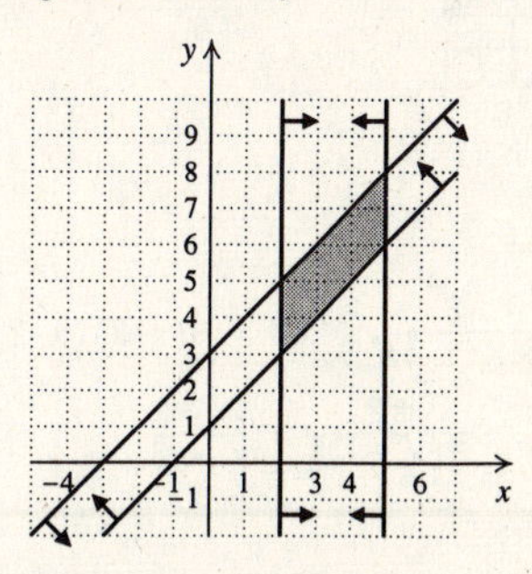

20.

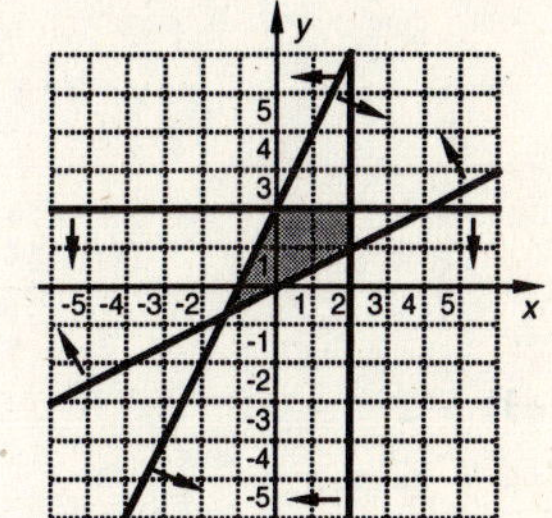

21. $y \leqslant x,$

$x \geqslant -2,$

$x \leqslant -y$

We graph the lines $y = x$, $x = -2$ and $x = -y$ using solid lines. We indicate the region for each inequality by the arrows at the ends of the lines. We shade the area where the regions overlap.

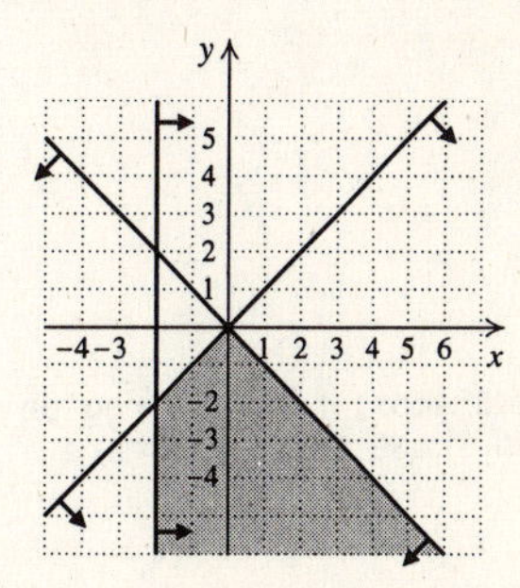

22. 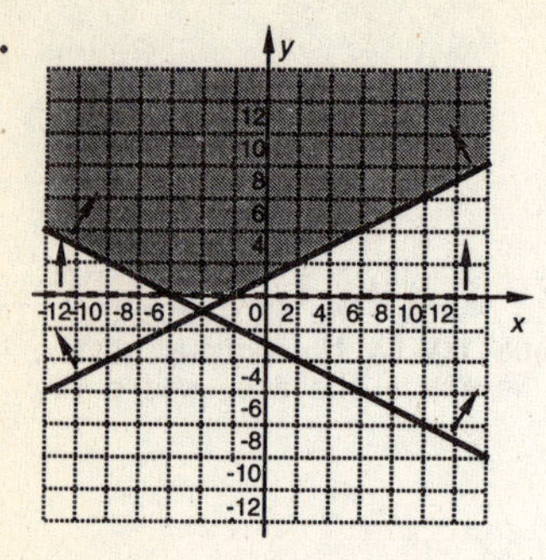

23. $\dfrac{3}{x^2 - 4} - \dfrac{2}{3x + 6}$

$= \dfrac{3}{(x + 2)(x - 2)} - \dfrac{2}{3(x + 2)}$, LCD = 3(x+2)(x-2)

$= \dfrac{3}{(x + 2)(x - 2)} \cdot \dfrac{3}{3} - \dfrac{2}{3(x + 2)} \cdot \dfrac{x - 2}{x - 2}$

$= \dfrac{9 - 2(x - 2)}{3(x + 2)(x - 2)}$

$= \dfrac{9 - 2x + 4}{3(x + 2)(x - 2)}$

$= \dfrac{13 - 2x}{3(x + 2)(x - 2)}$

24. $\dfrac{5a - 5 - 8a^2}{4a^2(a - 3)(a - 1)}$

25. $x^2 - 5x + 2 = (-3)^2 - 5(-3) + 2$

$= 9 + 15 + 2$

$= 26$

26. -153

27.

28.

29. $5a + 3b \geqslant 30,$

$2a + 3b \geqslant 21,$

$3a + 6b \geqslant 36,$

$a \geqslant 0,$

$b \geqslant 0$

We graph the related equations, find the region for each inequality, and shade the area where the regions overlap.

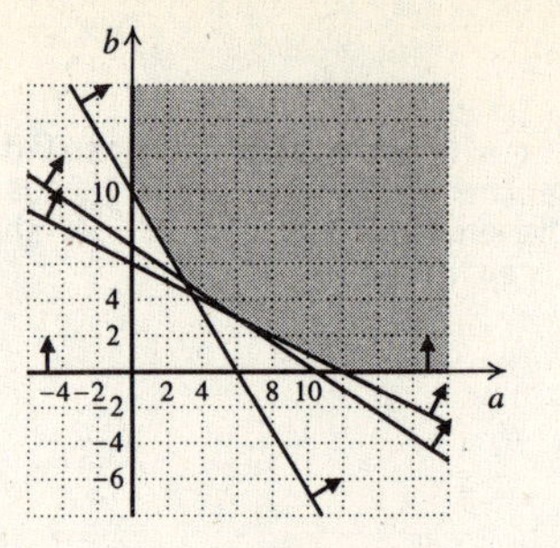

30. 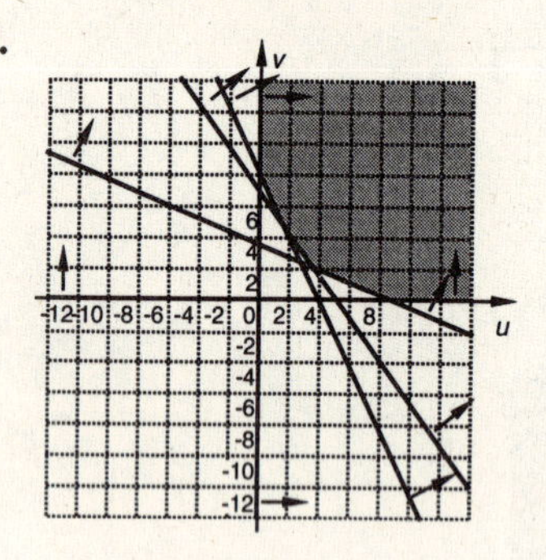

Exercise Set 9.1

1. The square roots of 1 are 1 and -1, because $1^2 = 1$ and $(-1)^2 = 1$.

2. 2, -2

3. The square roots of 16 are 4 and -4, because $4^2 = 16$ and $(-4)^2 = 16$.

4. 3, -3

5. The square roots of 49 are 7 and -7, because $7^2 = 49$ and $(-7)^2 = 49$.

6. 11, -11

7. The square roots of 169 are 13 and -13, because $13^2 = 169$ and $(-13)^2 = 169$.

8. 12, -12

9. $\sqrt{4} = 2$, finding the positive square root.

10. 1

11. $\sqrt{9} = 3$, so $-\sqrt{9} = -3$.

12. -5

13. $\sqrt{0} = 0$

14. -9

15. $\sqrt{121} = 11$, so $-\sqrt{121} = -11$.

16. 20

17. $\sqrt{361} = 19$, finding the positive square root.

18. 21

19. $\sqrt{144} = 12$, finding the positive square root.

20. 13

21. $\sqrt{625} = 25$, so $-\sqrt{625} = -25$.

22. -30

23. $a - 4$

24. $t + 3$

25. The radicand is the expression under the radical, $t^2 + 1$.

26. $x^2 + 5$

27. The radicand is the expression under the radical, $\frac{3}{x + 2}$.

28. $\frac{a}{a - b}$

29. $\sqrt{2}$ is irrational, since 2 is not a perfect square.

30. Irrational

31. $\sqrt{8}$ is irrational, since 8 is not a perfect square.

32. Irrational

33. $\sqrt{49}$ is rational, since 49 is a perfect square.

34. Rational

35. $\sqrt{98}$ is irrational, since 98 is not a perfect square.

36. Irrational

37. $-\sqrt{4}$ is rational, since 4 is a perfect square

38. Rational

39. $-\sqrt{12}$ is irrational, since 12 is not a perfect square.

40. Irrational

41. 2.236

42. 2.449

43. 4.123

44. 4.359

45. 9.644

46. 6.557

47. $\sqrt{t^2} = t$ Since t is assumed to be nonnegative

48. x

49. $\sqrt{9x^2} = \sqrt{(3x)^2} = 3x$ Since x is assumed to be nonnegative

50. 2a

51. $\sqrt{(ab)^2} = ab$

52. 6y

53. $\sqrt{(34d)^2} = 34d$

54. 53b

55. $\sqrt{(5ab)^2} = 5ab$

56. $7xy$

57. a) We substitute 25 into the formula:

$N = 2.5\sqrt{25} = 2.5(5) = 12.5 \approx 13$

b) We substitute 89 into the formula and use Table 1 or a calculator to find an approximation.

$N = 2.5\sqrt{89} \approx 2.5(9.434) = 23.585 \approx 24$

58. a) 20

b) 25

59. $m = \frac{\text{change in } y}{\text{change in } x} = \frac{-6 - 4}{5 - (-3)} = \frac{-10}{8} = -\frac{5}{4}$

60. $\frac{3}{5}$

61. Use the point-slope equation.

$y - y_1 = m(x - x_1)$

$y - 4 = 2(x - (-3))$

$y - 4 = 2(x + 3)$

$y - 4 = 2x + 6$

$y = 2x + 10$

62. $y = -\frac{5}{4}x + \frac{1}{4}$

63. ◈

64. ◈

65. We find the inner square root first.

$\sqrt{\sqrt{16}} = \sqrt{4} = 2$

66. 5

67. $-\sqrt{36} < -\sqrt{33} < -\sqrt{25}$, or $-6 < -\sqrt{33} < -5$

$-\sqrt{33}$ is between -6 and -5.

68. 64; answers may vary.

69. If $\sqrt{x^2} = 6$, then $x^2 = (6)^2$, or 36. Thus $x = 6$ or $x = -6$.

70. No solution

71. If $-\sqrt{x^2} = -3$, then $\sqrt{x^2} = 3$ and $x^2 = 3^2$, or 9. Thus, $x = 3$ or $x = -3$.

72. -7, 7

73. $\sqrt{(5a^2b)^2} = 5a^2b$ Since all variables represent positive numbers

74. $3a$

75. $\sqrt{\frac{144x^8}{36y^6}} = \sqrt{\frac{4x^8}{y^6}} = \sqrt{\left(\frac{2x^4}{y^3}\right)^2} = \frac{2x^4}{y^3}$

76. $\frac{y^6}{90}$

77. $\sqrt{\frac{169}{m^{16}}} = \sqrt{\left(\frac{13}{m^8}\right)^2} = \frac{13}{m^8}$

78. $\frac{p}{60}$

79. a) Locate 3 on the x-axis, move up vertically to the graph, and then move left horizontally to the y-axis to read the approximation.

$\sqrt{3} \approx 1.7$ (Answers may vary.)

b) Locate 5 on the x-axis, move up vertically to the graph, and then move left horizontally to the y-axis to read the approximation.

$\sqrt{5} \approx 2.2$ (Answers may vary.)

c) Locate 7 on the x-axis, move up vertically to the graph, and then move left horizontally to the y-axis to read the approximation.

$\sqrt{7} \approx 2.6$ (Answers may vary.)

80. $y_1 = \sqrt{x-2};\ y_2 = \sqrt{x+7};$
$y_3 = 5 + \sqrt{x};\ y_4 = -4 + \sqrt{x}$

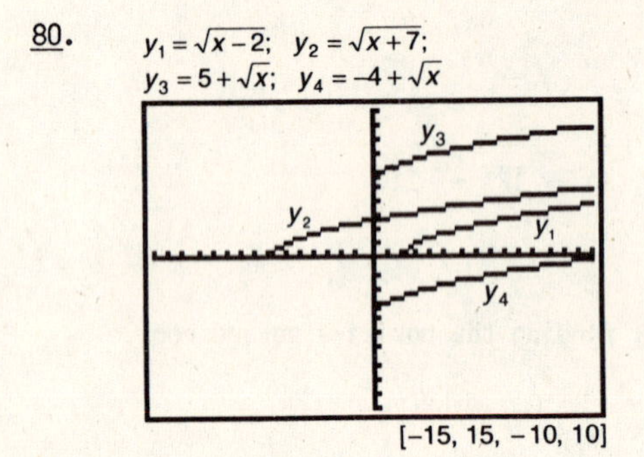

Exercise Set 9.2

1. $\sqrt{2}\sqrt{3} = \sqrt{2 \cdot 3} = \sqrt{6}$

2. $\sqrt{15}$

3. $\sqrt{4}\sqrt{3} = \sqrt{12}$, or

$\sqrt{4}\sqrt{3} = 2\sqrt{3}$ Taking the square root of 4

4. $\sqrt{18}$, or $3\sqrt{2}$

5. $\sqrt{\frac{2}{5}}\sqrt{\frac{3}{4}} = \sqrt{\frac{2 \cdot 3}{5 \cdot 4}} = \sqrt{\frac{3}{10}}$

6. $\sqrt{\frac{3}{40}}$, or $\frac{1}{2}\sqrt{\frac{3}{10}}$

7. $\sqrt{17}\ \sqrt{17} = \sqrt{17\cdot 17} = \sqrt{17^2} = 17$

8. 18

9. $\sqrt{25}\ \sqrt{3} = \sqrt{75}$, or

$\sqrt{25}\ \sqrt{3} = 5\sqrt{3}$ Taking the square root of 25

10. $\sqrt{72}$, or $6\sqrt{2}$

11. $\sqrt{2}\ \sqrt{x} = \sqrt{2\cdot x} = \sqrt{2x}$

12. $\sqrt{3a}$

13. $\sqrt{3}\ \sqrt{2x} = \sqrt{3\cdot 2x} = \sqrt{6x}$

14. $\sqrt{20x}$, or $2\sqrt{5x}$

15. $\sqrt{x}\ \sqrt{7y} = \sqrt{x\cdot 7y} = \sqrt{7xy}$

16. $\sqrt{10mn}$

17. $\sqrt{3a}\ \sqrt{2c} = \sqrt{3a\cdot 2c} = \sqrt{6ac}$

18. $\sqrt{3xyz}$

19. $\sqrt{12} = \sqrt{4\cdot 3}$ 4 is a perfect square

$= \sqrt{4}\ \sqrt{3}$ Factoring into a product of radicals

$= 2\sqrt{3}$ Taking the square root

20. $2\sqrt{2}$

21. $\sqrt{20} = \sqrt{4\cdot 5}$ 4 is a perfect square

$= \sqrt{4}\ \sqrt{5}$ Factoring into a product of radicals

$= 2\sqrt{5}$ Taking the square root

22. $3\sqrt{5}$

23. $\sqrt{200} = \sqrt{100\cdot 2}$ 100 is a perfect square

$= \sqrt{100}\ \sqrt{2}$

$= 10\sqrt{2}$

24. $10\sqrt{3}$

25. $\sqrt{9x} = \sqrt{9\cdot x} = \sqrt{9}\ \sqrt{x} = 3\sqrt{x}$

26. $2\sqrt{y}$

27. $\sqrt{75a} = \sqrt{25\cdot 3a} = \sqrt{25}\ \sqrt{3a} = 5\sqrt{3a}$

28. $2\sqrt{10m}$

29. $\sqrt{16a} = \sqrt{16\cdot a} = \sqrt{16}\ \sqrt{a} = 4\sqrt{a}$

30. $7\sqrt{b}$

31. $\sqrt{64y^2} = \sqrt{64}\ \sqrt{y^2} = 8y$, or

$\sqrt{64y^2} = \sqrt{(8y)^2} = 8y$

32. $3x$

33. $\sqrt{13x^2} = \sqrt{13}\ \sqrt{x^2} = \sqrt{13}\cdot x$, or $x\sqrt{13}$

34. $t\sqrt{29}$

35. $\sqrt{8t^2} = \sqrt{4\cdot t^2\cdot 2} = \sqrt{4}\ \sqrt{t^2}\ \sqrt{2} = 2t\sqrt{2}$

36. $5a\sqrt{5}$

37. $\sqrt{180} = \sqrt{36\cdot 5} = 6\sqrt{5}$

38. $7\sqrt{2}$

39. $\sqrt{288y} = \sqrt{144\cdot 2y} = \sqrt{144}\ \sqrt{2y} = 12\sqrt{2y}$

40. $11\sqrt{3p}$

41. $\sqrt{x^{20}} = \sqrt{(x^{10})^2} = x^{10}$

42. x^{15}

43. $\sqrt{x^{12}} = \sqrt{(x^6)^2} = x^6$

44. x^8

45. $\sqrt{x^5} = \sqrt{x^4x}$ One factor is a perfect square

$= \sqrt{x^4}\ \sqrt{x}$

$= \sqrt{(x^2)^2}\ \sqrt{x}$

$= x^2\sqrt{x}$

46. $x\sqrt{x}$

47. $\sqrt{t^{19}} = \sqrt{t^{18}t} = \sqrt{t^{18}}\ \sqrt{t} = \sqrt{(t^9)^2}\ \sqrt{t} = t^9\sqrt{t}$

48. $p^8\sqrt{p}$

49. $\sqrt{36m^3} = \sqrt{36\cdot m^2\cdot m} = \sqrt{36}\ \sqrt{m^2}\ \sqrt{m} = 6m\sqrt{m}$

50. $5y\sqrt{10y}$

51. $\sqrt{8a^5} = \sqrt{4a^4(2a)} = \sqrt{4(a^2)^2(2a)} =$

$\sqrt{4}\ \sqrt{(a^2)^2}\ \sqrt{2a} = 2a^2\sqrt{2a}$

52. $2b^3\sqrt{3b}$

53. $\sqrt{104p^{17}} = \sqrt{4p^{16}(26p)} = \sqrt{4(p^8)^2(26p)} =$

$\sqrt{4}\ \sqrt{(p^8)^2}\ \sqrt{26p} = 2p^8\sqrt{26p}$

54. $3m^{11}\sqrt{10m}$

55. $\sqrt{3}\sqrt{6} = \sqrt{18}$ Multiplying; note that 9 is a perfect-square factor of 18.

$= \sqrt{9\cdot 2}$ Factoring

$= \sqrt{9}\sqrt{2}$

$= 3\sqrt{2}$ Simplifying

56. $5\sqrt{2}$

57. $\sqrt{15}\sqrt{6} = \sqrt{90}$ Multiplying

$= \sqrt{9\cdot 10}$ Factoring

$= \sqrt{9}\sqrt{10}$

$= 3\sqrt{10}$

58. 9

59. $\sqrt{3x}\sqrt{12y} = \sqrt{36xy} = \sqrt{36}\sqrt{xy} = 6\sqrt{xy}$

60. $10\sqrt{xy}$

61. $\sqrt{10}\sqrt{10} = \sqrt{100} = 10$

62. $11\sqrt{x}$

63. $\sqrt{5b}\sqrt{15b} = \sqrt{75b^2} = \sqrt{25b^2\cdot 3} =$

$\sqrt{25}\sqrt{b^2}\sqrt{3} = 5b\sqrt{3}$

64. $6a\sqrt{3}$

65. $\sqrt{2t}\sqrt{2t} = \sqrt{4t^2} = 2t$

66. 3a

67. $\sqrt{ab}\sqrt{ac} = \sqrt{a^2bc} = \sqrt{a^2}\sqrt{bc} = a\sqrt{bc}$

68. $x\sqrt{yz}$

69. $\sqrt{2x}\sqrt{4x^5} = \sqrt{8x^6} = \sqrt{4x^6\cdot 2} = \sqrt{4}\sqrt{x^6}\sqrt{2} =$

$2x^3\sqrt{2}$

70. $5m^4\sqrt{3}$

71. $\sqrt{x^2y^3}\sqrt{xy^4} = \sqrt{x^3y^7} = \sqrt{x^2y^6\cdot xy} = \sqrt{x^2}\sqrt{y^6}\sqrt{xy} =$

$xy^3\sqrt{xy}$

72. $x^2y\sqrt{y}$

73. $\sqrt{50ab}\sqrt{10a^2b^4} = \sqrt{500a^3b^5} = \sqrt{100a^2b^4\cdot 5ab} =$

$\sqrt{100}\sqrt{a^2}\sqrt{b^4}\sqrt{5ab} = 10ab^2\sqrt{5ab}$

74. $5xy^2\sqrt{2xy}$

75. First we substitute 20 for L in the formula:

$r = 2\sqrt{5L} = 2\sqrt{5\cdot 20} = 2\sqrt{100} = 2\cdot 10 = 20$ mph

Then we substitute 150 for L:

$r = 2\sqrt{5\cdot 150} = 2\sqrt{750} = 2\sqrt{25\cdot 30} = 2\sqrt{25}\sqrt{30} =$

$2\cdot 5\sqrt{30} = 10\sqrt{30} \approx 10(5.477) \approx 54.77$ mph, or 54.8 mph (rounded to the nearest tenth)

76. 24.5 mph; 37.4 mph

77. Familiarize. We present the information in a table.

	d = Distance	r · Speed	t Time
First car	d	56	t
Second car	d	84	t - 1

Translate. From the rows of the table we get two equations:

$d = 56t,$

$d = 84(t - 1).$

Carry out. We use the substitution method.

$56t = 84(t - 1)$ Substituting 56t for d

$56t = 84t - 84$

$-28t = -84$

$t = 3$

The problem asks how far from Hereford the second car will overtake the first, so we need to find d. Substitute 3 for t in the first equation.

$d = 56t$

$d = 56\cdot 3$

$d = 168$

Check. If t = 3, then the first car travels 56·3, or 168 km, and the second car travels 84(3 - 1), or 84·2, or 168 km. Since the distances are the same, our answer checks.

State. The second car overtakes the first 168 km from Hereford.

78. r = 275 km/h

79.

80.

81. $\sqrt{0.01} = \sqrt{(0.1)^2} = 0.1$

82. 0.5

83. $\sqrt{0.0625} = \sqrt{(0.25)^2} = 0.25$

84. 0.001

85. ◈

86. $15 > 4\sqrt{14}$

87. $\sqrt{450} = \sqrt{225 \cdot 2} = 15\sqrt{2}$, so $15\sqrt{2} = \sqrt{450}$.

88. $16 > \sqrt{15}\sqrt{17}$

89. $3\sqrt{11} = \sqrt{9}\sqrt{11} = \sqrt{99}$ and
$7\sqrt{2} = \sqrt{49}\sqrt{2} = \sqrt{98}$, so
$3\sqrt{11} > 7\sqrt{2}$.

90. $5\sqrt{7} < 4\sqrt{11}$

91. Using a calculator or Table 1, we find $\sqrt{15} \approx 3.873$ and $\sqrt{17} \approx 4.123$, so $\sqrt{15} + \sqrt{17} \approx 3.873 + 4.123 \approx 7.996$. Then $8 > \sqrt{15} + \sqrt{17}$.

92. $18(x + 1)\sqrt{(x + 1)y}$

93. $\sqrt{18(x - 2)}\sqrt{20(x - 2)^3} = \sqrt{360(x - 2)^4} =$
$\sqrt{36(x - 2)^4 \cdot 10} = \sqrt{36}\sqrt{(x - 2)^4}\sqrt{10} =$
$6(x - 2)^2\sqrt{10}$

94. $2x^3\sqrt{5x}$

95. $\sqrt{2^{109}}\sqrt{x^{306}}\sqrt{x^{11}} = \sqrt{2^{109}x^{317}} =$
$\sqrt{2^{108} \cdot x^{316} \cdot 2x} = \sqrt{2^{108}}\sqrt{x^{316}}\sqrt{2x} = 2^{54}x^{158}\sqrt{2x}$

96. x^{4n}

97. $\sqrt{0.04x^{4n}} = \sqrt{(0.2x^{2n})^2} = 0.2x^{2n}$

98. $y^k\sqrt{y}$, where $k = \frac{n - 1}{2}$

Exercise Set 9.3

1. $\frac{\sqrt{18}}{\sqrt{2}} = \sqrt{\frac{18}{2}} = \sqrt{9} = 3$

2. 2

3. $\frac{\sqrt{60}}{\sqrt{15}} = \sqrt{\frac{60}{15}} = \sqrt{4} = 2$

4. 6

5. $\frac{\sqrt{75}}{\sqrt{15}} = \sqrt{\frac{75}{15}} = \sqrt{5}$

6. $\sqrt{6}$

7. $\frac{\sqrt{3}}{\sqrt{75}} = \sqrt{\frac{3}{75}} = \sqrt{\frac{1}{25}} = \frac{\sqrt{1}}{\sqrt{25}} = \frac{1}{5}$

8. $\frac{1}{4}$

9. $\frac{\sqrt{12}}{\sqrt{75}} = \sqrt{\frac{12}{75}} = \sqrt{\frac{4}{25}} = \frac{\sqrt{4}}{\sqrt{25}} = \frac{2}{5}$

10. $\frac{3}{4}$

11. $\frac{\sqrt{8x}}{\sqrt{2x}} = \sqrt{\frac{8x}{2x}} = \sqrt{4} = 2$

12. 3

13. $\frac{\sqrt{63y^3}}{\sqrt{7y}} = \sqrt{\frac{63y^3}{7y}} = \sqrt{9y^2} = 3y$

14. $4x$

15. $\frac{\sqrt{27x^5}}{\sqrt{3x}} = \sqrt{\frac{27x^5}{3x}} = \sqrt{9x^4} = 3x^2$

16. $2a^3$

17. $\frac{\sqrt{75x}}{\sqrt{3x^7}} = \sqrt{\frac{75x}{3x^7}} = \sqrt{\frac{25}{x^6}} = \frac{\sqrt{25}}{\sqrt{x^6}} = \frac{5}{x^3}$

18. $\frac{x^2}{3}$

19. $\frac{\sqrt{20a^{10}}}{\sqrt{10a^2}} = \sqrt{\frac{20a^{10}}{10a^2}} = \sqrt{2a^8} = \sqrt{a^8}\sqrt{2} = a^4\sqrt{2}$

20. $x^5\sqrt{5}$

21. $\sqrt{\frac{9}{49}} = \frac{\sqrt{9}}{\sqrt{49}} = \frac{3}{7}$

22. $\frac{4}{5}$

23. $\sqrt{\frac{1}{36}} = \frac{\sqrt{1}}{\sqrt{36}} = \frac{1}{6}$

24. $\frac{1}{2}$

25. $-\sqrt{\frac{16}{81}} = -\frac{\sqrt{16}}{\sqrt{81}} = -\frac{4}{9}$

26. $-\frac{5}{7}$

27. $\sqrt{\frac{64}{144}} = \sqrt{\frac{4}{9}} = \frac{\sqrt{4}}{\sqrt{9}} = \frac{2}{3}$

28. $\frac{9}{11}$

29. $\sqrt{\frac{1690}{1210}} = \sqrt{\frac{169\cdot 10}{121\cdot 10}} = \sqrt{\frac{169\cdot \cancel{10}}{121\cdot \cancel{10}}} = \frac{\sqrt{169}}{\sqrt{121}} = \frac{13}{11}$

30. $\frac{12}{25}$

31. $\sqrt{\frac{36}{a^2}} = \frac{\sqrt{36}}{\sqrt{a^2}} = \frac{6}{a}$

32. $\frac{5}{x}$

33. $\sqrt{\frac{9a^2}{625}} = \frac{\sqrt{9a^2}}{\sqrt{625}} = \frac{3a}{25}$

34. $\frac{xy}{12}$

35. $\sqrt{\frac{2}{5}} = \frac{\sqrt{2}}{\sqrt{5}} = \frac{\sqrt{2}}{\sqrt{5}} \cdot \frac{\sqrt{5}}{\sqrt{5}} = \frac{\sqrt{10}}{5}$

36. $\frac{\sqrt{14}}{7}$

37. $\sqrt{\frac{3}{8}} = \frac{\sqrt{3}}{\sqrt{8}} = \frac{\sqrt{3}}{\sqrt{4}\sqrt{2}} = \frac{\sqrt{3}}{2\sqrt{2}} = \frac{\sqrt{3}}{2\sqrt{2}} \cdot \frac{\sqrt{2}}{\sqrt{2}} = \frac{\sqrt{6}}{2\cdot 2} = \frac{\sqrt{6}}{4}$

38. $\frac{\sqrt{14}}{4}$

39. $\sqrt{\frac{7}{20}} = \frac{\sqrt{7}}{\sqrt{20}} = \frac{\sqrt{7}}{\sqrt{4}\sqrt{5}} = \frac{\sqrt{7}}{2\sqrt{5}} = \frac{\sqrt{7}}{2\sqrt{5}} \cdot \frac{\sqrt{5}}{\sqrt{5}} = \frac{\sqrt{35}}{2\cdot 5} = \frac{\sqrt{35}}{10}$

40. $\frac{\sqrt{3}}{6}$

41. $\sqrt{\frac{1}{18}} = \frac{\sqrt{1}}{\sqrt{18}} = \frac{\sqrt{1}}{\sqrt{9}\sqrt{2}} = \frac{1}{3\sqrt{2}} = \frac{1}{3\sqrt{2}} \cdot \frac{\sqrt{2}}{\sqrt{2}} = \frac{\sqrt{2}}{3\cdot 2} = \frac{\sqrt{2}}{6}$

42. $\frac{\sqrt{14}}{6}$

43. $\frac{3}{\sqrt{5}} = \frac{3}{\sqrt{5}} \cdot \frac{\sqrt{5}}{\sqrt{5}} = \frac{3\sqrt{5}}{5}$

44. $\frac{4\sqrt{3}}{3}$

45. $\sqrt{\frac{8}{3}} = \frac{\sqrt{8}}{\sqrt{3}} = \frac{\sqrt{4}\sqrt{2}}{\sqrt{3}} = \frac{2\sqrt{2}}{\sqrt{3}} = \frac{2\sqrt{2}}{\sqrt{3}} \cdot \frac{\sqrt{3}}{\sqrt{3}} = \frac{2\sqrt{6}}{3}$

46. $\frac{2\sqrt{15}}{5}$

47. $\sqrt{\frac{3}{x}} = \frac{\sqrt{3}}{\sqrt{x}} = \frac{\sqrt{3}}{\sqrt{x}} \cdot \frac{\sqrt{x}}{\sqrt{x}} = \frac{\sqrt{3x}}{x}$

48. $\frac{\sqrt{2x}}{x}$

49. $\sqrt{\frac{x}{y}} = \frac{\sqrt{x}}{\sqrt{y}} = \frac{\sqrt{x}}{\sqrt{y}} \cdot \frac{\sqrt{y}}{\sqrt{y}} = \frac{\sqrt{xy}}{y}$

50. $\frac{\sqrt{ab}}{b}$

51. $\frac{\sqrt{7}}{\sqrt{3}} = \frac{\sqrt{7}}{\sqrt{3}} \cdot \frac{\sqrt{3}}{\sqrt{3}} = \frac{\sqrt{21}}{3}$

52. $\frac{\sqrt{77}}{7}$

53. $\frac{\sqrt{9}}{\sqrt{8}} = \frac{\sqrt{9}}{\sqrt{4}\sqrt{2}} = \frac{3}{2\sqrt{2}} = \frac{3}{2\sqrt{2}} \cdot \frac{\sqrt{2}}{\sqrt{2}} = \frac{3\sqrt{2}}{2\cdot 2} = \frac{2\sqrt{3}}{4}$

54. $\frac{2\sqrt{3}}{9}$

55. $\frac{\sqrt{2}}{\sqrt{13}} = \frac{\sqrt{2}}{\sqrt{13}} \cdot \frac{\sqrt{13}}{\sqrt{13}} = \frac{\sqrt{26}}{13}$

56. $\frac{\sqrt{6}}{2}$

57. $\frac{2}{\sqrt{2}} = \frac{2}{\sqrt{2}} \cdot \frac{\sqrt{2}}{\sqrt{2}} = \frac{2\sqrt{2}}{2} = \sqrt{2}$

58. $\sqrt{3}$

59. $\frac{\sqrt{5}}{\sqrt{27}} = \frac{\sqrt{5}}{\sqrt{9}\sqrt{3}} = \frac{\sqrt{5}}{3\sqrt{3}} = \frac{\sqrt{5}}{3\sqrt{3}} \cdot \frac{\sqrt{3}}{\sqrt{3}} = \frac{\sqrt{15}}{3\cdot 3} = \frac{\sqrt{15}}{9}$

60. $\frac{\sqrt{77}}{11}$

61. $\frac{\sqrt{7}}{\sqrt{12}} = \frac{\sqrt{7}}{\sqrt{4}\sqrt{3}} = \frac{\sqrt{7}}{2\sqrt{3}} = \frac{\sqrt{7}}{2\sqrt{3}} \cdot \frac{\sqrt{3}}{\sqrt{3}} = \frac{\sqrt{21}}{2\cdot 3} = \frac{\sqrt{21}}{6}$

62. $\frac{\sqrt{10}}{6}$

63. $\frac{\sqrt{x}}{\sqrt{32}} = \frac{\sqrt{x}}{\sqrt{16}\sqrt{2}} = \frac{\sqrt{x}}{4\sqrt{2}} = \frac{\sqrt{x}}{4\sqrt{2}} \cdot \frac{\sqrt{2}}{\sqrt{2}} = \frac{\sqrt{2x}}{4\cdot 2} = \frac{\sqrt{2x}}{8}$

64. $\frac{\sqrt{10a}}{20}$

65. $\frac{\sqrt{8}}{\sqrt{18}} = \sqrt{\frac{8}{18}} = \sqrt{\frac{4\cdot 2}{9\cdot 2}} = \sqrt{\frac{4\cdot \cancel{2}}{9\cdot \cancel{2}}} = \frac{\sqrt{4}}{\sqrt{9}} = \frac{2}{3}$

66. $\frac{\sqrt{42}}{14}$

67. $\frac{\sqrt{3}}{\sqrt{x}} = \frac{\sqrt{3}}{\sqrt{x}} \cdot \frac{\sqrt{x}}{\sqrt{x}} = \frac{\sqrt{3x}}{x}$

68. $\frac{\sqrt{2y}}{y}$

69. $\frac{4y}{\sqrt{3}} = \frac{4y}{\sqrt{3}} \cdot \frac{\sqrt{3}}{\sqrt{3}} = \frac{4y\sqrt{3}}{3}$

70. $\frac{8x\sqrt{5}}{5}$

71. $\frac{\sqrt{6a}}{\sqrt{8}} = \sqrt{\frac{6a}{8}} = \sqrt{\frac{3a\cdot 2}{4\cdot 2}} = \sqrt{\frac{3a\cdot \cancel{2}}{4\cdot \cancel{2}}} = \frac{\sqrt{3a}}{\sqrt{4}} = \frac{\sqrt{3a}}{2}$

72. $\frac{\sqrt{x}}{3}$

73. $\frac{\sqrt{50}}{\sqrt{12x}} = \sqrt{\frac{50}{12x}} = \sqrt{\frac{25\cdot 2}{6x\cdot 2}} = \sqrt{\frac{25\cdot \cancel{2}}{6x\cdot \cancel{2}}} = \frac{\sqrt{25}}{\sqrt{6x}} = \frac{5}{\sqrt{6x}} =$

$\frac{5}{\sqrt{6x}} \cdot \frac{\sqrt{6x}}{\sqrt{6x}} = \frac{5\sqrt{6x}}{6x}$

74. $\frac{3\sqrt{10a}}{4a}$

75. $\frac{\sqrt{27c}}{\sqrt{32c^3}} = \sqrt{\frac{27c}{32c^3}} = \sqrt{\frac{27}{32c^2}} = \frac{\sqrt{9}\,\sqrt{3}}{\sqrt{16}\,\sqrt{c^2}\,\sqrt{2}} = \frac{3\sqrt{3}}{4c\sqrt{2}} =$

$\frac{3\sqrt{3}}{4c\sqrt{2}} \cdot \frac{\sqrt{2}}{\sqrt{2}} = \frac{3\sqrt{6}}{4c\cdot 2} = \frac{3\sqrt{6}}{8c}$

76. $\frac{x\sqrt{21}}{6}$

77. 2 ft: $T = 2\pi\sqrt{\frac{L}{32}} \approx 2(3.14)\sqrt{\frac{2}{32}} \approx 6.28\sqrt{\frac{1}{16}} \approx$

$6.28\left(\frac{1}{4}\right) \approx 1.57$ sec

8 ft: $T = 2\pi\sqrt{\frac{L}{32}} \approx 2(3.14)\sqrt{\frac{8}{32}} \approx 6.28\sqrt{\frac{1}{4}} \approx$

$6.28\left(\frac{1}{2}\right) = 3.14$ sec

64 ft: $T = 2\pi\sqrt{\frac{L}{32}} \approx 2(3.14)\sqrt{\frac{64}{32}} \approx 6.28\sqrt{2} \approx$

8.88 sec

100 ft: $T = 2\pi\sqrt{\frac{L}{32}} \approx 2(3.14)\sqrt{\frac{100}{32}} \approx 6.28\sqrt{\frac{50}{16}} \approx$

$\frac{6.28\sqrt{50}}{4} = 11.10$ sec

78. 0.262 sec

79. $T = 2\pi\sqrt{\frac{L}{32}} = 2\pi\sqrt{\frac{\frac{32}{\pi^2}}{32}} = 2\pi\sqrt{\frac{32}{\pi^2}\cdot\frac{1}{32}} = 2\pi\sqrt{\frac{1}{\pi^2}} = 2\pi\left(\frac{1}{\pi}\right) =$

2 sec

The time it takes the pendulum to swing from one side to the other and back is 2 sec, so it takes 1 sec to swing from one side to the other.

80. $\frac{3\sqrt{10}}{8}$ sec, or approximately 1.186 sec

81. $x = y + 2$,

$x + y = 6$

We first write both equations in the form $Ax + By = C$.

$x - y = 2$ (1)

$x + y = 6$ (2)

$2x = 8$ Adding

$x = 4$

Substitute 4 for x in one of the original equations and solve for y.

$4 + y = 6$ (2)

$y = 2$

The ordered pair (4,2) checks in both equations. It is the solution.

82. $\left(4,\frac{1}{3}\right)$

83. $y = kx$

$30 = k\cdot 9$ Substituting 30 for y and 9 for x

$\frac{30}{9} = k$

$\frac{10}{3} = k$ Simplifying

The equation of variation is $y = \frac{10}{3}x$.

84. $y = \frac{1}{4}x$

85.

86.

87. $\sqrt{\frac{5}{1600}} = \frac{\sqrt{5}}{\sqrt{1600}} = \frac{\sqrt{5}}{40}$

88. $\frac{\sqrt{30}}{100}$

89. $\sqrt{\frac{1}{5x^3}} = \frac{\sqrt{1}}{\sqrt{x^2}\,\sqrt{5x}} = \frac{1}{x\sqrt{5x}} = \frac{1}{x\sqrt{5x}} \cdot \frac{\sqrt{5x}}{\sqrt{5x}} = \frac{\sqrt{5x}}{x\cdot 5x} = \frac{\sqrt{5x}}{5x^2}$

90. $\frac{\sqrt{3xy}}{ax^2}$

91. $\sqrt{\frac{3a}{b}} = \frac{\sqrt{3a}}{\sqrt{b}} = \frac{\sqrt{3a}}{\sqrt{b}} \cdot \frac{\sqrt{b}}{\sqrt{b}} = \frac{\sqrt{3ab}}{b}$

92. $\frac{\sqrt{5z}}{5zw}$

93. $\sqrt{\frac{1}{x^2} - \frac{2}{xy} + \frac{1}{y^2}}$, LCD is x^2y^2

$= \sqrt{\frac{1}{x^2} \cdot \frac{y^2}{y^2} - \frac{2}{xy} \cdot \frac{xy}{xy} + \frac{1}{y^2} \cdot \frac{x^2}{x^2}}$

$= \sqrt{\frac{y^2 - 2xy + x^2}{x^2y^2}}$

$= \sqrt{\frac{(y - x)^2}{x^2y^2}}$

$= \frac{\sqrt{(y - x)^2}}{\sqrt{x^2y^2}}$

$= \frac{y - x}{xy}$

An alternate method of simplifying this expression is shown below.

$\sqrt{\frac{1}{x^2} - \frac{2}{xy} + \frac{1}{y^2}} = \sqrt{\left[\frac{1}{x} - \frac{1}{y}\right]^2}$

$= \frac{1}{x} - \frac{1}{y}$

The two answers are equivalent.

94. $\frac{(z^2 - 1)\sqrt{2}}{z^2}$

Exercise Set 9.4

1. $3\sqrt{2} + 4\sqrt{2} = (3 + 4)\sqrt{2}$ Using the distributive law

$= 7\sqrt{2}$

2. $11\sqrt{3}$

3. $7\sqrt{5} - 3\sqrt{5} = (7 - 3)\sqrt{5}$ Using the distributive law

$= 4\sqrt{5}$

4. $3\sqrt{2}$

5. $6\sqrt{x} + 7\sqrt{x} = (6 + 7)\sqrt{x} = 13\sqrt{x}$

6. $12\sqrt{y}$

7. $9\sqrt{x} - 11\sqrt{x} = (9 - 11)\sqrt{x} = -2\sqrt{x}$

8. $-8\sqrt{a}$

9. $5\sqrt{2a} + 3\sqrt{2a} = (5 + 3)\sqrt{2a} = 8\sqrt{2a}$

10. $9\sqrt{6x}$

11. $9\sqrt{10y} - \sqrt{10y} = (9 - 1)\sqrt{10y} = 8\sqrt{10y}$

12. $11\sqrt{14y}$

13. $5\sqrt{7} + 2\sqrt{7} + 4\sqrt{7} = (5 + 2 + 4)\sqrt{7} = 11\sqrt{7}$

14. $15\sqrt{5}$

15. $8\sqrt{2} - 11\sqrt{2} + 4\sqrt{2} = (8 - 11 + 4)\sqrt{2} = 1\sqrt{2}$, or $\sqrt{2}$

16. $\sqrt{10}$

17. $5\sqrt{3} + \sqrt{8} = 5\sqrt{3} + \sqrt{4\cdot 2}$ Factoring 8

$= 5\sqrt{3} + \sqrt{4}\sqrt{2}$

$= 5\sqrt{3} + 2\sqrt{2}$

$5\sqrt{3} + \sqrt{8}$, or $5\sqrt{3} + 2\sqrt{2}$, cannot be simplified further.

18. $5\sqrt{5}$

19. $\sqrt{x} - \sqrt{9x} = \sqrt{x} - \sqrt{9}\sqrt{x}$

$= \sqrt{x} - 3\sqrt{x}$

$= (1 - 3)\sqrt{x}$

$= -2\sqrt{x}$

20. $4\sqrt{a}$

21. $5\sqrt{8} + 15\sqrt{2} = 5\sqrt{4\cdot 2} + 15\sqrt{2}$

$= 5\cdot 2\sqrt{2} + 15\sqrt{2}$

$= 10\sqrt{2} + 15\sqrt{2}$

$= 25\sqrt{2}$

22. $8\sqrt{3}$

23. $\sqrt{27} - 2\sqrt{3} = \sqrt{9\cdot 3} - 2\sqrt{3}$

$= 3\sqrt{3} - 2\sqrt{3}$

$= (3 - 2)\sqrt{3}$

$= 1\sqrt{3}$

$= \sqrt{3}$

24. $32\sqrt{2}$

25. $\sqrt{45} - \sqrt{20} = \sqrt{9\cdot 5} - \sqrt{4\cdot 5}$

$= 3\sqrt{5} - 2\sqrt{5}$

$= (3 - 2)\sqrt{5}$

$= 1\sqrt{5}$

$= \sqrt{5}$

26. $\sqrt{3}$

27. $\sqrt{72} + \sqrt{98} = \sqrt{36\cdot 2} + \sqrt{49\cdot 2}$

$= 6\sqrt{2} + 7\sqrt{2}$

$= (6 + 7)\sqrt{2}$

$= 13\sqrt{2}$

28. $7\sqrt{5}$

29. $2\sqrt{12} + \sqrt{27} - \sqrt{48} = 2\sqrt{4\cdot 3} + \sqrt{9\cdot 3} - \sqrt{16\cdot 3}$

$= 2\cdot 2\sqrt{3} + 3\sqrt{3} - 4\sqrt{3}$

$= (4 + 3 - 4)\sqrt{3}$

$= 3\sqrt{3}$

30. $19\sqrt{2}$

31. $3\sqrt{18} - 2\sqrt{32} - 5\sqrt{50} = 3\sqrt{9\cdot 2} - 2\sqrt{16\cdot 2} - 5\sqrt{25\cdot 2}$

$= 3\cdot 3\sqrt{2} - 2\cdot 4\sqrt{2} - 5\cdot 5\sqrt{2}$

$= (9 - 8 - 25)\sqrt{2}$

$= -24\sqrt{2}$

32. $2\sqrt{2}$

33. $\sqrt{9x} + \sqrt{49x} - 9\sqrt{x} = 3\sqrt{x} + 7\sqrt{x} - 9\sqrt{x}$

$= (3 + 7 - 9)\sqrt{x}$

$= 1\sqrt{x}$

$= \sqrt{x}$

34. $5\sqrt{a}$

35. $\sqrt{3}(\sqrt{5} + \sqrt{7}) = \sqrt{3}\cdot\sqrt{5} + \sqrt{3}\cdot\sqrt{7}$
Using the distributive law

$= \sqrt{15} + \sqrt{21}$

36. $\sqrt{10} + \sqrt{55}$

37. $\sqrt{7}(\sqrt{6} - \sqrt{15}) = \sqrt{7}\cdot\sqrt{6} - \sqrt{7}\cdot\sqrt{15}$

$= \sqrt{42} - \sqrt{105}$

38. $3\sqrt{10} - \sqrt{42}$

39. $(3 + \sqrt{2})(5 + \sqrt{2}) = 3\cdot 5 + 3\cdot\sqrt{2} + \sqrt{2}\cdot 5 +$

$\sqrt{2}\cdot\sqrt{2}$ Using FOIL

$= 15 + 3\sqrt{2} + 5\sqrt{2} + 2$

$= 17 + 8\sqrt{2}$ Collecting like terms

40. $26 + 8\sqrt{11}$

41. $(\sqrt{6} - 2)(\sqrt{6} - 5) = \sqrt{6}\cdot\sqrt{6} - \sqrt{6}\cdot 5 -$

$2\cdot\sqrt{6} + 2(5)$ Using FOIL

$= 6 - 5\sqrt{6} - 2\sqrt{6} + 10$

$= 16 - 7\sqrt{6}$

42. $-18 - 3\sqrt{10}$

43. $(\sqrt{5} + 7)(\sqrt{5} - 7) = (\sqrt{5})^2 - 7^2$
Using $(A + B)(A - B) = A^2 - B^2$

$= 5 - 49$

$= -44$

44. -4

45. $(\sqrt{6} - \sqrt{3})(\sqrt{6} + \sqrt{3}) = (\sqrt{6})^2 - (\sqrt{3})^2$
Using $(A - B)(A + B) = A^2 - B^2$

$= 6 - 3$

$= 3$

46. -4

47. $(5 + 3\sqrt{2})(1 - \sqrt{2})$

$= 5\cdot 1 - 5\cdot\sqrt{2} + 3\sqrt{2}\cdot 1 - 3\sqrt{2}\cdot\sqrt{2}$ Using FOIL

$= 5 - 5\sqrt{2} + 3\sqrt{2} - 3\cdot 2$

$= 5 - 2\sqrt{2} - 6$

$= -1 - 2\sqrt{2}$

48. $-2 + 5\sqrt{7}$

49. $(6 + \sqrt{3})^2 = 6^2 + 2\cdot 6\cdot\sqrt{3} + (\sqrt{3})^2$
Using $(A + B)^2 = A^2 + 2AB + B^2$

$= 36 + 12\sqrt{3} + 3$

$= 39 + 12\sqrt{3}$

50. $9 + 4\sqrt{5}$

51. $(5 - 2\sqrt{3})^2 = 5^2 - 2\cdot 5\cdot 2\sqrt{3} + (2\sqrt{3})^2$
Using $(A - B)^2 = A^2 - 2AB + B^2$

$= 25 - 20\sqrt{3} + 4\cdot 3$

$= 25 - 20\sqrt{3} + 12$

$= 37 - 20\sqrt{3}$

52. $81 - 36\sqrt{5}$

53. $(\sqrt{x} - \sqrt{10})^2 = (\sqrt{x})^2 - 2\sqrt{x}\sqrt{10} + (\sqrt{10})^2$

$= x - 2\sqrt{10x} + 10$

54. $a - 2\sqrt{6a} + 6$

55. $\frac{5}{1+\sqrt{2}} = \frac{5}{1+\sqrt{2}} \cdot \frac{1-\sqrt{2}}{1-\sqrt{2}}$ Multiplying by 1

$= \frac{5(1-\sqrt{2})}{(1+\sqrt{2})(1-\sqrt{2})}$

$= \frac{5-5\sqrt{2}}{1^2-(\sqrt{2})^2}$

$= \frac{5-5\sqrt{2}}{1-2}$

$= \frac{5-5\sqrt{2}}{-1}$

$= \frac{-(5-5\sqrt{2})}{1}$ Since $\frac{a}{-b} = \frac{-a}{b}$

$= -5+5\sqrt{2}$

56. $\frac{3-\sqrt{5}}{2}$

57. $\frac{4}{2-\sqrt{5}} = \frac{4}{2-\sqrt{5}} \cdot \frac{2+\sqrt{5}}{2+\sqrt{5}}$

$= \frac{4(2+\sqrt{5})}{(2-\sqrt{5})(2+\sqrt{5})}$

$= \frac{8+4\sqrt{5}}{2^2-(\sqrt{5})^2}$

$= \frac{8+4\sqrt{5}}{4-5}$

$= \frac{8+4\sqrt{5}}{-1}$

$= \frac{-(8+4\sqrt{5})}{1}$ Since $\frac{a}{-b} = \frac{-a}{b}$

$= -8-4\sqrt{5}$

58. $\frac{21+3\sqrt{2}}{47}$

59. $\frac{2}{\sqrt{7}+3} = \frac{2}{\sqrt{7}+3} \cdot \frac{\sqrt{7}-3}{\sqrt{7}-3}$

$= \frac{2(\sqrt{7}-3)}{(\sqrt{7})^2-3^2}$

$= \frac{2\sqrt{7}-6}{7-9}$

$= \frac{2\sqrt{7}-6}{-2}$ All terms have a factor of 2

$= \frac{\cancel{2}(\sqrt{7}-3)}{\cancel{2}(-1)}$ Factoring and removing a factor of 1

$= \frac{\sqrt{7}-3}{-1}$

$= \frac{-(\sqrt{7}-3)}{1}$

$= -\sqrt{7}+3$

60. $-\frac{2\sqrt{10}-10}{5}$

61. $\frac{\sqrt{6}}{\sqrt{6}-5} = \frac{\sqrt{6}}{\sqrt{6}-5} \cdot \frac{\sqrt{6}+5}{\sqrt{6}+5} = \frac{\sqrt{6}\sqrt{6}+\sqrt{6}\cdot 5}{(\sqrt{6})^2-5^2} =$

$\frac{6+5\sqrt{6}}{6-25} = \frac{6+5\sqrt{6}}{-19} = -\frac{6+5\sqrt{6}}{19}$

62. $-\frac{10+7\sqrt{10}}{39}$

63. $\frac{\sqrt{5}}{\sqrt{5}-\sqrt{3}} = \frac{\sqrt{5}}{\sqrt{5}-\sqrt{3}} \cdot \frac{\sqrt{5}+\sqrt{3}}{\sqrt{5}+\sqrt{3}} =$

$\frac{\sqrt{5}\sqrt{5}+\sqrt{5}\sqrt{3}}{(\sqrt{5})^2-(\sqrt{3})^2} = \frac{5+\sqrt{15}}{5-3} = \frac{5+\sqrt{15}}{2}$

64. $\frac{7+\sqrt{35}}{2}$

65. $\frac{\sqrt{14}}{\sqrt{10}+\sqrt{14}} = \frac{\sqrt{14}}{\sqrt{10}+\sqrt{14}} \cdot \frac{\sqrt{10}-\sqrt{14}}{\sqrt{10}-\sqrt{14}} =$

$\frac{\sqrt{14}\sqrt{10}-\sqrt{14}\sqrt{14}}{(\sqrt{10})^2-(\sqrt{14})^2} = \frac{\sqrt{140}-14}{10-14} =$

$\frac{\sqrt{4\cdot 35}-14}{-4} = \frac{2\sqrt{35}-14}{-4} = \frac{\cancel{2}(\sqrt{35}-7)}{\cancel{2}(-2)} =$

$\frac{\sqrt{35}-7}{-2} = -\frac{\sqrt{35}-7}{2}$

66. $\frac{\sqrt{21}+3}{4}$

67. $\frac{\sqrt{3}-\sqrt{2}}{\sqrt{3}+\sqrt{2}} = \frac{\sqrt{3}-\sqrt{2}}{\sqrt{3}+\sqrt{2}} \cdot \frac{\sqrt{3}-\sqrt{2}}{\sqrt{3}-\sqrt{2}}$

$= \frac{(\sqrt{3}-\sqrt{2})^2}{(\sqrt{3}+\sqrt{2})(\sqrt{3}-\sqrt{2})}$

$= \frac{(\sqrt{3})^2-2\sqrt{3}\sqrt{2}+(\sqrt{2})^2}{(\sqrt{3})^2-(\sqrt{2})^2}$

$= \frac{3-2\sqrt{6}+2}{3-2} = \frac{5-2\sqrt{6}}{1}$

$= 5-2\sqrt{6}$

68. $4+\sqrt{15}$

69. $\frac{\sqrt{6}+\sqrt{5}}{\sqrt{6}-\sqrt{5}} = \frac{\sqrt{6}+\sqrt{5}}{\sqrt{6}-\sqrt{5}} \cdot \frac{\sqrt{6}+\sqrt{5}}{\sqrt{6}+\sqrt{5}} =$

$= \frac{(\sqrt{6}+\sqrt{5})^2}{(\sqrt{6}-\sqrt{5})(\sqrt{6}+\sqrt{5})}$

$= \frac{(\sqrt{6})^2+2\sqrt{6}\sqrt{5}+(\sqrt{5})^2}{(\sqrt{6})^2-(\sqrt{5})^2}$

$= \frac{6+2\sqrt{30}+5}{6-5} = \frac{11+2\sqrt{30}}{1}$

$= 11+2\sqrt{30}$

70. $\frac{17 - 2\sqrt{70}}{3}$

71. $\frac{1 - \sqrt{7}}{3 + \sqrt{7}} = \frac{1 - \sqrt{7}}{3 + \sqrt{7}} \cdot \frac{3 - \sqrt{7}}{3 - \sqrt{7}} =$

$\frac{1\cdot 3 - 1 \cdot \sqrt{7} - \sqrt{7}\cdot 3 + \sqrt{7}\sqrt{7}}{3^2 - (\sqrt{7})^2} =$

$\frac{3 - \sqrt{7} - 3\sqrt{7} + 7}{9 - 7} = \frac{10 - 4\sqrt{7}}{2} = \frac{\not{2}(5 - 2\sqrt{7})}{\not{2}\cdot 1} =$

$5 - 2\sqrt{7}$

72. $\frac{13 + 6\sqrt{5}}{11}$

73. Familiarize and Translate. The problem states that we have inverse variation between t and r. Thus, an equation $t = \frac{k}{r}$ applies.

Carry out.

First find an equation of variation.

$t = \frac{k}{r}$

$2 = \frac{k}{40}$ Substituting 2 for t and 40 for r

$80 = k$ Multiplying by 40

The equation of variation is $t = \frac{80}{r}$.

When $r = 60$,

$t = \frac{80}{r}$

$t = \frac{80}{60}$ Substituting 60 for r

$t = \frac{4}{3}$

Check. Let us do some reasoning about the answer. The time decreased from 2 hr to $\frac{4}{3}$ hr when the speed increased from 40 mph to 60 mph. This is what we would expect with inverse variation. We can also do the calculation again. Furthermore, we note that the products $2\cdot 40$ and $\frac{4}{3} \cdot 60$ are both 80.

State. It will take $\frac{4}{3}$ hr to travel the fixed distance at 60 mph.

Since $t = \frac{80}{r}$, or $rt = 80$, the variation constant is the fixed distance, 80 mi. (Speed·Time = Distance)

74. 16 gal of A, 64 gal of B

75. ◈

76. ◈

77. $\sqrt{10} + \sqrt{50} = \sqrt{10} + \sqrt{10}\sqrt{5} = \sqrt{10}(1 + \sqrt{5})$

$\sqrt{10} + \sqrt{50} = \sqrt{10} + \sqrt{25\cdot 2} = \sqrt{10} + 5\sqrt{2}$

$\sqrt{10} + \sqrt{50} = \sqrt{2}\sqrt{5} + \sqrt{2}\sqrt{25} =$

$\sqrt{2}(\sqrt{5} + \sqrt{25}) = \sqrt{2}(\sqrt{5} + 5)$, or $\sqrt{2}(5 + \sqrt{5})$

All three are correct.

78. Any pairs of numbers a, b such that $a = 0, b \geqslant 0$ or $a \geqslant 0, b = 0$

79. $\sqrt{125} - \sqrt{45} + 2\sqrt{5} = \sqrt{25\cdot 5} - \sqrt{9\cdot 5} + 2\sqrt{5}$

$= 5\sqrt{5} - 3\sqrt{5} + 2\sqrt{5}$

$= (5 - 3 + 2)\sqrt{5}$

$= 4\sqrt{5}$

80. $16\sqrt{2}$

81. $\frac{3}{5}\sqrt{24} + \frac{2}{5}\sqrt{150} - \sqrt{96} = \frac{3}{5}\sqrt{4\cdot 6} + \frac{2}{5}\sqrt{25\cdot 6} - \sqrt{16\cdot 6}$

$= \frac{3}{5} \cdot 2\sqrt{6} + \frac{2}{5} \cdot 5\sqrt{6} - 4\sqrt{6}$

$= \frac{6}{5}\sqrt{6} + 2\sqrt{6} - 4\sqrt{6}$

$= \left(\frac{6}{5} + 2 - 4\right)\sqrt{6}$

$= -\frac{4}{5}\sqrt{6}$, or $-\frac{4\sqrt{6}}{5}$

82. $11\sqrt{3} - 10\sqrt{2}$

83. $\sqrt{ab^6} + b\sqrt{a^3} + a\sqrt{a} = \sqrt{b^6\cdot a} + b\sqrt{a^2\cdot a} + a\sqrt{a}$

$= b^3\sqrt{a} + ab\sqrt{a} + a\sqrt{a}$

$= (b^3 + ab + a)\sqrt{a}$

84. 0

85. $7x\sqrt{12xy^2} - 9y\sqrt{27x^3} + 5\sqrt{300x^3y^2}$

$= 7x\sqrt{4y^2\cdot 3x} - 9y\sqrt{9x^2\cdot 3x} + 5\sqrt{100x^2y^2\cdot 3x}$

$= 7x\cdot 2y\sqrt{3x} - 9y\cdot 3x\sqrt{3x} + 5\cdot 10xy\sqrt{3x}$

$= 14xy\sqrt{3x} - 27xy\sqrt{3x} + 50xy\sqrt{3x}$

$= (14xy - 27xy + 50xy)\sqrt{3x}$

$= 37xy\sqrt{3x}$

86. $\frac{x + 1}{x}\sqrt{x}$

87. Substitute 30 for T and 25 for v.

$T_w = 91.4 - \frac{(10.45+6.68\sqrt{25} - 0.447\cdot 25)(457-5\cdot 30)}{110}$

$= 91.4 - \frac{(10.45 + 33.4 - 11.175)(457 - 150)}{110}$

$= 91.4 - \frac{(32.675)(307)}{110} \approx 91.4 - 91.2 \approx 0.2$

$\approx 0°$ To the nearest degree

88. -10°

89. Substitute 20 for T and 40 for v.

$T_w = 91.4 - \frac{(10.45+6.68\sqrt{40} - 0.447\cdot 40)(457-5\cdot 20)}{110}$

$\approx 91.4 - \frac{(10.45 + 42.25 - 17.88)(457 - 100)}{110}$

$\approx 91.4 - \frac{(34.82)(357)}{110} \approx 91.4 - 113.01 \approx -21.61$

$\approx -22°$ To the nearest degree

90. -64°

Exercise Set 9.5

1. $\sqrt{x} = 5$
$(\sqrt{x})^2 = 5^2$
$x = 25$

Check: $\sqrt{x} = 5$; $\sqrt{25}$? 5; 5 | 5 TRUE

2. 49

3. $\sqrt{x} = 1.5$
$(\sqrt{x})^2 = (1.5)^2$
$x = 2.25$

Check: $\sqrt{x} = 1.5$; $\sqrt{2.25}$? 1.5; 1.5 | 1.5 TRUE

4. 18.49

5. $\sqrt{x + 3} = 8$
$(\sqrt{x + 3})^2 = 8^2$
$x + 3 = 64$
$x = 61$

Check: $\sqrt{x + 3} = 8$; $\sqrt{61 + 3}$? 8; $\sqrt{64}$; 8 | 8 TRUE

6. 117

7. $\sqrt{2x + 4} = 9$
$(\sqrt{2x + 4})^2 = 9^2$
$2x + 4 = 81$
$2x = 77$
$x = \frac{77}{2}$

Check: $\sqrt{2x + 4} = 9$; $\sqrt{2 \cdot \frac{77}{2} + 4}$? 9; $\sqrt{77 + 4}$; $\sqrt{81}$; 9 | 9 TRUE

8. 84

9. $3 + \sqrt{x - 1} = 5$
$\sqrt{x - 1} = 2$
$(\sqrt{x - 1})^2 = 2^2$
$x - 1 = 4$
$x = 5$

Check: $3 + \sqrt{x - 1} = 5$; $3 + \sqrt{5 - 1}$? 5; $3 + \sqrt{4}$; $3 + 2$; 5 | 5 TRUE

10. 52

11. $6 - 2\sqrt{3n} = 0$
$6 = 2\sqrt{3n}$
$6^2 = (2\sqrt{3n})^2$
$36 = 4\cdot 3n$
$36 = 12n$
$3 = n$

Check: $6 - 2\sqrt{3n} = 0$; $6 - 2\sqrt{3\cdot 3}$? 0; $6 - 2\cdot 3$; $6 - 6$; 0 | 0 TRUE

12. $\frac{4}{5}$

13. $\sqrt{5x - 7} = \sqrt{x + 10}$
$(\sqrt{5x - 7})^2 = (\sqrt{x + 10})^2$
$5x - 7 = x + 10$
$4x = 17$
$x = \frac{17}{4}$

Check: $\sqrt{5x - 7} = \sqrt{x + 10}$; $\sqrt{5 \cdot \frac{17}{4} - 7}$? $\sqrt{\frac{17}{4} + 10}$; $\sqrt{\frac{85}{4} - \frac{28}{4}}$ | $\sqrt{\frac{57}{4}}$; $\sqrt{\frac{57}{4}}$ | $\sqrt{\frac{57}{4}}$ TRUE

14. $\frac{14}{3}$

15. $\sqrt{x} = -7$
There is no solution. The principal square root of x cannot be negative.

16. No solution

17. $\sqrt{2y + 6} = \sqrt{2y - 5}$
$(\sqrt{2y + 6})^2 = (\sqrt{2y - 5})^2$
$2y + 6 = 2y - 5$
$6 = -5$
The equation 6 = -5 is false; there is no solution.

18. No real-number solution

19. $x - 7 = \sqrt{x - 5}$
$(x - 7)^2 = (\sqrt{x - 5})^2$
$x^2 - 14x + 49 = x - 5$
$x^2 - 15x + 54 = 0$
$(x - 9)(x - 6) = 0$
$x - 9 = 0$ or $x - 6 = 0$
$x = 9$ or $x = 6$

Check: For 9:

$$\begin{array}{c|c} x - 7 = \sqrt{x - 5} & \\ \hline 9 - 7 \ ? & \sqrt{9 - 5} \\ 2 & \sqrt{4} \\ 2 & 2 \end{array} \quad \text{TRUE}$$

For 6:

$$\begin{array}{c|c} x - 7 = \sqrt{x - 5} & \\ \hline 6 - 7 \ ? & \sqrt{6 - 5} \\ -1 & \sqrt{1} \\ -1 & 1 \end{array} \quad \text{FALSE}$$

The number 9 checks, but 6 does not. The solution is 9.

20. 9

21. $\sqrt{x + 18} = x - 2$

$(\sqrt{x + 18})^2 = (x - 2)^2$

$x + 18 = x^2 - 4x + 4$

$0 = x^2 - 5x - 14$

$0 = (x - 7)(x + 2)$

$x - 7 = 0$ or $x + 2 = 0$

$x = 7$ or $x = -2$

Check: For 7:

$$\begin{array}{c|c} \sqrt{x + 18} = x - 2 & \\ \hline \sqrt{7 + 18} \ ? & 7 - 2 \\ \sqrt{25} & 5 \\ 5 & 5 \end{array} \quad \text{TRUE}$$

For -2:

$$\begin{array}{c|c} \sqrt{x + 18} = x - 2 & \\ \hline \sqrt{-2 + 18} \ ? & -2 - 2 \\ \sqrt{16} & -4 \\ 4 & -4 \end{array} \quad \text{FALSE}$$

The number 7 checks, but -2 does not. The solution is 7.

22. 12

23. $2\sqrt{x - 1} = x - 1$

$(2\sqrt{x - 1})^2 = (x - 1)^2$

$4(x - 1) = x^2 - 2x + 1$

$4x - 4 = x^2 - 2x + 1$

$0 = x^2 - 6x + 5$

$0 = (x - 5)(x - 1)$

$x - 5 = 0$ or $x - 1 = 0$

$x = 5$ or $x = 1$

Both numbers check. The solutions are 5 and 1.

24. 0, 8

25. $\sqrt{5x + 21} = x + 3$

$(\sqrt{5x + 21})^2 = (x + 3)^2$

$5x + 21 = x^2 + 6x + 9$

$0 = x^2 + x - 12$

$0 = (x + 4)(x - 3)$

$x + 4 = 0$ or $x - 3 = 0$

$x = -4$ or $x = 3$

Check: For -4:

$$\begin{array}{c|c} \sqrt{5x + 21} = x + 3 & \\ \hline \sqrt{5(-4) + 21} \ ? & -4 + 3 \\ \sqrt{1} & -1 \\ 1 & -1 \end{array} \quad \text{FALSE}$$

For 3:

$$\begin{array}{c|c} \sqrt{5x + 21} = x + 3 & \\ \hline \sqrt{5 \cdot 3 + 21} \ ? & 3 + 3 \\ \sqrt{36} & 6 \\ 6 & 6 \end{array} \quad \text{TRUE}$$

The number 3 checks, but -4 does not. The solution is 3.

26. 6

27. $x = 1 + 6\sqrt{x - 9}$

$x - 1 = 6\sqrt{x - 9}$ Isolating the radical

$(x - 1)^2 = (6\sqrt{x - 9})^2$

$x^2 - 2x + 1 = 36(x - 9)$

$x^2 - 2x + 1 = 36x - 324$

$x^2 - 38x + 325 = 0$

$(x - 13)(x - 25) = 0$

$x - 13 = 0$ or $x - 25 = 0$

$x = 13$ or $x = 25$

Both numbers check.

28. 5

29. $\sqrt{x^2 + 6} - x + 3 = 0$

$\sqrt{x^2 + 6} = x - 3$ Isolating the radical

$(\sqrt{x^2 + 6})^2 = (x - 3)^2$

$x^2 + 6 = x^2 - 6x + 9$

$-3 = -6x$ Adding $-x^2$ and -9

$\frac{1}{2} = x$

Check:

$$\begin{array}{c|c} \sqrt{x^2 + 6} - x + 3 = 0 & \\ \hline \sqrt{\left(\frac{1}{2}\right)^2 + 6} - \frac{1}{2} + 3 \ ? & 0 \\ \sqrt{\frac{25}{4}} - \frac{1}{2} + 3 & \\ \frac{5}{2} - \frac{1}{2} + 3 & \\ 5 & 0 \end{array} \quad \text{FALSE}$$

The number $\frac{1}{2}$ does not check. There are no solutions.

30. No solution

31. $\sqrt{(p + 6)(p + 1)} - 2 = p + 1$

$\sqrt{(p + 6)(p + 1)} = p + 3$ Isolating the radical

$\left[\sqrt{(p + 6)(p + 1)}\right]^2 = (p + 3)^2$

$(p + 6)(p + 1) = p^2 + 6p + 9$

$p^2 + 7p + 6 = p^2 + 6p + 9$

$p = 3$

The number 3 checks. It is the solution.

32. 5

33. $\sqrt{2 - x} = \sqrt{3x - 7}$

$(\sqrt{2 - x})^2 = (\sqrt{3x - 7})^2$

$2 - x = 3x - 7$

$9 = 4x$

$\frac{9}{4} = x$

Check: $\sqrt{2 - x} = \sqrt{3x - 7}$

$\sqrt{2 - \frac{9}{4}} \;?\; \sqrt{3 \cdot \frac{9}{4} - 7}$

$\sqrt{-\frac{1}{4}} \;\big|\; \sqrt{-\frac{1}{4}}$

Since $\sqrt{-\frac{1}{4}}$ is not a real number, there are no solutions that yield real-number values for the radical expressions.

34. No real-number solution

35. $1 + \sqrt{1 - x} = x$

$\sqrt{1 - x} = x - 1$ Isolating the radical

$(\sqrt{1 - x})^2 = (x - 1)^2$

$1 - x = x^2 - 2x + 1$

$0 = x^2 - x$

$0 = x(x - 1)$

$x = 0$ or $x - 1 = 0$

$x = 0$ or $x = 1$

Check: For $x = 0$: $1 + \sqrt{1 - x} = x$

$1 + \sqrt{1 - 0} \;?\; 0$

$1 + \sqrt{1}$

$1 + 1$

$2 \;\big|\; 0$ FALSE

For $x = 1$: $1 + \sqrt{1 - x} = x$

$1 + \sqrt{1 - 1} \;?\; 1$

$1 + \sqrt{0}$

$1 + 0$

$1 \;\big|\; 1$ TRUE

Since 1 checks but 0 does not, the solution is 1.

36. 3

37. Familiarize and Translate. We use the equation $V = 3.5\sqrt{h}$.

$17.5 = 3.5\sqrt{h}$ Substituting 17.5 for V

Carry out. We solve the equation.

$17.5 = 3.5\sqrt{h}$

$5 = \sqrt{h}$ Dividing by 3.5

$5^2 = (\sqrt{h})^2$

$25 = h$

Check. We go over the computation.

State. The altitude of the steeplejack's eyes is 25 m.

38. 400 m

39. Familiarize and Translate. We use the equation $V = 3.5\sqrt{h}$.

$371 = 3.5\sqrt{h}$ Substituting 371 for V

Carry out. We solve the equation $371 = 3.5\sqrt{h}$.

$106 = \sqrt{h}$ Dividing by 3.5

$106^2 = (\sqrt{h})^2$

$11{,}236 = h$

Check. We go over the computation.

State. The airplane is 11,236 m high.

40. 806.56 m

41. $r = 2\sqrt{5L}$

$50 = 2\sqrt{5L}$ Substituting 50 for r

$25 = \sqrt{5L}$

$25^2 = (\sqrt{5L})^2$

$625 = 5L$

$125 = L$

The car will skid 125 ft at 50 mph.

$70 = 2\sqrt{5L}$ Substituting 70 for r

$35 = \sqrt{5L}$

$35^2 = (\sqrt{5L})^2$

$1225 = 5L$

$245 = L$

The car will skid 245 ft at 70 mph.

42. 180 ft; 500 ft

43. Familiarize. Let x represent the number.

Translate. We reword the problem.

Two times the square root of a number is 14.

$2 \cdot \sqrt{x} = 14$

Carry out.

$2\sqrt{x} = 14$

$\sqrt{x} = 7$

$(\sqrt{x})^2 = 7^2$

$x = 49$

Check. The principal square root of 49 is 7 and twice 7 is 14.

State. The number is 49.

44. 121

45. Familiarize. Let x represent the number. Then 4 more than 5 times the number is represented by 5x + 4.

Translate.

The square root of 5x + 4 is 8.

$\sqrt{5x + 4} = 8$

Carry out.

$\sqrt{5x + 4} = 8$

$(\sqrt{5x + 4})^2 = 8^2$

$5x + 4 = 64$

$5x = 60$

$x = 12$

Check. Four more than five times 12 is 5·12 + 4, or 64. The principal square root of 64 is 8.

State. The number is 12.

46. 32

47. $T = 2\pi\sqrt{\frac{L}{32}}$

$1.6 = 2(3.14)\sqrt{\frac{L}{32}}$ Substituting 1.6 for T and 3.14 for π

$1.6 = 6.28\sqrt{\frac{L}{32}}$

$\frac{1.6}{6.28} = \sqrt{\frac{L}{32}}$

$\left(\frac{1.6}{6.28}\right)^2 = \left(\sqrt{\frac{L}{32}}\right)^2$

$0.065 \approx \frac{L}{32}$

$2.08 \approx L$

The pendulum is about 2.08 ft long.

48. About 7.30 ft

49. $\frac{7x^9}{27} \cdot \frac{9}{7x^3} = \frac{63x^9}{189x^3}$

$= \frac{63x^3 \cdot x^6}{63x^3 \cdot 3}$

$= \frac{\cancel{63x^3} \cdot x^6}{\cancel{63x^3} \cdot 3}$

$= \frac{x^6}{3}$

50. $\frac{x - 3}{4(x + 3)}$

51. $\frac{x}{x + 5} + \frac{x^2 - 20}{x + 5} = \frac{x + x^2 - 20}{x + 5}$

$= \frac{x^2 + x - 20}{x + 5}$ Rearranging in the numerator

$= \frac{(x + 5)(x - 4)}{x + 5}$

$= \frac{\cancel{(x + 5)}(x - 4)}{\cancel{(x + 5)} \cdot 1}$

$= x - 4$

52. $\frac{13}{a - 1}$

53. ◈

54. ◈

55. $\sqrt{x + 9} = 1 + \sqrt{x}$

$(\sqrt{x + 9})^2 = (1 + \sqrt{x})^2$

$x + 9 = 1 + 2 \cdot 1 \cdot \sqrt{x} + (\sqrt{x})^2$

$x + 9 = 1 + 2\sqrt{x} + x$

$8 = 2\sqrt{x}$ Isolating the radical

$4 = \sqrt{x}$ Multiplying by $\frac{1}{2}$

$4^2 = (\sqrt{x})^2$

$16 = x$

The number 16 checks. It is the solution.

56. 9

57. $\sqrt{3x+1} = 1 - \sqrt{x+4}$

$(\sqrt{3x+1})^2 = (1 - \sqrt{x+4})^2$

$3x + 1 = 1 - 2\cdot1\cdot\sqrt{x+4} + (\sqrt{x+4})^2$

$3x + 1 = 1 - 2\sqrt{x+4} + x + 4$

$3x + 1 = 5 - 2\sqrt{x+4} + x$

$2x - 4 = -2\sqrt{x+4}$ Isolating the radical

$2(x - 2) = -2\sqrt{x+4}$

$x - 2 = -\sqrt{x+4}$ Multiplying by $\frac{1}{2}$

$(x - 2)^2 = (-\sqrt{x+4})^2$

$x^2 - 4x + 4 = x + 4$

$x^2 - 5x = 0$

$x(x - 5) = 0$

$x = 0$ or $x - 5 = 0$

$x = 0$ or $x = 5$

Check:

For 0:

$\sqrt{3x+1}$	$=$	$1 - \sqrt{x+4}$	
$\sqrt{3\cdot0+1}$	?	$1 - \sqrt{0+4}$	
$\sqrt{1}$		$1 - 2$	
1		-1	FALSE

For 5:

$\sqrt{3x+1}$	$=$	$1 - \sqrt{x+4}$	
$\sqrt{3\cdot5+1}$	?	$1 - \sqrt{5+4}$	
$\sqrt{16}$		$1 - 3$	
4		-2	FALSE

Neither number checks. There is no solution.

58. 1

59. $4 + \sqrt{19-x} = 6 + \sqrt{4-x}$

$\sqrt{19-x} = 2 + \sqrt{4-x}$ Isolating one radical

$(\sqrt{19-x})^2 = (2 + \sqrt{4-x})^2$

$19 - x = 4 + 4\sqrt{4-x} + (4 - x)$

$19 - x = 4\sqrt{4-x} + 8 - x$

$11 = 4\sqrt{4-x}$

$11^2 = (4\sqrt{4-x})^2$

$121 = 16(4 - x)$

$121 = 64 - 16x$

$57 = -16x$

$-\frac{57}{16} = x$

$-\frac{57}{16}$ checks, so it is the solution.

60. 3

61. $2\sqrt{x-1} - \sqrt{3x-5} = \sqrt{x-9}$

$(2\sqrt{x-1} - \sqrt{3x-5})^2 = (\sqrt{x-9})^2$

$4(x - 1) - 4\sqrt{(x-1)(3x-5)} + (3x - 5) = x - 9$

$4x - 4 - 4\sqrt{3x^2 - 8x + 5} + 3x - 5 = x - 9$

$7x - 9 - 4\sqrt{3x^2 - 8x + 5} = x - 9$

$-4\sqrt{3x^2 - 8x + 5} = -6x$

$2\sqrt{3x^2 - 8x + 5} = 3x$

$(2\sqrt{3x^2 - 8x + 5})^2 = (3x)^2$

$4(3x^2 - 8x + 5) = 9x^2$

$12x^2 - 32x + 20 = 9x^2$

$3x^2 - 32x + 20 = 0$

$(3x - 2)(x - 10) = 0$

$3x - 2 = 0$ or $x - 10 = 0$

$3x = 2$ or $x = 10$

$x = \frac{2}{3}$ or $x = 10$

The number 10 checks, but $\frac{2}{3}$ does not. The solution is 10.

62. 0, 4

63. Familiarize. We will use the formula $V = 3.5\sqrt{h}$. We present the information in a table.

	Height	Distance to the horizon
First sighting	h	V
Second sighting	h + 100	V + 20

Translate. The rows of the table give us two equations.

$V = 3.5\sqrt{h}$, (1)

$V + 20 = 3.5\sqrt{h+100}$ (2)

Carry out. We substitute $3.5\sqrt{h}$ for V in equation (2) and solve for h.

$3.5\sqrt{h} + 20 = 3.5\sqrt{h+100}$

$(3.5\sqrt{h} + 20)^2 = (3.5\sqrt{h+100})^2$

$12.25h + 140\sqrt{h} + 400 = 12.25(h + 100)$

$12.25h + 140\sqrt{h} + 400 = 12.25h + 1225$

$140\sqrt{h} = 825$

$28\sqrt{h} = 165$ Multiplying by $\frac{1}{5}$

$(28\sqrt{h})^2 = (165)^2$

$784h = 27{,}225$

$h \approx 34.726$

Check. When $h \approx 34.726$, then $V \approx 3.5\sqrt{34.726} \approx 20.625$ km. When $h \approx 100 + 34.726$, or 134.726, then $V \approx 3.5\sqrt{134.726} \approx 40.625$ km. This is 20 km more than 20.625. The answer checks.

State. The climber was at a height of about 34.726 m when the first computation was made.

64. $b = \dfrac{a}{A^4 - 2A^2 + 1}$

65. Graph $y = \sqrt{x}$.

We make a table of values. Note that we must choose nonnegative values of x in order to have a nonnegative radicand.

x	y
0	0
1	1
2	1.414
4	2
5	2.236

We plot these points and connect them with a smooth curve.

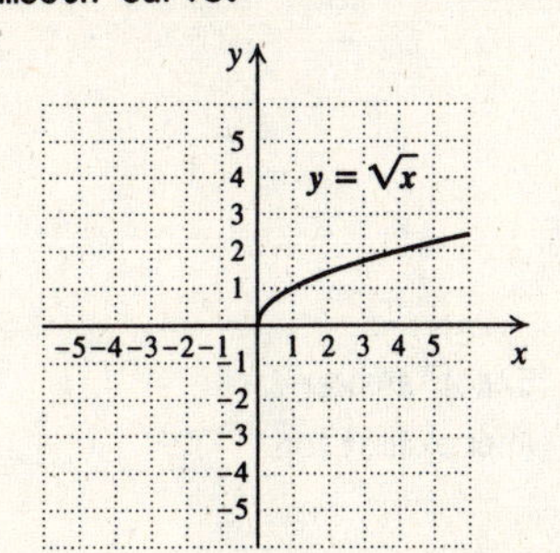

66.

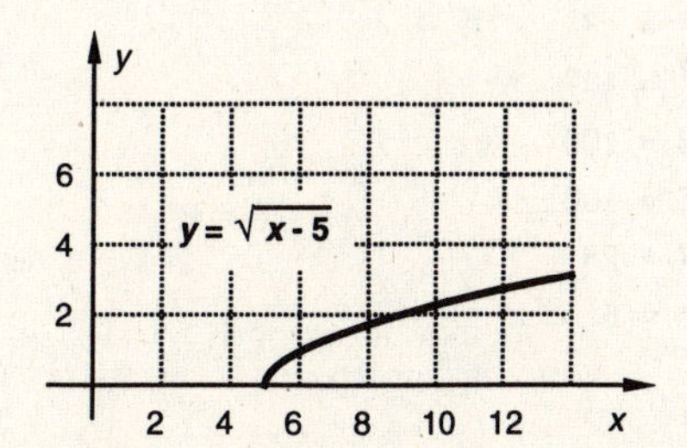

67. Graph $y = \sqrt{x - 1}$.

We make a table of values. Note that we must choose values for x that are greater than or equal to 1 in order to have a nonnegative radicand.

x	y
1	0
2	1
3	1.414
4	1.732
5	2

We plot these points and connect them with a smooth curve.

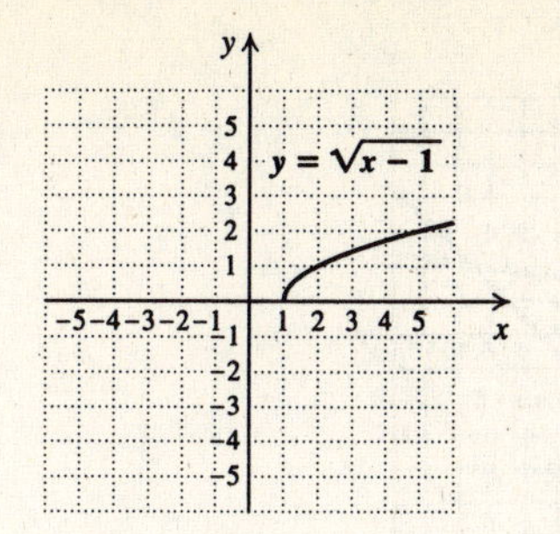

68.

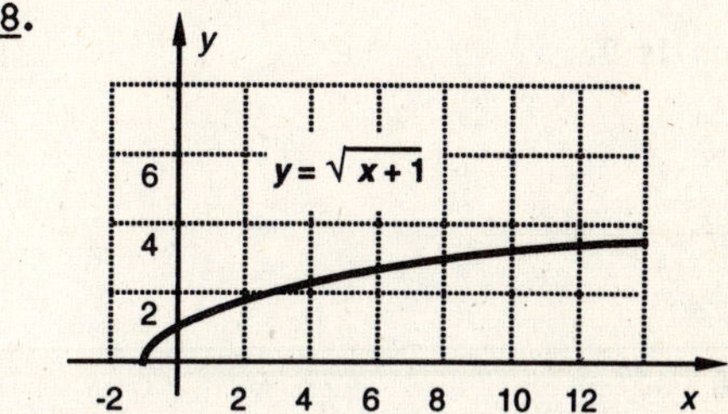

69. We can graph $y = x - 7$ using the intercepts, (0,-7) and (7,0).

We make a table of values for $y = \sqrt{x - 5}$.

x	y
5	0
6	1
7	1.414
8	1.732
9	2

We plot these points and connect them with a smooth curve.

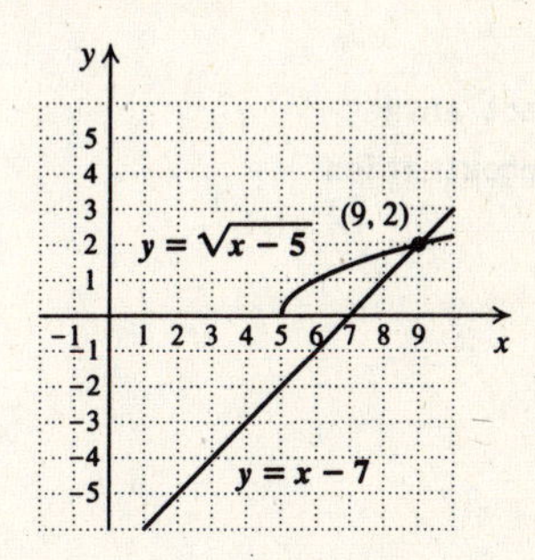

The graphs intersect at (9,2), so the solution of $x - 7 = \sqrt{x - 5}$ is 9.

70.

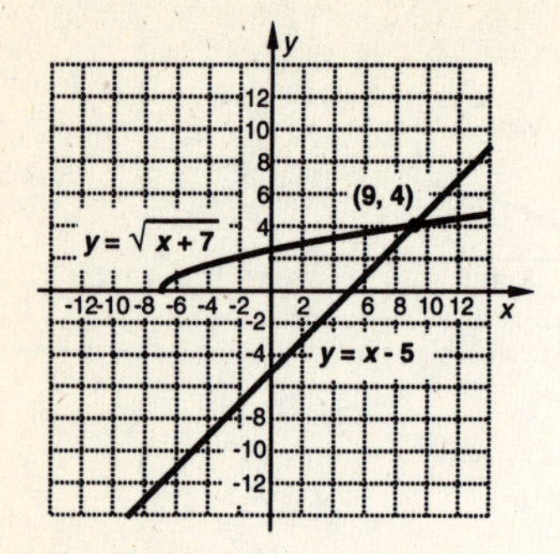

The solution is 9.

71. 1.57

72. -0.32

Exercise Set 9.6

1. $a^2 + b^2 = c^2$
$8^2 + 15^2 = c^2$ Substituting
$64 + 225 = c^2$
$289 = c^2$
$\sqrt{289} = c$
$17 = c$

2. $\sqrt{34} \approx 5.831$

3. $a^2 + b^2 = c^2$
$4^2 + 4^2 = c^2$ Substituting
$16 + 16 = c^2$
$32 = c^2$
$\sqrt{32} = c$ Exact answer
$5.657 \approx c$ Approximation

4. $\sqrt{98} \approx 9.899$

5. $a^2 + b^2 = c^2$
$5^2 + b^2 = 13^2$
$25 + b^2 = 169$
$b^2 = 144$
$b = 12$

6. 5

7. $a^2 + b^2 = c^2$
$a^2 + (4\sqrt{3})^2 = 8^2$
$a^2 + 16\cdot3 = 64$
$a^2 + 48 = 64$
$a^2 = 16$
$a = 4$

8. $\sqrt{31} \approx 5.568$

9. $a^2 + b^2 = c^2$
$10^2 + 24^2 = c^2$
$100 + 576 = c^2$
$676 = c^2$
$26 = c$

10. 13

11. $a^2 + b^2 = c^2$
$9^2 + b^2 = 15^2$
$81 + b^2 = 225$
$b^2 = 144$
$b = 12$

12. 24

13. $a^2 + b^2 = c^2$
$a^2 + 1^2 = (\sqrt{5})^2$
$a^2 + 1 = 5$
$a^2 = 4$
$a = 2$

14. 1

15. $a^2 + b^2 = c^2$
$1^2 + b^2 = (\sqrt{3})^2$
$1 + b^2 = 3$
$b^2 = 2$
$b = \sqrt{2}$ Exact answer
$b \approx 1.414$ Approximation

16. $\sqrt{8} \approx 2.828$

17. $a^2 + b^2 = c^2$
$a^2 + (5\sqrt{3})^2 = 10^2$
$a^2 + 25\cdot3 = 100$
$a^2 + 75 = 100$
$a^2 = 25$
$a = 5$

18. $\sqrt{50} \approx 7.071$

19. Familiarize. We first make a drawing. We label the unknown height h.

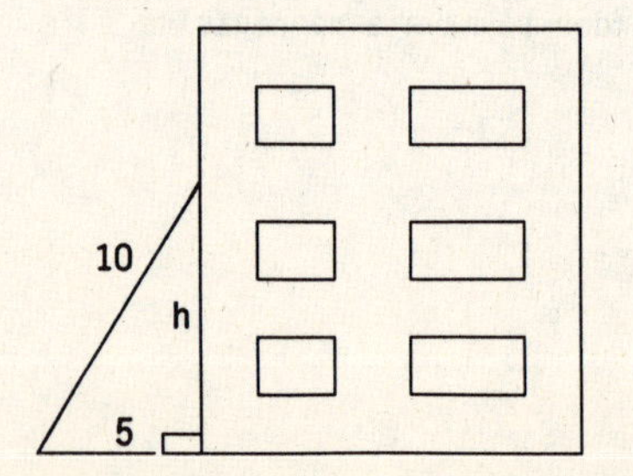

Translate. We use the Pythagorean theorem.

$a^2 + b^2 = c^2$

$5^2 + h^2 = 10^2$ Substituting 5 for a, h for b, and 10 for c

Carry out. We solve the equation.

$5^2 + h^2 = 10^2$

$25 + h^2 = 100$

$h^2 = 75$

$h = \sqrt{75}$ Exact answer

$h \approx 8.660$ Approximation

Check. We check by substituting 5, $\sqrt{75}$, and 10 into the Pythagorean equation:

$a^2 + b^2 = c^2$	
$5^2 + (\sqrt{75})^2$?	10^2
$25 + 75$	100
100	100 TRUE

State. The top of the ladder is $\sqrt{75}$ or about 8.660 m from the ground.

20. $\sqrt{18} \approx 4.243$ cm

21. Familiarize. We first make a drawing. We label the unknown length w.

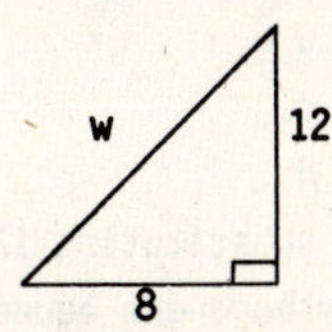

Translate. We use the Pythagorean theorem.

$a^2 + b^2 = c^2$

$8^2 + 12^2 = w^2$ Substituting 8 for a, 12 for b, and w for c.

Carry out. We solve the equation.

$8^2 + 12^2 = w^2$

$64 + 144 = w^2$

$208 = w^2$

$\sqrt{208} = w$

$4\sqrt{13} = w$ Exact answer

$14.422 \approx w$ Approximation

Check. We check by substituting 8, 12, and $\sqrt{208}$ into the Pythagorean equation:

$a^2 + b^2 = c^2$	
$8^2 + 12^2$?	$(\sqrt{208})^2$
$64 + 144$	208
208	208 TRUE

State. The pipe should be $4\sqrt{13}$ or about 14.422 feet long.

22. $\sqrt{250} \approx 15.811$ m

23. Familiarize. We first make a drawing. We label the diagonal d.

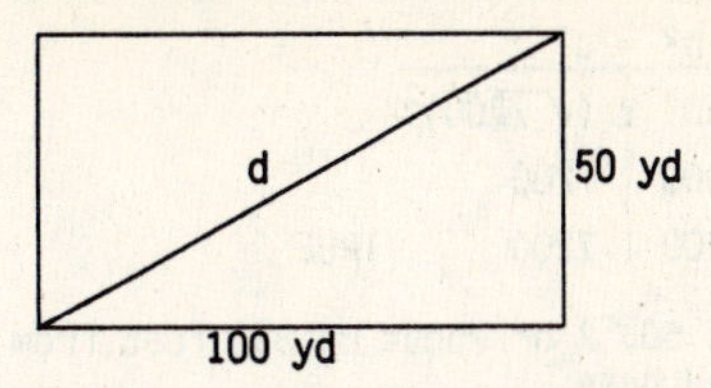

Translate. We use the Pythagorean theorem, substituting 50 for a, 100 for b, and d for c.

$a^2 + b^2 = c^2$

$50^2 + 100^2 = d^2$

Carry out. We solve the equation.

$50^2 + 100^2 = d^2$

$2500 + 10,000 = d^2$

$12,500 = d^2$

$\sqrt{12,500} = d$ Exact answer

$111.803 \approx d$ Approximation

Check. We check by substituting 50,100, and $\sqrt{12,500}$ into the Pythagorean equation:

$a^2 + b^2 = c^2$	
$50^2 + 100^2$?	$(\sqrt{12,500})^2$
$2500 + 10,000$	$12,500$
$12,500$	$12,500$ TRUE

State. The length of a diagonal is $\sqrt{12,500}$ or about 111.803 yd.

24. $\sqrt{26,900} \approx 164.012$ yd

25. Familiarize. We first make a drawing. We label the length from home to second base d.

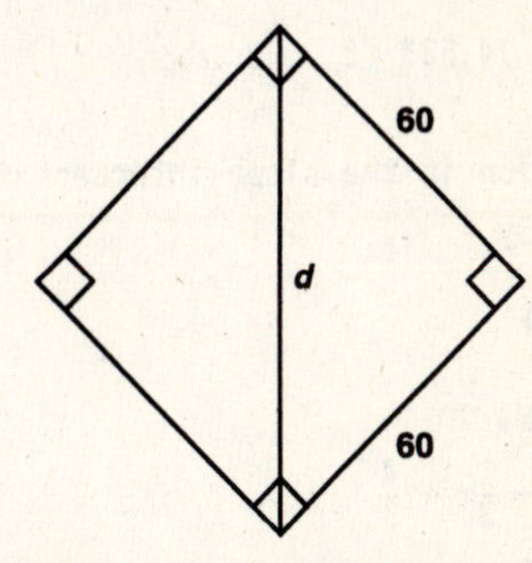

Translate. We use the Pythagorean theorem.

$a^2 + b^2 = c^2$

$60^2 + 60^2 = d^2$ Substituting 60 for a, 60 for b and d for c

Carry out. We solve the equation.

$60^2 + 60^2 = d^2$

$3600 + 3600 = d^2$

$7200 = d^2$

$\sqrt{7200} = d$

$60\sqrt{2} = d$ Exact answer

$84.853 \approx d$ Approximation

Check. We check by substituting 60, 60, and $\sqrt{7200}$ into the Pythagorean equation:

$$\begin{array}{c|l} \multicolumn{2}{c}{a^2 + b^2 = c^2} \\ \hline 60^2 + 60^2 & ? \ (\sqrt{7200})^2 \\ 3600 + 3600 & 7200 \\ 7200 & 7200 \quad \text{TRUE} \end{array}$$

State. It is $60\sqrt{2}$ or about 84.853 feet from home to second base.

26. $\sqrt{16{,}200} \approx 127.279$ ft

27. Familiarize. Referring to the drawing in the text, we let d represent the distance from P to R. This is the length of the lake.

Translate. We use the Pythagorean theorem.

$a^2 + b^2 = c^2$

$25^2 + 35^2 = d^2$ Substituting 25 for a, 35 for b, and d for c

Carry out. We solve the equation.

$$\begin{aligned} 25^2 + 35^2 &= d^2 \\ 625 + 1225 &= d^2 \\ 1850 &= d^2 \\ \sqrt{1850} &= d \\ 43 &\approx d \end{aligned}$$

Check. We check by substituting 25, 35, and $\sqrt{1850}$ into the Pythagorean equation:

$$\begin{array}{c|l} \multicolumn{2}{c}{a^2 + b^2 = c^2} \\ \hline 25^2 + 35^2 & ? \ (\sqrt{1850})^2 \\ 625 + 1225 & 1850 \\ 1850 & 1850 \quad \text{TRUE} \end{array}$$

State. The length of the lake is about 43 km.

28. $\sqrt{211{,}200{,}000} \approx 14{,}533$ ft

29. Write the equation in the slope-intercept form.

$$\begin{aligned} 4 - x &= 3y \\ \frac{1}{3}(4 - x) &= y \\ \frac{4}{3} - \frac{1}{3}x &= y, \text{ or} \\ y &= -\frac{1}{3}x + \frac{4}{3} \end{aligned}$$

The slope is $-\frac{1}{3}$.

30. $\frac{5}{8}$

31. $-\frac{3}{5}x < 15$

$x > -\frac{5}{3} \cdot 15$ Multiplying by $-\frac{5}{3}$ and reversing the inequality

$x > -25$

The solution set is $\{x | x > -25\}$.

32. $\{x | x \leqslant 1\}$

33.

34.

35. Familiarize. We make a drawing. The outfielder is 40 ft + 90 ft, or 130 ft, from home plate. We label the unknown distance d.

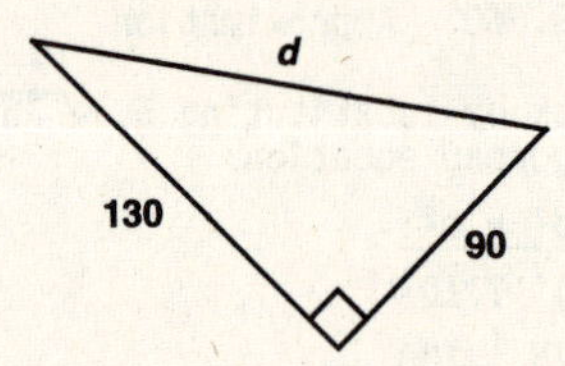

Translate. We use the Pythagorean theorem.

$a^2 + b^2 = c^2$

$130^2 + 90^2 = d^2$ Substituting 130 for a, 90 for b, and d for c

Carry out. We solve the equation.

$$\begin{aligned} 130^2 + 90^2 &= d^2 \\ 16{,}900 + 8100 &= d^2 \\ 25{,}000 &= d^2 \\ \sqrt{25{,}000} &= d \\ 50\sqrt{10} &= d \\ 158 &\approx d \end{aligned}$$

Check. We check by substituting 130, 90, and $\sqrt{25{,}000}$ into the Pythagorean equation.

$$\begin{array}{c|l} \multicolumn{2}{c}{a^2 + b^2 = c^2} \\ \hline 130^2 + 90^2 & ? \ (\sqrt{25{,}000})^2 \\ 16{,}900 + 8100 & 25{,}000 \\ 25{,}000 & 25{,}000 \quad \text{TRUE} \end{array}$$

State. The outfielder will have to throw the ball $50\sqrt{10}$ or about 158 ft.

36. 8 ft

37. Familiarize. Let s = the length of a side of the square. Recall that the area A of a square with side s is given by $A = s^2$.

Translate. We substitute 7 for A in the formula.

$7 = s^2$

Carry out. We solve the equation.

$$\begin{aligned} 7 &= s^2 \\ \sqrt{7} &= s \\ 2.646 &\approx s \end{aligned}$$

Check. If the length of a side of a square is $\sqrt{7}$ m, then the area of the square is $(\sqrt{7}\text{ m})^2$, or 7 m^2. The result checks.

State. The length of a side of the square is $\sqrt{7}$ or about 2.646 m.

38. 3, 4, 5

39.

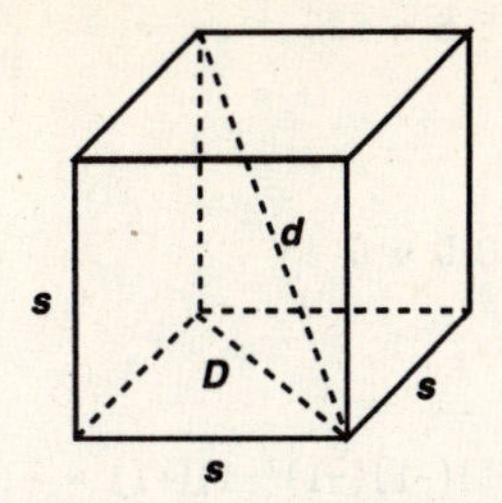

From the drawing we see that the diagonal of the cube is the hypotenuse of a right triangle with one leg of length s, where s is the length of a side of the cube, and the other leg of length D, where D is the length of the diagonal of the base of the cube. First we find D:

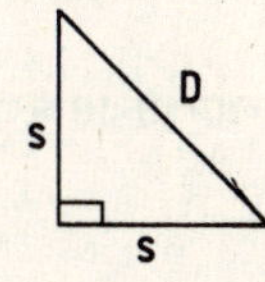

Using the Pythagorean theorem we have:

$$s^2 + s^2 = D^2$$
$$2s^2 = D^2$$
$$\sqrt{2s^2} = D$$
$$s\sqrt{2} = D$$

Then we find d:

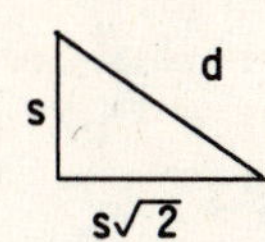

Using the Pythagorean theorem again we have:

$$s^2 + (s\sqrt{2})^2 = d^2$$
$$s^2 + 2s^2 = d^2$$
$$3s^2 = d^2$$
$$\sqrt{3s^2} = d$$
$$s\sqrt{3} = d$$

40. $\frac{2}{3}$

41. Using the Pythagorean theorem we get:

$$h^2 + \left(\frac{a}{2}\right)^2 = a^2$$
$$h^2 + \frac{a^2}{4} = a^2$$
$$h^2 = a^2 - \frac{a^2}{4}$$
$$h^2 = \frac{3a^2}{4}$$
$$h = \sqrt{\frac{3a^2}{4}}$$
$$h = \frac{a\sqrt{3}}{2}$$

42. $A = \frac{a^2\sqrt{3}}{4}$

43. Familiarize. We first make a drawing. We let x represent the width; then x + 1 represents the length.

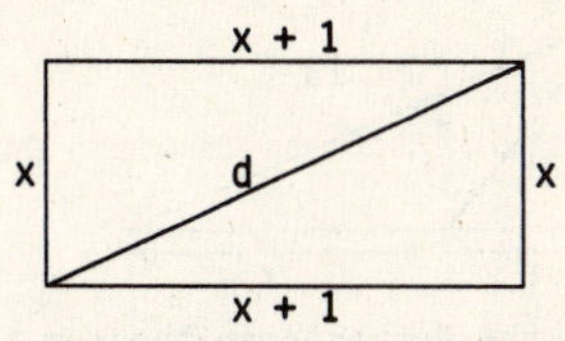

Using an area formula we can determine the value of x. Then we can use the Pythagorean theorem to determine the length of the diagonal labeled d.

Translate.

The area of the rectangle is 90 cm².

This translates to $(x + 1)x = 90$.

Carry out. We solve the equation.

$$x^2 + x = 90$$
$$x^2 + x - 90 = 0$$
$$(x + 10)(x - 9) = 0$$

$x + 10 = 0$ or $x - 9 = 0$

$x = -10$ or $x = 9$

Check. Since the width of the rectangle cannot be negative, we only check x = 9. If the width is 9 cm, then the length is 10 cm and the area is 9·10, or 90 cm².

We repeat the previous three steps to determine the length of the diagonal.

We repeat the previous three steps to determine the length of the diagonal.

Translate. We use the Pythagorean theorem.

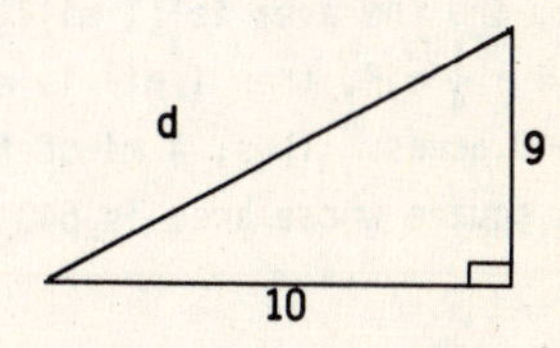

$$a^2 + b^2 = c^2$$

$10^2 + 9^2 = d^2$ Substituting 10 for a, 9 for b, and d for c

Carry out.

$$100 + 81 = d^2$$
$$181 = d^2$$
$$\sqrt{181} = d$$

$13.454 \approx d$ Using a calculator

Check. We check by substituting 10, 9, and $\sqrt{181}$ into the Pythagorean equation:

$a^2 + b^2 = c^2$	
$10^2 + 9^2$? $(\sqrt{181})^2$	
100 + 81	181
181	181 TRUE

State. The length of the diagonal of the rectangle is $\sqrt{181}$, or about 13.454 cm.

44. $\sqrt{1525} \approx 39.051$ mi

45.

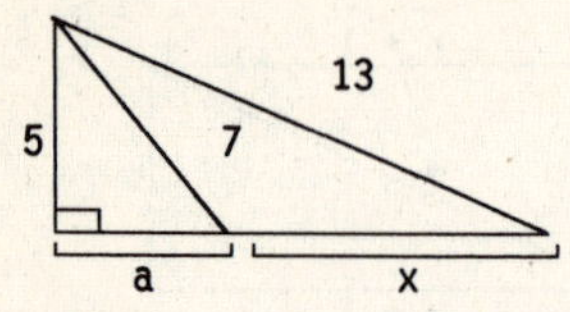

First we use the Pythagorean theorem to find a.

$a^2 + 5^2 = 7^2$

$a^2 + 25 = 49$

$a^2 = 24$

$a = \sqrt{24}$, or $2\sqrt{6}$

Now we use the Pythagorean theorem again to find x.

$(a + x)^2 + 5^2 = 13^2$

$(2\sqrt{6} + x)^2 + 5^2 = 13^2$ Substituting $2\sqrt{6}$ for a

$(2\sqrt{6} + x)^2 + 25 = 169$

$(2\sqrt{6} + x)^2 = 144$

$2\sqrt{6} + x = 12$

$x = 12 - 2\sqrt{6}$

46. 6

47. The perimeter of the smaller square plot is 2 mi, so each side is $\frac{1}{4} \cdot 2$, or $\frac{1}{2}$ mi, and the area is $\left(\frac{1}{2}\text{ mi}\right)^2$, or $\frac{1}{4}$ mi². This tells us that $\frac{1}{4}$ mi² is is equivalent to 160 acres. The perimeter of the larger square plot is 4 mi, so each side is $\frac{1}{4} \cdot 4$, or 1 mi, and the area is $(1\text{ mi})^2$, or 1 mi². Since 1 mi² = $4 \cdot \frac{1}{4}$ mi², then 1 mi² is equivalent to 4·160, or 640 acres. Thus, 4 mi of fencing will enclose a square whose area is 640 acres.

48. 50 ft²

Exercise Set 9.7

1. $\sqrt[3]{-8} = -2$ $(-2)^3 = (-2)(-2)(-2) = -8$

2. 4

3. $\sqrt[3]{1000} = 10$ $10^3 = 10 \cdot 10 \cdot 10 = 1000$

4. -3

5. $-\sqrt[3]{125} = -5$ $\sqrt[3]{125} = 5$, so $-\sqrt[3]{125} = -5$

6. -2

7. $\sqrt[3]{216} = 6$ $6^3 = 6 \cdot 6 \cdot 6 = 216$

8. -7

9. $\sqrt[4]{625} = 5$ $5^4 = 5 \cdot 5 \cdot 5 \cdot 5 = 625$

10. 3

11. $\sqrt[5]{0} = 0$ $0^5 = 0 \cdot 0 \cdot 0 \cdot 0 \cdot 0 = 0$

12. 1

13. $\sqrt[5]{-1} = -1$ $(-1)^5 = (-1)(-1)(-1)(-1)(-1) = -1$

14. -3

15. $\sqrt[4]{-81}$ is not a real number, because it is an even root of a negative number.

16. Not a real number

17. $\sqrt[4]{10{,}000} = 10$ $10^4 = 10 \cdot 10 \cdot 10 \cdot 10 = 10{,}000$

18. 10

19. $\sqrt[3]{5^3} = 5$ $5^3 = 5 \cdot 5 \cdot 5$

20. 7

21. $\sqrt[6]{64} = 2$ $2^6 = 2 \cdot 2 \cdot 2 \cdot 2 \cdot 2 \cdot 2 = 64$

22. 1

23. $\sqrt[3]{x^3} = x$ $x^3 = x \cdot x \cdot x$

24. t

25. $\sqrt[3]{32} = \sqrt[3]{8 \cdot 4} = \sqrt[3]{8}\,\sqrt[3]{4} = 2\sqrt[3]{4}$

26. $3\sqrt[3]{2}$

27. $\sqrt[4]{48} = \sqrt[4]{16 \cdot 3} = \sqrt[4]{16}\,\sqrt[4]{3} = 2\sqrt[4]{3}$

28. $2\sqrt[5]{5}$

29. $\sqrt[3]{\frac{27}{64}} = \frac{\sqrt[3]{27}}{\sqrt[3]{64}} = \frac{3}{4}$

30. $\frac{5}{4}$

31. $\sqrt[4]{\frac{256}{625}} = \frac{\sqrt[4]{256}}{\sqrt[4]{625}} = \frac{4}{5}$

32. $\frac{3}{2}$

33. $\sqrt[3]{\frac{17}{8}} = \frac{\sqrt[3]{17}}{\sqrt[3]{8}} = \frac{\sqrt[3]{17}}{2}$

34. $\frac{\sqrt[5]{11}}{2}$

35. $\sqrt[4]{\frac{13}{81}} = \frac{\sqrt[4]{13}}{\sqrt[4]{81}} = \frac{\sqrt[4]{13}}{3}$

36. $\frac{\sqrt[3]{10}}{3}$

37. $25^{1/2} = \sqrt{25} = 5$

38. 3

39. $1000^{1/3} = \sqrt[3]{1000} = 10$

40. 5

41. $16^{1/4} = \sqrt[4]{16} = 2$

42. 2

43. $16^{5/4} = (16^{1/4})^5 = (\sqrt[4]{16})^5 = 2^5 = 32$

44. 16

45. $4^{5/2} = (4^{1/2})^5 = (\sqrt{4})^5 = 2^5 = 32$

46. 27

47. $64^{2/3} = (64^{1/3})^2 = (\sqrt[3]{64})^2 = 4^2 = 16$

48. 4

49. $8^{2/3} = (8^{1/3})^2 = (\sqrt[3]{8})^2 = 2^2 = 4$

50. 8

51. $25^{5/2} = (25^{1/2})^5 = (\sqrt{25})^5 = 5^5 = 3125$

52. $\frac{1}{2}$

53. $36^{-1/2} = \frac{1}{36^{1/2}} = \frac{1}{\sqrt{36}} = \frac{1}{6}$

54. $\frac{1}{2}$

55. $256^{-1/4} = \frac{1}{256^{1/4}} = \frac{1}{\sqrt[4]{256}} = \frac{1}{4}$

56. $\frac{1}{1000}$

57. $81^{-3/4} = \frac{1}{81^{3/4}} = \frac{1}{(\sqrt[4]{81})^3} = \frac{1}{3^3} = \frac{1}{27}$

58. $\frac{1}{11}$

59. $144^{-1/2} = \frac{1}{144^{1/2}} = \frac{1}{\sqrt{144}} = \frac{1}{12}$

60. $\frac{1}{8}$

61. $81^{-5/4} = \frac{1}{81^{5/4}} = \frac{1}{(\sqrt[4]{81})^5} = \frac{1}{3^5} = \frac{1}{243}$

62. $\frac{1}{4}$

63. $8^{-2/3} = \frac{1}{8^{2/3}} = \frac{1}{(\sqrt[3]{8})^2} = \frac{1}{2^2} = \frac{1}{4}$

64. $\frac{1}{125}$

65. Familiarize and Translate. This problem states that we have direct variation between F and I. Thus, an equation $F = kI$ applies.

Carry out.

First find an equation of variation.

$F = kI$

$7644 = k \cdot 29{,}400$ Substituting 7644 for F and 29,400 for I

$\frac{7644}{29{,}400} = k$

$0.26 = k$

The equation of variation is $F = 0.26I$.

When $I = \$41{,}000$,

$F = 0.26I$

$F = 0.26(\$41{,}000)$ Substituting \$41,000 for I

$F = \$10{,}660$

Check. Let us do some reasoning about the answer. The income increased from \$29,400 to \$41,000. Similarly, the amount spent on food increased from \$7644 to \$10,660. This is what we would expect with direct variation. We can also redo the calculations.

State. The amount spent on food is \$10,660.

66. $y = 0.625x$

67. $\frac{a + 5}{a^2 - 25} - \frac{3a + 15}{a^2 - 25} = \frac{a + 5 - (3a + 15)}{a^2 - 25}$

$= \frac{a + 5 - 3a - 15}{a^2 - 25}$

$= \frac{-2a - 10}{a^2 - 25}$

$= \frac{-2\cancel{(a + 5)}}{\cancel{(a + 5)}(a - 5)}$

$= \frac{-2}{a - 5}$

68. $\frac{x - 1}{3x^3(x + 3)}$

69.

70.

71. Enter 10, press the power key, enter 0.6 (or $3 \div 5$), and then press $\boxed{=}$.

$10^{3/5} \approx 3.981$

72. 2.213

73. Enter 10, press the power key, enter 1.5 (or 3 ÷ 2), and then press [=].

$$10^{3/2} \approx 31.623$$

74. 9.391

75. $(x^{2/3})^{5/3} = x^{\frac{2}{3} \cdot \frac{5}{3}}$ Multiplying exponents

$= x^{\frac{10}{9}}$

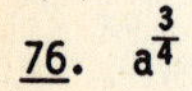

76. $a^{\frac{3}{4}}$

77. $\frac{p^{4/5}}{p^{2/3}} = p^{\frac{4}{5} - \frac{2}{3}}$ Subtracting exponents

$= p^{\frac{12}{15} - \frac{10}{15}} = p^{\frac{2}{15}}$

78. $m^{\frac{7}{12}}$

79. Graph $y = \sqrt[3]{x}$.

We make a table of values.

x	y
-8	-2
-1	-1
0	0
1	1
8	2

We plot these points and connect them with a smooth curve.

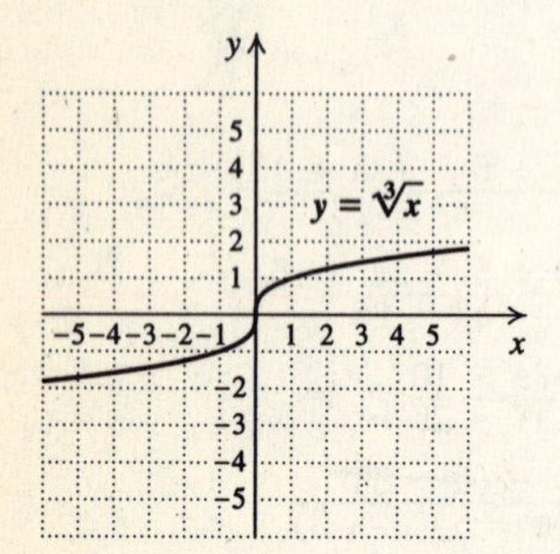

80.

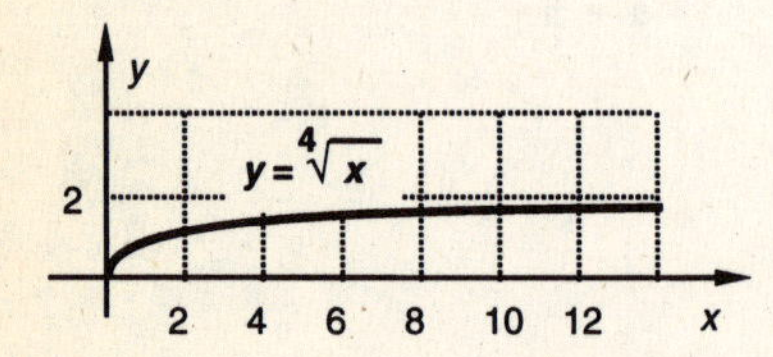

81.

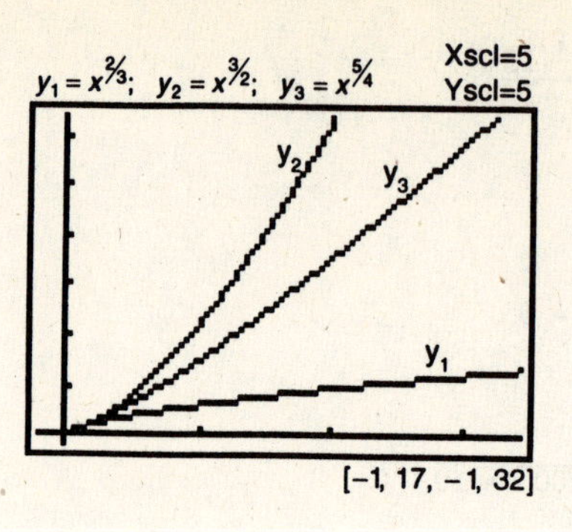

Exercise Set 10.1

1. $x^2 = 25$

$x = \sqrt{25}$ or $x = -\sqrt{25}$ Using the principle of square roots

$x = 5$ or $x = -5$

Check: For 5: $x^2 = 25$; $5^2 \mathrel{?} 25$; $25 \mid 25$ TRUE

For -5: $x^2 = 25$; $(-5)^2 \mathrel{?} 25$; $25 \mid 25$ TRUE

The solutions are 5 and -5.

2. 6, -6

3. $a^2 = 81$

$a = \sqrt{81}$ or $a = -\sqrt{81}$ Using the principle of square roots

$a = 9$ or $a = -9$

Both numbers check. The solutions are 9 and -9.

4. 11, -11

5. $m^2 = 15$

$m = \sqrt{15}$ or $m = -\sqrt{15}$

Both numbers check. The solutions are $\sqrt{15}$ and $-\sqrt{15}$.

6. $\sqrt{10}$, $-\sqrt{10}$

7. $x^2 = 19$

$x = \sqrt{19}$ or $x = -\sqrt{19}$

Both numbers check. The solutions are $\sqrt{19}$ and $-\sqrt{19}$.

8. $\sqrt{29}$, $-\sqrt{29}$

9. $5a^2 = 35$

$a^2 = 7$ Dividing by 5

$a = \sqrt{7}$ or $a = -\sqrt{7}$ Using the principle of square roots

Both numbers check. The solutions are $\sqrt{7}$ and $-\sqrt{7}$.

10. $2\sqrt{7}$, $-2\sqrt{7}$

11. $7x^2 = 140$

$x^2 = 20$ Dividing by 7

$x = \sqrt{20}$ or $x = -\sqrt{20}$

$x = \sqrt{4\cdot5}$ or $x = -\sqrt{4\cdot5}$ } Simplifying

$x = 2\sqrt{5}$ or $x = -2\sqrt{5}$

Both numbers check. The solutions are $2\sqrt{5}$ and $-2\sqrt{5}$.

12. $2\sqrt{2}$, $-2\sqrt{2}$

13. $4t^2 - 25 = 0$

$4t^2 = 25$

$t^2 = \frac{25}{4}$

$t = \sqrt{\frac{25}{4}}$ or $t = -\sqrt{\frac{25}{4}}$

$t = \frac{\sqrt{25}}{\sqrt{4}}$ or $t = -\frac{\sqrt{25}}{\sqrt{4}}$

$t = \frac{5}{2}$ or $t = -\frac{5}{2}$

Both numbers check. The solutions are $\frac{5}{2}$ and $-\frac{5}{2}$.

14. $\frac{2}{3}$, $-\frac{2}{3}$

15. $3x^2 - 49 = 0$

$3x^2 = 49$

$x^2 = \frac{49}{3}$

$x = \sqrt{\frac{49}{3}}$ or $x = -\sqrt{\frac{49}{3}}$

$x = \frac{\sqrt{49}}{\sqrt{3}}$ or $x = -\frac{\sqrt{49}}{\sqrt{3}}$

$x = \frac{7}{\sqrt{3}}$ or $x = -\frac{7}{\sqrt{3}}$

$x = \frac{7}{\sqrt{3}} \cdot \frac{\sqrt{3}}{\sqrt{3}}$ or $x = -\frac{7}{\sqrt{3}} \cdot \frac{\sqrt{3}}{\sqrt{3}}$ Rationalizing the denominators

$x = \frac{7\sqrt{3}}{3}$ or $x = -\frac{7\sqrt{3}}{3}$

Both numbers check. The solutions are $\frac{7\sqrt{3}}{3}$ and $-\frac{7\sqrt{3}}{3}$.

16. $\frac{4\sqrt{5}}{5}$, $-\frac{4\sqrt{5}}{5}$

17. $4y^2 - 3 = 9$

$4y^2 = 12$

$y^2 = 3$

$y = \sqrt{3}$ or $y = -\sqrt{3}$

The solutions are $\sqrt{3}$ and $-\sqrt{3}$.

18. $\frac{2\sqrt{5}}{7}$, $-\frac{2\sqrt{5}}{7}$

19. $25x^2 - 35 = 0$

$25x^2 = 35$

$x^2 = \frac{35}{25}$

$x = \sqrt{\frac{35}{25}}$ or $x = -\sqrt{\frac{35}{25}}$

$x = \frac{\sqrt{35}}{\sqrt{25}}$ or $x = -\frac{\sqrt{35}}{\sqrt{25}}$

$x = \frac{\sqrt{35}}{5}$ or $x = -\frac{\sqrt{35}}{5}$

The solutions are $\frac{\sqrt{35}}{5}$ and $-\frac{\sqrt{35}}{5}$.

20. $2\sqrt{6}$, $-2\sqrt{6}$

21. $(x - 2)^2 = 49$

$x - 2 = 7$ or $x - 2 = -7$ Principle of square roots

$x = 9$ or $x = -5$

The solutions are 9 and -5.

22. 4, -6

23. $(x + 3)^2 = 36$

$x + 3 = 6$ or $x + 3 = -6$ Principle of square roots

$x = 3$ or $x = -9$

The solutions are 3 and -9.

24. 13, -5

25. $(m + 3)^2 = 21$

$m + 3 = \sqrt{21}$ or $m + 3 = -\sqrt{21}$

$m = -3 + \sqrt{21}$ or $m = -3 - \sqrt{21}$

The solutions are $-3 + \sqrt{21}$ and $-3 - \sqrt{21}$, or $-3 \pm \sqrt{21}$.

26. $3 \pm \sqrt{6}$

27. $(a + 13)^2 = 8$

$a + 13 = \sqrt{8}$ or $a + 3 = -\sqrt{8}$

$a + 13 = 2\sqrt{2}$ or $a + 13 = -2\sqrt{2}$

$a = -13 + 2\sqrt{2}$ or $a = -13 - 2\sqrt{2}$

The solutions are $-13 + 2\sqrt{2}$ and $-13 - 2\sqrt{2}$, or $-13 \pm 2\sqrt{2}$.

28. 21, 5

29. $(x - 7)^2 = 12$

$x - 7 = \sqrt{12}$ or $x - 7 = -\sqrt{12}$

$x - 7 = 2\sqrt{3}$ or $x - 7 = -2\sqrt{3}$

$x = 7 + 2\sqrt{3}$ or $x = 7 - 2\sqrt{3}$

The solutions are $7 + 2\sqrt{3}$ and $7 - 2\sqrt{3}$, or $7 \pm 2\sqrt{3}$.

30. $-1 \pm \sqrt{14}$

31. $(x + 9)^2 = 34$

$x + 9 = \sqrt{34}$ or $x + 9 = -\sqrt{34}$

$x = -9 + \sqrt{34}$ or $x = -9 - \sqrt{34}$

The solutions are $-9 + \sqrt{34}$ and $-9 - \sqrt{34}$, or $-9 \pm \sqrt{34}$.

32. 3, -7

33. $\left(x + \frac{3}{2}\right)^2 = \frac{7}{2}$

$x + \frac{3}{2} = \sqrt{\frac{7}{2}}$ or $x + \frac{3}{2} = -\sqrt{\frac{7}{2}}$

$x = -\frac{3}{2} + \sqrt{\frac{7}{2}}$ or $x = -\frac{3}{2} - \sqrt{\frac{7}{2}}$

$x = -\frac{3}{2} + \frac{\sqrt{7}}{\sqrt{2}}$ or $x = -\frac{3}{2} - \frac{\sqrt{7}}{\sqrt{2}}$

$x = -\frac{3}{2} + \frac{\sqrt{7}}{\sqrt{2}} \cdot \frac{\sqrt{2}}{\sqrt{2}}$ or $x = -\frac{3}{2} - \frac{\sqrt{7}}{\sqrt{2}} \cdot \frac{\sqrt{2}}{\sqrt{2}}$

$x = -\frac{3}{2} + \frac{\sqrt{14}}{2}$ or $x = -\frac{3}{2} - \frac{\sqrt{14}}{2}$

$x = \frac{-3 + \sqrt{14}}{2}$ or $x = \frac{-3 - \sqrt{14}}{2}$

The solutions are $\frac{-3 \pm \sqrt{14}}{2}$.

34. $\frac{3 \pm \sqrt{17}}{4}$

35. $x^2 - 6x + 9 = 64$

$(x - 3)^2 = 64$ Factoring the left side

$x - 3 = 8$ or $x - 3 = -8$ Principle of square roots

$x = 11$ or $x = -5$

The solutions are 11 and -5.

36. 15, -5

37. $y^2 + 14y + 49 = 4$

$(y + 7)^2 = 4$ Factoring the left side

$y + 7 = 2$ or $y + 7 = -2$ Principle of square roots

$y = -5$ or $y = -9$

The solutions are -5 and -9.

38. -3, -5

39. $m^2 - 2m + 1 = 5$

$(m - 1)^2 = 5$

$m - 1 = \sqrt{5}$ or $m - 1 = -\sqrt{5}$

$m = 1 + \sqrt{5}$ or $m = 1 - \sqrt{5}$

The solutions are $1 \pm \sqrt{5}$.

40. $-3 \pm \sqrt{13}$

41. $x^2 + 4x + 4 = 12$

$(x + 2)^2 = 12$

$x + 2 = \sqrt{12}$ or $x + 2 = -\sqrt{12}$

$x = -2 + \sqrt{12}$ or $x = -2 - \sqrt{12}$

$x = -2 + \sqrt{4 \cdot 3}$ or $x = -2 - \sqrt{4 \cdot 3}$

$x = -2 + 2\sqrt{3}$ or $x = -2 - 2\sqrt{3}$

The solutions are $-2 \pm 2\sqrt{3}$.

42. $6 \pm 3\sqrt{2}$

43. Write each equation in slope-intercept form.

$y + 5 = 2x$ $\qquad$ $y - 2x = 7$

$y = 2x - 5$ $\qquad$ $y = 2x + 7$

Since the slopes are the same ($m = 2$) and the y-intercept are different, the equations represent parallel lines.

44. No

45. Write each equation in slope-intercept form.

$3x - 2y = 9$ $\qquad$ $4y - 6x = 8$

$-2y = -3x + 9$ $\qquad$ $4y = 6x + 8$

$y = \frac{3}{2}x - \frac{9}{2}$ $\qquad$ $y = \frac{3}{2}x + 2$

Since the slopes are the same $\left(m = \frac{3}{2}\right)$ and the y-intercepts are different, the equations represent parallel lines.

46. Yes

47.

48.

49. $x^2 + 5x + \frac{25}{4} = \frac{13}{4}$

$\left(x + \frac{5}{2}\right)^2 = \frac{13}{4}$

$x + \frac{5}{2} = \frac{\sqrt{13}}{2}$ or $x + \frac{5}{2} = -\frac{\sqrt{13}}{2}$

$x = -\frac{5}{2} + \frac{\sqrt{13}}{2}$ or $x = -\frac{5}{2} - \frac{\sqrt{13}}{2}$

$x = \frac{-5 + \sqrt{13}}{2}$ or $x = \frac{-5 - \sqrt{13}}{2}$

The solutions are $\frac{-5 \pm \sqrt{13}}{2}$.

50. $\frac{7 \pm \sqrt{7}}{6}$

51. $m^2 - \frac{3}{2}m + \frac{9}{16} = \frac{17}{16}$

$\left(m - \frac{3}{4}\right)^2 = \frac{17}{16}$

$m - \frac{3}{4} = \frac{\sqrt{17}}{4}$ or $m - \frac{3}{4} = -\frac{\sqrt{17}}{4}$

$m = \frac{3 + \sqrt{17}}{4}$ or $m = \frac{3 - \sqrt{17}}{4}$

The solutions are $\frac{3 \pm \sqrt{17}}{4}$.

52. 2, -5

53. $x^2 + 0.5x + 0.0625 = 13.69$

$(x + 0.25)^2 = 13.69$

$x + 0.25 = 3.7$ or $x + 0.25 = -3.7$

$x = 3.45$ or $x = -3.95$

The solutions are 3.45 and -3.95.

54. 1.85, -4.35

55. $a^2 - 3.8a + 3.61 = 27.04$

$(a - 1.9)^2 = 27.04$

$a - 1.9 = 5.2$ or $a - 1.9 = -5.2$

$x = 7.1$ or $a = -3.3$

The solutions are 7.1 and -3.3.

56. 9.9, -4.7

Exercise Set 10.2

1. $x^2 - 2x$

$\left(\frac{-2}{2}\right)^2 = (-1)^2 = 1$ Taking half the x-coefficient and squaring

$\downarrow$

$x^2 - 2x + 1$

The trinomial $x^2 - 2x + 1$ is the square of $x - 1$.

2. $x^2 - 4x + 4$

3. $x^2 + 18x$ $\left[\frac{18}{2}\right]^2 = 9^2 = 81$

The trinomial $x^2 + 18x + 81$ is the square of $x + 9$.

4. $x^2 + 22x + 121$

5. $x^2 - x$ $\left[\frac{-1}{2}\right]^2 = \left[-\frac{1}{2}\right]^2 = \frac{1}{4}$

The trinomial $x^2 - x + \frac{1}{4}$ is the square of $x - \frac{1}{2}$.

6. $x^2 + x + \frac{1}{4}$

7. $t^2 + 5t$ $\left[\frac{5}{2}\right]^2 = \frac{25}{4}$

The trinomial $t^2 + 5t + \frac{25}{4}$ is the square of $t + \frac{5}{2}$.

8. $y^2 - 9y + \frac{81}{4}$

9. $x^2 - \frac{3}{2}x$ $\left[\frac{-\frac{3}{2}}{2}\right]^2 = \left[-\frac{3}{4}\right]^2 = \frac{9}{16}$

The trinomial $x^2 - \frac{3}{2}x + \frac{9}{16}$ is the square of $x - \frac{3}{4}$.

10. $x^2 + \frac{4}{3}x + \frac{4}{9}$

11. $m^2 + \frac{9}{2}m$ $\left[\frac{\frac{9}{2}}{2}\right]^2 = \left[\frac{9}{4}\right]^2 = \frac{81}{16}$

The trinomial $m^2 + \frac{9}{2}m + \frac{81}{16}$ is the square of $m + \frac{9}{4}$.

12. $r^2 - \frac{2}{5}r + \frac{1}{25}$

13. $x^2 - 6x - 16 = 0$

$x^2 - 6x = 16$ Adding 16

$x^2 - 6x + 9 = 16 + 9$ Adding 9: $\left[\frac{-6}{2}\right]^2 = (-3)^2 = 9$

$(x - 3)^2 = 25$

$x - 3 = 5$ or $x - 3 = -5$ Principle of square roots

$x = 8$ or $x = -2$

The solutions are 8 and -2.

14. -3, -5

15. $x^2 + 22x + 21 = 0$

$x^2 + 22x = -21$ Adding -21

$x^2 + 22x + 121 = -21 + 121$ Adding 121: $\left[\frac{22}{2}\right]^2 = 11^2 = 121$

$(x + 11)^2 = 100$

$x + 11 = 10$ or $x + 11 = -10$ Principle of square roots

$x = -1$ or $x = -21$

The solutions are -1 and -21.

16. 1, -15

17. $3x^2 - 6x - 15 = 0$

$\frac{1}{3}(3x^2 - 6x - 15) = \frac{1}{3} \cdot 0$ Multiplying by $\frac{1}{3}$ to make the x^2-coefficient 1

$x^2 - 2x - 5 = 0$

$x^2 - 2x = 5$

$x^2 - 2x + 1 = 5 + 1$ Adding 1: $\left[\frac{-2}{2}\right]^2 = (-1)^2 = 1$

$(x - 1)^2 = 6$

$x - 1 = \sqrt{6}$ or $x - 1 = -\sqrt{6}$

$x = 1 + \sqrt{6}$ or $x = 1 - \sqrt{6}$

The solutions are $1 \pm \sqrt{6}$.

18. $2 \pm \sqrt{15}$

19. $x^2 - 22x + 102 = 0$

$x^2 - 22x = -102$

$x^2 - 22x + 121 = -102 + 121$

$(x - 11)^2 = 19$

$x - 11 = \sqrt{19}$ or $x - 11 = -\sqrt{19}$

$x = 11 + \sqrt{19}$ or $x = 11 - \sqrt{19}$

The solutions are $11 \pm \sqrt{19}$.

20. $9 \pm \sqrt{7}$

21. $x^2 + 10x - 4 = 0$

$x^2 + 10x = 4$

$x^2 + 10x + 25 = 4 + 25$

$(x + 5)^2 = 29$

$x + 5 = \sqrt{29}$ or $x + 5 = -\sqrt{29}$

$x = -5 + \sqrt{29}$ or $x = -5 - \sqrt{29}$

The solutions are $-5 \pm \sqrt{29}$.

22. $5 \pm \sqrt{29}$

23. $x^2 - 7x - 2 = 0$

$x^2 - 7x = 2$

$x^2 - 7x + \frac{49}{4} = 2 + \frac{49}{4}$ Adding $\frac{49}{4}$: $\left(\frac{-7}{2}\right)^2 = \frac{49}{4}$

$\left(x - \frac{7}{2}\right)^2 = \frac{8}{4} + \frac{49}{4} = \frac{57}{4}$

$x - \frac{7}{2} = \frac{\sqrt{57}}{2}$ or $x - \frac{7}{2} = -\frac{\sqrt{57}}{2}$

$x = \frac{7}{2} + \frac{\sqrt{57}}{2}$ or $x = \frac{7}{2} - \frac{\sqrt{57}}{2}$

$x = \frac{7 + \sqrt{57}}{2}$ or $x = \frac{7 - \sqrt{57}}{2}$

The solutions are $\frac{7 \pm \sqrt{57}}{2}$.

24. $\frac{-7 \pm \sqrt{57}}{2}$

25. $2x^2 + 6x - 56 = 0$

$\frac{1}{2}(2x^2 + 6x - 56) = \frac{1}{2} \cdot 0$ Multiplying by $\frac{1}{2}$ to make the x^2-coefficient 1

$x^2 + 3x - 28 = 0$

$x^2 + 3x = 28$

$x^2 + 3x + \frac{9}{4} = 28 + \frac{9}{4}$ Adding $\frac{9}{4}$: $\left(\frac{3}{2}\right)^2 = \frac{9}{4}$

$\left(x + \frac{3}{2}\right)^2 = \frac{121}{4}$

$x + \frac{3}{2} = \frac{11}{2}$ or $x + \frac{3}{2} = -\frac{11}{2}$

$x = \frac{8}{2}$ or $x = -\frac{14}{2}$

$x = 4$ or $x = -7$

The solutions are 4 and -7.

26. 7, -4

27. $x^2 + \frac{3}{2}x - \frac{1}{2} = 0$

$x^2 + \frac{3}{2}x = \frac{1}{2}$

$x^2 + \frac{3}{2}x + \frac{9}{16} = \frac{1}{2} + \frac{9}{16}$ Adding $\frac{9}{16}$: $\left[\frac{\frac{3}{2}}{2}\right]^2 = \left(\frac{3}{4}\right)^2 = \frac{9}{16}$

$\left(x + \frac{3}{4}\right)^2 = \frac{17}{16}$

$x + \frac{3}{4} = \frac{\sqrt{17}}{4}$ or $x + \frac{3}{4} = -\frac{\sqrt{17}}{4}$

$x = -\frac{3}{4} + \frac{\sqrt{17}}{4}$ or $x = -\frac{3}{4} - \frac{\sqrt{17}}{4}$

$x = \frac{-3 + \sqrt{17}}{4}$ or $x = \frac{-3 - \sqrt{17}}{4}$

The solutions are $\frac{-3 \pm \sqrt{17}}{4}$.

28. $\frac{3 \pm \sqrt{41}}{4}$

29. $2x^2 + 3x - 17 = 0$

$\frac{1}{2}(2x^2 + 3x - 17) = \frac{1}{2} \cdot 0$ Multiplying by $\frac{1}{2}$ to make the x^2-coefficient 1

$x^2 + \frac{3}{2}x - \frac{17}{2} = 0$

$x^2 + \frac{3}{2}x = \frac{17}{2}$

$x^2 + \frac{3}{2}x + \frac{9}{16} = \frac{17}{2} + \frac{9}{16}$ Adding $\frac{9}{16}$: $\left[\frac{\frac{3}{2}}{2}\right]^2 = \left(\frac{3}{4}\right)^2 = \frac{9}{16}$

$\left(x + \frac{3}{4}\right)^2 = \frac{145}{16}$

$x + \frac{3}{4} = \frac{\sqrt{145}}{4}$ or $x + \frac{3}{4} = -\frac{\sqrt{145}}{4}$

$x = \frac{-3 + \sqrt{145}}{4}$ or $x = \frac{-3 - \sqrt{145}}{4}$

The solutions are $\frac{-3 \pm \sqrt{145}}{4}$.

30. $\frac{3 \pm \sqrt{65}}{4}$

31. $3x^2 + 4x - 1 = 0$

$\frac{1}{3}(3x^2 + 4x - 1) = \frac{1}{3} \cdot 0$

$x^2 + \frac{4}{3}x - \frac{1}{3} = 0$

$x^2 + \frac{4}{3}x = \frac{1}{3}$

$x^2 + \frac{4}{3}x + \frac{4}{9} = \frac{1}{3} + \frac{4}{9}$

$\left(x + \frac{2}{3}\right)^2 = \frac{7}{9}$

$x + \frac{2}{3} = \frac{\sqrt{7}}{3}$ or $x + \frac{2}{3} = -\frac{\sqrt{7}}{3}$

$x = \frac{-2 + \sqrt{7}}{3}$ or $x = \frac{-2 - \sqrt{7}}{3}$

The solutions are $\frac{-2 \pm \sqrt{7}}{3}$.

32. $\frac{2 \pm \sqrt{13}}{3}$

33. $2x^2 = 9x + 5$

$$2x^2 - 9x - 5 = 0$$

$$\frac{1}{2}(2x^2 - 9x - 5) = \frac{1}{2} \cdot 0$$

$$x^2 - \frac{9}{2}x - \frac{5}{2} = 0$$

$$x^2 - \frac{9}{2}x = \frac{5}{2}$$

$$x^2 - \frac{9}{2}x + \frac{81}{16} = \frac{5}{2} + \frac{81}{16}$$

$$\left(x - \frac{9}{4}\right)^2 = \frac{121}{16}$$

$$x - \frac{9}{4} = \frac{11}{4} \quad \text{or} \quad x - \frac{9}{4} = -\frac{11}{4}$$

$$x = \frac{20}{4} \quad \text{or} \quad x = -\frac{2}{4}$$

$$x = 5 \quad \text{or} \quad x = -\frac{1}{2}$$

The solutions are 5 and $-\frac{1}{2}$.

34. $4, -\frac{3}{2}$

35. $4x^2 + 12x = 7$

$$4x^2 + 12x - 7 = 0$$

$$\frac{1}{4}(4x^2 + 12x - 7) = \frac{1}{4} \cdot 0$$

$$x^2 + 3x - \frac{7}{4} = 0$$

$$x^2 + 3x = \frac{7}{4}$$

$$x^2 + 3x + \frac{9}{4} = \frac{7}{4} + \frac{9}{4}$$

$$\left(x + \frac{3}{2}\right)^2 = \frac{16}{4} = 4$$

$$x + \frac{3}{2} = 2 \quad \text{or} \quad x + \frac{3}{2} = -2$$

$$x = \frac{1}{2} \quad \text{or} \quad x = -\frac{7}{2}$$

The solutions are $\frac{1}{2}$ and $-\frac{7}{2}$.

36. $\frac{2}{3}, -\frac{5}{2}$

37. $y - x = 5,$ (1)
$y + 2x = 7$ (2)

We solve equation (1) for y.

$y - x = 5$ (1)
$y = x + 5$ (3)

Then we substitute $x + 5$ for y in equation (2) and solve for x.

$y + 2x = 7$ (2)
$x + 5 + 2x = 7$
$3x + 5 = 7$
$3x = 2$
$x = \frac{2}{3}$

Substitute $\frac{2}{3}$ for x in equation (3).

$$y = x + 5 = \frac{2}{3} + 5 = \frac{17}{3}.$$

The solution is $\left(\frac{2}{3}, \frac{17}{3}\right)$.

38. (-2,4)

39. Graph $y = \frac{3}{5}x - 1$.

First we plot the y-intercept (0,-1). The slope is $\frac{3}{5}$. Starting at (0,-1), we move up 3 units and right 5 units to another point on the graph, (5,2). We can also think of the slope as $\frac{-3}{-5}$. Starting at (0,-1) again, we move down 3 units and left 5 units to a third point on the graph, (-5,-4). We draw the line through these points.

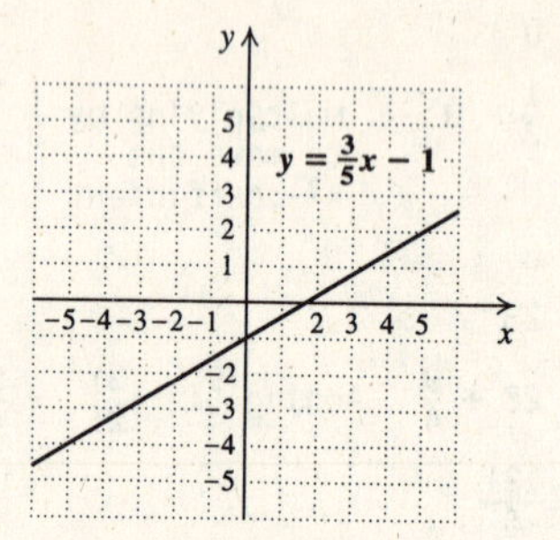

40.

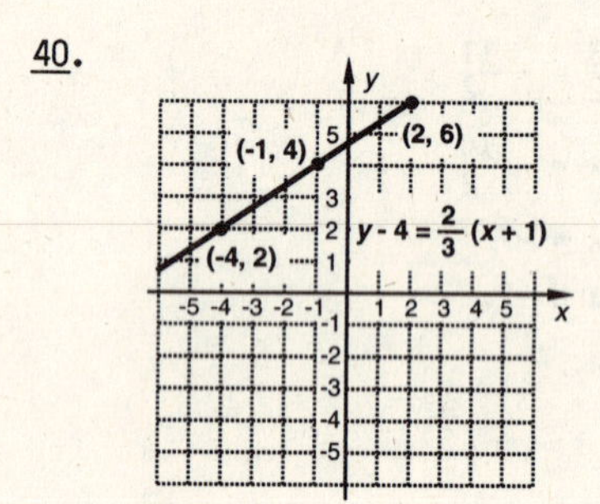

41.

42.

43. $x^2 + bx + 36$

The trinomial is a square if the square of one-half the x-coefficient is equal to 36. Thus we have:

$$\left(\frac{b}{2}\right)^2 = 36$$

$$\frac{b^2}{4} = 36$$

$$b^2 = 144$$

$b = 12$ or $b = -12$ Principle of square roots

44. $2\sqrt{55}, -2\sqrt{55}$

45. $x^2 + bx + 128$

The trinomial is a square if the square of one-half the x-coefficient is equal to 128. Thus we have:

$$\left(\frac{b}{2}\right)^2 = 128$$

$$\frac{b^2}{4} = 128$$

$$b^2 = 512$$

$b = \sqrt{512}$ or $b = -\sqrt{512}$

$b = 16\sqrt{2}$ or $b = -16\sqrt{2}$

46. 16, -16

47. $x^2 + bx + c$

The trinomial is a square if the square of one-half the x-coefficient is equal to c. Thus we have:

$$\left(\frac{b}{2}\right)^2 = c$$

$$\frac{b^2}{4} = c$$

$$b^2 = 4c$$

$b = \sqrt{4c}$ or $b = -\sqrt{4c}$

$b = 2\sqrt{c}$ or $b = -2\sqrt{c}$

48. $2\sqrt{ac}$, $-2\sqrt{ac}$

49. 8.00, 2.00

50. 2.00, -8.00

51. -0.39, -7.61

52. 7.41, 4.59

53. 7.27, -0.27

54. 0.27, -7.27

55. -0.50, 5.00

56. 4.00, -1.50

Exercise Set 10.3

1. $x^2 - 4x = 21$

$x^2 - 4x - 21 = 0$ Standard form

We can factor.

$x^2 - 4x - 21 = 0$

$(x - 7)(x + 3) = 0$

$x - 7 = 0$ or $x + 3 = 0$

$x = 7$ or $x = -3$

The solutions are 7 and -3.

2. -9, 2

3. $x^2 = 6x - 9$

$x^2 - 6x + 9 = 0$ Standard form

We can factor.

$x^2 - 6x + 9 = 0$

$(x - 3)(x - 3) = 0$

$x - 3 = 0$ or $x - 3 = 0$

$x = 3$ or $x = 3$

The solution is 3.

4. 4

5. $3y^2 - 2y - 8 = 0$

We can factor.

$3y^2 - 2y - 8 = 0$

$(3y + 4)(y - 2) = 0$

$3y + 4 = 0$ or $y - 2 = 0$

$3y = -4$ or $y = 2$

$y = -\frac{4}{3}$ or $y = 2$

The solutions are $-\frac{4}{3}$ and 2.

6. $\frac{4}{3}$, 1

7. $4x^2 + 12x = 7$

$4x^2 + 12x - 7 = 0$

We can factor.

$4x^2 + 12x - 7 = 0$

$(2x - 1)(2x + 7) = 0$

$2x - 1 = 0$ or $2x + 7 = 0$

$2x = 1$ or $2x = -7$

$x = \frac{1}{2}$ or $x = -\frac{7}{2}$

The solutions are $\frac{1}{2}$ and $-\frac{7}{2}$.

8. $-\frac{5}{2}$, $\frac{3}{2}$

9. $x^2 - 9 = 0$ Difference of squares

$(x + 3)(x - 3) = 0$

$x + 3 = 0$ or $x - 3 = 0$

$x = -3$ or $x = 3$

The solutions are -3 and 3.

10. -2, 2

11. $x^2 - 2x - 2 = 0$
We use the quadratic formula.
$a = 1$, $b = -2$, $c = -2$

$$x = \frac{-(-2) \pm \sqrt{(-2)^2 - 2\cdot1\cdot(-2)}}{2\cdot1}$$

$$x = \frac{2 \pm \sqrt{12}}{2} = \frac{2 \pm \sqrt{4\cdot3}}{2}$$

$$x = \frac{2 \pm 2\sqrt{3}}{2} = \frac{2(1 \pm \sqrt{3})}{2}$$

$$x = 1 \pm \sqrt{3}$$

12. $2 \pm \sqrt{11}$

13. $y^2 - 10y + 22 = 0$
We use the quadratic formula.
$a = 1$, $b = -10$, $c = 22$

$$y = \frac{-(-10) \pm \sqrt{(-10)^2 - 4\cdot1\cdot22}}{2\cdot1}$$

$$y = \frac{10 \pm \sqrt{12}}{2} = \frac{10 \pm \sqrt{4\cdot3}}{2}$$

$$y = \frac{10 \pm 2\sqrt{3}}{2} = \frac{2(5 \pm \sqrt{3})}{2}$$

$$y = 5 \pm \sqrt{3}$$

14. $-3 \pm \sqrt{10}$

15. $x^2 + 4x + 4 = 7$
$x^2 + 4x - 3 = 0$ Adding -7 to get standard form
We use the quadratic formula.
$a = 1$, $b = 4$, $c = -3$

$$x = \frac{-4 \pm \sqrt{4^2 - 4\cdot1\cdot(-3)}}{2\cdot1}$$

$$x = \frac{-4 \pm \sqrt{28}}{2} = \frac{-4 \pm \sqrt{4\cdot7}}{2}$$

$$x = \frac{-4 \pm 2\sqrt{7}}{2} = \frac{2(-2 \pm \sqrt{7})}{2}$$

$$x = -2 \pm \sqrt{7}$$

16. $1 \pm \sqrt{5}$

17. $3x^2 + 8x + 2 = 0$
We use the quadratic formula.
$a = 3$, $b = 8$, $c = 2$

$$x = \frac{-8 \pm \sqrt{8^2 - 4\cdot3\cdot2}}{2\cdot3}$$

$$x = \frac{-8 \pm \sqrt{40}}{6} = \frac{-8 \pm \sqrt{4\cdot10}}{6}$$

$$x = \frac{-8 \pm 2\sqrt{10}}{6} = \frac{2(-4 \pm \sqrt{10})}{2\cdot3}$$

$$x = \frac{-4 \pm \sqrt{10}}{3}$$

18. $\frac{2 \pm \sqrt{10}}{3}$

19. $2x^2 - 5x = 1$
$2x^2 - 5x - 1 = 0$ Adding -1 to get standard form
We use the quadratic formula.
$a = 2$, $b = -5$, $c = -1$

$$x = \frac{-(-5) \pm \sqrt{(-5)^2 - 4\cdot2\cdot(-1)}}{2\cdot2}$$

$$x = \frac{5 \pm \sqrt{33}}{4}$$

20. $\frac{-1 \pm \sqrt{7}}{2}$

21. $4y^2 - 4y - 1 = 0$
We use the quadratic formula.
$a = 4$, $b = -4$, $c = -1$

$$y = \frac{-(-4) \pm \sqrt{(-4)^2 - 4\cdot4\cdot(-1)}}{2\cdot4}$$

$$y = \frac{4 \pm \sqrt{32}}{8} = \frac{4 \pm \sqrt{16\cdot2}}{8}$$

$$y = \frac{4 \pm 4\sqrt{2}}{8} = \frac{4(1 \pm \sqrt{2})}{4\cdot2}$$

$$y = \frac{1 \pm \sqrt{2}}{2}$$

22. $\frac{-1 \pm \sqrt{2}}{2}$

23. $2t^2 + 6t + 5 = 0$
We use the quadratic formula.
$a = 2$, $b = 6$, $c = 5$

$$t = \frac{-6 \pm \sqrt{6^2 - 4\cdot2\cdot5}}{2\cdot2}$$

$$t = \frac{-6 \pm \sqrt{-4}}{4}$$

Since the radicand, -4, is negative, there are no real-number solutions.

24. No real-number solutions

25. $3x^2 = 5x + 4$
$3x^2 - 5x - 4 = 0$
We use the quadratic formula.
$a = 3$, $b = -5$, $c = -4$

$$x = \frac{-(-5) \pm \sqrt{(-5)^2 - 4\cdot3\cdot(-4)}}{2\cdot3}$$

$$x = \frac{5 \pm \sqrt{73}}{6}$$

26. $\frac{-3 \pm \sqrt{17}}{4}$

27. $2y^2 - 6y = 10$

$2y^2 - 6y - 10 = 0$

$y^2 - 3y - 5 = 0$ Multiplying by $\frac{1}{2}$ to simplify

We use the quadratic formula.

$a = 1,\ b = -3,\ c = -5$

$$y = \frac{-(-3) \pm \sqrt{(-3)^2 - 4\cdot1\cdot(-5)}}{2\cdot1}$$

$$y = \frac{3 \pm \sqrt{29}}{2}$$

28. $\frac{11 \pm \sqrt{181}}{10}$

29. $6x^2 - 9x = 0$ We can factor.

$3x(2x - 3) = 0$

$3x = 0$ or $2x - 3 = 0$

$x = 0$ or $2x = 3$

$x = 0$ or $x = \frac{3}{2}$

30. No real-number solutions

31. $5t^2 - 7t = -4$

$5t^2 - 7t + 4 = 0$

We use the quadratic formula.

$a = 5,\ b = -7,\ c = 4$

$$t = \frac{-(-7) \pm \sqrt{(-7)^2 - 4\cdot5\cdot4}}{2\cdot5}$$

$$t = \frac{7 \pm \sqrt{-31}}{10}$$

Since the radicand, -31, is negative, there are no real-number solutions.

32. $0,\ -\frac{2}{3}$

33. $4x^2 = 90$

$x^2 = \frac{90}{4}$

$x = \sqrt{\frac{90}{4}}$ or $x = -\sqrt{\frac{90}{4}}$ Principle of square roots

$x = \frac{\sqrt{90}}{2}$ or $x = -\frac{\sqrt{90}}{2}$

$x = \frac{3\sqrt{10}}{2}$ or $x = -\frac{3\sqrt{10}}{2}$ $(\sqrt{90} = \sqrt{9\cdot10} = 3\sqrt{10})$

34. $2\sqrt{5},\ -2\sqrt{5}$

35. $x^2 - 4x - 7 = 0$

$a = 1,\ b = -4,\ c = -7$

$$x = \frac{-(-4) \pm \sqrt{(-4)^2 - 4\cdot1\cdot(-7)}}{2\cdot1}$$

$$x = \frac{4 \pm \sqrt{16 + 28}}{2} = \frac{4 \pm \sqrt{44}}{2}$$

$$x = \frac{4 \pm \sqrt{4\cdot11}}{2} = \frac{4 \pm 2\sqrt{11}}{2}$$

$$x = \frac{2(2 \pm \sqrt{11})}{2} = 2 \pm \sqrt{11}$$

Using a calculator or Table 1, we see that $\sqrt{11} \approx 3.317$:

$2 + \sqrt{11} \approx 2 + 3.317$ or $2 - \sqrt{11} \approx 2 - 3.317$

≈ 5.317 or ≈ -1.317

The approximate solutions, to the nearest thousandth, are 5.317 and -1.317.

36. 0.732, -2.732

37. $y^2 - 6y - 1 = 0$

$a = 1,\ b = -6,\ c = -1$

$$y = \frac{-(-6) \pm \sqrt{(-6)^2 - 4\cdot1\cdot(-1)}}{2\cdot1}$$

$$y = \frac{6 \pm \sqrt{36 + 4}}{2} = \frac{6 \pm \sqrt{40}}{2}$$

$$y = \frac{6 \pm \sqrt{4\cdot10}}{2} = \frac{6 \pm 2\sqrt{10}}{2}$$

$$y = \frac{2(3 \pm \sqrt{10})}{2} = 3 \pm \sqrt{10}$$

Using a calculator or Table 1, we see that $\sqrt{10} \approx 3.162$:

$3 + \sqrt{10} \approx 3 + 3.162$ or $3 - \sqrt{10} \approx 3 - 3.162$

≈ 6.162 or ≈ -0.162

The approximate solutions, to the nearest thousandth, are 6.162 and -0.162.

38. -3.268, -6.732

39. $4x^2 + 4x = 1$

$4x^2 + 4x - 1 = 0$ Standard form

$a = 4,\ b = 4,\ c = -1$

$$x = \frac{-4 \pm \sqrt{4^2 - 4\cdot4\cdot(-1)}}{2\cdot4}$$

$$x = \frac{-4 \pm \sqrt{16 + 16}}{8} = \frac{-4 \pm \sqrt{32}}{8}$$

$$x = \frac{-4 \pm \sqrt{16\cdot2}}{8} = \frac{-4 \pm 4\sqrt{2}}{8}$$

$$x = \frac{4(-1 \pm \sqrt{2})}{4\cdot2} = \frac{-1 \pm \sqrt{2}}{2}$$

Using a calculator or Table 1, we see that $\sqrt{2} \approx 1.414$:

$$\frac{-1+\sqrt{2}}{2} \approx \frac{-1+1.414}{2} \quad \text{or} \quad \frac{-1-\sqrt{2}}{2} \approx \frac{-1-1.414}{2}$$

$$\approx \frac{0.414}{2} \quad \text{or} \quad \approx \frac{-2.414}{2}$$

$$\approx 0.207 \quad \text{or} \quad \approx -1.207$$

The approximate solutions, to the nearest thousandth, are 0.207 and -1.207.

40. 1.207, -0.207

41. Familiarize. We will use the formula

$$d = \frac{n^2 - 3n}{2},$$

where d is the number of diagonals and n is the number of sides.

Translate. We substitute 8 for n.

$$d = \frac{8^2 - 3\cdot 8}{2}$$

Carry out. We do the computation.

$$d = \frac{8^2 - 3\cdot 8}{2} = \frac{64 - 24}{2} = \frac{40}{2} = 20$$

Check. We can substitute 20 for d in the original formula and see if this yields n = 8. This is left to the student.

State. An octagon has 20 diagonals.

42. 35

43. Familiarize. We will use the formula

$$d = \frac{n^2 - 3n}{2},$$

where d is the number of diagonals and n is the number of sides.

Translate. We substitute 14 for d.

$$14 = \frac{n^2 - 3n}{2}$$

Carry out. We solve the equation.

$$\frac{n^2 - 3n}{2} = 14$$

$n^2 - 3n = 28$ Multiplying by 2

$$n^2 - 3n - 28 = 0$$

$$(n - 7)(n + 4) = 0$$

$$n - 7 = 0 \quad \text{or} \quad n + 4 = 0$$

$$n = 7 \quad \text{or} \quad n = -4$$

Check. Since the number of sides cannot be negative, -4 cannot be a solution. To check 7, we substitute 7 for n in the original formula and determine if this yields d = 14. This is left to the student.

State. The polygon has 7 sides.

44. 5

45. Familiarize. We will use the formula $s = 16t^2$.

Translate. We substitute 1136 for s.

$$1136 = 16t^2$$

Carry out. We solve the equation.

$$1136 = 16t^2$$

$71 = t^2$ Dividing by 16

$\sqrt{71} = t$ or $-\sqrt{71} = t$ Principle of square roots

$8.43 \approx t$ or $-8.43 \approx t$ Using a calculator and rounding to the nearest hundredth

Check. The number -8.43 cannot be a solution, because time cannot be negative in this situation. We substitute 8.43 in the original equation.

$$s = 16(8.43)^2 = 16(71.0649) = 1137.0384$$

This is close. Remember that we approximated a solution. Thus we have a check.

State. It takes about 8.43 sec for an object to fall to the ground from the top of the World Trade Center.

46. About 7.95 sec

47. Familiarize. We will use the formula $s = 16t^2$.

Translate. We substitute 175 for s.

$$175 = 16t^2$$

Carry out. We solve the equation.

$$175 = 16t^2$$

$$\frac{175}{16} = t^2$$

$$\sqrt{\frac{175}{16}} = t \quad \text{or} \quad -\sqrt{\frac{175}{16}} = t$$

$$\frac{\sqrt{175}}{4} = t \quad \text{or} \quad -\frac{\sqrt{175}}{4} = t$$

$\frac{5\sqrt{7}}{4} = t$ or $-\frac{5\sqrt{7}}{4} = t$ $(\sqrt{175} = \sqrt{25\cdot 7} = 5\sqrt{7})$

$$3.31 \approx t \quad \text{or} \quad -3.31 \approx t$$

Check. The number -3.31 cannot be a solution because time cannot be negative in this situation. We substitute 3.31 in the original equation.

$$s = 16(3.31)^2 = 16(10.9561) = 175.2976$$

This is close. Remember that we approximated a solution. Thus we have a check.

State. The fall took about 3.31 sec.

48. About 4.41 sec

49. Familiarize. We first make a drawing and label it. We let x represent the length of one leg. Then x + 1 represents the length of the other leg.

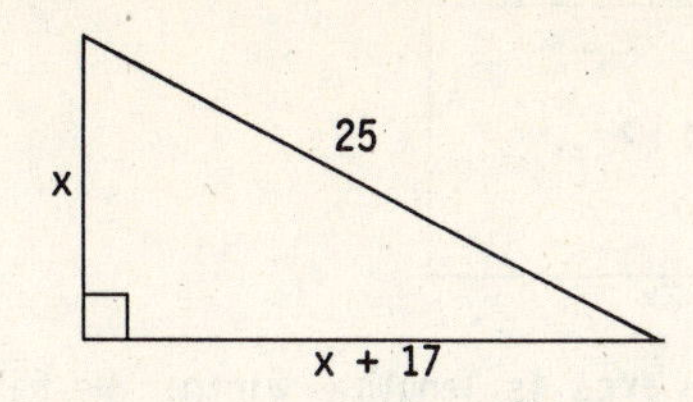

Translate. We use the Pythagorean equation.

$x^2 + (x + 17)^2 = 25^2$

Carry out. We solve the equation.

$x^2 + x^2 + 34x + 289 = 625$

$2x^2 + 34x - 336 = 0$

$x^2 + 17x - 168 = 0$ Multiplying by $\frac{1}{2}$

$(x - 7)(x + 24) = 0$

$x - 7 = 0$ or $x + 24 = 0$

$x = 7$ or $x = -24$

Check. Since the length of a leg cannot be negative, -24 does not check. But 7 does check. If the smaller leg is 7, the other leg is 7 + 17, or 24. Then, $7^2 + 24^2 = 49 + 576 = 625$, and $\sqrt{625} = 25$, the length of the hypotenuse.

State. The legs measure 7 ft and 24 ft.

50. 10 yd, 24 yd

51. Familiarize. We consider the drawing in the text, where w represents the width of the rectangle and w + 2 represents the length.

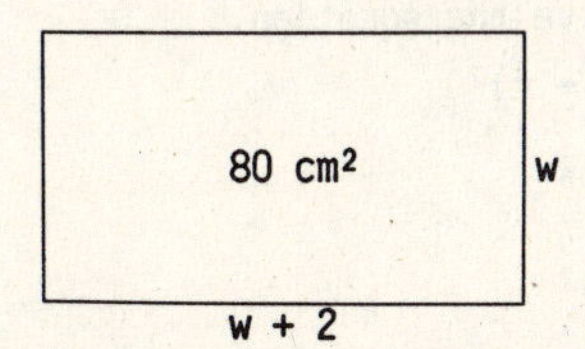

Translate. The area is length × width. Thus, we have two expressions for the area of the rectangle: (w + 2)w and 80. This gives us a translation.

$(w + 2)w = 80$

Carry out. We solve the equation.

$w^2 + 2w = 80$

$w^2 + 2w - 80 = 0$

$(w + 10)(w - 8) = 0$

$w + 10 = 0$ or $w - 8 = 0$

$w = -10$ or $w = 8$

Check. Since the length of a side cannot be negative, -10 does not check. But 8 does check. If the width is 8, then the length is 8 + 2, or 10. The area is 10 × 8, or 80. This checks.

State. The length is 10 cm, and the width is 8 cm.

52. Length: 10 m, width: 7 m

53. Familiarize. We make a drawing. We let w = the width of the yard. Then w + 5 = the length.

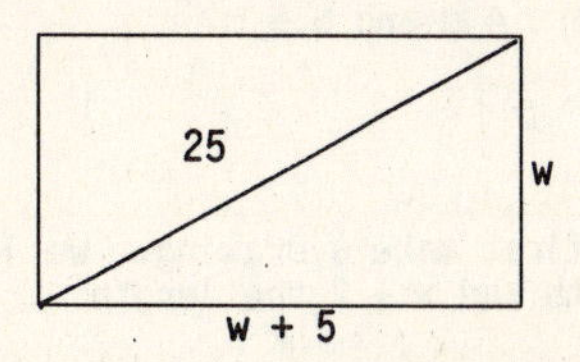

Translate. We use the Pythagorean equation.

$w^2 + (w + 5)^2 = 25^2$

Carry out. We solve the equation.

$w^2 + w^2 + 10w + 25 = 625$

$2w^2 + 10w - 600 = 0$

$w^2 + 5w - 300 = 0$

$(w + 20)(w - 15) = 0$

$w + 20 = 0$ or $w - 15 = 0$

$w = -20$ or $w = 15$

Check. Since the width cannot be negative, -20 does not check. But 15 does check. If the width is 15, then the length is 15 + 5, or 20, and $15^2 + 20^2 = 225 + 400 = 625 = 25^2$.

State. The yard is 15m by 20 m.

54. 24 ft

55. Familiarize. We make a drawing. Let x = the shorter leg of the right triangle. Then x + 2.5 = the longer leg.

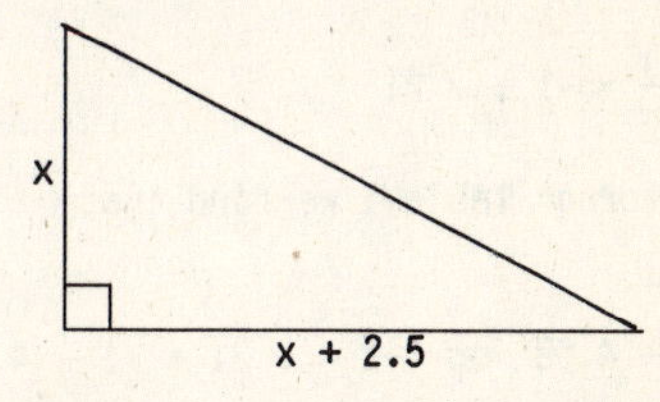

Translate. Using the formula $A = \frac{1}{2}bh$, we substitute 13 for A, x + 2.5 for b, and x for h.

$13 = \frac{1}{2}(x + 2.5)(x)$

Carry out. We solve the equation.

$13 = \frac{1}{2}(x + 2.5)(x)$

$26 = (x + 2.5)(x)$ Multiplying by 2

$26 = x^2 + 2.5x$

$0 = x^2 + 2.5x - 26$

$0 = 10x^2 + 25x - 260$ Multiplying by 10 to clear the decimal

$0 = 5(2x^2 + 5x - 52)$

$0 = 5(2x + 13)(x - 4)$

$2x + 13 = 0 \quad \text{or} \quad x - 4 = 0$

$2x = -13 \quad \text{or} \quad x = 4$

$x = -6.5 \quad \text{or} \quad x = 4$

Check. Since the length cannot be negative, -6.5 does not check. But 4 does check. If the shorter leg is 4, then the longer leg is 4 + 2.5, or 6.5, and $A = \frac{1}{2}(6.5)(4) = 13$.

State. The legs are 4 m and 6.5 m.

56. 5 cm, 6.2 cm

57. Familiarize. We first make a drawing. We let x represent the width and x + 2 the length.

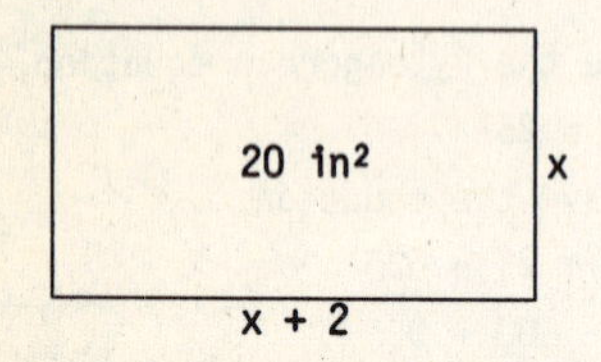

Translate. The area is length × width. We have two expressions for the area of the rectangle: $(x + 2)x$ and 20. This gives us a translation.

$(x + 2)x = 20$

Carry out. We solve the equation.

$x^2 + 2x = 20$

$x^2 + 2x - 20 = 0$

$a = 1,\ b = 2,\ c = -20$

$$x = \frac{-2 \pm \sqrt{2^2 - 4\cdot1\cdot(-20)}}{2\cdot1}$$

$$x = \frac{-2 \pm \sqrt{4 + 80}}{2} = \frac{-2 \pm \sqrt{84}}{2}$$

$$x = \frac{-2 \pm \sqrt{4\cdot21}}{2} = \frac{-2 \pm 2\sqrt{21}}{2}$$

$$x = \frac{2(-1 \pm \sqrt{21})}{2} = -1 \pm \sqrt{21}$$

Using a calculator or Table 1 we find that $\sqrt{21} \approx 4.58$:

$-1 + \sqrt{21} \approx -1 + 4.58 \quad \text{or} \quad -1 - \sqrt{21} \approx -1 - 4.58$

$\approx 3.58 \quad \text{or} \quad \approx -5.58$

Check. Since the length of a side cannot be negative, -5.58 does not check. But 3.58 does check. If the width is 3.58, then the length is 3.58 + 2, or 5.58. The area is 5.58(3.58), or 19.9764 ≈ 20.

State. The length is about 5.58 in., and the width is about 3.58 in.

58. Length: 5.65 ft, width: 2.65 ft

59. Familiarize. We first make a drawing. We let x represent the width and 2x the length.

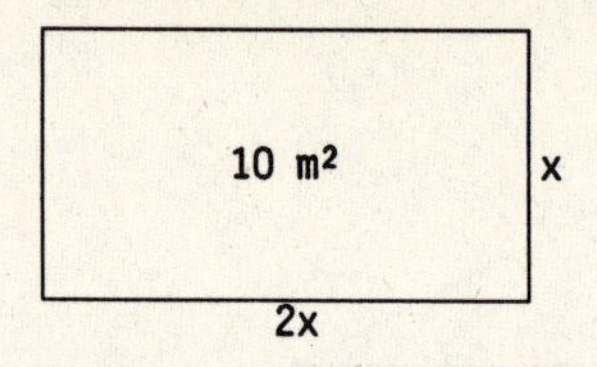

Translate. The area is length × width. We have two expressions for the area of the rectangle: $2x\cdot x$ and 10. This gives us a translation.

$2x\cdot x = 10$

Carry out. We solve the equation.

$2x^2 = 10$

$x^2 = 5$

$x = \sqrt{5} \quad \text{or} \quad x = -\sqrt{5}$

$x \approx 2.24 \quad \text{or} \quad x \approx -2.24$ Using a calculator or Table 1

Check. Since the length cannot be negative, -2.24 does not check. But 2.24 does check. If the width is $\sqrt{5}$ m, then the length is $2(\sqrt{5})$ or 4.47 m. The area is (4.47)(2.24), or 10.0128 ≈ 10.

State. The length is about 4.47 m, and the width is about 2.24 m.

60. Length: 6.32 cm, width: 3.16 cm

61. Familiarize. We will use the formula $A = P(1 + r)^t$ where P is \$1000, A is \$1210 and t is 2.

Translate. We make the substitutions.

$A = P(1 + r)^t$

$1210 = 1000(1 + r)^2$

Carry out. We solve the equation.

$1210 = 1000(1 + r)^2$

$\frac{1210}{1000} = (1 + r)^2$

$\frac{121}{100} = (1 + r)^2$

$\frac{11}{10} = 1 + r \quad \text{or} \quad -\frac{11}{10} = 1 + r$ Principle of square roots

$\frac{11}{10} - \frac{10}{10} = r \quad \text{or} \quad -\frac{11}{10} - \frac{10}{10} = r$

$\frac{1}{10} = r \quad \text{or} \quad -\frac{21}{10} = r$

Check. Since the interest rate cannot be negative,

$\frac{1}{10} = r$

$0.1 = r$

$10\% = r.$

We check 10%, or 0.1, in the formula.

$1000(1 + 0.1)^2 = 1000(1.1)^2 = 1000(1.21) = 1210$

Our answer checks.

State. The interest rate is 10%.

62. 20%

63. Familiarize. We will use the formula $A = P(1 + r)^t$ where P is \$2560, A is \$3610, and t is 2.

Translate. We make the substitutions.

$$A = P(1 + r)^t$$
$$3610 = 2560(1 + r)^2$$

Carry out. We solve the equation.

$$3610 = 2560(1 + r)^2$$
$$\frac{3610}{2560} = (1 + r)^2$$
$$\frac{361}{256} = (1 + r)^2$$
$$\frac{19}{16} = 1 + r \quad \text{or} \quad -\frac{19}{16} = 1 + r \quad \text{Principle of square roots}$$
$$\frac{19}{16} - \frac{16}{16} = r \quad \text{or} \quad -\frac{19}{16} - \frac{16}{16} = r$$
$$\frac{3}{16} = r \quad \text{or} \quad -\frac{35}{16} = r$$

Check. Since the interest rate cannot be negative,

$$\frac{3}{16} = r$$
$$0.1875 = r$$
$$18.75\% = r.$$

We check 18.75%, or 0.1875, in the formula.

$$2560(1 + 0.1875)^2 = 2560(1.1875)^2 = 3610$$

Our answer checks.

State. The interest rate is 18.75%.

64. 5%

65. Familiarize. We will use the formula $A = P(1 + r)^t$ where P is \$6250, A is \$7290, and t is 2.

Translate. We make the substitutions.

$$A = P(1 + r)^t$$
$$7290 = 6250(1 + r)^2$$

Carry out. We solve the equation.

$$7290 = 6250(1 + r)^2$$
$$\frac{7290}{6250} = (1 + r)^2$$
$$\frac{729}{625} = (1 + r)^2$$
$$\frac{27}{25} = 1 + r \quad \text{or} \quad -\frac{27}{25} = 1 + r$$
$$\frac{27}{25} - \frac{25}{25} = r \quad \text{or} \quad -\frac{27}{25} - \frac{25}{25} = r$$
$$\frac{2}{25} = r \quad \text{or} \quad -\frac{52}{25} = r$$

Check. Since the interest rate cannot be negative,

$$\frac{2}{25} = r$$
$$0.08 = r$$
$$8\% = r.$$

We check 8%, or 0.08, in the formula.

$$6250(1 + 0.08)^2 = 6250(1.08)^2 = 6250(1.1664) = 7290$$

Our answer checks.

State. The interest rate is 8%.

66. 4%

67. Familiarize. Let d = the diameter (or width) of the flower garden. Then $\frac{d}{2}$ = the radius. We will use the formula for the area of a circle, $A = \pi r^2$.

Translate. We substitute 250 for A, 3.14 for π, and $\frac{d}{2}$ for r in the formula.

$$250 = 3.14\left(\frac{d}{2}\right)^2$$
$$250 = 3.14\left(\frac{d^2}{4}\right)$$

Carry out. We solve the equation.

$$250 = 3.14\left(\frac{d^2}{4}\right)$$
$$250 = 0.785d^2 \qquad \left[\frac{3.14}{4} = 0.785\right]$$
$$\frac{250}{0.785} = d^2$$
$$17.85 \approx d \quad \text{or} \quad -17.85 \approx d$$

Check. Since the diameter cannot be negative, -17.85 cannot be a solution. If d = 17.85, then $A = 3.14\left(\frac{17.85}{2}\right)^2 = 250.1186625 \approx 250$. The answer checks.

State. The width of the largest circular garden Laura can cover with the mulch is about 17.85 ft.

68. About 159.62 m

69.
$$5(2x - 3) + 4x = 9 - 6x$$
$$10x - 15 + 4x = 9 - 6x \quad \text{Removing parentheses}$$
$$14x - 15 = 9 - 6x$$
$$20x = 24 \quad \text{Adding } -14x \text{ and } 6x$$
$$x = \frac{24}{20}$$
$$x = \frac{6}{5} \quad \text{Simplifying}$$

70. -20

71. $\sqrt{40} - 2\sqrt{10} + \sqrt{90} = \sqrt{4\cdot 10} - 2\sqrt{10} + \sqrt{9\cdot 10}$
$= \sqrt{4}\sqrt{10} - 2\sqrt{10} + \sqrt{9}\sqrt{10}$
$= 2\sqrt{10} - 2\sqrt{10} + 3\sqrt{10}$
$= (2 - 2 + 3)\sqrt{10}$
$= 3\sqrt{10}$

72. $30x^5\sqrt{10}$

73. ◈

74. ◈

75. $5x + x(x - 7) = 0$
$5x + x^2 - 7x = 0$
$x^2 - 2x = 0$ We can factor.
$x(x - 2) = 0$

$x = 0$ or $x - 2 = 0$
$x = 0$ or $x = 2$
The solutions are 0 and 2.

76. $0, -\frac{4}{3}$

77. $3 - x(x - 3) = 4$
$3 - x^2 + 3x = 4$
$0 = x^2 - 3x + 1$ Standard form
We use the quadratic formula.
$a = 1, b = -3, c = 1$

$x = \frac{-(-3) \pm \sqrt{(-3)^2 - 4\cdot 1\cdot 1}}{2\cdot 1}$

$x = \frac{3 \pm \sqrt{5}}{2}$

78. $\frac{7 \pm \sqrt{69}}{10}$

79. $(y + 4)(y + 3) = 15$
$y^2 + 7y + 12 = 15$
$y^2 + 7y - 3 = 0$ Standard form
We use the quadratic formula.
$a = 1, b = 7, c = -3$

$y = \frac{-7 \pm \sqrt{7^2 - 4\cdot 1\cdot(-3)}}{2\cdot 1}$

$y = \frac{-7 \pm \sqrt{61}}{2}$

80. $\frac{-2 \pm \sqrt{10}}{2}$

81. $(x + 2)^2 + (x + 1)^2 = 0$
$x^2 + 4x + 4 + x^2 + 2x + 1 = 0$
$2x^2 + 6x + 5 = 0$ Standard form
We use the quadratic formula.
$a = 2, b = 6, c = 5$

$x = \frac{-6 \pm \sqrt{6^2 - 4\cdot 2\cdot 5}}{2\cdot 2}$

$x = \frac{-6 \pm \sqrt{-4}}{4}$

Since the radicand, -4, is negative, there are no real-number solutions.

82. No real-number solutions

83. $\frac{x^2}{x - 4} - \frac{7}{x - 4} = 0$, LCD is $x - 4$

$(x - 4)\left[\frac{x^2}{x - 4} - \frac{7}{x - 4}\right] = (x - 4)\cdot 0$

$x^2 - 7 = 0$
$x^2 = 7$

$x = \sqrt{7}$ or $x = -\sqrt{7}$ Principle of square roots

Both numbers check. The solutions are $\sqrt{7}$ and $-\sqrt{7}$.

84. $\sqrt{5}, -\sqrt{5}$

85. $\frac{1}{x} + \frac{1}{x + 6} = \frac{1}{5}$, LCD is $5x(x + 6)$

$5x(x + 6)\left[\frac{1}{x} + \frac{1}{x + 6}\right] = 5x(x + 6)\cdot\frac{1}{5}$
$5(x + 6) + 5x = x(x + 6)$
$5x + 30 + 5x = x^2 + 6x$
$10x + 30 = x^2 + 6x$
$0 = x^2 - 4x - 30$
We use the quadratic formula.
$a = 1, b = -4, c = -30$

$x = \frac{-(-4) \pm \sqrt{(-4)^2 - 4\cdot 1\cdot(-30)}}{2\cdot 1}$

$x = \frac{4 \pm \sqrt{136}}{2} = \frac{4 \pm \sqrt{4\cdot 34}}{2}$

$x = \frac{4 \pm 2\sqrt{34}}{2} = \frac{2(2 \pm \sqrt{34})}{2}$

$x = 2 \pm \sqrt{34}$
Both numbers check. The solutions are $2 + \sqrt{34}$ and $2 - \sqrt{34}$.

86. $\frac{5 \pm \sqrt{37}}{2}$

87. Familiarize. From the drawing in the text we see that we have a right triangle whose legs are both r and whose hypotenuse is r + 1.

Translate. We use the Pythagorean equation.

$r^2 + r^2 = (r + 1)^2$

Carry out. We solve the equation.

$2r^2 = r^2 + 2r + 1$

$r^2 - 2r - 1 = 0$

$a = 1, b = -2, c = -1$

$r = \frac{-(-2) \pm \sqrt{(-2)^2 - 4\cdot1\cdot(-1)}}{2\cdot1}$

$r = \frac{2 \pm \sqrt{4 + 4}}{2} = \frac{2 \pm \sqrt{8}}{2}$

$r = \frac{2 \pm \sqrt{4\cdot2}}{2} = \frac{2 \pm 2\sqrt{2}}{2}$

$r = \frac{2(1 \pm \sqrt{2})}{2} = 1 \pm \sqrt{2}$

Check. Since $1 - \sqrt{2}$ is negative, it cannot be the length of a leg. Thus it cannot be a solution. If the length of a leg is $1 + \sqrt{2}$, then the length of the hypotenuse is $2 + \sqrt{2}$. We check using the Pythagorean equation.

$(1 + \sqrt{2})^2 + (1 + \sqrt{2})^2$	$(2 + \sqrt{2})^2$
$= 2(1 + 2\sqrt{2} + 2)$	$= 4 + 4\sqrt{2} + 2$
$= 2(3 + 2\sqrt{2})$	$= 6 + 4\sqrt{2}$
$= 6 + 4\sqrt{2}$	

Thus, $(1 + \sqrt{2})^2 + (1 + \sqrt{2})^2 = (2 + \sqrt{2})^2$. The value checks.

State. In the figure, $r = 1 + \sqrt{2} \approx 2.41$ cm.

88. 12 and 13 or -13 and -12

89. From the drawing we see that we have a problem similar to Exercise 87.

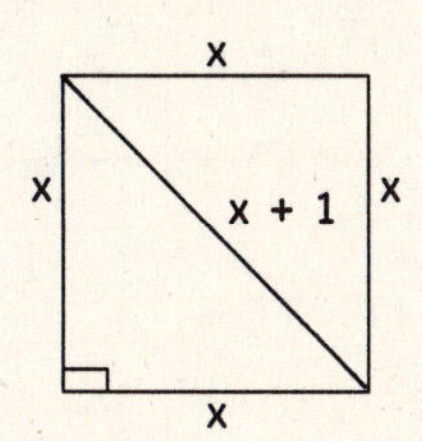

$x^2 + x^2 = (x + 1)^2$

From Exercise 87, we know that $x = 1 + \sqrt{2}$. Then the area is $(1 + \sqrt{2})(1 + \sqrt{2}) = 1 + 2\sqrt{2} + 2 = 3 + 2\sqrt{2} \approx 5.828$ square units.

90. 7.5 ft from the bottom

91. Familiarize. The radius of a 10-in. pizza is $\frac{10}{2}$, or 5 in. The radius of a d-in. pizza is $\frac{d}{2}$ in. The area of a circle is πr^2.

Translate.

Area of d-in. pizza = Area of 10-in. pizza + Area of 10-in. pizza

$\pi\left(\frac{d}{2}\right)^2 = \pi\cdot5^2 + \pi\cdot5^2$

Carry out. We solve the equation.

$\frac{d^2}{4}\pi = 25\pi + 25\pi$

$\frac{d^2}{4}\pi = 50\pi$

$\frac{d^2}{4} = 50$ Multiplying by $\frac{1}{\pi}$

$d^2 = 200$

$d = \sqrt{200}$ or $d = -\sqrt{200}$

$d = 10\sqrt{2}$ or $d = -10\sqrt{2}$

$d \approx 14.14$ or $d \approx -14.14$ Using a calculator or Table 1

Check. Since the diameter cannot be negative, -14.14 is not a solution. If $d = 10\sqrt{2}$, or 14.14, then $r = 5\sqrt{2}$ and the area is $\pi(5\sqrt{2})^2$, or 50π. The area of the two 10" pizzas is $2\cdot\pi\cdot5^2$, or 50π. The value checks.

State. The diameter of the pizza should be $10\sqrt{2}$, or about 14.14 in.

The area of two 10" pizzas is approximately the same as a 14" pizza. Thus, you get more to eat with two 10" pizzas than with a 13-in. pizza.

92. $3 + 3\sqrt{2}$, or about 7.2 cm

93. Familiarize. We will use the formula $A = P(1 + r)^t$ where P is $4000, A is $5267.03, and t is 2.

Translate. We make the substitutions.

$A = P(1 + r)^t$

$5267.03 = 4000(1 + r)^2$

Carry out. We solve the equation.

$5267.03 = 4000(1 + r)^2$

$\frac{5267.03}{4000} = (1 + r)^2$

$\sqrt{\frac{5267.03}{4000}} = 1 + r$ or $-\sqrt{\frac{5267.03}{4000}} = 1 + r$

$1.1475 \approx 1 + r$ or $-1.1475 \approx 1 + r$ Using a calculator

$0.1475 \approx r$ or $-2.1475 \approx r$

Check. Since the interest rate cannot be negative, we check 0.1475 in the formula.

$4000(1 + 0.1475)^2 = 4000(1.1475)^2 = 5267.03$

Our answer checks.

State. The interest rate is about 0.1475, or 14.75%.

94. $2239.13

95. Familiarize. The area of the actual strike zone is 15(40), so the area of the enlarged zone is 15(40) + 0.4(15)(40), or 1.4(15)(40). From the drawing in the text we see that the dimensions of the enlarged strike zone are 15 + 2x by 40 + 2x.

Translate. Using the formula $A = \ell w$, we write an equation for the area of the enlarged strike zone.

$$1.4(15)(40) = (15 + 2x)(40 + 2x)$$

Carry out. We solve the equation.

$$1.4(15)(40) = (15 + 2x)(40 + 2x)$$

$$840 = 600 + 110x + 4x^2 \quad \text{Multiplying on both sides}$$

$$0 = 4x^2 + 110x - 240$$

$$0 = 2x^2 + 55x - 120 \quad \text{Dividing by 2}$$

$a = 2$, $b = 55$, $c = -120$

$$x = \frac{-55 \pm \sqrt{55^2 - 4 \cdot 2 \cdot (-120)}}{2 \cdot 2}$$

$$x = \frac{-55 \pm \sqrt{3985}}{4}$$

$x \approx 2.0$ or $x \approx -29.5$

Check. Since the measurement cannot be negative, -29.5 cannot be a solution. If x = 2, then the dimensions of the enlarged strike zone are $15 + 2 \cdot 2$, or 19, by $40 + 2 \cdot 2$, or 44, and the area is $19 \cdot 44 = 836 \approx 840$. The answer checks.

State. The dimensions of the enlarged strike zone are 19 in. by 44 in.

96. 

Exercise Set 10.4

1. $\sqrt{-1} = i$

2. $3i$

3. $\sqrt{-49} = \sqrt{-1 \cdot 49} = \sqrt{-1} \cdot \sqrt{49} = i \cdot 7 = 7i$

4. $4i$

5. $\sqrt{-8} = \sqrt{-1 \cdot 8} = \sqrt{-1} \cdot \sqrt{8} = i \cdot 2\sqrt{2} = 2i\sqrt{2}$

6. $2i\sqrt{5}$

7. $-\sqrt{-12} = -\sqrt{-1 \cdot 12} = -\sqrt{-1} \cdot \sqrt{12} = -i \cdot 2\sqrt{3} = -2i\sqrt{3}$

8. $-3i\sqrt{5}$

9. $-\sqrt{-27} = -\sqrt{-1 \cdot 27} = -\sqrt{-1} \cdot \sqrt{27} = -i \cdot 3\sqrt{3} = -3i\sqrt{3}$

10. $-2i$

11. $\sqrt{-50} = \sqrt{-1 \cdot 50} = \sqrt{-1}\sqrt{50} = i \cdot 5\sqrt{2} = 5i\sqrt{2}$

12. $-6i\sqrt{2}$

13. $-\sqrt{-300} = -\sqrt{-1 \cdot 300} = -\sqrt{-1} \cdot \sqrt{300} = -i \cdot 10\sqrt{3} = -10i\sqrt{3}$

14. $9i\sqrt{2}$

15. $7 + \sqrt{-16} = 7 + \sqrt{-1 \cdot 16} = 7 + \sqrt{-1} \cdot \sqrt{16} = 7 + i \cdot 4 = 7 + 4i$

16. $-8 - 6i$

17. $3 - \sqrt{-98} = 3 - \sqrt{-1 \cdot 98} = 3 - \sqrt{-1} \cdot \sqrt{98} = 3 - i \cdot 7\sqrt{2} = 3 - 7i\sqrt{2}$

18. $-2 + 5i\sqrt{5}$

19. $x^2 + 4 = 0$

$x^2 = -4$

$x = \sqrt{-4}$ or $x = -\sqrt{-4}$ Principle of square roots

$x = i\sqrt{4}$ or $x = -i\sqrt{4}$

$x = 2i$ or $x = -2i$

The solutions are $2i$ and $-2i$, or $\pm 2i$.

20. $\pm 3i$

21. $x^2 = -12$

$x = \sqrt{-12}$ or $x = -\sqrt{-12}$ Principle of square roots

$x = i\sqrt{12}$ or $x = -i\sqrt{12}$

$x = 2i\sqrt{3}$ or $x = -2i\sqrt{3}$

The solutions are $2i\sqrt{3}$ and $-2i\sqrt{3}$, or $\pm 2i\sqrt{3}$.

22. $\pm 4i\sqrt{3}$

23. $x^2 - 4x + 6 = 0$

$a = 1$, $b = -4$, $c = 6$

$$x = \frac{-b \pm \sqrt{b^2 - 4ac}}{2a}$$

$$x = \frac{-(-4) \pm \sqrt{(-4)^2 - 4 \cdot 1 \cdot 6}}{2 \cdot 1}$$

$$x = \frac{4 \pm \sqrt{-8}}{2}$$

$$x = \frac{4 \pm \sqrt{-1}\sqrt{8}}{2} = \frac{4 \pm i \cdot 2\sqrt{2}}{2}$$

$$x = \frac{4}{2} \pm \frac{2\sqrt{2}}{2}i \quad \text{Writing in the form } a + bi$$

$$x = 2 \pm \sqrt{2}i$$

The solutions are $2 \pm \sqrt{2}i$.

24. $-2 \pm i$

25. $(x - 2)^2 = -16$

$x - 2 = \sqrt{-16}$ or $x - 2 = -\sqrt{-16}$ Principle of square roots

$x - 2 = i\sqrt{16}$ or $x - 2 = -i\sqrt{16}$

$x - 2 = 4i$ or $x - 2 = -4i$

$x = 2 + 4i$ or $x = 2 - 4i$

The solutions are $2 + 4i$ and $2 - 4i$, or $2 \pm 4i$.

26. $-1 \pm 5i$

27. $x^2 + 2x + 2 = 0$

$a = 1, b = 2, c = 2$

$$x = \frac{-b \pm \sqrt{b^2 - 4ac}}{2a}$$

$$x = \frac{-2 \pm \sqrt{2^2 - 4\cdot1\cdot2}}{2\cdot1}$$

$$x = \frac{-2 \pm \sqrt{-4}}{2} = \frac{-2 \pm i\sqrt{4}}{2} = \frac{-2 \pm 2i}{2}$$

$$x = \frac{-2}{2} \pm \frac{2i}{2} = -1 \pm i$$

The solutions are $-1 \pm i$.

28. $1 \pm i$

29. $x^2 + 7 = 4x$

$x^2 - 4x + 7 = 0$ Standard form

$a = 1, b = -4, c = 7$

$$x = \frac{-b \pm \sqrt{b^2 - 4ac}}{2a}$$

$$x = \frac{-(-4) \pm \sqrt{(-4)^2 - 4\cdot1\cdot7}}{2\cdot1}$$

$$x = \frac{4 \pm \sqrt{-12}}{2} = \frac{4 \pm i\sqrt{12}}{2}$$

$$x = \frac{4 \pm 2i\sqrt{3}}{2} = \frac{4}{2} \pm \frac{2\sqrt{3}}{2}i = 2 \pm \sqrt{3}i$$

The solutions are $2 \pm \sqrt{3}i$.

30. $-2 \pm \sqrt{3}i$

31. $2t^2 + 6t + 5 = 0$

$a = 2, b = 6, t = 5$

$$t = \frac{-b \pm \sqrt{b^2 - 4ac}}{2a}$$

$$t = \frac{-6 \pm \sqrt{6^2 - 4\cdot2\cdot5}}{2\cdot2}$$

$$t = \frac{-6 \pm \sqrt{-4}}{4} = \frac{-6 \pm i\sqrt{4}}{4} = \frac{-6 \pm 2i}{4}$$

$$t = \frac{-6}{4} \pm \frac{2}{4}i = -\frac{3}{2} \pm \frac{1}{2}i$$

The solutions are $-\frac{3}{2} \pm \frac{1}{2}i$.

32. $-\frac{3}{8} \pm \frac{\sqrt{23}}{8}i$

33. $1 + 2m + 3m^2 = 0$

$3m^2 + 2m + 1 = 0$ Standard form

$a = 3, b = 2, c = 1$

$$m = \frac{-b \pm \sqrt{b^2 - 4ac}}{2a}$$

$$m = \frac{-2 \pm \sqrt{2^2 - 4\cdot3\cdot1}}{2\cdot3}$$

$$m = \frac{-2 \pm \sqrt{-8}}{6} = \frac{-2 \pm i\sqrt{8}}{6} = \frac{-2 \pm 2i\sqrt{2}}{6}$$

$$m = \frac{-2}{6} \pm \frac{2\sqrt{2}}{6}i = -\frac{1}{3} \pm \frac{\sqrt{2}}{3}i$$

The solutions are $-\frac{1}{3} \pm \frac{\sqrt{2}}{3}i$.

34. $\frac{3}{4} \pm \frac{\sqrt{3}}{4}i$

35. $$\frac{3}{1 - \sqrt{2}} = \frac{3}{1 - \sqrt{2}} \cdot \frac{1 + \sqrt{2}}{1 + \sqrt{2}} = \frac{3(1 + \sqrt{2})}{(1 - \sqrt{2})(1 + \sqrt{2})} =$$

$$\frac{3 + 3\sqrt{2}}{1^2 - (\sqrt{2})^2} = \frac{3 + 3\sqrt{2}}{1 - 2} = \frac{3 + 3\sqrt{2}}{-1} = \frac{-(3 + 3\sqrt{2})}{1} =$$

$-3 - 3\sqrt{2}$

36. $\frac{20 - 5\sqrt{7}}{9}$

37. $$\frac{\sqrt{2} + \sqrt{5}}{\sqrt{2} - \sqrt{5}} = \frac{\sqrt{2} + \sqrt{5}}{\sqrt{2} - \sqrt{5}} \cdot \frac{\sqrt{2} + \sqrt{5}}{\sqrt{2} + \sqrt{5}} =$$

$$\frac{(\sqrt{2})^2 + 2\sqrt{2}\,\sqrt{5} + (\sqrt{5})^2}{(\sqrt{2})^2 - (\sqrt{5})^2} = \frac{2 + 2\sqrt{10} + 5}{2 - 5} =$$

$$\frac{7 + 2\sqrt{10}}{-3} = -\frac{7 + 2\sqrt{10}}{3}$$

38. $-6 + \sqrt{35}$

39.

40.

41. $(x + 2)^2 + (x + 1)^2 = 0$

$(x^2 + 4x + 4) + (x^2 + 2x + 1) = 0$

$2x^2 + 6x + 5 = 0$

This is the equation we solved in Exercise 31 (using the variable t instead of x). The solutions are $-\frac{3}{2} \pm \frac{1}{2}i$.

42. $-2 \pm i$

43. $\frac{2x - 1}{5} - \frac{2}{x} = \frac{x}{2}$

We multiply by 10x, the LCD.

$$10x\left[\frac{2x - 1}{5} - \frac{2}{x}\right] = 10x \cdot \frac{x}{2}$$

$$2x(2x - 1) - 10\cdot 2 = 5x\cdot x$$

$$4x^2 - 2x - 20 = 5x^2$$

$$0 = x^2 + 2x + 20$$

$a = 1$, $b = 2$, $c = 20$

$$x = \frac{-b \pm \sqrt{b^2 - 4ac}}{2a}$$

$$x = \frac{-2 \pm \sqrt{2^2 - 4\cdot 1\cdot 20}}{2\cdot 1}$$

$$x = \frac{-2 \pm \sqrt{-76}}{2} = \frac{-2 \pm i\sqrt{76}}{2} = \frac{-2 \pm 2i\sqrt{19}}{2}$$

$$x = \frac{-2}{2} \pm \frac{2\sqrt{19}}{2}i = -1 \pm \sqrt{19}i$$

The solutions are $-1 \pm \sqrt{19}i$.

44. $\frac{1}{2} \pm \frac{\sqrt{3}}{6}i$

45. 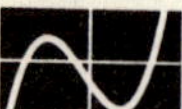

Exercise Set 10.5

1. $y = x^2 - 1$

We first find the vertex. The x-coordinate is

$$-\frac{b}{2a} = -\frac{0}{2\cdot 1} = 0.$$

We substitute into the equation to find the second coordinate of the vertex.

$$y = x^2 - 1 = 0^2 - 1 = -1$$

The vertex is (0,-1). The line of symmetry is x = 0, the y-axis.

We choose some x-values on both sides of the vertex and graph the parabola.

When $x = 1$, $y = 1^2 - 1 = 1 - 1 = 0$.

When $x = -1$, $y = (-1)^2 - 1 = 1 - 1 = 0$.

When $x = 2$, $y = 2^2 - 1 = 4 - 1 = 3$.

When $x = -2$, $y = (-2)^2 - 1 = 4 - 1 = 3$.

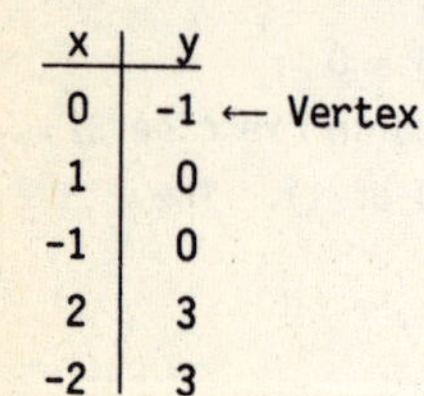

x	y
0	-1 ← Vertex
1	0
-1	0
2	3
-2	3

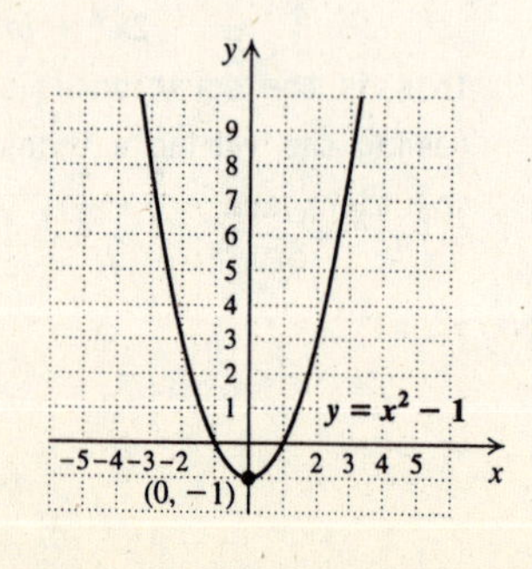

2. 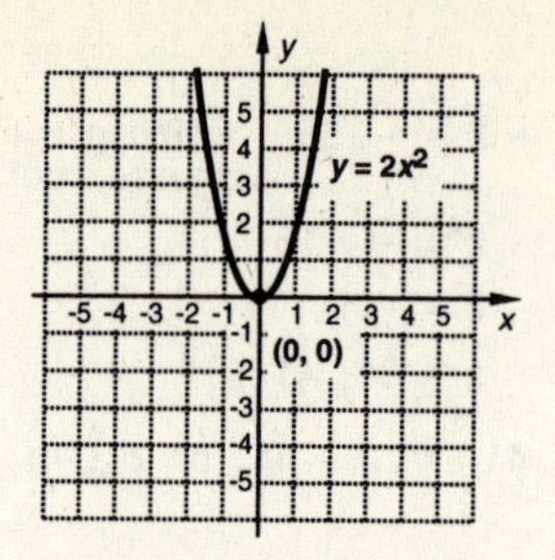

3. $y = -1\cdot x^2$

Find the vertex. The x-coordinate is

$$-\frac{b}{2a} = -\frac{0}{2(-1)} = 0.$$

The y-coordinate is

$$y = -1\cdot x^2 = -1\cdot 0^2 = 0.$$

The vertex is (0,0). The line of symmetry is x = 0, the y-axis.

Choose some x-values on both sides of the vertex and graph the parabola.

When $x = -2$, $y = -1\cdot(-2)^2 = -1\cdot 4 = -4$.

When $x = -1$, $y = -1\cdot(-1)^2 = -1\cdot 1 = -1$.

When $x = 1$, $y = -1\cdot 1^2 = -1\cdot 1 = -1$.

When $x = 2$, $y = -1\cdot 2^2 = -1\cdot 4 = -4$.

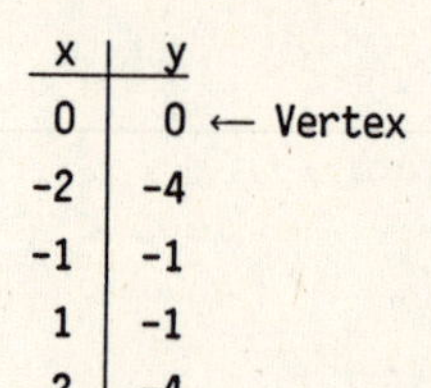

x	y
0	0 ← Vertex
-2	-4
-1	-1
1	-1
2	-4

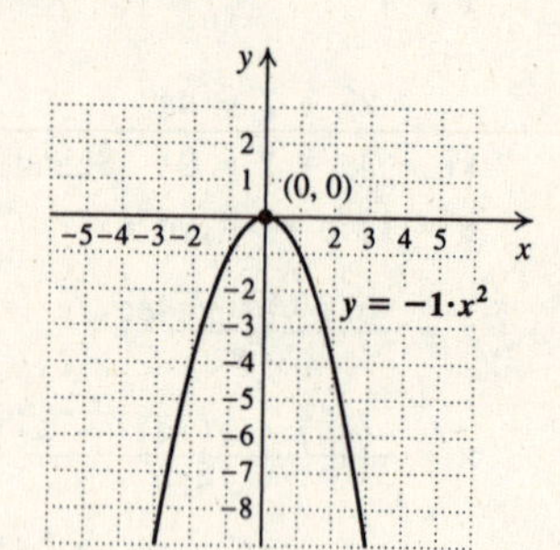

4. 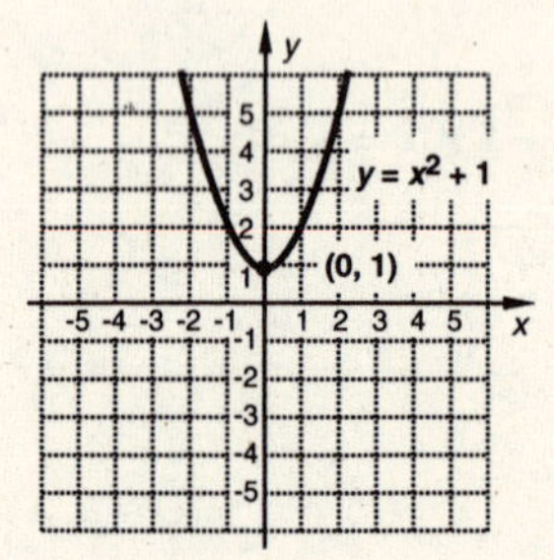

5. $y = -x^2 + 2x$

Find the vertex. The x-coordinate is

$$-\frac{b}{2a} = -\frac{2}{2(-1)} = -\frac{2}{-2} = 1.$$

The y-coordinate is

$$y = -x^2 + 2x = -(1)^2 + 2\cdot 1 = -1 + 2 = 1.$$

The vertex is (1,1).

We choose some x-values on both sides of the vertex and graph the parabola. We make sure we find y when x = 0. This gives us the y-intercept.

x	y	
1	1	← Vertex
0	0	← y-intercept
-1	-3	
2	0	
3	-3	

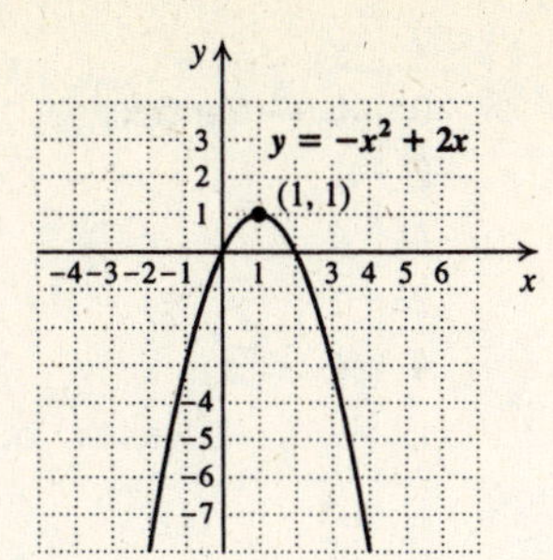

6.

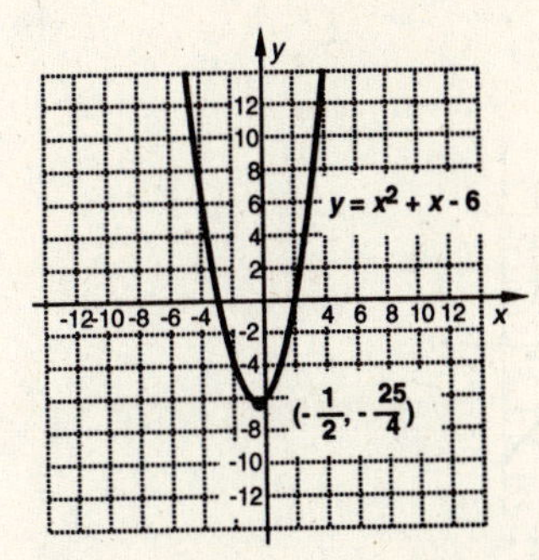

7. $y = 8 - 6x - x^2$, or $y = -x^2 - 6x + 8$

Find the vertex. The x-coordinate is

$$-\frac{b}{2a} = -\frac{-6}{2(-1)} = -3.$$

The y-coordinate is

$$y = 8 - 6x - x^2 = 8 - 6(-3) - (-3)^2 =$$
$$8 + 18 - 9 = 17.$$

The vertex is (-3,17).

We choose some x-values on both sides of the vertex and graph the parabola.

x	y	
-3	17	← Vertex
-2	16	
-1	13	
-4	16	
-5	13	

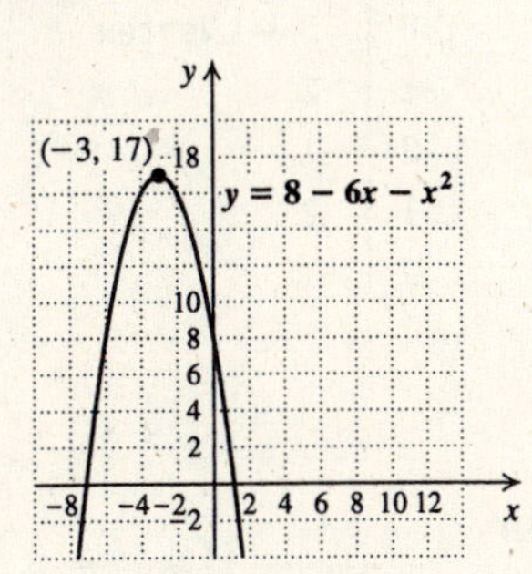

8.

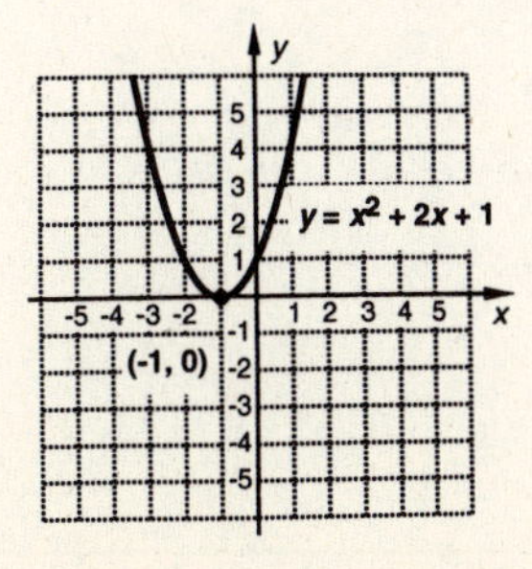

9. $y = x^2 - 2x + 1$

Find the vertex. The x-coordinate is

$$-\frac{b}{2a} = -\frac{-2}{2\cdot1} = -(-1) = 1.$$

The y-coordinate is

$$y = x^2 - 2x + 1 = 1^2 - 2\cdot1 + 1 = 1 - 2 + 1 = 0.$$

The vertex is (1,0).

We choose some x-values on both sides of the vertex and graph the parabola.

x	y	
1	0	← Vertex
0	1	← y-intercept
-1	4	
2	1	
3	4	

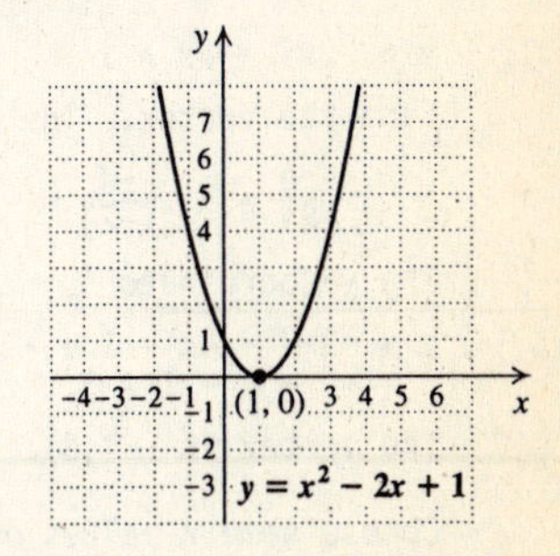

10.

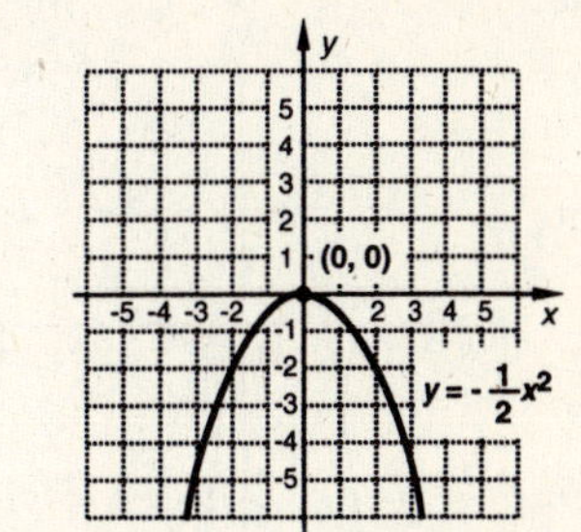

11. $y = -x^2 + 2x + 3$

Find the vertex. The x-coordinate is

$$-\frac{b}{2a} = -\frac{2}{2(-1)} = -(-1) = 1.$$

The y-coordinate is

$$y = -x^2 + 2x + 3 = -(1)^2 + 2\cdot1 + 3 = -1 + 2 + 3 = 4.$$

The vertex is (1,4).

Choose some x-values on both sides of the vertex and graph the parabola.

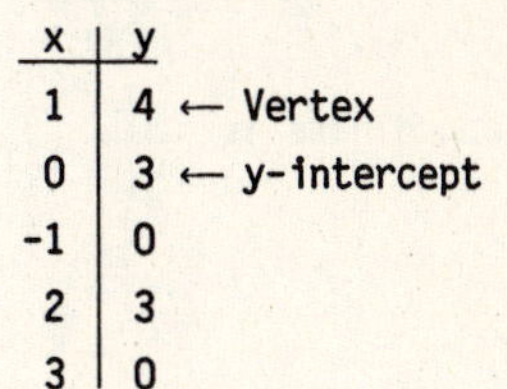

x	y	
1	4	← Vertex
0	3	← y-intercept
-1	0	
2	3	
3	0	

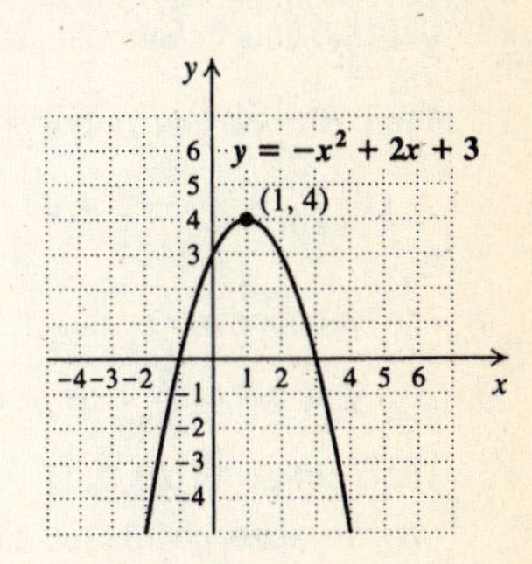

12.

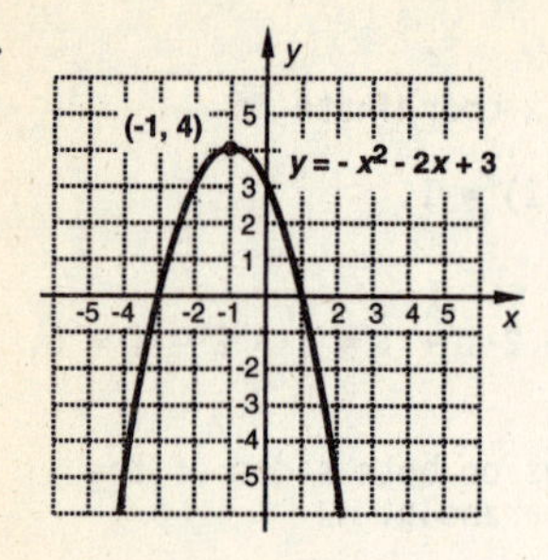

13. $y = -2x^2 - 4x + 1$

Find the vertex. The x-coordinate is

$$-\frac{b}{2a} = -\frac{-4}{2(-2)} = -1.$$

The y-coordinate is

$y = -2x^2 - 4x + 1 = -2(-1)^2 - 4(-1) + 1 =$
$-2 + 4 + 1 = 3.$

The vertex is (-1,3).

Choose some x-values on both sides of the vertex and graph the parabola.

x	y
-1	3 ← Vertex
0	1 ← y-intercept
1	-5
-2	1
-3	-5

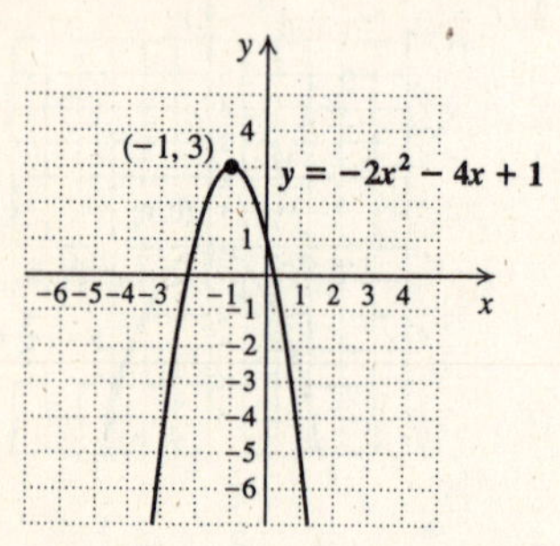

14.

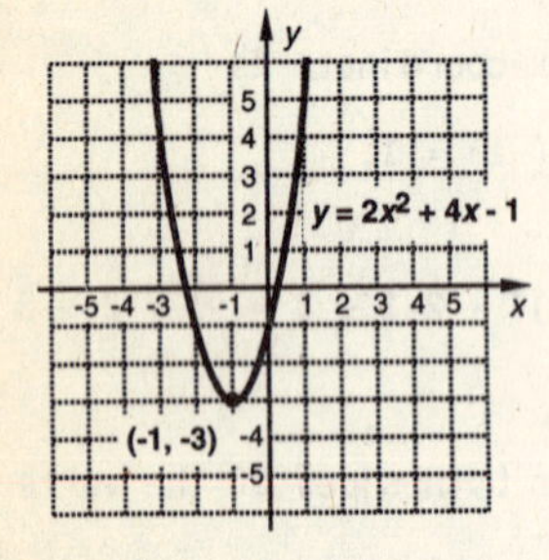

15. $y = \frac{1}{4}x^2$

Find the vertex. The x-coordinate is

$$-\frac{b}{2a} = -\frac{0}{2\left(\frac{1}{4}\right)} = 0.$$

The y-coordinate is

$$y = \frac{1}{4}x^2 = \frac{1}{4} \cdot 0^2 = 0.$$

The vertex is (0,0).

Choose some points on both sides of the vertex and graph the parabola.

x	y
0	0 ← Vertex
-2	1
-4	4
2	1
4	4

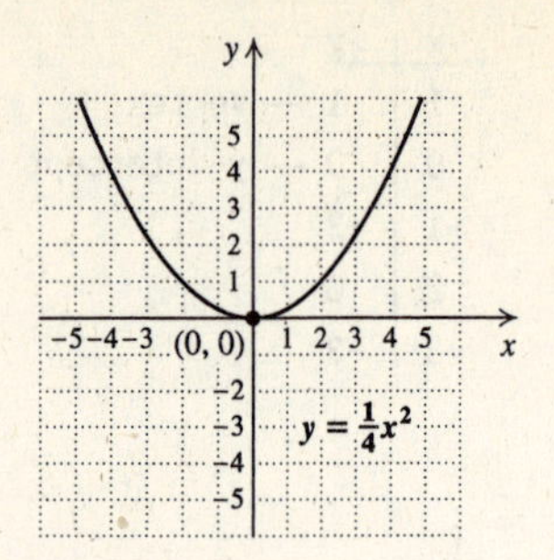

16.

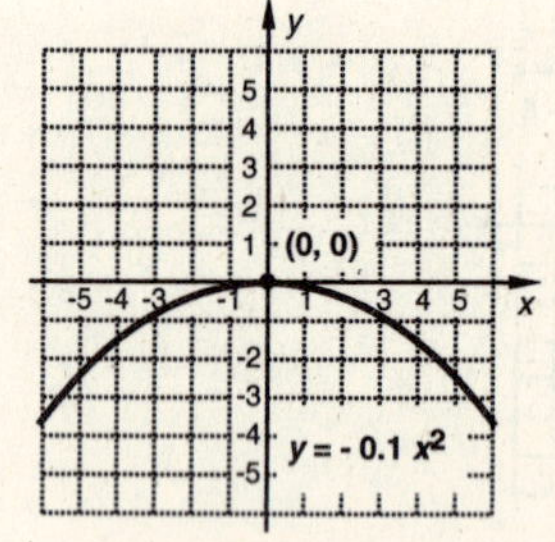

17. $y = 3 - x^2$, or $y = -x^2 + 3$

Find the vertex. The x-coordinate is

$$-\frac{b}{2a} = -\frac{0}{2(-1)} = 0.$$

The y-coordinate is

$$y = 3 - x^2 = 3 - 0^2 = 3.$$

The vertex is (0,3).

Choose some x-values on both sides of the vertex and graph the parabola.

x	y
0	3 ← Vertex
-1	2
-2	-1
1	2
2	-1

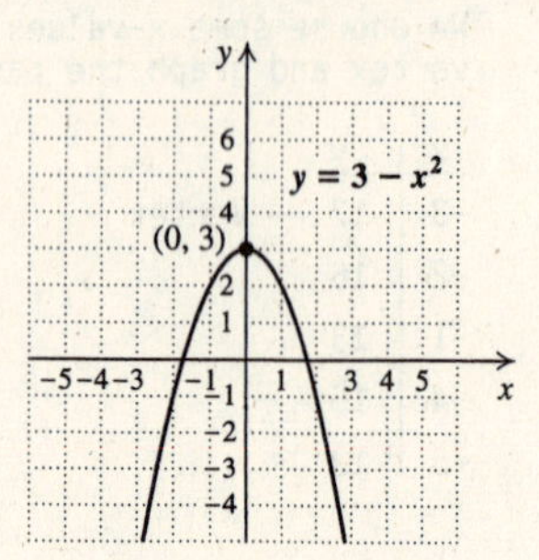

18.

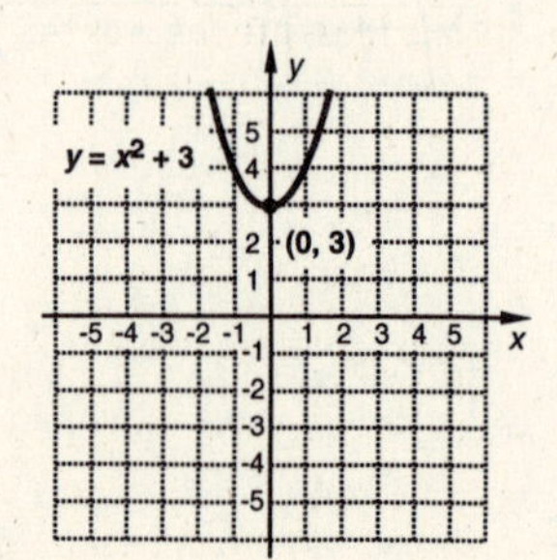

19. $y = -x^2 + x - 1$

Find the vertex. The x-coordinate is

$$-\frac{b}{2a} = -\frac{1}{2(-1)} = -\left(-\frac{1}{2}\right) = \frac{1}{2}.$$

The y-coordinate is

$$y = -x^2 + x - 1 = -\left(\frac{1}{2}\right)^2 + \frac{1}{2} - 1 = -\frac{1}{4} + \frac{1}{2} - 1 = -\frac{3}{4}.$$

The vertex is $\left(\frac{1}{2}, -\frac{3}{4}\right)$.

Choose some x-values on both sides of the vertex and graph the parabola.

x	y	
$\frac{1}{2}$	$-\frac{3}{4}$	← Vertex
0	-1	← y-intercept
-1	-3	
1	-1	
2	-3	

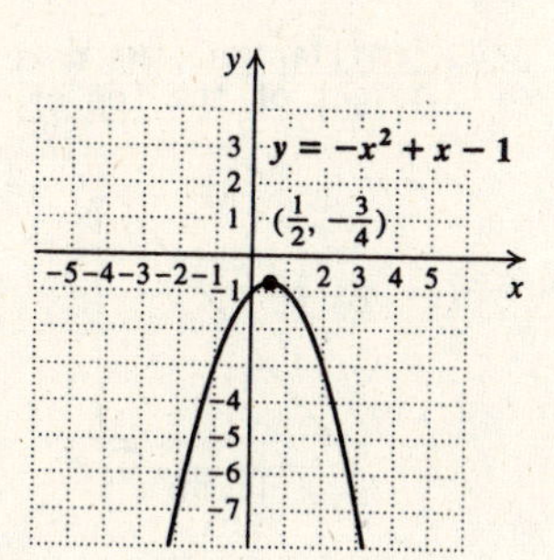

20.

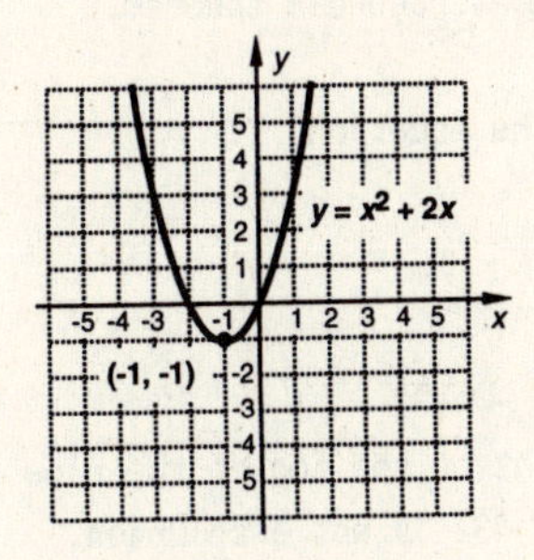

21. $y = -2x^2$

Find the vertex. The x-coordinate is

$$-\frac{b}{2a} = -\frac{0}{2(-2)} = 0.$$

The y-coordinate is

$$y = -2x^2 = -2 \cdot 0^2 = 0.$$

The vertex is (0,0).

Choose some x-values on both sides of the vertex and graph the parabola.

x	y	
0	0	← Vertex
-1	-2	
-2	-8	
1	-2	
2	-8	

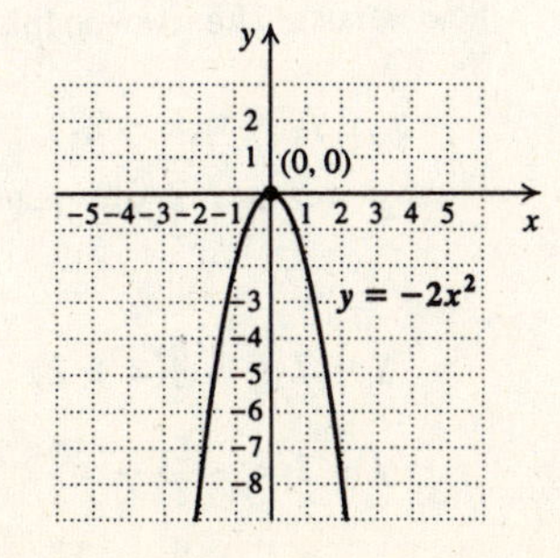

22.

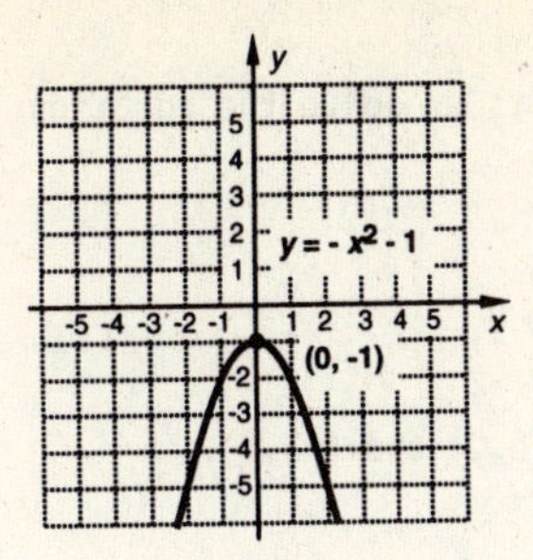

23. $y = x^2 - x - 6$

Find the vertex. The x-coordinate is

$$-\frac{b}{2a} = -\frac{-1}{2 \cdot 1} = -\left(-\frac{1}{2}\right) = \frac{1}{2}.$$

The y-coordinate is

$$y = x^2 - x - 6 = \left(\frac{1}{2}\right)^2 - \frac{1}{2} - 6 = \frac{1}{4} - \frac{1}{2} - 6 = -\frac{25}{4}.$$

The vertex is $\left(\frac{1}{2}, -\frac{25}{4}\right)$.

Choose some x-values on both sides of the vertex and graph the parabola.

x	y	
$\frac{1}{2}$	$-\frac{25}{4}$	← Vertex
0	-6	← y-intercept
-1	-4	
1	-6	
2	-4	

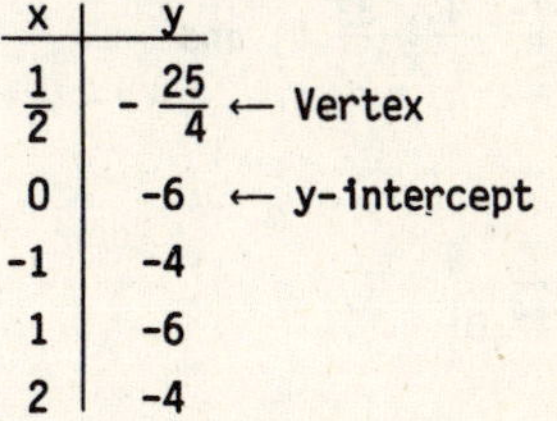
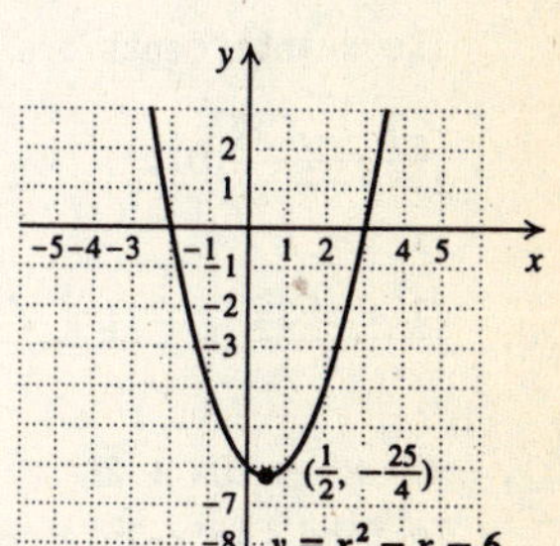

24.

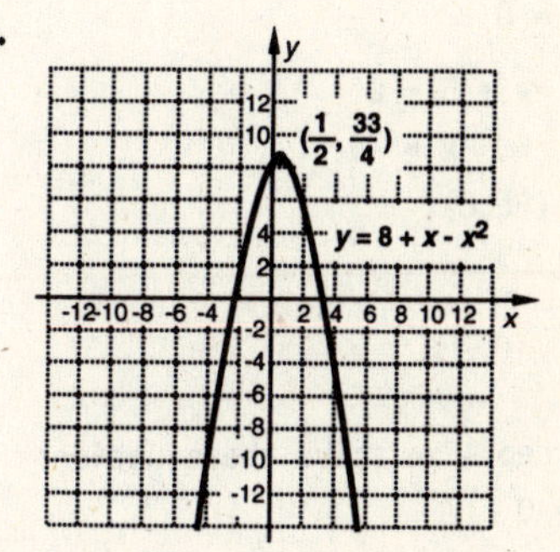

25. $y = x^2 - 5$

To find the x-intercepts we solve the equation:

$$x^2 - 5 = 0$$
$$x^2 = 5$$
$$x = \sqrt{5} \text{ or } x = -\sqrt{5} \quad \text{Principle of square roots}$$

The x-intercepts are $(\sqrt{5},0)$ and $(-\sqrt{5},0)$.

26. $(\sqrt{3},0)$, $(-\sqrt{3},0)$

27. $y = x^2 + 2x$

To find the x-intercepts we solve the equation:

$x^2 + 2x = 0$

$x(x + 2) = 0$

$x = 0$ or $x + 2 = 0$

$x = 0$ or $x = -2$

The x-intercepts are (0,0) and (-2,0).

28. (0,0), (2,0)

29. $y = 8 - x - x^2$

To find the intercepts we solve the equation:

$8 - x - x^2 = 0$

$0 = x^2 + x - 8$ Standard form

$a = 1,\ b = 1,\ c = -8$

$$x = \frac{-1 \pm \sqrt{1^2 - 4\cdot 1\cdot(-8)}}{2\cdot 1}$$

$$x = \frac{-1 \pm \sqrt{33}}{2}$$

The x-intercepts are $\left(\frac{-1 + \sqrt{33}}{2}, 0\right)$ and $\left(\frac{-1 - \sqrt{33}}{2}, 0\right)$.

30. $\left(\frac{1 + \sqrt{33}}{2}, 0\right)$, $\left(\frac{1 - \sqrt{33}}{2}, 0\right)$

31. $y = x^2 + 10x + 25$

To find the x-intercepts we solve the equation:

$x^2 + 10x + 25 = 0$

$(x + 5)(x + 5) = 0$

$x + 5 = 0$ or $x + 5 = 0$

$x = -5$ or $x = -5$

The x-intercept is (-5,0).

32. (4,0)

33. $y = -2x^2 - 4x + 1$

To find the x-intercepts we solve the equation:

$-2x^2 - 4x + 1 = 0$

$2x^2 + 4x - 1 = 0$ Standard form

$a = 2,\ b = 4,\ c = -1$

$$x = \frac{-4 \pm \sqrt{4^2 - 4\cdot 2\cdot(-1)}}{2\cdot 2}$$

$$x = \frac{-4 \pm \sqrt{24}}{4} = \frac{-4 \pm \sqrt{4\cdot 6}}{4} = \frac{-4 \pm 2\sqrt{6}}{4}$$

$$x = \frac{2(-2 \pm \sqrt{6})}{2\cdot 2} = \frac{-2 \pm \sqrt{6}}{2}$$

The x-intercepts are $\left(\frac{-2 + \sqrt{6}}{2}, 0\right)$ and $\left(\frac{-2 - \sqrt{6}}{2}, 0\right)$.

34. $\left(\frac{-2 + \sqrt{6}}{2}, 0\right)$, $\left(\frac{-2 - \sqrt{6}}{2}, 0\right)$

35. $y = x^2 + 5$

To find the x-intercepts we solve the equation:

$x^2 + 5 = 0$

$x^2 = -5$

The negative number -5 has no real-number square roots. Thus there are no x-intercepts. (The graph does not cross the x-axis.)

36. None

37. Familiarize. We make a drawing. Let h = the height of the top of the pipe.

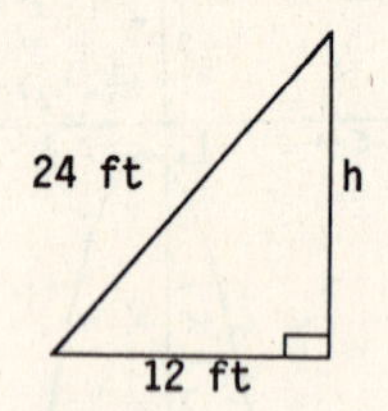

Translate. We use the Pythagorean theorem.

$12^2 + h^2 = 24^2$

Carry out. We solve the equation.

$12^2 + h^2 = 24^2$

$144 + h^2 = 576$

$h^2 = 432$

$h = \sqrt{432}$ or $h = -\sqrt{432}$

Check. Since the height of the top of the pipe cannot be negative, $-\sqrt{432}$ is not a solution. If $h = \sqrt{432}$, then $12^2 + (\sqrt{432})^2 = 144 + 432 = 576 = 24^2$ and $\sqrt{432}$ is the solution.

State. The height of the top of the pipe is $\sqrt{432} \approx 20.78$ ft.

38. $\sqrt{11{,}336} \approx 106.47$ ft

39. First we find the slope.

$$m = \frac{\text{change in } y}{\text{change in } x} = \frac{-3 - 7}{4 - (-2)} = \frac{-10}{6} = -\frac{5}{3}$$

Now we use the two-point equation and solve for y.

$y - y_1 = m(x - x_1)$

$y - 7 = -\frac{5}{3}(x - (-2))$ Substituting $-\frac{5}{3}$ for m, -2 for x_1, and 7 for y_1

$y - 7 = -\frac{5}{3}(x + 2)$

$y - 7 = -\frac{5}{3}x - \frac{10}{3}$

$y = -\frac{5}{3}x + \frac{11}{3}$ Slope-intercept equation

40. -3

41. 

42.

43.

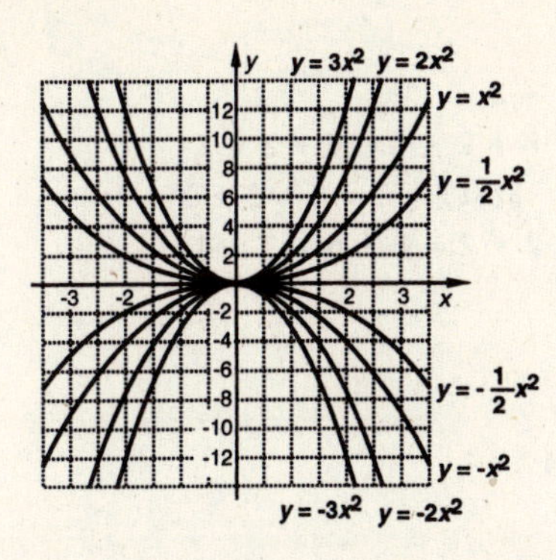

As |a| increases, the graph of $y = ax^2$ is stretched vertically.

44.

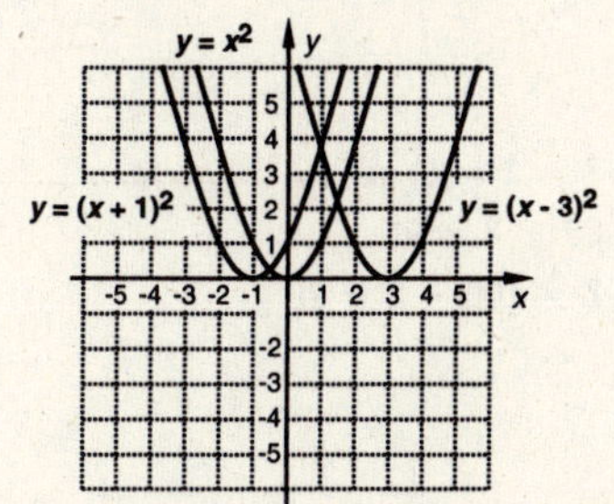

We can move the graph of $y = x^2$ to the right h units if $h > 0$ or to the left |h| units if $h < 0$ to obtain the graph of $y = (x - h)^2$.

45. $S(p) = p^2 + p + 10$

p	S
0	10
1	12
2	16
3	22
4	30
5	40
6	52

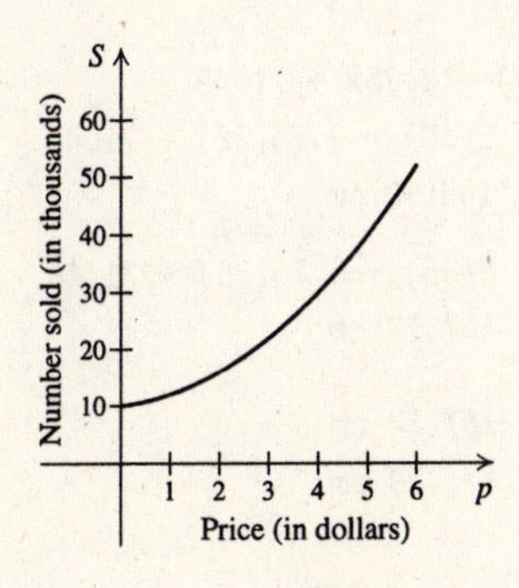

46.

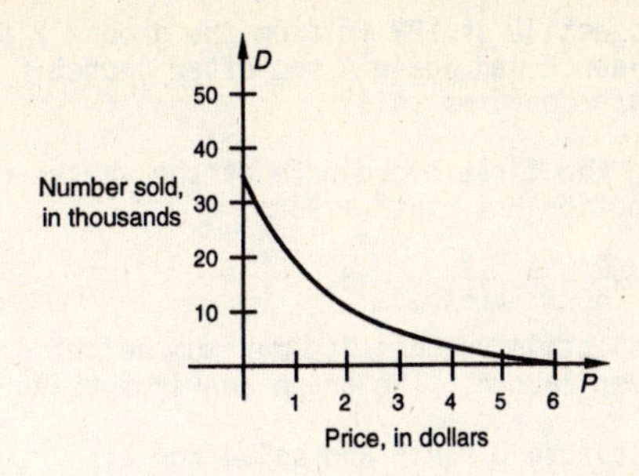

47.

$$D = S$$
$$(p - 6)^2 = p^2 + p + 10$$
$$p^2 - 12p + 36 = p^2 + p + 10$$
$$26 = 13p$$
$$2 = p$$

For p = \$2, D = S.

When p = \$2, $D(p) = (p - 6)^2 = (2 - 6)^2 = 16$, so 16,000 will be sold at that price.

48. a) 56.25 ft, 120 ft, 206.25 ft, 276.25 ft, 356.25 ft, 600 ft

b)

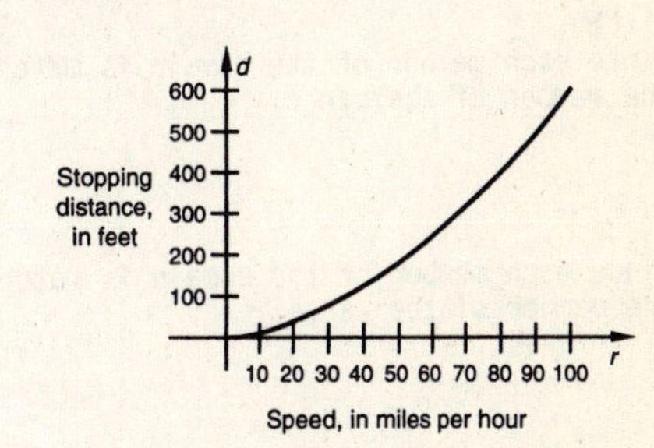

49.

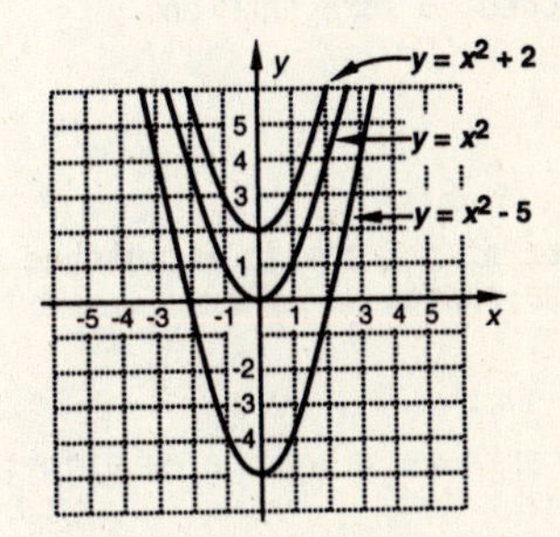

We can move the graph of $y = x^2$ up k units if $k > 0$ or down |k| units if $k < 0$ to obtain the graph of $y = x^2 + k$.

50. 2.2361

51. a) We substitute 128 for H and solve for t:

$$128 = -16t^2 + 96t$$
$$16t^2 - 96t + 128 = 0$$
$$16(t^2 - 6t + 8) = 0$$
$$16(t - 2)(t - 4) = 0$$
$$t - 2 = 0 \quad \text{or} \quad t - 4 = 0$$
$$t = 2 \quad \text{or} \quad t = 4$$

The projectile is 128 ft from the ground 2 sec after launch and again 4 sec after launch. The graph confirms this.

b) We find the first coordinate of the vertex of the function $H = -16t^2 + 96t$:

$$-\frac{b}{2a} = -\frac{96}{2(-16)} = 3$$

The projectile reaches its maximum height 3 sec after launch. The graph confirms this.

c) We substitute 0 for H and solve for t:

$$0 = -16t^2 + 96t$$
$$0 = -16t(t - 6)$$
$$-16t = 0 \quad \text{or} \quad t - 6 = 0$$
$$t = 0 \quad \text{or} \quad t = 6$$

At $t = 0$ sec the projectile has not yet been launched. Thus, we use $t = 6$. The projectile returns to the ground 6 sec after launch. The graph confirms this.

Exercise Set 10.6

1. Yes, since each member of the domain is matched to only one member of the range.

2. Yes

3. Yes, since each member of the domain is matched to only one member of the range.

4. No

5. No, the members of the domain New York and Los Angeles are each matched to more than one member of the range.

6. Yes

7. Yes, since each member of the domain is matched to only one member of the range.

8. Yes

9. $f(x) = x + 5$
$f(3) = 3 + 5 = 8$
$f(7) = 7 + 5 = 12$
$f(-9) = -9 + 5 = -4$

10. -6, 0, 12

11. $h(p) = 3p$
$h(-2) = 3(-2) = -6$
$h(5) = 3\cdot 5 = 15$
$h(24) = 3\cdot 24 = 72$

12. -24, 2, -80

13. $g(s) = 2s + 4$
$g(1) = 2\cdot 1 + 4 = 2 + 4 = 6$
$g(-7) = 2(-7) + 4 = -14 + 4 = -10$
$g(6.7) = 2(6.7) + 4 = 13.4 + 4 = 17.4$

14. 19, 19, 19

15. $F(x) = 2x^2 - 3x + 2$
$F(0) = 2\cdot 0^2 - 3\cdot 0 + 2 = 0 - 0 + 2 = 2$
$F(-1) = 2(-1)^2 - 3(-1) + 2 = 2 + 3 + 2 = 7$
$F(2) = 2\cdot 2^2 - 3\cdot 2 + 2 = 8 - 6 + 2 = 4$

16. 5, 21, 26

17. $h(x) = |x|$
$h(-4) = |-4| = 4$
$h\left(\frac{2}{3}\right) = \left|\frac{2}{3}\right| = \frac{2}{3}$
$h(-3.8) = |-3.8| = 3.8$

18. $1, \frac{13}{4}$

19. $f(x) = |x| - 2$
$f(-3) = |-3| - 2 = 3 - 2 = 1$
$f(93) = |93| - 2 = 93 - 2 = 91$
$f(-100) = |-100| - 2 = 100 - 2 = 98$

20. -122, 3

21. $h(x) = x^4 - 3$
$h(0) = 0^4 - 3 = 0 - 3 = -3$
$h(-1) = (-1)^4 - 3 = 1 - 3 = -2$
$h(3) = 3^4 - 3 = 81 - 3 = 78$

22. $-1, 4, \frac{2}{3}, \frac{1}{5}$

23. $E(x) = 2.75x + 71.48$

a) $E(30) = 2.75(30) + 71.48 = 82.50 + 71.48 = 153.98$ cm

b) $E(35) = 2.75(35) + 71.48 = 96.25 + 71.48 = 167.73$ cm

24. a) 157.34 cm
b) 171.79 cm

25. $P(d) = 1 + \frac{d}{33}$
$P(20) = 1 + \frac{20}{33} = 1\frac{20}{33} \approx 1.606$ atmospheres
$P(30) = 1 + \frac{30}{33} = 1\frac{10}{11} \approx 1.909$ atmospheres
$P(100) = 1 + \frac{100}{33} = 1 + 3\frac{1}{33} = 4\frac{1}{33} \approx 4.03$ atmospheres

26. $166\frac{2}{3}$, $666\frac{2}{3}$, $833\frac{1}{3}$

27. T(d) = 10d + 20
T(5) = 10·5 + 20 = 50 + 20 = 70° C
T(20) = 10·20 + 20 = 200 + 20 = 220° C
T(1000) = 10·1000 + 20 = 10,000 + 20 = 10,020° C

28. 1.792 cm, 2.8 cm, 11.2 cm

29. D(p) = -2.7p + 16.3
D(1) = -2.7(1) + 16.3 = -2.7 + 16.3 = 13.6 million
D(2) = -2.7(2) + 16.3 = -5.4 + 16.3 = 10.9 million
D(3) = -2.7(3) + 16.3 = -8.1 + 16.3 = 8.2 million
D(4) = -2.7(4) + 16.3 = -10.8 + 16.3 = 5.5 million
D(5) = -2.7(5) + 16.3 = -13.5 + 16.3 = 2.8 million

30. 1.1 million, 8.1 million, 15.1 million, 22.1 million, 29.1 million

31. Graph f(x) = x + 4.
Make a list of function values in a table.
When x = -2, f(-2) = -2 + 4 = 2.
When x = 0, f(0) = 0 + 4 = 4.
When x = 1, f(1) = 1 + 4 = 5.

x	f(x)
-2	2
0	4
1	5

We plot these points and connect them.

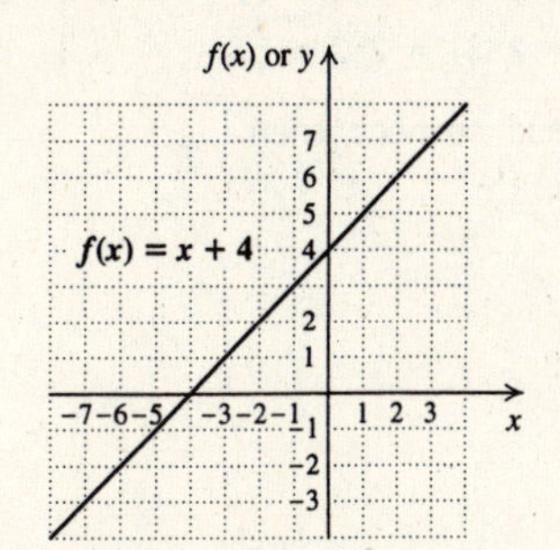

32.

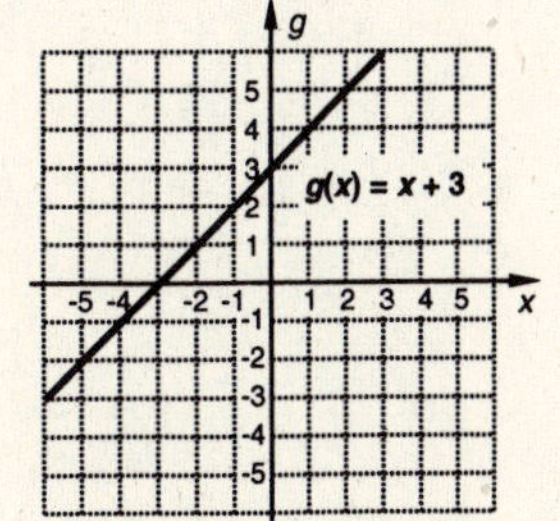

33. Graph h(x) = 2x - 3
Make a list of function values in a table.
When x = -1, h(-1) = 2(-1) - 3 = -2 - 3 = -5.
When x = 0, h(0) = 2·0 - 3 = 0 - 3 = -3.
When x = 3, h(3) = 2·3 - 3 = 6 - 3 = 3.

x	h(x)
-1	-5
0	-3
3	3

We plot these points and connect them.

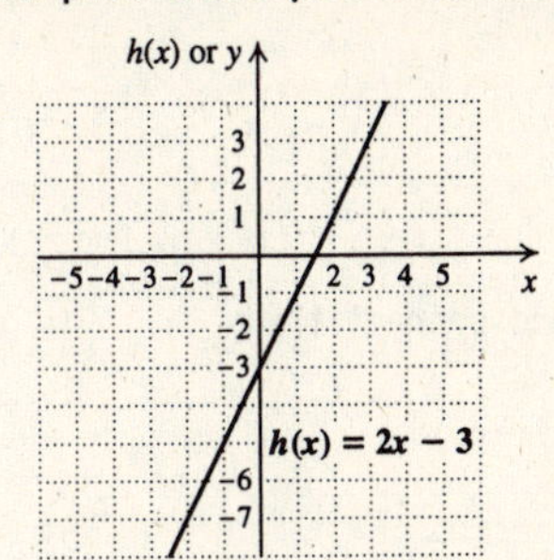

34.

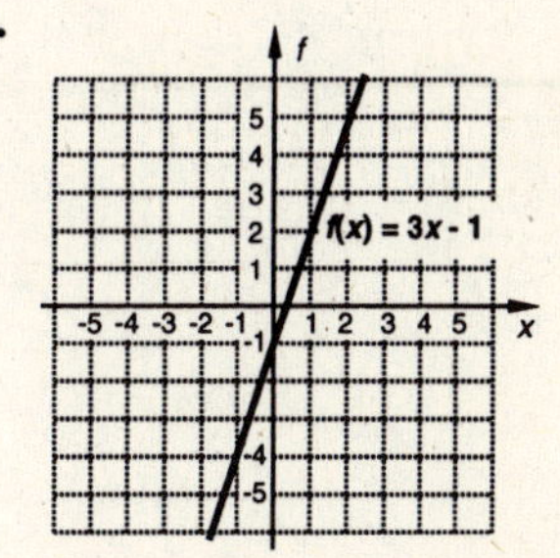

35. Graph g(x) = x - 6.
Make a list of function values in a table.
When x = 0, g(0) = 0 - 6 = -6.
When x = 2, g(2) = 2 - 6 = -4.
When x = 6, g(6) = 6 - 6 = 0.

x	g(x)
0	-6
2	-4
6	0

We plot these points and connect them.

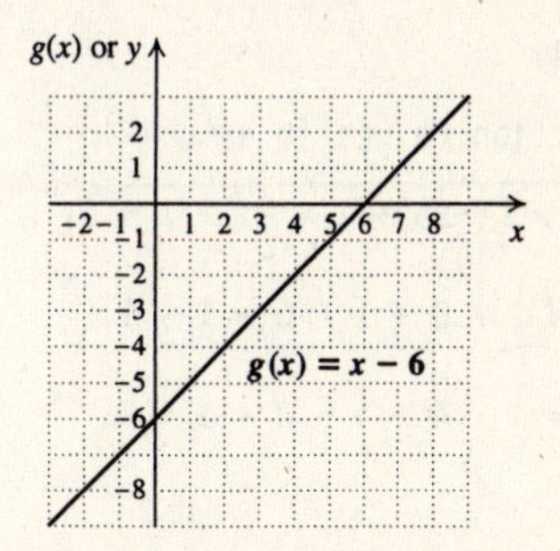

36.

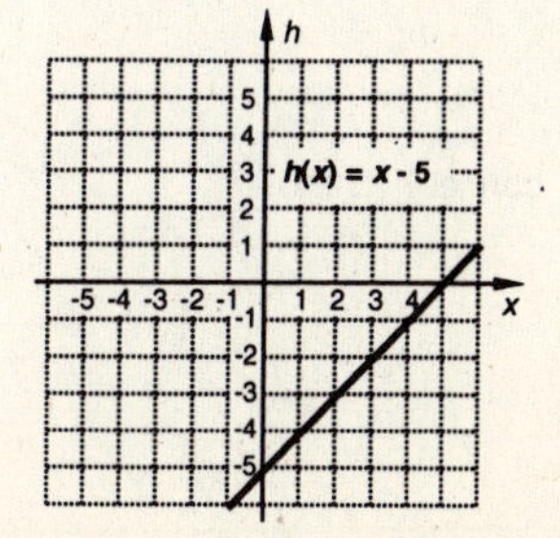

37. Graph f(x) = 2x - 7.

Make a list of function values in a table.

When x = 1, f(1) = 2·1 - 7 = 2 - 7 = -5.

When x = 2, f(2) = 2·2 - 7 = 4 - 7 = -3.

When x = 5, f(5) = 2·5 - 7 = 10 - 7 = 3.

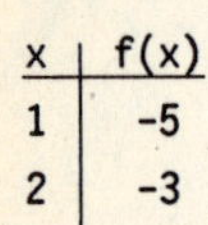

x	f(x)
1	-5
2	-3
5	3

Plot these points and connect them.

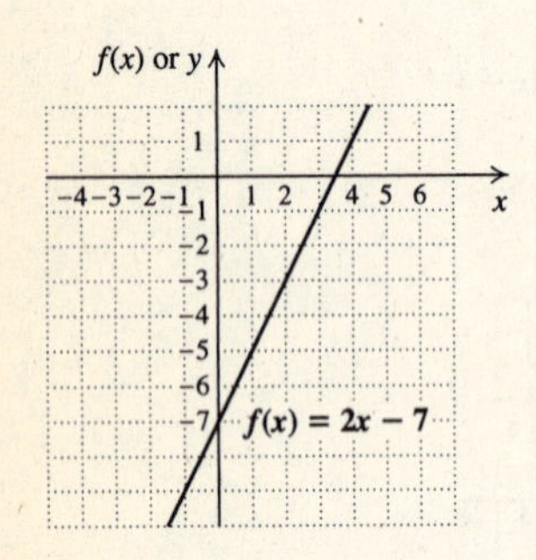

38.

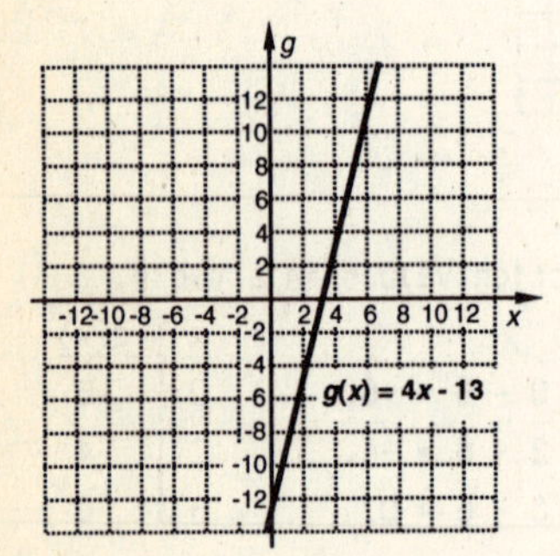

39. Graph $f(x) = \frac{1}{2}x + 1$.

Make a list of function values in a table.

When x = -2, $f(-2) = \frac{1}{2}(-2) + 1 = -1 + 1 = 0$.

When x = 0, $f(0) = \frac{1}{2} \cdot 0 + 1 = 0 + 1 = 1$.

When x = 4, $f(4) = \frac{1}{2} \cdot 4 + 1 = 2 + 1 = 3$.

x	f(x)
-2	0
0	1
4	3

Plot these points and connect them.

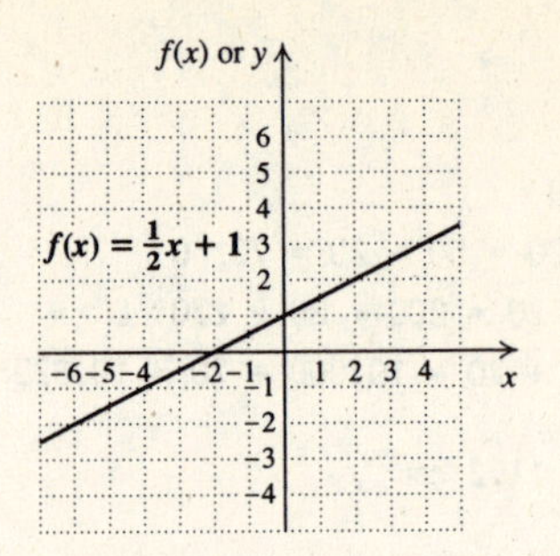

40.

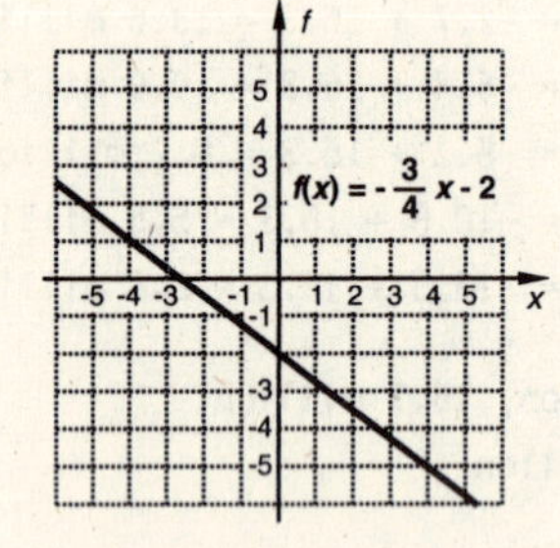

41. Graph g(x) = 2|x|.

Make a list of function values in a table.

When x = -3, g(-3) = 2|-3| = 2·3 = 6.

When x = -1, g(-1) = 2|-1| = 2·1 = 2.

When x = 0, g(0) = 2|0| = 2·0 = 0.

When x = 1, g(1) = 2|1| = 2·1 = 2.

When x = 3, g(3) = 2|3| = 2·3 = 6.

x	g(x)
-3	6
-1	2
0	0
1	2
3	6

Plot these points and connect them.

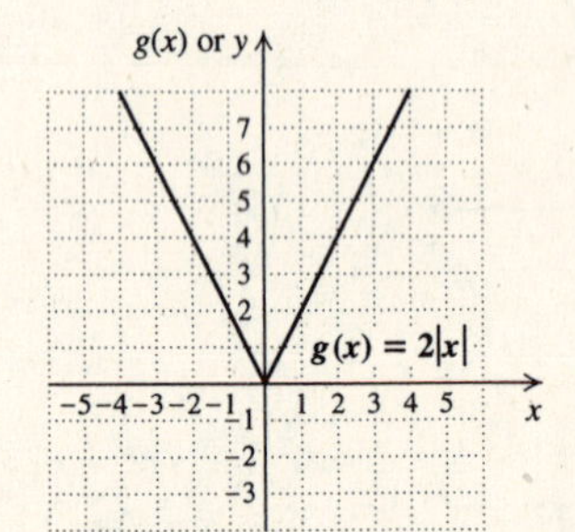

42.

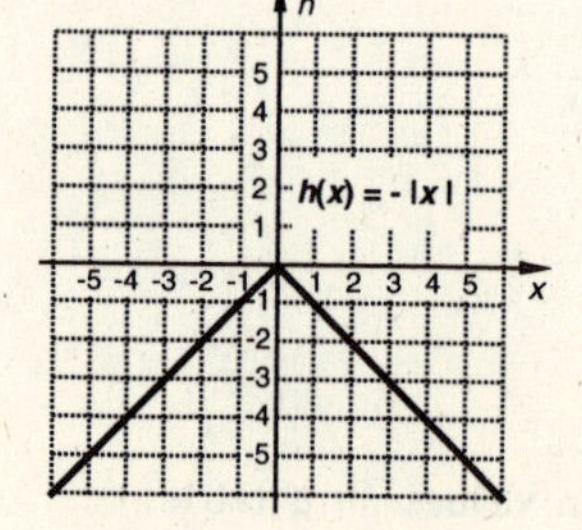

43. Graph $g(x) = x^2$.

Recall from Section 10.5 that the graph is a parabola. Make a list of function values in a table.

When $x = -2$, $g(-2) = (-2)^2 = 4$.

When $x = -1$, $g(-1) = (-1)^2 = 1$.

When $x = 0$, $g(0) = 0^2 = 0$

When $x = 1$, $g(1) = 1^2 = 1$.

When $x = 2$, $g(2) = 2^2 = 4$.

x	g(x)
-2	4
-1	1
0	0
1	1
2	4

Plot these points and connect them.

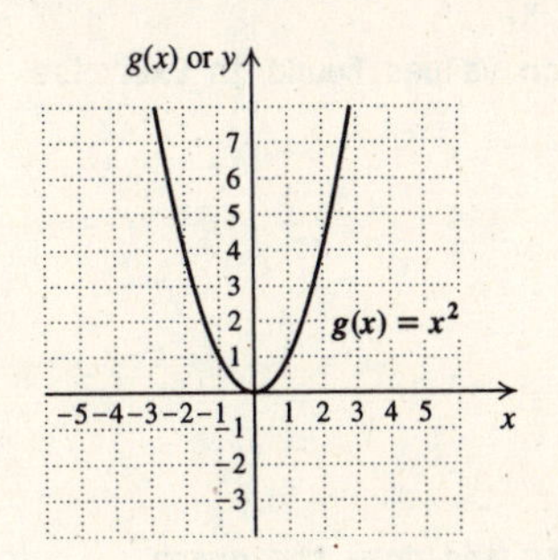

44.

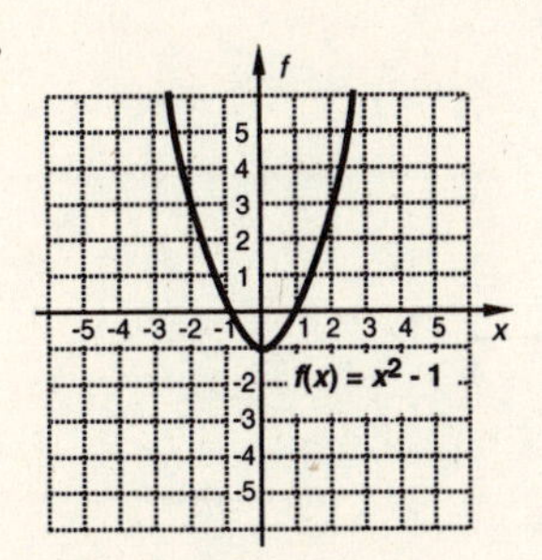

45. Graph $f(x) = x^2 - x - 2$.

Recall from Section 10.5 that the graph is a parabola. Make a list of function values in a table.

When $x = -1$, $f(-1) = (-1)^2 - (-1) - 2 = 1 + 1 - 2 = 0$.

When $x = 0$, $f(0) = 0^2 - 0 - 2 = -2$.

When $x = 1$, $f(1) = 1^2 - 1 - 2 = 1 - 1 - 2 = -2$.

When $x = 2$, $f(2) = 2^2 - 2 - 2 = 4 - 2 - 2 = 0$.

x	f(x)
-1	0
0	-2
1	-2
2	0

Plot these points and connect them.

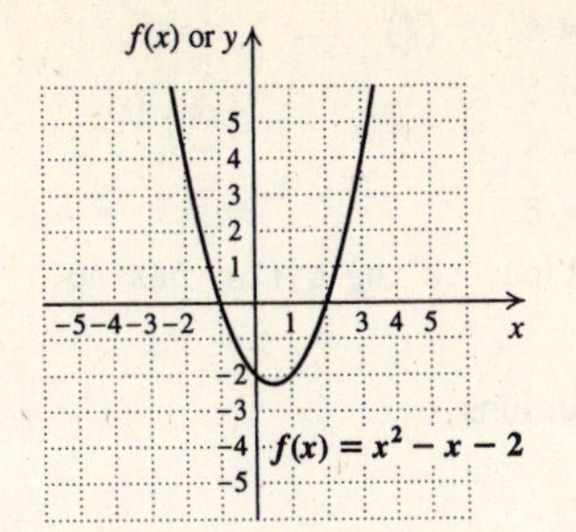

46.

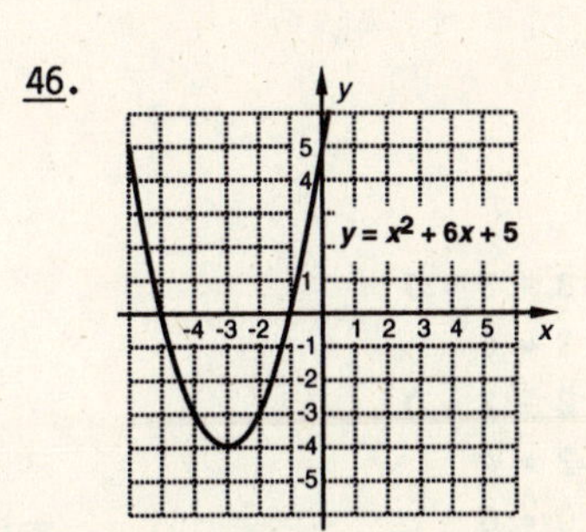

47. The graph is that of a function because no vertical line can cross the graph at more than one point.

48. No

49. The graph is not that of a function because a vertical line, say $x = 1$, crosses the graph at more than one point.

50. Yes

51. The graph is that of function because no vertical line can cross the graph at more than one point.

52. No

53. The first equation is in slope-intercept form:

$$y = \frac{3}{4}x - 7,\ m = \frac{3}{4}$$

We write the second equation in slope-intercept form.

$$3x + 4y = 7$$
$$4y = -3x + 7$$
$$y = -\frac{3}{4}x + \frac{7}{4},\ m = -\frac{3}{4}$$

Since the slopes are different, the equations do not represent parallel lines.

54. Yes

55. $2x - y = 6,\quad (1)$

$4x - 2y = 5\quad (2)$

We solve equation (1) for y.

$2x - y = 6\quad (1)$

$2x - 6 = y\quad$ Adding y and -6

Substitute $2x - 6$ for y in equation (2) and solve for x.

$$4x - 2y = 5 \quad (2)$$
$$4x - 2(2x - 6) = 5$$
$$4x - 4x + 12 = 5$$
$$12 = 5$$

We get a false equation, so the system has no solution.

56. Infinitely many solutions

57. ◈

58. ◈

59. $g(x) = |x| + x$

$g(-3) = |-3| + (-3) = 3 - 3 = 0$

$g(3) = |3| + 3 = 3 + 3 = 6$

$g(-2) = |-2| + (-2) = 2 - 2 = 0$

$g(2) = |2| + 2 = 2 + 2 = 4$

$g(0) = |0| + 0 = 0 + 0 = 0$

60. 6, 0, 4, 0, 0

61. Graph $f(x) = \frac{|x|}{x}$.

We list some function values in a table.

When $x = -3$, $f(-3) = \frac{|-3|}{-3} = \frac{3}{-3} = -1$.

When $x = -2$, $f(-2) = \frac{|-2|}{-2} = \frac{2}{-2} = -1$.

When $x = 0$, $f(0) = \frac{|0|}{0}$, undefined

When $x = 2$, $f(2) = \frac{|2|}{2} = \frac{2}{2} = 1$.

When $x = 3$, $f(3) = \frac{|3|}{3} = \frac{3}{3} = 1$.

x	f(x)
-3	-1
-2	-1
0	undefined
2	1
3	1

We plot these points and draw the graph. We use open circles on the y-axis to indicate that the function is undefined for x = 0.

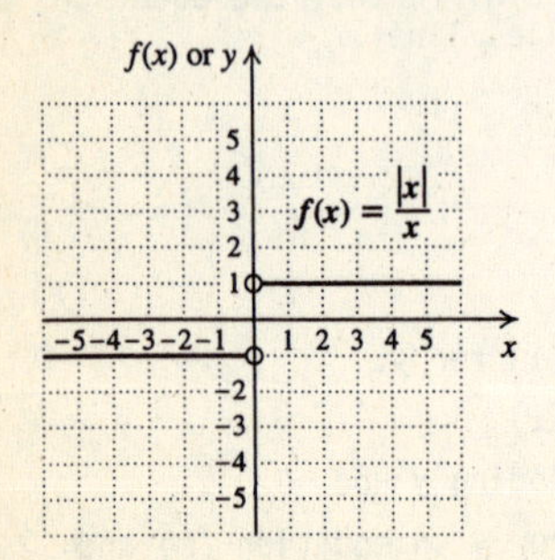

62.

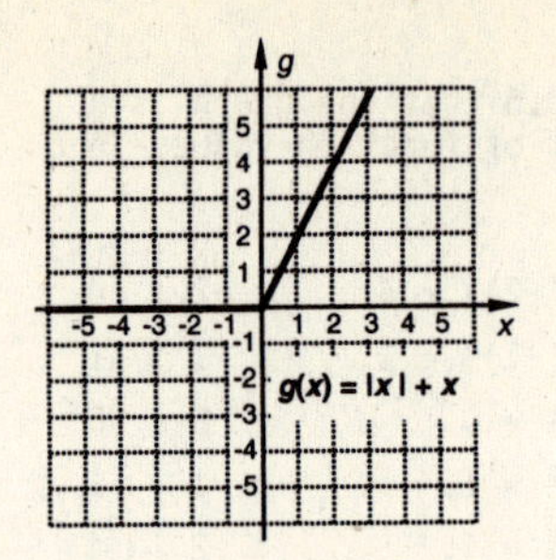

63. Graph $h(x) = |x| - x$.

We list the function values found in Exercise 60 in a table.

x	h(x)
-3	6
-2	4
0	0
2	0
3	0

We plot these points and draw the graph.

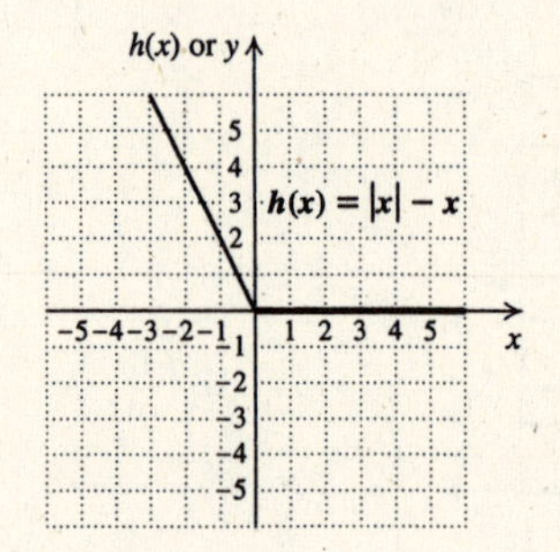

64. Answers may vary.

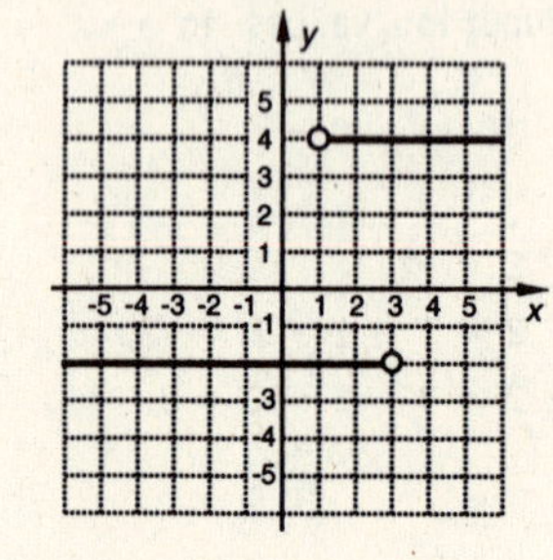

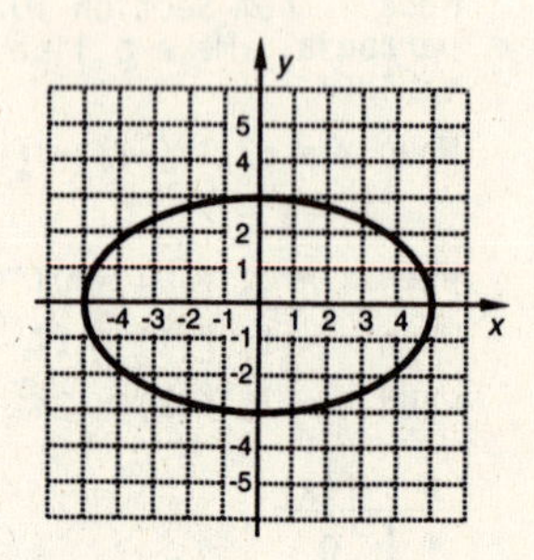

65. Graph $|y| = x$.

We list some function values in a table. For this function it is perhaps best to choose any value for y and then find the corresponding x value. Note that the ordered pairs are still of the form (x,y) even though we choose y first.

	x	y
When $y = -3$, $x = \|-3\| = 3$.	3	-3
When $y = -1$, $x = \|-1\| = 1$.	1	-1
When $y = 0$, $x = \|0\| = 0$.	0	0
When $y = 1$, $x = \|1\| = 1$.	1	1
When $y = 3$, $x = \|3\| = 3$.	3	3

We plot these points and draw the graph.

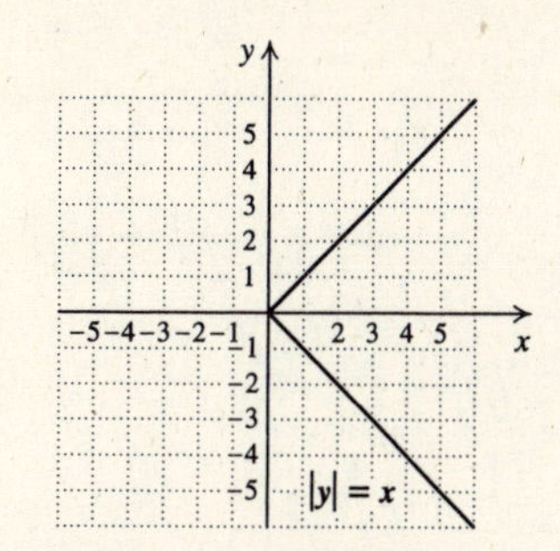

This is not the graph of a function because a vertical line, say $x = 2$, crosses the graph at more than one point.

66.

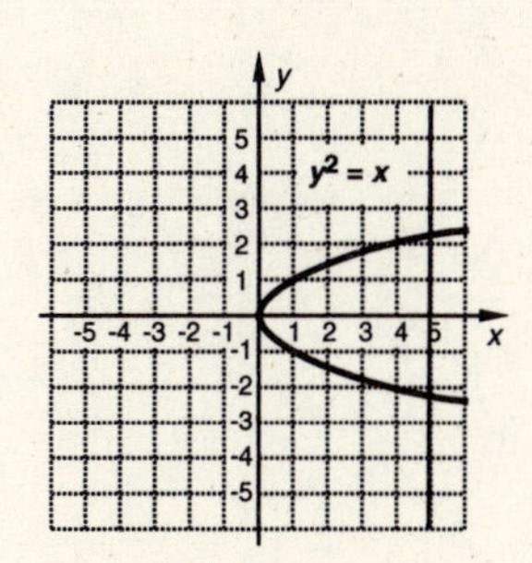

No

67. $f(-1) = -7$ gives us the ordered pair $(-1,-7)$.

$f(3) = 8$ gives us the ordered pair $(3,8)$.

The slope of the line determined by these points is

$$m = \frac{-7 - 8}{-1 - 3} = \frac{-15}{-4} = \frac{15}{4}.$$

We use the two-point equation to find the equation of the line with slope $\frac{15}{4}$ and containing the point (3,8).

$$y - y_1 = m(x - x_1)$$

$$y - 8 = \frac{15}{4}(x - 3)$$

$$y - 8 = \frac{15}{4}x - \frac{45}{4}$$

$$y = \frac{15}{4}x - \frac{13}{4}$$

68. $g(x) = x^2 - 4$

69. $f(x) = 3x + 5$

The domain is the set {0, 1, 2, 3}.

$f(0) = 3\cdot0 + 5 = 0 + 5 = 5$

$f(1) = 3\cdot1 + 5 = 3 + 5 = 8$

$f(2) = 3\cdot2 + 5 = 6 + 5 = 11$

$f(3) = 3\cdot3 + 5 = 9 + 5 = 14$

The range is the set {5, 8, 11, 14}.

70. {-5, -4, -1, 4}

71. $h(x) = |x| - x$

The domain is the set {-1,0,1,2,3,4,5,6,7,8,9, 10,11,12,13,14,15,16,17,18,19}.

$h(-1) = |-1| - (-1) = 1 + 1 = 2$

$h(0) = |0| - 0 = 0 - 0 = 0$

$h(1) = |1| - 1 = 1 - 1 = 0$

$h(2) = |2| - 2 = 2 - 2 = 0$

$h(3) = |3| - 3 = 3 - 3 = 0$

.

.

.

$h(19) = |19| - 19 = 19 - 19 = 0$

The range is the set {0, 2}.

72. {-7, 0, 1, 2, 9}

73. (-3.00, 0), (-1.00, 0), (2.00, 0), (3.00, 0)

Appendix A

1. $\{3,4,5,6,7,8\}$ 2. $\{101,102,103,104,105,106,107\}$

3. $\{41,43,45,47,49\}$ 4. $\{15,20,25,30,35\}$

5. $\{-3,3\}$ 6. $\{0.008\}$ 7. False 8. True

9. True 10. True 11. True 12. True

13. True 14. True 15. True 16. False

17. False 18. True 19. $\{c,d,e\}$ 20. $\{u,i\}$

21. $\{1,10\}$ 22. $\{0,1\}$ 23. $\emptyset$ 24. $\emptyset$

25. The system has no solution. The lines are parallel. Their intersection is empty. 26. The system has no solution. The lines are parallel. Their intersection is empty. 27. $\{a,e,i,o,u,q,c,k\}$

28. $\{a,b,c,d,e,f,g\}$ 29. $\{0,1,2,5,7,10\}$

30. $\{1,2,5,10,0,7\}$ 31. $\{a,e,i,o,u,m,n,f,g,h\}$

32. $\{1,2,5,10,a,b\}$ 33. The solution set is $\{3,-5\}$. This set is the union of the solution sets of the equations $x - 3 = 0$ and $x + 5 = 0$, which are $\{3\}$ and $\{-5\}$. 34. The solution set is $\{-3,1\}$. This set is the union of the solution sets of the equations $x + 3 = 0$ and $x - 1 = 0$, which are $\{-3\}$ and $\{1\}$.

35. The set of integers 36. $\emptyset$ 37. The set of real numbers 38. The set of positive even integers 39. $\emptyset$ 40. The set of integers 41. a) A, b) A, c) A, d) $\emptyset$ 42. a) Yes, b) No, c) No, d) Yes, e) Yes, f) No 43. True

Appendix B

1. $(t + 3)(t^2 - 3t + 9)$ 2. $(p + 2)(p^2 - 2p + 4)$

3. $(a - 1)(a^2 + a + 1)$ 4. $(w - 4)(w^2 + 4w + 16)$

5. $(z + 5)(z^2 - 5z + 25)$ 6. $(x + 1)(x^2 - x + 1)$

7. $(2a - 1)(4a^2 + 2a + 1)$ 8. $(3x - 1)(9x^2 + 3x + 1)$

9. $(y - 3)(y^2 + 3y + 9)$ 10. $(p - 2)(p^2 + 2p + 4)$

11. $(4 + 5x)(16 - 20x + 25x^2)$

12. $(2 + 3b)(4 - 6b + 9b^2)$

13. $(5p - 1)(25p^2 + 5p + 1)$

14. $(4w - 1)(16w^2 + 4w + 1)$

15. $(3m + 4)(9m^2 - 12m + 16)$

16. $(2t + 3)(4t^2 - 6t + 9)$

17. $(p - q)(p^2 + pq + q^2)$ 18. $(a + b)(a^2 - ab + b^2)$

19. $\left[x + \frac{1}{2}\right]\left[x^2 - \frac{1}{2}x + \frac{1}{4}\right]$ 20. $\left[y + \frac{1}{3}\right]\left[y^2 - \frac{1}{3}y + \frac{1}{9}\right]$

21. $2(y - 4)(y^2 + 4y + 16)$

22. $3(z - 1)(z^2 + z + 1)$

23. $3(2a + 1)(4a^2 - 2a + 1)$

24. $2(3x + 1)(9x^2 - 3x + 1)$

25. $r(s + 4)(s^2 - 4s + 16)$

26. $a(b + 5)(b^2 - 5b + 25)$

27. $5(x - 2z)(x^2 + 2xz + 4z^2)$

28. $2(y - 3z)(y^2 + 3yz + 9z^2)$

29. $(x + 0.1)(x^2 - 0.1x + 0.01)$

30. $(y + 0.5)(y^2 - 0.5y + 0.25)$

31. $8(2x^2 - t^2)(4x^4 + 2x^2t^2 + t^4)$

32. $(5c^2 - 2d^2)(25c^4 + 10c^2d^2 + 4d^4)$

33. $(x - 0.3)(x^2 + 0.3x + 0.09)$

34. $(y + 0.2)(y^2 - 0.2y + 0.04)$

35. $(x^{2a} + y^b)(x^{4a} - x^{2a}y^b + y^{2b})$

36. $(ax - by)(a^2x^2 + abxy + b^2y^2)$

37. $3(x^a + 2y^b)(x^{2a} - 2x^ay^b + 4y^{2b})$

38. $\left[\frac{2}{3}x + \frac{1}{4}y\right]\left[\frac{4}{9}x^2 - \frac{1}{6}xy + \frac{1}{16}y^2\right]$

39. $\frac{1}{3}\left[\frac{1}{2}xy + z\right]\left[\frac{1}{4}x^2y^2 - \frac{1}{2}xyz + z^2\right]$

40. $\frac{1}{2}\left[\frac{1}{2}x^a + y^{2a}z^{3b}\right]\left[\frac{1}{4}x^{2a} - \frac{1}{2}x^ay^{2a}z^{3b} + y^{4a}z^{6b}\right]$